Krause · Jäger (Eds.)

High Performance Computing in Science and Engineering '99

Springer

Berlin
Heidelberg
New York
Barcelona
Hong Kong
London
Milan
Paris
Singapore
Tokyo

E. Krause W. Jäger (Eds.)

High Performance Computing in Science and Engineering '99

Transactions of the High Performance Computing Center Stuttgart (HLRS) 1999

With 281 Figures, 99 in Color and 28 Tables

Springer

Egon Krause
Aerodynamisches Institut der RWTH Aachen
Wuellnerstraße zw. 5 u. 7
52062 Aachen
Germany
e-mail: ek@aia.rwth-aachen.de

Willi Jäger
Interdisziplinäres Zentrum für Wissenschaftliches Rechnen
Universität Heidelberg
Im Neuenheimer Feld 368
69120 Heidelberg
Germany
e-mail: jaeger@iwr.uni-heidelberg.de

Front cover figure: "Concentration of gas and temperature distribution in a turbulent flame" – The Interdisciplinary Center for Scientific Computing at the University of Heidelberg.

Libary of Congress Cataloging-in-Publication Data applied for

Die Deutsche Bibliothek – CIP-Einheitsaufnahme
High performance computing in science and engineering '99: transactions of the High Performance Computing Center Stuttgart (HLRS) 1999 / E. Krause ; W. Jäger (ed.). – Berlin ; Heidelberg ; New York ; Barcelona ; Hong Kong ; London ; Mailand ; Paris ; Singapur ; Tokio ; Springer, 2000

Mathematics Subject Classification (1991): 65Cxx, 65C99, 68U20

ISBN 978-3-642-64084-1 ISBN 978-3-642-59686-5 (eBook)
DOI 10.1007/978-3-642-59686-5

Cover Design: design & production, 69121 Heidelberg
Production: Pro-Edit GmbH, 69126 Heidelberg, Germany
Typeset by the Authors
SPIN 10724101 46/3143 – 5 4 3 2 1 0 – Printed on acid free paper

Preface

Prof. Dr. Egon Krause
Aerodynamisches Institut, RWTH Aachen
Wüllnerstr. 5 u. 7, D-52062 Aachen

Prof. Dr. Willi Jäger
Interdisziplinäres Zentrum für Wissenschaftliches Rechnen
Universität Heidelberg
Im Neuenheimer Feld 368, D-69120 Heidelberg

High Performance Computing is progressing as a discipline providing important tools for research and development in science and industry. The High-Performance Computing Center Stuttgart (HLRS) is not only providing the facilities, hard- and software for a growing community of researchers and developers, but it also promotes the know-how to use supercomputers efficiently. Regular exchange of information, of ideas and methods is essential in improving the proper use of the facilities,and their performance as well as the application of algorithms and of simulation techniques.

A Second Result and Review Workshop on High-Performance Computing in Science and Engineering, (October 4 - 6,1999) was organized by the HLRS in order to give an overview of the scientific work carried out during the past year and to demonstrate the state of the art in the various fields. In 1998 the Land Baden-Württemberg decided to extend the responsibilities of the Steering Committee of the HLRS and therewith also the rules of access to its Scientific Supercomputing Center (SSC) Karlsruhe. That center was recently upgraded with the IBM RS 6000 SP, thereby significantly increasing the attractivity of the two centers, since the joint portfolio of computer-architectures now covers most of the application-profile of their users.

In the second half of 1998 and the first half of 1999 the Scientific Steering Committee accepted altogether 190 projects for computation on the parallel vector system NEC SX4/40H2 and the massively parallel systems CRAY T3E-900/512 and IBM RS/6000 SP/256. Scientists from 42 universities and research centers in Germany were involved in these projects. About one half of the total available machine power was allocated for simulating processes in fluid dynamics and reacting flows, 37 percent for physical processes and those in solid state physics, 10 percent for chemical processes and the rest for processes in life sciences, structural mechanics, electrical engineering, and computer science.

The present volume contains the 43 representative contributions, that were reviewed and selected by the referees from the projects processed in both centers and communicated at the workshop. They are arranged according to the nine disciplines mentioned above. A wide range of problems is covered in the articles. All of them require high-performance computers for their solution. However, large-scale research projects demanding an interdisciplinary joint approach by different scientific disciplines are still in minority. Computational science and engineering had enormous success during recent years. In order to treat complex real life processes without reducing their dimensionality or dealing only with partial aspects, high-performance computing and interdisciplinary research have to be promoted and appropriate steps be taken. Recently, the Academic Strategic Alliance Program (ASAP) was conceived and implemented in USA. The primary goals of this initiative are the following: to establish within the academic community the practices of large-scale modeling and simulation, to accelerate advances in basic sciences, mathematics and computer science for computational science and engineering, and to establish coupling of such projects between the partners of the strategic alliance. The influence of this program on the selected projects and on science in general has to be analyzed. However, already now it can be stated: European activities for initializing and funding similar interdisciplinary research are long overdue.

In order to maintain international competitiveness of the HLRS the Steering Committee in June 1998 recommended to increase the capacity of the parallel vector system. The proposal was immediately approved by the Land Baden-Württemberg and also by the Deutsche Forschungsgemeinschaft, so that the necessary funds could soon be allocated, strongly supported also by the industrial partners, the debis Systemhaus and the Porsche AG. The new machine, the NEC SX-5/32M2e increases the computational power of the parallel vector system by a factor of three. With a memory of 80 GB, of which up to 48 GB can be used as shared-memory, the machine at the present time offers the largest memory installed in Germany for applications that have to rely on a uniformly accessible common address system.

Many articles published in this volume confirm that such a large memory is urgently needed for projects simulated on Parallel Vector Systems. Especially flow simulations and related topics need large storage capacities, as for example the simulation of two-phase flows, of crystal growth, of the process of laminar-turbulent transition, of flow separation, and of direct and large-eddy simulation of turbulent flows. The requirement for large storage capacities is even more pronounced for flows of chemically reacting gases. In addition to the fluid dynamical properties as velocity, pressure, temperature, and density, the changes of the chemical species have to be taken into account, enlarging the required storage capacity in many cases by a factor of ten and more.

The contributions in this volume clearly show, that efficiency in high-performance computing can only be achieved, if computer-platforms and algorithms are carefully chosen for the solution processes considered. It is easy to visualize, that with further increase of compute-power, the machine architectures as well as the further development of algorithms will have to be geared to scientific concentrations in the application-profile. Stronger coordination between HPC-centers will also be necessary to keep computational costs low and provide the service required by the scientific disciplines. It is hoped that the articles of this volume again are demonstrating the importance of high-performance computing. The editors strongly believe that this discipline needs further strengthening. Scientific and technological developments have to be imbedded in an appropriate computational environment in order to achieve essential progress and to stay internationally competitive. All disciplines involved have to join their potential in an integrated approach to solve important large-scale problems.

Stuttgart, October 1999

W. Jäger E. Krause

Contents

Physics

Prof. Dr. Werner Hanke
Physikalisches Institut, Lehrstuhl für Theoretische Physik I
Universität Würzburg
Am Hubland, D-97074 Würzburg

Prof. Dr. Hanns Ruder
Institut für Theoretische Astrophysik
Universität Tübingen
Auf der Morgenstelle 10, D-72076 Tübingen

In almost all fields of physics, such as atomic and molecular physics, solid state physics, plasma physics, hydrodynamics, electrodynamics, quantum mechanics, quantum chromodynamics, high-energy physics, astro- and geophysics, fundamental new results were achieved by means of High Performance Computing. It provides the possibility of analyzing complex systems by the aid of simulation. This procedure is as important as theoretical investigations and closely interacts with the latter. Numerical simulation is more and more becoming a pillar equivalent to the two classical pillars of gaining knowledge, namely the theoretical investigation and the experiment. Corresponding to the problems involved and the methods and tools used, "Wissenschaftliches Rechnen" - the German designation - has a strongly interdisciplinary character by integrating contributions from different fields of natural sciences, applied and numerical mathematics as well as informatics. In the following, out of numerous projects in the field of physics currently running at the HLRS, 14 projects were selected to illustrate the scientific progress which can be achieved with the supercomputer resources of the HLRS with examples from geophysics, astrophysics, molecular dynamics and solid state physics. Out of these excellent projects, two topics concerning the subject of solid physics are discussed in some detail. One concerns the theme "Vibronic studies of absorbate-covered semiconductor surfaces with the help of HPC" by U. Freking, A. Mazur and J. Pollmann. This project concerns technologically relevant surface systems with silver or selenium atoms on germanium surfaces. These adsorbate systems are prototypes for the passivation of semiconductor surfaces. In recent years, the electronic and structural properties have been studied by ab-initio methods, in particular the local density version of the density functional theory. On the other hand, the description of vibronic properties of the surface atoms was limited to semi-empirical methods. The use of high performance computing for the first time allowed the calculation of surface lattice vibrations for these particular systems from first principle. The efficiency achieved with the parallelization is very high and independent of both the thickness of the super cell used in the repeated slab model and the number of plane waves taken into account in the electronic basis set. The reason for this efficiency lies in the fact that almost the whole computation depends on special points, and the loops over these points can be parallelized comparably easily by using the message passing interface. A comparable efficiency seems to be reachable also for other surfaces, so that the density functional technique in a parallelized version facilitates an ab-initio determination of surface phonons for many interesting semi-conductor surfaces, both clean and adsorbate-covered. Because of the technological importance of semi-conductor nanostructures in fields such as micro-electronics, the description of these vibronic properties is of great interest for future studies.

The second project which should be mentioned in more detail is titled "Towards the limits of present-day supercomputers: Exact diagonalization of strongly correlated electron-phonon systems" by G. Wellein and H. Fehske. This project aims at the exact diagonalization of strongly correlated electron-phonon systems such as the high-Tc superconducting compounds, the non-metallic nickelates, and the magnetoresistance manganites. Again, these compounds are very interesting from a material science point of view and are potential candidates for a variety of technological applications. In these systems, in contrast to the before-mentioned semiconductor surface-systems, the challenge is the simultaneous interplay of the vibronic degrees of freedom with the strong electronic correlations, which can not be treated with perturbation theory. Employing the exact diagonalization technique in combination with the kernel polynomial moment expansion and the maximum entropy method on massive parallel computers, the authors extracted ground state properties and spectral functions for a variety of models of electrons or spins strongly interacting with quantum phonons. In the numerical work, the full dynamics of all the phonon modes was taken into account on finite lattices up to 16 sites. The efficiency and accuracy of the phonon Hilbert space used in the Lanczos diagonalization was demonstrated in this project. The numerical techniques presented in this work were applied successfully also to various other issues, such as the study of self-trapping phenomena of electrons and excitons and to the investigation of polaronic superlattices.

These two selected topics in the area of solid-state physics are examples of exciting projects which was dealt with at the RUS, using the HLRS supercomputing resources. The same certainly is true for every other project within this chapter concerning actual problems in geophysics, astrophysics and molecular dynamics.

Finite difference modelling of elastic wave propagation in the Earth's uppermost mantle

Trond Ryberg[1], Marc Tittgemeyer[2], and Friedemann Wenzel[2]

[1] GeoForschungsZentrum, Telegrafenberg, 14473 Potsdam, Germany.
[2] Geophysikalisches Institut, Universität Karlsruhe, Hertz Str. 16, 76187 Karlsruhe, Germany.

Abstract. Most models of the Earth's upper mantle had previously assumed a homogeneous elastic structure. In contrast many seismological data sets show conclusive evidence for strong scattering. For instance scattering within the uppermost mantle is prominently documented in the so-called high-frequency teleseismic P_n phase, generated by mantle velocity fluctuations. This phase, and its correspondence S_n, is seen globally in data sets from active and passive seismology.

We demonstrate that a wave guide, which is caused by random fluctuations of the mantle's elastic properties can explain the main features of the teleseismic P_n. We focus on the statistical properties of these fluctuations acting as scatterers. To test the hypothesis of an upper mantle scattering wave guide we calculate synthetic seismograms and compare them with observations. To compute realistic seismograms we solved the elastic wave equation numerically. Using a 2D finite difference scheme we calculate synthetic seismograms for a variety of very large models (larger than 1000 wave length). The size of the models employed and the number of time steps computed are unprecedented so far and inconceivable without modern high-performance computing. We developed and optimized an efficient code with High Performance FORTRAN (HPF) for a massive parallel computer system. We discuss the influence of the vertical and horizontal correlation length, RMS velocity fluctuations, and thickness of the heterogeneous layer on the scattering properties of the upper mantle and on the propagation mechanism of the teleseismic P_n.

1 Introduction

The analysis of the propagation of elastic waves (seismic phases) plays an essential role for understanding important questions concerning the structure, composition and evolution of the Earth. Since direct observation of the Earth deeper interior is not possible, elastic waves generated from earthquakes or artificial sources are recorded with seismographs, usually located as array at the Earth's surface, and are analyzed for seismic phases. These phases carry important information about the velocity structure within the Earth. In most cases the direct derivation of a velocity model of the Earth is not possible. To interpret the recorded seismic data and to understand the propagation of the seismic wave field through the Earth, synthetic seismograms are calculated for an initial starting model. These synthetic seismograms are then compared to the observations and the initial model is modified. Again, synthetics are calculated, until the synthetic and observed seismograms are similar enough to stop this iterative process.

In this paper we discuss the long-known, but not yet fully understood, unusual teleseismic P_n phase, propose an alternative velocity model for the Earth's uppermost mantle and test our hypothesis by computing synthetic seismograms. Mantle P_n and S_n phases (compressional and shear waves) with unusually high frequencies and extremely long coda emergent from earthquakes and other sources have been observed in many different areas of the world [5,7,13] as early as 1935. These phases are called oceanic P_n, P_0 and long-range or teleseismic P_n, propagate with a group velocity of ~8.0 km/s and are observed occasionally to a distance of more than 4000 km. It is remarkable that these high-frequency signals (typically >15 Hz, occasionally as high as 35 Hz) are reported for both continental and oceanic paths.

Several explanations for high-frequency teleseismic P_n and S_n have been proposed in the past. All of them can only explain some properties of these phases. We suggested an alternative model of the Earth upper mantle which includes a wave guide with random velocity fluctuations which act as scatterers. Early simulations of wave propagation with model seismics [6] by means of down-scaled models made from copper, steel and plastic, show that scattering effects could explain the coda phenomenon of the teleseismic P_n phase. Our final goal is to quantify our models by computing complete synthetic seismograms and comparing them with observations. The seismic recordings along long-range profiles in Russia using Peaceful Nuclear Explosions (PNEs) as source of elastic energy are a unique data set to encipher this phase. The properties of the high-frequency P_n led us conclude that strong scattering along the propagation path plays an important role [9–11,8]. Our suggested model bases on a global Earth model: IASP91 by [4]. While IASP91 was derived mainly from travel time information of seismic body and surface waves, we tried to include dynamic information (amplitudes, frequency content) of the recorded wave field into the derivation of our model. It contains relatively weak velocity fluctuations with specific spatial properties overlaying the IASP91 model.

2 Wave equation

To test our hypothesis of mantle scattering we calculate synthetic seismograms for different models and compare them with observations. Unfortunately, there is no theory available that predicts wave propagation in heterogeneous media without essential restrictions. We therefore have to apply numerical modelling techniques in order to understand key features of wave propagation for the class of models we believe are relevant to the data. A quite standard approach to calculate the seismic wave field is to solve the wave equation numerically. While this is computationally expensive, especially for realistic models, it has the advantage that complete synthetic seismograms are calculated and no approximations have to be done. Routinely used high-frequency approximations of the wave equation would not be applicable to describe the wave propagation adequately, because the characteristic spatial properties of heterogeneities a are of the order of the dominant wavenumber k ($ka \simeq 1$) and the expected multiple scattering in the proposed wave guide [10]. Solving the wave equation numerically for our large models is challenging and can only be carried out on modern massive parallel computers. Due to the integration of the elastic wave equation over a long wave path ($> 1000\lambda$), we have to restrict ourselfs to two-dimensional modelling.

The wave equation bases on Newton's (force balance) and Hooke's laws (elastic material properties). It describes the temporal and spatial development of the field of motion (wave field) in an elastic solid body with respect to a given velocity model for both components, U and W:

$$\rho \frac{\partial^2 U}{\partial t^2} = \frac{\partial}{\partial x}\left[\lambda\left(\frac{\partial U}{\partial x} + \frac{\partial W}{\partial z}\right) + 2\mu\frac{\partial U}{\partial x} + \frac{\partial}{\partial z}\left(\mu\frac{\partial W}{\partial x} + \frac{\partial U}{\partial z}\right)\right], \tag{1}$$

$$\rho \frac{\partial^2 W}{\partial t^2} = \frac{\partial}{\partial z}\left[\lambda\left(\frac{\partial U}{\partial x} + \frac{\partial W}{\partial z}\right) + 2\mu\frac{\partial W}{\partial z} + \frac{\partial}{\partial x}\left(\mu\frac{\partial W}{\partial x} + \frac{\partial U}{\partial z}\right)\right]. \tag{2}$$

Here U and W represent the vertical and horizontal component of the wave field and μ and λ are the elastic constants, which describe the elastic properties (velocities of propagation) of the model. Generally they are spatial varying functions.

3 Finite difference method

To solve equations (1) and (2) numerically for a given model we have to resample it on a grid, and replace the spatial and temporal partial derivatives by their finite difference equivalents. Replacing the partial derivatives in equation (1) by their second-order in time and space ($O(\Delta t^2, \Delta x^2)$) approximations the equation (1) can be written as:

$$U_{i,j}^{n+1} = F(U_{i,j}^{n-1}, U_{i,j}^{n}, U_{i\pm1,j}^{n}, U_{i,j\pm1}^{n}, W_{i\pm1,j\pm1}^{n}) \tag{3}$$

here $U_{i,j}^n$ represents the vertical component for the n-th time step at the location i, j on the grid. Function F is a simple linear function of their arguments with coefficients which depend on the local values of λ and μ (i.e. the local velocities). Equation (3) gives the rule to calculate the n+1 time step of $U_{i,j}$ from its past values $U_{i,j}^n$ and $U_{i,j}^{n-1}$ and from the spatial neighbours $U_{i\pm1,j\pm1}^n$ and $W_{i\pm1,j\pm1}^n$. A similar scheme holds for $W_{i,j}^{n+1}$:

$$W_{i,j}^{n+1} = F'(W_{i,j}^{n-1}, W_{i,j}^{n}, W_{i\pm1,j}^{n}, W_{i,j\pm1}^{n}, U_{i\pm1,j\pm1}^{n}) \tag{4}$$

Both expressions (3) and (4) can easily be and efficiently calculated on a massive parallel computer system. Of course, special care has to be taken at the surface and at the model boundaries. A point source is approximated by an initial wave field at time $T = 0$ and the synthetic seismograms are recorded with receivers at the surface (Z=0).

We choose a second-order approximation in time and space ($O(\Delta t^2, \Delta x^2)$) of the derivative of the wave equation [3] for the elastic and isotropic case. Numerical grid dispersion for this low-order approximation occurs as one of the most significant numerical problem, which can be overcome by sampling the model finer. A sampling increment ($\Delta x = \Delta z$) which fullfills the relation $\lambda/\Delta x > 10$ for the smallest source wave length λ turned out to be fine enough to avoid numerical dispersion. The advantage to use a second-order (time and space) approximation over the established more sophisticated FD schemes is, that it can be easily implemented on a massively parallel supercomputer (Cray T3E), where sufficient direct-access memory is not an issue. To take advantage of the multiple CPUs the finite difference schemes had to be code in parallel by using High Performance FORTRAN (HPF). Efficient programming can only be done if the tasks (solving the wave equation) are distributed over the individual CPUs uniformly. This is generally very effective for the central part of our models, while the special treatment of the wave equation at model boundaries can not be coded efficiently in parallel. We therefore decided to neglect the left, right, and lower model boundaries, but instead extending our model by a buffer zone. At the model boundaries we did not solve the wave equations (3) and (4) correctly. This led to the generation of parasitic waves emerging from the boundaries. The thickness of that buffer zone was chosen to minimize the interference of these numerical phases with the seismic phases, which we endeavor to study. Benchmark testing has shown, that solving the wave equation on a massive parallel computer system for large models surrounded by these buffer zones can be done much faster than implementing realistic (absorbing, non-reflecting) boundary conditions. Figure 1 shows the computational efficiency of the parallel programming for different model sizes and numbers of CPUs used. It shows the obtained performance expressed in floating point operations per CPU. To update the wave field according to equations (1) and (2) we need ~ 74 floating point operation per grid point (additions and multiplications). Chosing the appropriate number of CPUs a sustained performance of > 100 MFLOPS can be reached easily on a T3E-900/512. Special care had to be taken to avoid caching problems by properly allocating the memory. Best numerical performance was obtained when using the following HPF statements for memory allocation and CPU distribution:

```
      REAL*4 A(0:8,0:35000,-1:5000)
$HPF PROCESSORS VEC(256)
$HPF DISTRIBUTE (*,BLOCK,*) ONTO VEC :: A
```

The strip-like distribution `(*,BLOCK,*)` was chosen because of the necessary special treatment of the upper model boundary (Earth's surface).

To solve the elastic wave equation for models of the size 2100 × 300 km at frequencies around 5 Hz we need a spatial and temporal discretization of

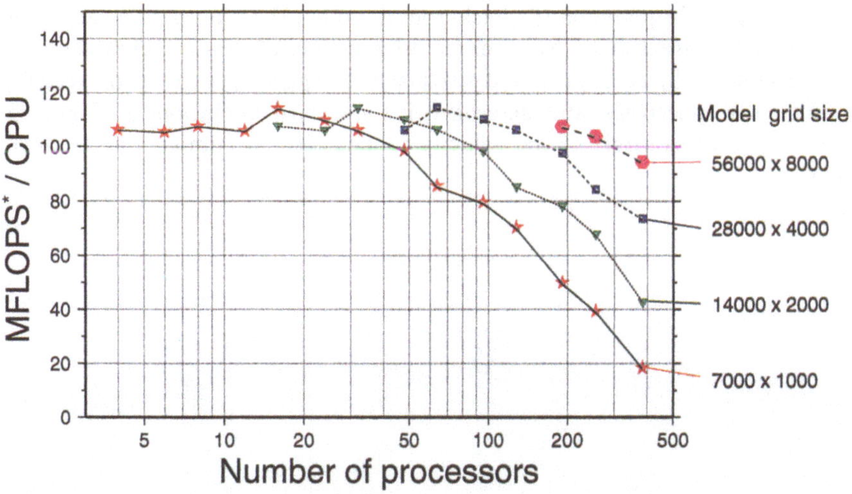

Fig. 1. Performance test for different model sizes and processor numbers. Increasing the model size will led to an increased computational efficiency, especially for large numbers of CPU used.

60 m and 5.6 ms, respectively. This results in a model of 35000 by 5000 grid points and an integration over ~ 50000 time steps. To compute the wave field for only one model more than $6 \cdot 10^{14}$ floating point operations had to be carried out, the required memory was >25 GBytes. Model input and seismogram output increased the computational time by not more than 5%. As a source we choose a simple, relatively broadband (2 – 8 Hz) wavelet with the main frequency centered at 5 Hz. This is similar to the average frequency content of the high-frequency P_n phase in the observations along the PNE profile Quartz [9].

4 Model generation

When calculating synthetic seismograms we use a rather simple crustal structure. It consists of a homogeneous crust with a P-wave velocity of 6.2 km/s and a transition zone between 30 km and 35 km depth with a P-wave velocity value of 7.5 km/s reached at the base of this zone. The crust-mantle boundary (Moho) is marked by a first order discontinuity at 35 km depth with a velocity contrast from 7.5 km/s to 8.04 km/s. The upper 100 km of the mantle are characterized by a gradient zone, which is subsequently underlain by a homogeneous halfspace. In the sub-Moho zone we randomize the P velocity. The spatial distribution of the velocity perturbations is generated in the two-dimensional wavenumber domain by filtering an array of random numbers with a filter operator chosen to generate a random velocity field with a Gaussian distribution [12]. This random number field with zero mean, was scaled and superimposed on the background model. We generated a variety of models, each characterized by a unique description in terms of heterogeneous layer thickness, L, horizontal and vertical heterogeneity correlation length, a_x and a_z, respectively, and heterogeneity standard deviation, σ. Figure 2 shows a typical example of the velocity model with the upper mantle wave guide.

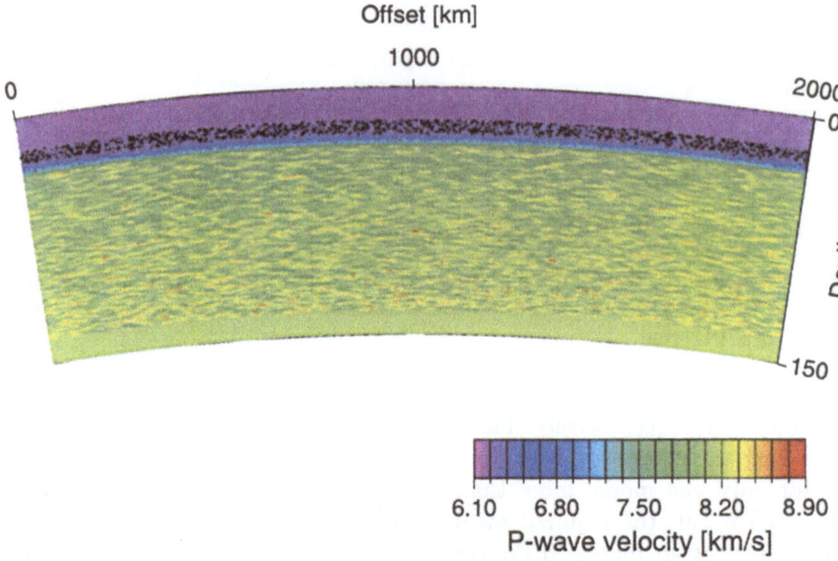

Fig. 2. Long distance wave propagation of P_n and S_n is achieved by statistical fluctuations of elastic parameters in the uppermost mantle, which differs remarkable in the scale of structure and in the magnitude of variation from the overlaying lower crust. A typical lower crustal structure is generated according to [2]. Mantle heterogeneities have a gaussian distribution with a horizontal and vertical correlation length of 20 km and 0.5 km, respectively. The standard deviation of the velocity fluctuations is 2 %. Both targets, lower crust and upper mantle provide a generic description of geologic heterogeneity. (The figure is plotted with a radial exaggeration of 1:4)

5 Parameter scan

The aim of our investigation consists firstly in the verification whether two-dimensional velocity fluctuations in the upper mantle could explain the observation of high-frequency P_n phases. We perform a grid search to scan the possible range of layer thickness, L, horizontal and vertical heterogeneity correlation lengths, a_x and a_z, heterogeneity standard deviation, σ, and other properties. Considering the effort of calculating synthetic seismograms for these huge models no extensive and complete grid search can be done.

We started our studies with the analysis of the wave propagation in the background model. Figure 3a shows the vertical component, unfiltered record section for the background model. These calculations were kinematically verified by the one-dimensional reflectivity method [1] to test the implementation of the finite difference code. The record section is mainly characterized by the Moho reflection ($P_M P$), and the phase diving into the upper mantle (P). The secondary wave field consists of their respective multiple reflected and converted waves generated at the free surface (P') and the Moho. The mantle refraction remains weak between the critical distance and of $P_M P$ and 750 km. It splits into two phases at a distance around 1000 km, the fast being the direct wave diving deeper into the upper mantle. The second phase has a lower apparent velocity ($\sim 8.0\,\mathrm{km/s}$) and represents the branches of the whispering-gallery phase (WG) at the crust/mantle boundary. From this simple testing (Fig. 3a) it is quite obvious that the whispering-gallery phase, although having arrival times and an apparent velocity similar to the observations, does not have a long coda. The background model and its seismograms served as a reference for the further calculations, which include mantle velocity fluctuations.

By perturbing the smooth upper mantle gradient with weak velocity fluctuations the corresponding wave field gets more distorted (Fig. 2). The introduction of random velocity fluctuations provokes a redistribution of energy on individual phases, see Fig. 3b. With σ=2% the main energy is travelling as a secondary phase: the high-frequency P_n. A teleseismic P_n separates from the refracted mantle phase P at distances $\geq$ 1000 km. While this phase is clearly low frequency, the teleseismic P_n is dominated by a higher frequency content due to the frequency selective multiple scattering process. This behaviour is very similar to the observed data (Fig. 3c) and was formerly shown to be a characteristic feature of teleseismic P_n observations [9,11]. It was also quite instructive to look at the spatial distribution of the wave field at several times. Figure 4 shows the horizontal component of the wave field at T=24.64 s and T=168 s time of wave propagation. The snapshot for the early time shows no scattering of the seismic phases. As expected, at later times the snapshot is dominated by strong scattering phases which appear as spatial distributed coda.

We performed a simple scan of the parameter space (L, σ, a_x, a_z) and compared the synthetic seismograms with the observations. A qualitative comparison with seismic recordings (e.g. Fig. 3c) had shown that only models with specific properties of the velocity structure are capable to propagate these phases efficiently. Our favorite model for the Russian PNE profiles includes a 100 to 150 km thick zone of scatterers below crust-mantle boundary, containing velocity heterogeneities with a horizontal correlation-length of 20 km, a vertical correlation-length of 0.5 km and a RMS velocity perturbation of 2%, while keeping the P- and S-wave velocities correlated.

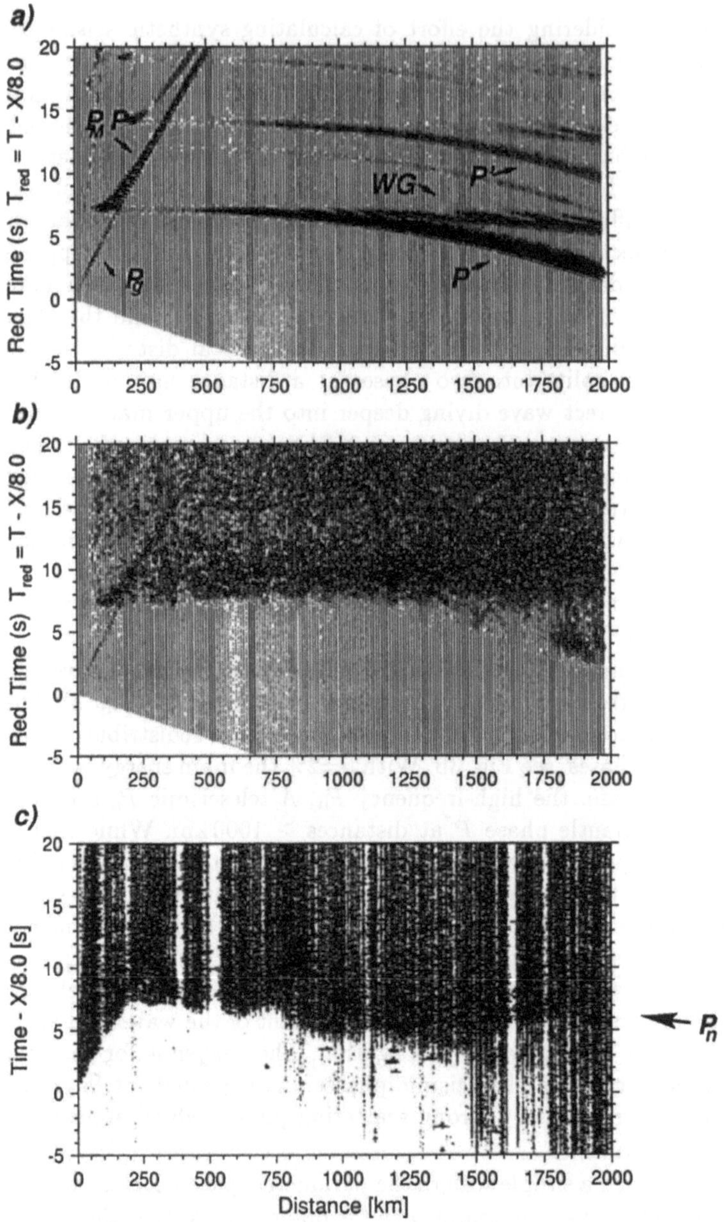

Fig. 3. a,b) Synthetic seismograms computed for different models of upper mantle structure. The model without scatterers for the seismograms in a) was taken as reference. This model is characterized by a simple and smooth, velocity structure. Thus its seismic response can be described in terms of discrete phases (see text). The underlying model for the seismogram section in b) is the same as depicted in Fig. 2. Note the obvious resemblance of the seismograms to the observations c). c) shows an example of observed (recorded) data along the russian long-range profile Ruby.

Several interpretations for the cause of velocity fluctuations are possible: Conceivable variations in the mantle mineralogy (olivine, ortho- and clinopyroxene, and garnet content) can easily explain variations of the P and S velocities of 2%. This holds true for both typical peridotitic and eclogitic mantle material. An alternative to the compositional interpretation of the velocity fluctuation consists of a velocity model with spatial varying intrinsic anisotropy.

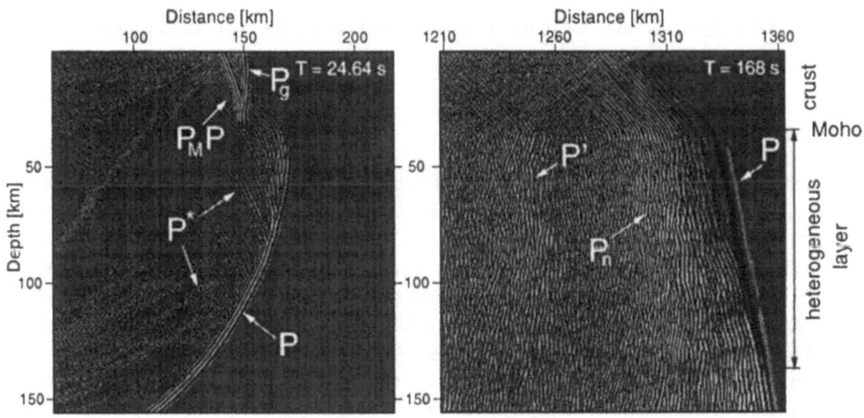

Fig. 4. Snapshots of the horizontal component of the wave field at 24.64 s (left panel) and 168 s (right panel) travel time.

6 Conclusion

Numerical simulations of wave propagation in the upper mantle requires 2D modelling. We carried out such calculations in order to explain the propagation of a guided wave in the uppermost mantle, the so-called teleseismic P_n phase. The propagation mechanism of this phase is under discussion since its first observation in 1935. As the teleseismic P_n carries important information about the structure of the upper mantle the phase is thus relevant for the understanding of tectonic processes and of the accretion of continents. Theoretical studies of wave propagation in heterogeneous media can not be done without essential restrictions and thus realistic forward modelling of the teleseismic P_n has been a challenge since decades, but was impossible so far. The exploitation of modern high-performance computing now allows to determine and decide on the physics of wave propagation for realistic models of the Earth's upper mantle.

To test the hypothesis of an upper mantle scattering we carried out calculations of upper mantle models and compared them with observations. The best fit to the simulated high-frequency teleseismic P_n coda is obtained for a model consisting of a scattering zone of 100 km thickness, with velocity fluctuations of 0.5 by 20 km scale, and a RMS velocity perturbation of 2 %.

Acknowledgements. We are grateful to Karl Fuchs for the many discussions concerning wave propagation in the Earth, as to his comments and suggestions which improved earlier versions of this manuscript. The numerical calculations of the synthetic seismograms has been supported by the High-Performance Computer-Center in Stuttgart and has been carried out on a Cray T3E-900/512. All figures were generated using the GMT software of [14].

References

1. Fuchs, K., Müller, G.: Computation of synthetic seismograms with the reflectivity method and comparison with observations. Geophys. J. R. Astron. Soc. **23** (1971) 417–433
2. Holliger, K., Levander, A., Carbonell, R., Hobbs, R.: Some attributes of wavefields scattered from Ivrea-type lower crust. Tectonophysics **119** (1994) 497–510.
3. Kelly, K. R., Ward, R. W., Treitel, S., Alford, R. M.: Synthetic seismograms: a finite-difference approach. Geophysics **41** (1976) 2–27
4. Kennett, B. L. N., Engdahl, E. R.: Traveltimes for global earthquake location and phase identification. Geophys. J. Int., **105** (1991) 429–465.
5. Linehan, D.: Earthquakes in the West Indian region. EOS Trans. Am. Geophys. Union **21** (1940) 229–232.
6. Menke, W. H., Richards, P. G.: The horizontal propagation of P waves through scattering media: Analog model studies relevant to long range P_n propagation. Bull. Seismol. Soc. Am. **73** (1983) 125–142
7. Molnar, P. and Oliver, J.: Lateral variations of attenuation in the upper mantle and discontinuities in the lithosphere. J. Geophys. Res. **74** (1969) 2.648–2.682
8. Ryberg, T., Wenzel, F.: High-frequency wave propagation in the uppermost mantle. J. Geophys. Res. (to appear)
9. Ryberg, T., Fuchs, K., Egorkin, A. V., Solodilov, L.: Observation of high-frequency teleseismic P_n waves on the long-range Quartz profile across Northern Eurasia. J. Geophys. Res. **100** (1995) 18.151–18.163
10. Tittgemeyer, M., Wenzel, F., Fuchs, K., Ryberg, T.: Wave propagation in a multiple-scattering upper mantle — observations and modelling, Geophys. J. Int. **127** (1996) 492–502
11. Tittgemeyer, M., Ryberg, T., Fuchs, K., Wenzel, F.: Observation of teleseismic P_n/S_n on super long-range seismic profiles in northern Eurasia and their implications for the structure of the lithosphere. In Upper Mantle Heterogeneities from Active and Passive Seismology edited by K. Fuchs NATO ASI Series (1997) 63–73 Kluwer Academic Publishers Dordecht /Boston /London
12. Tittgemeyer, M., Wenzel, F., Ryberg, T., Fuchs, K.: Scales of heterogeneities in the continental crust and upper mantle. Pure Appl. Geophys. (in press)
13. Walker, D. A.: High-frequency P_n and S_n phases recorded in the western Pacific. J. Geophys. Res. **82** (1977) 3.350–3.360
14. Wessel, P. and Smith, W.: Free software helps map and display data. EOS Trans. Am. Geophys. Union **72** (1991) 441,445–446

Direct Simulation of Seismic Wave Propagation

Martin Karrenbach[1]

Geophysical Institute, Karlsruhe University, 76187 Karlsruhe, Germany

Abstract. Full wave form modeling techniques in 3D complex heterogeneous elastic media are computationally expensive, even on today's largest supercomputers. In contrast asymptotic techniques only compute a small subset of the wave field, but can be very fast. Using asymptotic methods in combination with full wave form techniques can speed up the overall computation. I use a particular asymptotic method to compute dynamically the progressingly active computational domain for a parallel full wave form finite difference technique. This leads to an overall decreased computational runtime. This computational technique is applied to simulation in a industrially oriented oil/gas bearing subsruface model as well as to simulation of earthquake related scientific scenarios.

1 Introduction

Traditional full wave form seismic simulation techniques are well established such as pseudo-spectral and high-order finite difference methods and others. Although improvements are continuously ongoing, the major limitation so far has been the huge expense associated with realistic 3D computation for complex heterogeneous subsurface models. Full wave form techniques aim at producing exact seismic wave form solutions in complex 3D subsurface models for a prescribed recording period. One of the biggest challenges in 3D full wave form modeling is to make the computational effort more economical. Some attempts promise to be successful, such as fixed-geometry based domain decomposition and the use of unstructured grids.

The method of active domain decomposition which I describe later, can be used in conjunction with traditional Finite Difference techniques as well as with the previously mentioned novel approaches. Active domain decomposition techniques as well as hybrid simulation techniques can be implemented using various asymptotic methods. However, one critical issue is efficiency and speed. If the asymptotic method is slower than computing the full wave form solution, then the asymptotic methods is not practical, unless certain wave field subsets need to be computed only.

Fast marching level set methods are particularly suitable for the approach presented here, because they can be efficiently implemented to solve 3D problems. Given an equilateral computational grid with n grid points along each axis, full wave form finite-difference solutions show an operational count of the order $O(n^3)$ per time step while fast marching methods, which track the wave front only, have a substantially reduced operational count $O(logn^3)$ for the entire problem.

Several years ago seismic modeling initiatives [1,3] were pursued by SEG, EAGE, Industry and National Laboratories to numerically calculate acoustic 3D seismic data. The 3D models were chosen to represent realistic geologic settings. However, due to computer memory limitations and time constraints the seismic data were modeled purely acoustically, and thus lack certain real-world effects. Nevertheless, even up to now, repeating the same simulation for an elastic subsurface model, has been hampered by the availability of computers that can provide enough memory and compute power.

2 Finite Difference Techniques

For computing the full wave form solution, I am using high-order optimized finite difference operators to approximate partial derivatives in space and time as described in more detail in [4,15,5,6]. I am solving basic anisotropic wave equations of the form

$$\underset{\approx}{\mathbf{a}}(\underset{\sim}{\mathbf{x}})\ \underset{\sim}{\nabla}\ \underset{\approx}{\mathbf{b}}(\underset{\sim}{\mathbf{x}})\ \underset{\sim}{\nabla}^t\ \underset{\sim}{\mathbf{u}}(\underset{\sim}{\mathbf{x}},t)\ -\ \frac{\partial^2}{\partial t^2}\underset{\sim}{\mathbf{u}}(\underset{\sim}{\mathbf{x}},t) = \underset{\sim}{\mathbf{f}}(\underset{\sim}{\mathbf{x}}_0,t)\ , \tag{1}$$

which can be easily extended to visco-elastic and more complicated cases and where $\underset{\sim}{\mathbf{u}}$ is an arbitrary wave field (scalar or vector) and $\underset{\sim}{\mathbf{f}}$ is the force applied at source locations $\underset{\sim}{\mathbf{x}}_0$. ∇^t is a general gradient operator and $\underset{\sim}{\nabla}$ the associated divergence operator applied to the wave field components in three dimensions. $\underset{\approx}{\mathbf{a}}$ and $\underset{\approx}{\mathbf{b}}$ are medium property fields such as density, velocity or stiffnesses.

I solve equation 1 numerically as set of first order coupled equations. This allows to freely impose boundary and initial conditions and to extract all desired wave field quantities, such as pressure, particle displacement, acceleration, stress, strain and force. Such a flexible simulation method, easily produces wave fields for acoustic, elastic, anisotropic and viscous media, with and without free surface effects and allows a large degree of freedom for variable recording geometry and observable. The computational complexity increases typically to the third power with increasing computational 3D volume.

3 Static Domain Decomposition

The modeling code proceeds to parallelize the entire computation by decomposing the computational domain into a set of subdomains. All subdomains can be computed concurrently using the same computational stencils.

The logical steps of the parallel algorithm are:

- Determine processors layout depending on 2D or 3D input model files or user specified input

- Determine local coordinates in subdomains
- Set some specific subdomain properties
- Read model grid on master processor and distribute in subdomains
- for each timestep
  - compute the new complete wave field according to the non-parallel FD algorithm
  - exchange boundary values of the wave field with neighbours
  - collect seismogram information from subdomains and write out
- repeat for next source location, leave model in memory

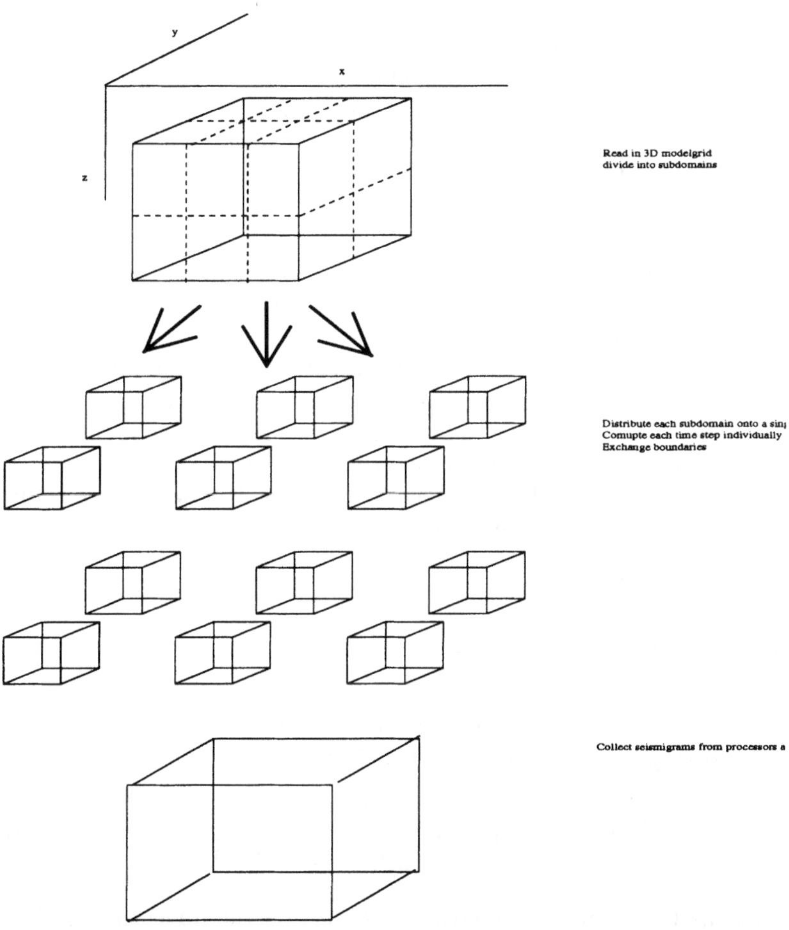

Data flow using the parallel domain decomposition implementation.

The available processors are arranged in Cartesian 2D or 3D grids. The user can specify how many processors are used along each individual spatial dimension, however there is an optimal layout that balances computational load and boundary exchange. Each processor determines its nearest neighbours in each dimension, as well as the diagonal neighbours for the diagonal exchange. Then each processor determines how the global input space is dissected and how it is assigned to a processor. This particular information is needed when the elastic parameters are distributed or when the boundaries of the wave fields are exchanged at each timestep. Thus each processor contains a unique part of the global input space. All local input spaces consist of roughly the same number of grid points. This is a current restriction and cannot be easily overcome without loss off efficiency.

The master processor reads in all parameters and input files. It is responsible for distributing all information to the individual processors. The input model, which can easily be on the order of 1GB to 10GB, is read in sequential manner, thus creating a bottleneck. For large scale modeling runs this initial loading time represents only a small fraction of the total runtime. The input elastic parameter file is read and portions are distributed into subdomains on the individual processors. It uses a double-buffering mechanism to perform the distribution efficiently.

A global stability check for the numerical method is carried out on each subdomain and results are collected on the master processors. There the decision is made if the modeling run is stable and non-dispersive, or if the run should be aborted. The algorithm proceeds to convert global source and receiver coordinates into local subdomain coordinates. Following these setup procedures, the simulation proceeds to compute the full wave field solution for a given number of timesteps.

4 Fast Marching Methods

Fast marching level set methods [11] are relatively recent contributions to numerically solving Eikonal equations of the form

$$\left(\frac{\partial t}{\partial x}\right)^2 + \left(\frac{\partial t}{\partial y}\right)^2 + \left(\frac{\partial t}{\partial y}\right)^2 = \frac{1}{v^2(x,y,z)}\,, \tag{2}$$

where x, y, z are Cartesian coordinates and $t(x, y, z)$ is the travel time field at each point is space. In the early 1990s methods were developed and became popular that solved the Eikonal equation by using finite-difference approximations and produced travel time maps for fixed source seismic experiments. These traveltime maps were usually used successfully in the context of seismic imaging [14,12]. Fast marching methods aim at producing viscosity solutions by using upwind finite-difference schemes. Thus, they compute the continuous first arrivals of the travel time field. By choosing particular finite-difference stencils the method can be made numerically extremely stable. Using a particularly structured algorithm, the method can be made extremely fast. The numerical complexity is orders of magnitudes simpler than for a full wave form finite-difference computation in the same volume.

In this paper I am using a fast marching algorithm to compute the progressively active computational domain for the full wave form finite-difference method. The fast marching algorithm tracks very efficiently the limiting wave fronts and thus decomposes the computational domain dynamically into subdomains. Within these smaller subdomains the full wave form solutions can be computed faster while maintaining identical accuracy.

5 Seismic Simulation in the SEG/EAGE Salt Model

In the following I apply the active domain decomposition by a fast marching method to the computational grid of a high-order finite difference technique.

In the 3D model, Fig. 1, I employed a surface recording geometry. The model consists of a smooth background velocity, with a shallow soft sea floor, that exhibits gentle slopes. There are several interfaces and faults that present some structural complexity. A anomalous geopressure zone is incorporated in the model. Details can be found in [10] and in workshops [13] held during the SEG. Previously I generated [5] an elastic model out of the original purely acoustic subsurface model and compared synthetic seismic data in form of snapshots and seismograms for these two scenarios. In this paper I improve on the computational efficiency.

Figure 1 shows the original 3D subsurface model. The dynamic domain decomposition was computed by a fast marching method and Figure 2 shows the outlines of individual subdomains. The dark contour lines represent the boundaries of the subdomains at given instances in time. The shaded colors within those subdomains illustrate the progression of the active domain over time. The domain boundaries are heavily influenced by the 3D subsurface p-wave velocity model. The salt body in particular dramatically increases the computational domain. Figure 3 shows a snapshot of the vertical component of the elastic wave field in this model. An explosive source was used at the surface and we see the complicated wave field generated in the layers on the top of the salt and less energetic penetration through the salt. P-wave to S-wave conversions play a major role in this energy partitioning. The full wave form solution computed using the fixed entire 3D volume is identical to the solution computed with the dynamic domain decomposition, except for numerical round-off errors.

6 Direct Seismic Simulation at the San Andreas Fault Zone System

Earthquake mechanisms and related wave propagation phenomena are still not fully understood and represent challenging problems to the geoscientific

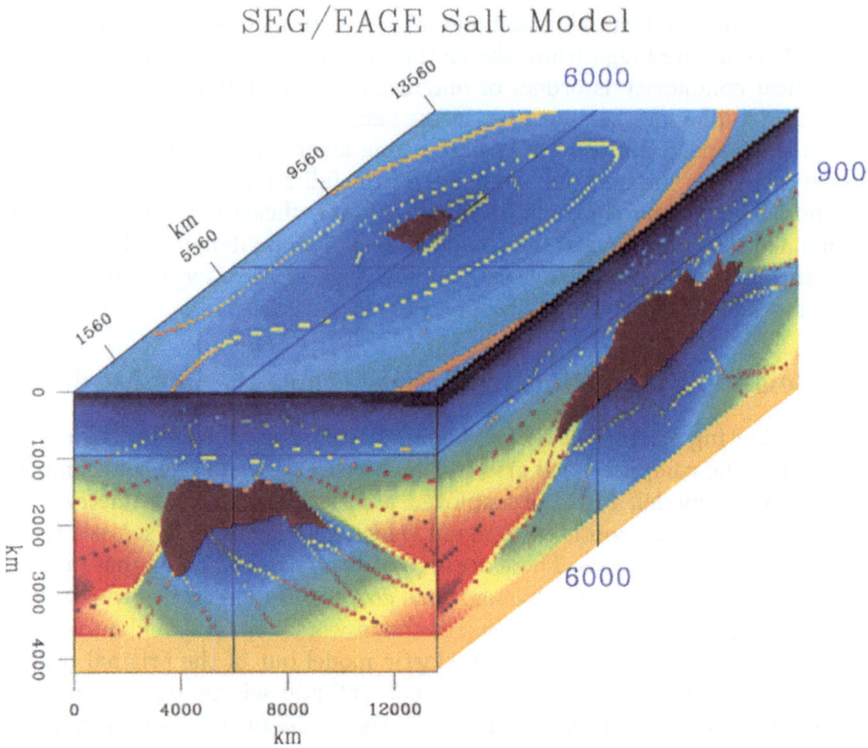

Fig. 1. The compressional wave velocity model of the SEG/EAGE salt model.

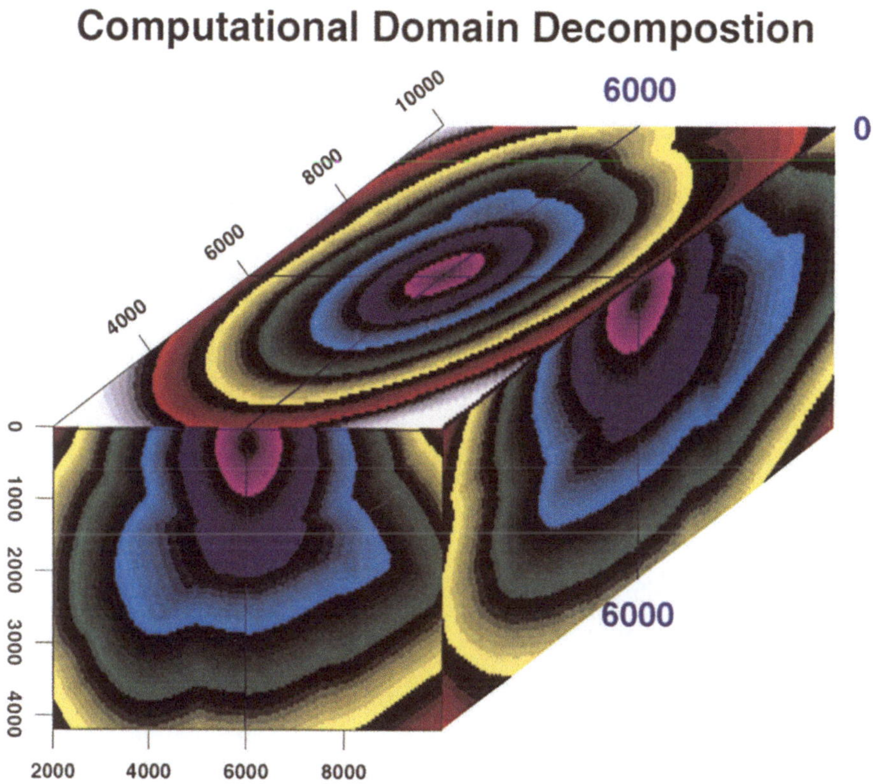

Fig. 2. In the SEG/EAGE salt model Fast Marching Level Set Methods can efficiently estimate the active computational domains dynamically. The determination of the active areas in the 3D model can be followed by computing full wave form solutions only in subdomains of reduced size.

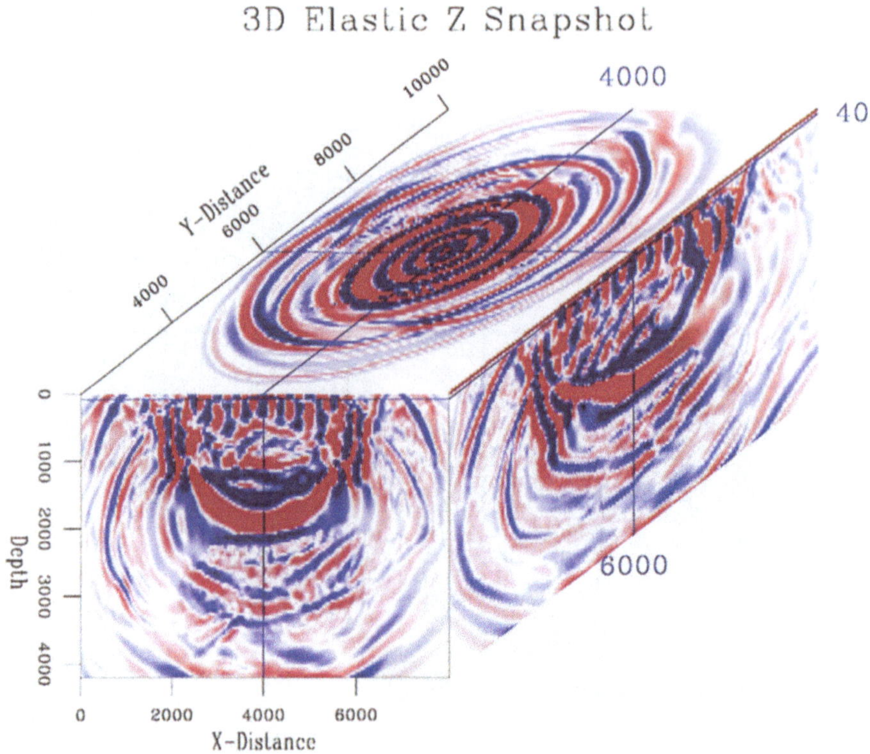

Fig. 3. Snapshot of the elastic full wave form solution computed through the SEG/EAGE salt model using the a Fast Marching Level Set Method to bound the active computational domain for each time step. The result is identical to the one computed on the full 3D grid.

community. A scientific investigation is planned at the San Andreas Fault Zone System in the Parkfield Region, California, USA. In this scientific endeavor a hole might be drilled into an active earthquake fault zone (Zoback et al, 1997). This effort, although initiated by US researchers is an international effort with the aim of determining physical and chemical composition of seismogenic fault zones. Klemperer et al. (1998) propose a 3D reflexion experiment, with which the local 1D information gained in the borehole itself, can be extrapolated into a large 3D volume encompassing a representative portion of the fault zone.

In this investigation 3D seismic wave fields are simulated in complex fault zones. Using these full wave field data can help in interpreting the measured 3D data. On the other hand the influence of petrophysical parameters of the complex fault zone on the seismic wave field can be examined in such simulations.

Using these simulated data sets special processing methods, such as 3D Prestack Depth Migration, Amplitude versus Angle/Azimuth Analysis can be optimized and calibrated. One can test the applicability of these methods in seismogenic fault zones, or one can determine the sensitivity of the wave field towards changes of petrophysical parameters, such as porosity and fluid content. During the simulation an acquisition geometry can be chosen, such that a variety of standard acquisition methods can be compared with respect to the information content which the data embody.

In order to obtain realistic 3D background velocity models, the results of two tomographic inversions (Michelini and McEvilly, 1991; Eberhardt-Phillips and Michael, 1993) were combined into one single 3D subsurface model.

The fault zone itself was approximated by extrapolating the surface geology into depth. and was compared to the fault zone geometry obtained by averaging local earthquake hypocenter locations as given in the southern California Earthquake Cataloque (http://scec.gps.caltech.edu/catalog-search.html). In Figure 4 both subsurface models are combined into one for better comparisons and the location of a possible borehole is plotted into the picture, indicating where it should penetrate the fault zone at depth.

Seismic wave field simulations were carried out in various frequency ranges, in order to mimic local earthquake scenarios as well as artificial sources of a reflexion seismic acquisition. Figure 5 shows a snapshot of the vertical displacement component of the 3D wave field at 1.8 sec after initiating the source.

7 Implementation

The individual runtimes are extremely model and acquisition geometry dependent. A realistic small simulation scenario runs in parallel on 121 processors of the T3E taking about 67 minutes for about 1000 timesteps. The

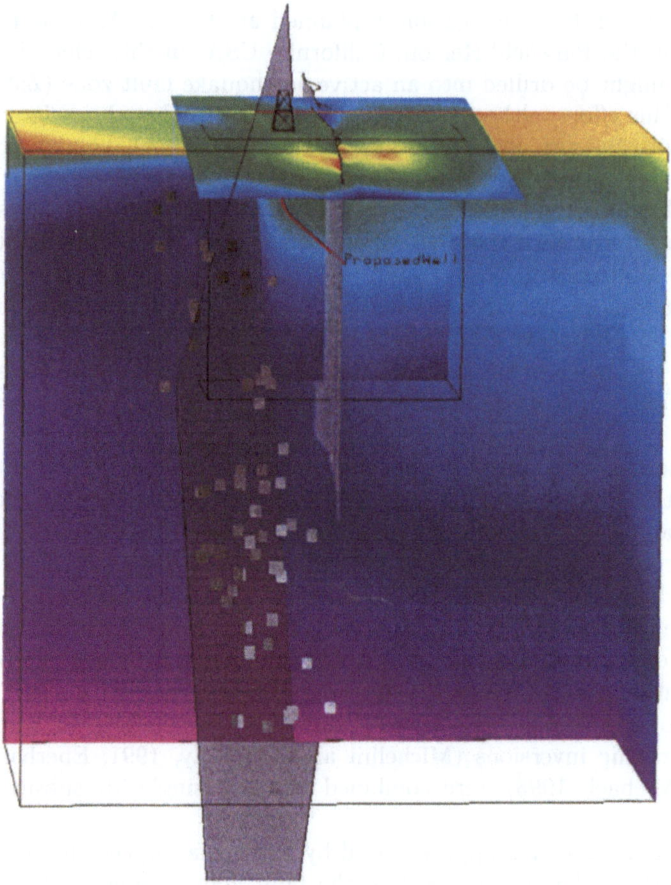

Fig. 4. *Comparison of the fault zones in both 3D subsurface models: The fault surface obtained from the local earthquake are displaced about 1 km to the West in comparison to the fault surface obtained using the surface geology. The colors indicate variations in p-wave velocities.*

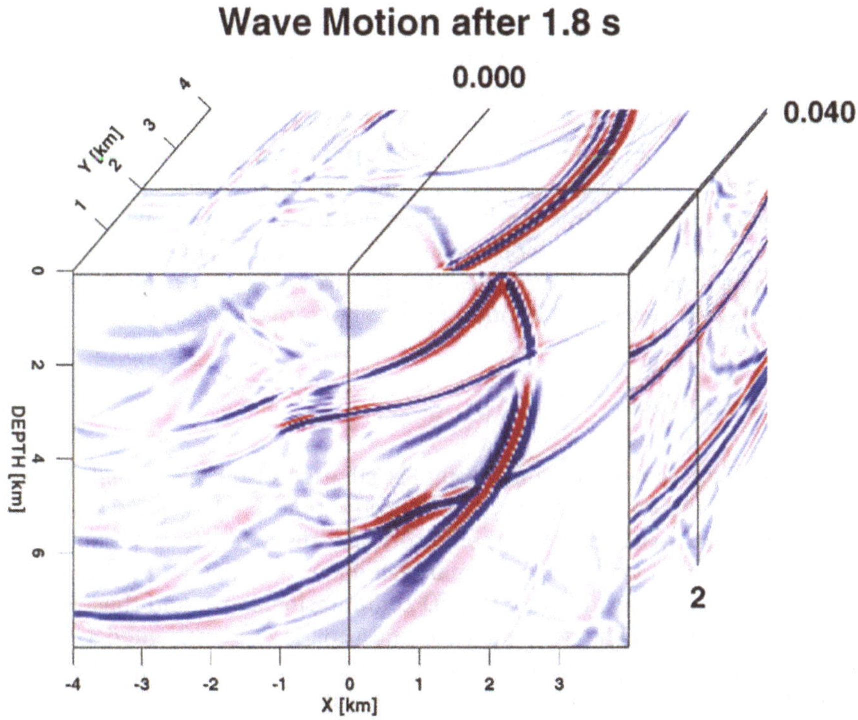

Fig. 5. *Snapshot vertical displacement of the seismic wave field after 1.8 sec. The source was located at the surface and used a dominant frequency of 20 Hz. A low velocity zone in the vicinity of the fault causes a caustic in wave phases.*

memory load per processor is about 110 MB each using a global 210x550x550 3D grid. Typical realistic runs would require between 3000-15000 time steps and the grid dimensions to be increased by a factor of two in each of the three dimension, requiring about 968 processors with 128 MB main memory each; or alternatively more main memory per processor using a smaller number of processors and tolerating increased runtime. We would like to run full wave form simulations in realistic earth models, but the computational problem is memory-bound as well as compute-bound. Thus direct simulations of seismic wave propagation at realistic scales still requires adequate compute power, which is often not available at one single computer center.

8 Conclusions

3D elastic (or visco elastic, anisotropic) simulation of full wave fields is useful for survey design, processing and interpretation. However, there are still challenges, that need to be overcome to make them practical. I show here the combination of two computational techniques, that allows to change dynamically the active grid size of the 3D computational grid and thus reduces the overall runtime. The dynamic decomposition is extremely model, source and receiver geometry dependent. This method can be used together with other hybridization techniques to simulate within a reasonable amount of time realistic sized 3D surveys and to produce 3D full wave fields in complex heterogeneous reservoirs or to simulate earthquake wave propagation in 3D earth models. These small modeling examples demonstrate the practicality of the numerical modeling method seismic exploration settings as wells as for earthquake wave propagation related scenarios. More elaborate modeling scenarios are planned for the Collaborative Research Project SFB 461 and other international collaborative projects at Karlsruhe University.

References

1. F. Aminzadeh, N. Burkhard, J. Long, T. Kunz, and P. Duclos. Three dimensional SEG/EAEG models — an update. *The Leading Edge*, 15(2):131–134, 1996.
2. Donna Eberhardt-Phillips and Andrew J. Micheal. Three-Dimensional Velocity Structure, Seismicity, and Fault Structure in the Parkfield Region, Central California. *gpr*, 98(98):15,737–15,758, 1993.
3. L. House, M. Fehler, J. Bahren, F. Aminzadeh, and S. Larsen. A national laboratory-industry collaboration to use SEG/EAEG model data sets. *The Leading Edge*, 15(2):135–136, 1996.
4. M. Karrenbach. Elastic tensor wave fields. *Ph.D. Thesis Stanford University*, 1995.
5. M. Karrenbach. Full wave form modeling in complex media. 1998.
6. M. Karrenbach. Modeling physical systems, 1999.

7. S.L. Klemperer, J.A. Hole, and M. Talwani. San Andreas Scientific Drilling Project: 3D Seismic Reflection to Extrapolate 1D Drilling into the 3D-Heterogeneous Fault-Zone, 1998. submitted to the National Scientific Foundation (NSF).
8. Gary F. Margrave and University of Calgary The CREWES Project. Seismic acquisition parameter considerations for a linear velocity medium. In *Soc.Expl.Geophys., Extended Abstracts*, pages 47–50, 1997.
9. A. Michelini and T.V. McEvilly. Seismological studies at Parkfield I: Simultaneous Inversion for Velocity Structure and Hypocenters using cubic B-splines Parametrization. *BSSA*, 81(2):524–552, 1991.
10. M. L. O'Brien and S. Gray. Can we image beneath salt ? *The Leading Edge*, 15(1):17–22, 1996.
11. J. A. Sethian. *Level set methods: evolving interfaces in geometry, fluid mechanics, computer vision and material sciences.* Cambridge University Press, 1996.
12. J. van Trier and W. W. Symes. Upwind finite-difference calculation of traveltimes. *Geophysics*, 56(6):812–821, 1991.
13. R. Versteeg and G. Grau. Practical aspects of seismic data inversion, the Marmousi experience. In *Proceedings of 1990 EAEG Workshop*, pages 1–194. Eur. Assoc. Expl. Geophys., 1990.
14. J. E. Vidale. Finite-difference calculations of traveltimes in three dimensions. *Geophysics*, 55(5):521–526, 1990.
15. J. Virieux. P-SV wave propagation in heterogeneous media - velocity-stress finite -difference method. *Geophysics*, 51(4):889–901, 1986.
16. Mark D. Zoback, Stephen H. Hickmann, and William L . Ellsworth. Scientific Drilling Into The San Andreas Fault at Parkfield, Ca: Project Overview and Operational Plan, 1997. submitted to the National Scientific Foundation (NSF).

Summary of Project 11172

Erik H. Saenger[1], Peter Hubral[1], Andreas Kirchner[1], Kai-Uwe Vieth[1], Tobias Müller[1], and Serge A. Shapiro[2]

[1] Geophysical Institute, Karlsruhe University, 76187 Karlsruhe, Germany
[2] Fachrichtung Geophysik, Freie Universität Berlin, 12249 Berlin, Germany

Summary

The project 11172 is embedded in the Collaborative Research Center 381 'Charakterisierung des Schädigungsverlaufes in Faserverbundwerkstoffen mittels zerstörungsfreier Prüfung' (SFB 381). The SFB 381 consists of theoretical and experimental groups in Stuttgart and one theoretical group in Karlsruhe. With help of the CrayT3E we perform large-scale numerical solutions of the elastodynamic wave equation in 2-D and 3-D. These solutions are required to get a better understanding of complex wave phenomena and hence to develop new non-destructive testing methods. Our work on the CrayT3E is based on the program Ultimod [1,4]. The parallel HPF-implementation has been done in the Cray-projects PARSEP and ERDMOD in Stuttgart described in [2,3]. In one part of our project we have modified this program in order to solve our specific problems. This is described in the article 'Modeling of elastic waves in fractured media' in this report. The other three articles are related to SFB 381 topics that are in connection with our Cray-project and rely very much on synthetic data generated within this project.
The resource requirements of our program strongly depend on the given problem (model-size, timesteps to calculate, ...). Until now, we restrict ourselves to relatively small jobs (N64med-queue), but with our new approach - high contrast 3D-FD-modeling - we are going to calculate very large-scale problems.

References

1. M. Karrenbach. *Modeling Physical Systems*. Karlsruhe University (1998).
2. M. Karrenbach. *Finite Difference Modeling of Earthquakes*. Ph.D. thesis, Stanford University (1995).
3. M. Karrenbach, P. Knopf, and M. Jacob. *Parallel Seismic Data Processing*. High Performance Computing in Science and Engineering '98, Springer (1999).
4. M. Karrenbach. *Elastic tensor wavefields*. High Performance Computing in Science and Engineering '98, Springer (1999).

Modeling of elastic waves in fractured media

Erik H. Saenger[1] and Serge A. Shapiro[2]

[1] Geophysical Institute, Karlsruhe University, 76187 Karlsruhe, Germany
[2] Fachrichtung Geophysik, Freie Universität Berlin, 12249 Berlin, Germany

Abstract. *The modeling of elastic waves with an explicit Finite Difference (FD) scheme on a staggered grid causes instability problems when the medium possesses high contrast discontinuities (cracks). We have derived a rotated staggered grid in 2D and 3D where all medium parameters are defined at appropriate positions within an elementary cell for the essential operations. Using this modified grid it is possible to simulate the propagation of elastic waves in a medium containing cracks, pores or free surfaces without applying boundary conditions. In this paper we show the von Neumann stability criterion and the dispersion error for the rotated staggered grid. Additionally, we show a synthetic example to demonstrate the rotated staggered grid in modeling fractured media.*

1 Introduction

Finite Difference (FD) methods discretize the wave equation on a grid. They replace spatial derivatives by FD operators using neighboring points. The wavefield is also discretized in time, and the wavefield for the next timestep is calculated in general by using a Taylor expansion. Elastic FD methods can be separated in time domain FD methods (e.g. [1]), using the 2nd order wave equation, and velocity stress methods, solving two coupled first order equations (e.g. [2]). The main idea using a staggered grid is to calculate spatial derivatives halfway between two gridpoints. Hence, some modeling parameters are defined on intergrid locations, such that they have to be averaged or the grid values halfway between two gridpoints have to be used. This yields in both methods mentioned above inaccurate results or instability problems when the propagation of waves in media with strong fluctuations of the elastic parameters (cracks) is simulated although the von Neumann stability is fulfilled (see e.g. [3]). In such situations, different types of Finite Element (FE) methods or Finite Integral (FI) methods seem to be more appropriate. In this paper, however, we propose an approach to FD staggered grid modeling of the elastic wave propagation in arbitrary heterogeneous and, in principle, arbitrary anisotropic media.

2 Basic numerical procedures

The propagation of elastic waves is described by the elastodynamic wave equation (e.g. [4]):

$$\rho(\mathbf{r})\ddot{u}_i(\mathbf{r}) = (c_{ijkl}(\mathbf{r})u_{k,l}(\mathbf{r}))_{,j} + f(\mathbf{r}). \quad (1)$$

For modeling elastic waves with finite-differences, it is necessary to discretize the stiffness tensor c_{ijkl}, the density ρ and the wavefield u_i on a grid. Calculation of

$u(t+\Delta t)$ from $u(t)$ and $u(t-\Delta t)$ and maybe from the wavefield at timesteps before, when applying a Taylor time update of higher order than two, requires the following steps (when neglecting outer forces and boundary conditions) :

- Calculation of the strain $\epsilon_{kl} = u_{k,l}$.
- Calculation of the stress $\sigma_{ij} = c_{ijkl}u_{k,l}$.
- Calculation of the divergence of stress $\sigma_{ij,j}$.
- Calculation of the acceleration $\ddot{u}_i = \frac{1}{\rho}\sigma_{ij,j}$.
- Performing the time update.

3 The standard staggered grid

A standard way of discretization (e.g. [1]) is shown in Figure 1a. The main reason for using this method is that small scale medium fluctuations are not suppressed, because no averaging (or smoothing) of medium parameters has to be done. The Figure 1a shows the elementary cell, replacing, for example, the density at the left and the lower side by the density of the center and replacing the shear modulus at the lower left corner by the shear modulus at the center.
There is only one density location and one location for the Lamé parameter μ in an elementary cell. The calculation of the stress component σ_{xz} has to be done by multiplying the values of strain and stiffness defined at different positions. The same problem arises for the calculation of the acceleration, since the density has changed its location completely. When the wavefield hits the inhomogeneities with high density contrasts (cracks), stability problems can also occur. We obtain an unstable modeling of a wavefield diffraction on a crack. Note, that such stability problems exist although the von Neumann stability criterion is fulfilled.

4 The rotated staggered grid

All these problems described above can be avoided by choosing another configuration of the grid. Placing all components of the stiffness tensor at the same position within the elementary grid cell (e.g., the center), the positions of the modeling parameters are found directly as shown in Figure 1b.
The grid in Figure 1b satisfies all conditions stated in the previous Section with respect to the five operations that are necessary to perform a timestep. Parameters that have to be multiplied are defined at the same location and derivatives are defined between the parameters that have to be differentiated. Since the density is not located at the same position as the stiffness tensor elements, a density averaging (using the four surrounding cells) has also to be done for this grid. In the case of homogeneous cells or a linear behavior of the density between the stiffness locations, the density coincides with the exact density after averaging.
The directions of spatial derivatives have changed from $\mathbf{x}$ and $\mathbf{z}$ to $\tilde{\mathbf{x}}$ and $\tilde{\mathbf{z}}$. In the following Section, we describe how the derivatives must be performed in the new system of coordinates.

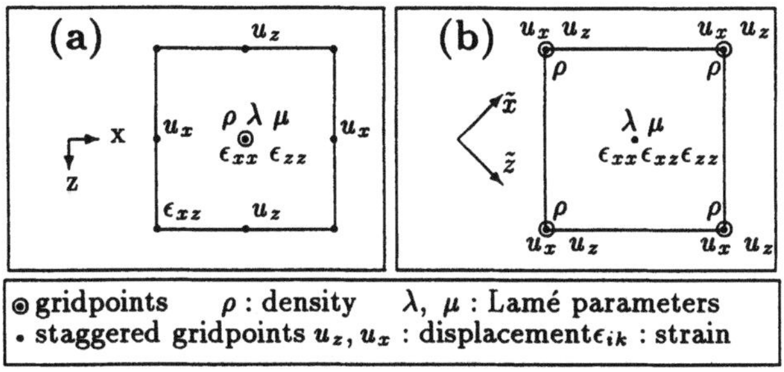

Fig. 1. Elementary cells of different staggered grids. Locations where strains, displacement and elastic parameters are defined. (a) Locations on a standard staggered grid (e.g.. [1]) if no averaging of medium parameters is performed. (b) Elementary cell of the new rotated staggered grid. Spatial derivatives are performed along the $\tilde{x}$- and $\tilde{z}$-axes. The wave equation and the elements of the stiffness tensor are the same as in (a).

5 Discretization of derivatives of the rotated staggered grid in 2D and 3D

In the case of a grid with rectangular elementary cells of length Δz and Δx, one yields for the new derivative directions (in two dimensions):

$$\tilde{\mathbf{z}} = \frac{\Delta x}{\Delta r}\mathbf{x} + \frac{\Delta z}{\Delta r}\mathbf{z}, \tag{2}$$

$$\tilde{\mathbf{x}} = \frac{\Delta x}{\Delta r}\mathbf{x} - \frac{\Delta z}{\Delta r}\mathbf{z}, \tag{3}$$

$$\Delta r = \sqrt{\Delta z^2 + \Delta x^2}.$$

In order to perform numerical simulations, one must define differentiation operators $D_{\tilde{x}}$ and $D_{\tilde{z}}$ for the derivatives $\frac{\partial}{\partial \tilde{x}}$ and $\frac{\partial}{\partial \tilde{z}}$ in $\tilde{\mathbf{x}}$- and $\tilde{\mathbf{z}}$-direction:

$$D_{\tilde{x}}u(z,x,t) = \frac{1}{\Delta r}[u(z - \frac{\Delta z}{2}, x + \frac{\Delta x}{2}, t) - u(z + \frac{\Delta z}{2}, x - \frac{\Delta x}{2}, t)], \tag{4}$$

$$D_{\tilde{z}}u(z,x,t) = \frac{1}{\Delta r}[u(z + \frac{\Delta z}{2}, x + \frac{\Delta x}{2}, t) - u(z - \frac{\Delta z}{2}, x - \frac{\Delta x}{2}, t)]. \tag{5}$$

The spatial derivatives in $\mathbf{z}$- and $\mathbf{x}$-direction in the new grid (i.e., using the new derivative directions) are:

$$\frac{\partial}{\partial z}u(z,x,t) \approx \frac{\Delta r}{2\Delta z}\left(D_{\tilde{z}}u(z,x,t) - D_{\tilde{x}}u(z,x,t)\right), \tag{6}$$

$$\frac{\partial}{\partial x}u(z,x,t) \approx \frac{\Delta r}{2\Delta x}\left(D_{\tilde{z}}u(z,x,t) + D_{\tilde{x}}u(z,x,t)\right). \tag{7}$$

Instead of a 2nd order differentiation operator the use of higher order operators is possible. For the 3-dimensional case the necessary derivative directions are:

$$\tilde{\mathbf{d}_1} = \frac{\Delta x}{\Delta l}\mathbf{x} + \frac{\Delta z}{\Delta l}\mathbf{z} + \frac{\Delta y}{\Delta l}\mathbf{y}, \tag{8}$$

$$\tilde{\mathbf{d}_2} = \frac{\Delta x}{\Delta l}\mathbf{x} - \frac{\Delta z}{\Delta l}\mathbf{z} + \frac{\Delta y}{\Delta l}\mathbf{y}, \tag{9}$$

$$\tilde{\mathbf{d}_3} = \frac{\Delta x}{\Delta l}\mathbf{x} + \frac{\Delta z}{\Delta l}\mathbf{z} - \frac{\Delta y}{\Delta l}\mathbf{y}, \tag{10}$$

$$\tilde{\mathbf{d}_4} = \frac{\Delta x}{\Delta l}\mathbf{x} - \frac{\Delta z}{\Delta l}\mathbf{z} - \frac{\Delta y}{\Delta l}\mathbf{y}, \tag{11}$$

$$\Delta l = \sqrt{\Delta z^2 + \Delta x^2 + \Delta y^2}.$$

With an analogous calculation as in the 2D case we receive the following result for the derivatives in z-, x- and y-direction:

$$\frac{\partial}{\partial x} u(z,x,y,t) \approx \frac{\Delta l}{4\Delta x} \left(D_{\tilde{d}_1} + D_{\tilde{d}_2} + D_{\tilde{d}_3} + D_{\tilde{d}_4}\right) u(z,x,y,t), \tag{12}$$

$$\frac{\partial}{\partial y} u(z,x,y,t) \approx \frac{\Delta l}{4\Delta y} \left(D_{\tilde{d}_1} + D_{\tilde{d}_2} - D_{\tilde{d}_3} - D_{\tilde{d}_4}\right) u(z,x,y,t), \tag{13}$$

$$\frac{\partial}{\partial z} u(z,x,y,t) \approx \frac{\Delta l}{4\Delta z} \left(D_{\tilde{d}_1} - D_{\tilde{d}_2} + D_{\tilde{d}_3} - D_{\tilde{d}_4}\right) u(z,x,y,t). \tag{14}$$

Therefore, each derivative in z-, x- and y-direction is replaced by a linear combination of four derivatives.

6 Stability and dispersion

Since FD modeling approximates derivatives by numerical operators and uses Taylor polynomials to perform the time update, inaccuracies occur, especially for coarse grids. One can separate these numerical errors into amplitude and phase errors. For a plane wave propagating through an infinite, isotropic and homogeneous medium, the amplitude must be conserved, and the velocity of propagation is not frequency-dependent. In FD modeling, it is possible that the amplitude increases exponentially with every timestep. In this case, the modeling scheme is said to be unstable. Frequency-dependent velocity errors, also called numerical dispersion, cannot be excluded completely but can be estimated and, therefore, reduced to a known and acceptable degree. The goal of this Section is to give the results of a standard von Neumann style analysis (see e.g. [3]) for the rotated staggered grid. Note, that the von Neumann stability criterion is not in connection with the stability problems for high contrast inclusions.

6.1 Neumann Stability

For the rotated grid the stability criterion for the 3D case and the 2D case are the same. We obtain:

$$\frac{\Delta t v_p}{\Delta h} \leq 1/(\sum_{k=1}^{n} |c_k|). \tag{15}$$

In this equation c_k denotes the difference coefficients (e.g. Central Limit coefficients [5], v_p the compressional wave velocity, Δt the time increment, and Δh the gridspacing. This result yields the von Neumann stability criterion for the rotated grid for all wavenumbers in the case of homogeneous media and for 2nd order operator in time.

6.2 Dispersion

To determine the dispersion relation, the 2nd order time update and a time harmonic plane wave are used. For the case of 2nd order spatial differentiations, we obtain for the rotated staggered grid:

$$\begin{aligned} &\sin^2\left(\tfrac{\hat{\omega}\Delta t}{2}\right) = \\ &\tfrac{\Delta t^2 v_{p,s}^2}{\Delta z^2} \sin^2\left(\tfrac{k_z \Delta z}{2}\right) \cos^2\left(\tfrac{k_x \Delta x}{2}\right) \cos^2\left(\tfrac{k_y \Delta y}{2}\right) + \\ &\tfrac{\Delta t^2 v_{p,s}^2}{\Delta x^2} \cos^2\left(\tfrac{k_z \Delta z}{2}\right) \sin^2\left(\tfrac{k_x \Delta x}{2}\right) \cos^2\left(\tfrac{k_y \Delta y}{2}\right) \\ &+ \tfrac{\Delta t^2 v_{p,s}^2}{\Delta y^2} \cos^2\left(\tfrac{k_z \Delta z}{2}\right) \cos^2\left(\tfrac{k_x \Delta x}{2}\right) \sin^2\left(\tfrac{k_y \Delta y}{2}\right). \end{aligned} \tag{16}$$

For the dispersion analysis (similar to [3]) we define a dispersion parameter γ given by

$$\gamma = v_{ph} \frac{\Delta t}{\Delta h}. \tag{17}$$

The maximum value for γ is associated with the stability criterion. It is given by the absolute value of the right-hand side of inequality (15). The quantity H given by

$$H = \Delta h / \lambda = \frac{k \; \Delta h}{2\pi} \tag{18}$$

controls the number of nodes per wavelength of the plane wave. The maximum value of H is 0.5 given by the Nyquist-theorem. To obtain the relative accuracy of the phase velocity A_{rel} as a function of the quantity H, the dispersion parameter γ and the direction of wave propagation (ϑ, φ), we define:

$$A_{rel}(H, \gamma, \varphi, \vartheta) := \frac{v_{ph,numerical}}{v_{ph}} = \frac{\hat{\omega}\,(kx, ky, kz, \Delta h, v_{ph}, \Delta t, c_1, c_2, ..., c_n)}{v_{ph} |\boldsymbol{k}|} \tag{19}$$

with:

$$\begin{aligned} kx = H \tfrac{2\pi}{\Delta h} \sin\varphi \sin\vartheta, kz = H \tfrac{2\pi}{\Delta h} \cos\varphi \sin\vartheta, \\ ky = H \tfrac{2\pi}{\Delta h} \cos\vartheta, \varphi \in [0, 2\pi], \vartheta \in [0, \pi], \Delta t = \gamma \tfrac{\Delta h}{v_{ph}}. \end{aligned}$$

Note that $\hat{\omega}\,(kx, ky, kz, \Delta h, v_{ph}, \Delta t, c_1, c_2, ..., c_n)$ is given from the dispersion relation. The numerical dispersion error is illustrated for different cases in the Figures 2 and 3.

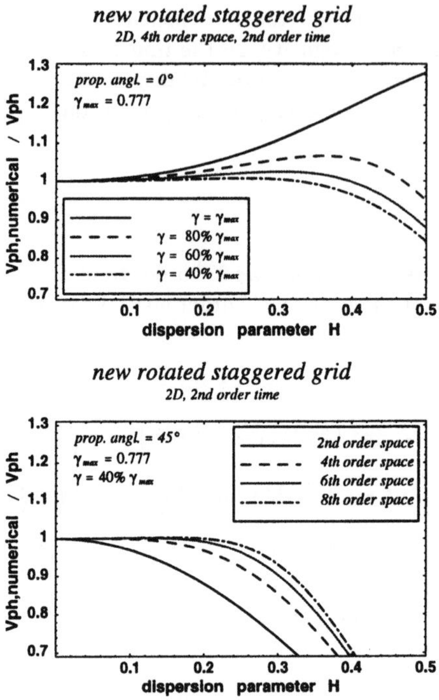

Fig. 2. Curves showing modeling errors caused by numerical dispersion for the rotated staggered grid for the 2D case ($\vartheta = \pi/2$). In the bottom row the angle between the direction of wave propagation and the axes is 45 degrees ($\varphi = \pi/4$). For the top row the angle is 0 degree ($\varphi = 0$). In the top plot γ varies from γ_{max} (top) to 40% of γ_{max} (bottom) in steps of 20%. All this curves are calculated for 8th order spatial differentiations using Central Limit coefficients [5] and 2nd order time update. In the bottom plot the order of the spatial differentiations varies from 2 (bottom) to 8 (top) in steps of 2.

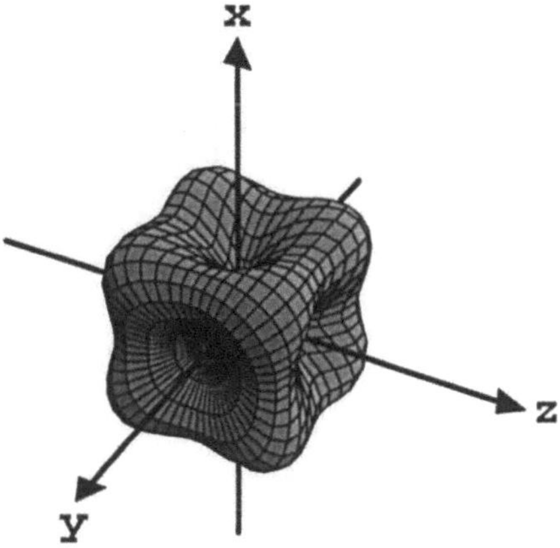

Fig. 3. The spatial variation of the relative error of the phase velocity for the new rotated grid is depicted. The distance between the curves and the origin has the value $|A_{rel} - 1|$ of the corresponding direction. The computation is done for 2nd order spatial differentiations and 2nd order time update ($\gamma = 80\%\ \gamma_{max}$, $H = 0.1$). Note that the maximum phase velocity error of the new rotated staggered grid shown in this figure is 2.82%.

7 Crack modeling

To demonstrate that the rotated grid can model cracks and high contrasts we created the 2D model shown in Figure 4. The model (gridspacing $\Delta h = 0.0002m$) consists of a $10cm \times 10cm$ area in which the compressional and shear wave velocity is set to $3950\frac{m}{s}$ and $2244\frac{m}{s}$; the density is $2.172\frac{g}{cm^3}$. Above this area and in the cracks in this area we set the the velocities to zero and the density to $0.000001\frac{g}{cm^3}$ which shall approximate vacuum. A bodyforce linesource ($f_{fund} = 8 \cdot 10^5 \frac{1}{s}$, $f_{max} \approx 2.5 \cdot 10^6 \frac{1}{s}$, $\Delta t = 7 \cdot 10^{-9} s$) in z-direction is placed at the interface rock/vacuum. The modeling is done with 2nd order time update and with a combination of 2nd and 8th order spatial differentiation operators [6].

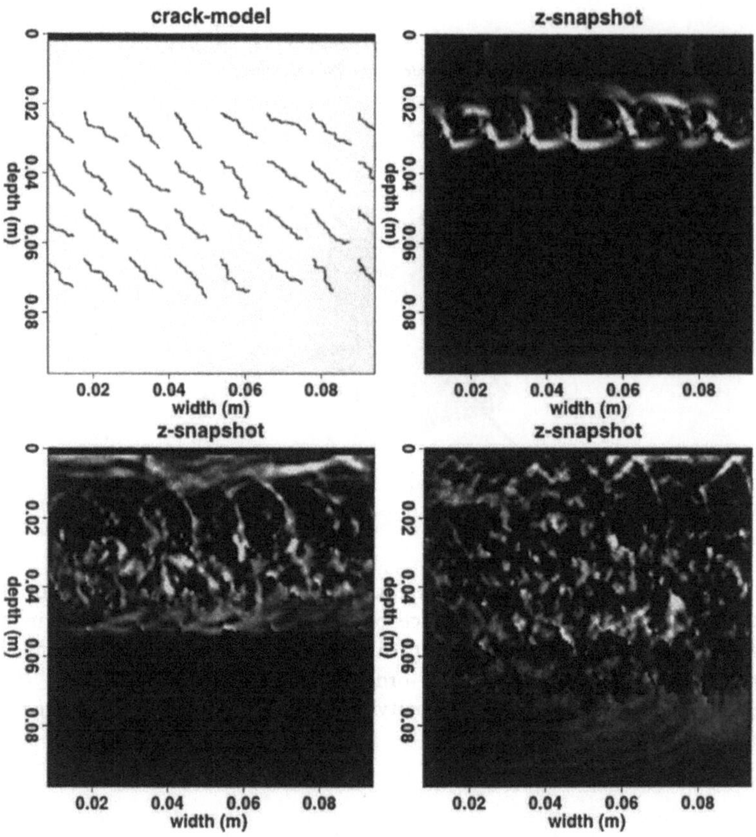

Fig. 4. The 2D crack model and three snapshots at different moments of time. The snapshots show the propagation of a plane wave from top through a region including some cracks.

8 Conclusions

Since FD modeling discretizes the medium and the wavefield on a grid and not by finite volumes (like, e.g., Finite Element schemes), it requires very few assumptions. If the discretization is done correctly, FD modeling is very fast and accurate. We find that the problem of arbitrary high contrasts of the medium parameters cannot be solved with a standard staggered grid. The grid modifications found in this paper make it possible. Moreover, it enables one to model arbitrary inhomogeneities without boundary conditions. In contrast to a standard staggered grid high-contrast inclusions do not cause instability problems for our rotated staggered grid. With a von Neumann style analysis we showed that our rotated staggered grid has accuracy criteria and limitations similar to those of the standard staggered grid. We demonstrated in our example that our modification of the grid can model fractured media very well.

References

1. G. Kneib and C. Kerner. "Accurate and efficient seismic modelling in random media." *Geophysics* 58, 576–588 (1993).
2. A. R. Levander. "Fourth-order finite-difference P-SV seismograms." *Geophysics* 53, 1425–1436 (1988).
3. E. Crase. "High-order (space and time) finite-difference modeling of the elastic wave equation." *60th Annual Internat. Mtg., Soc. Expl. Geophys., Expanded Abstracts*, volume 90, 987–991 (1990).
4. K. Aki and P. G. Richards. *Quantitative Seismology, Theory and Methods*. W.H. Freeman and Comp., San Fransisco (1980).
5. M. Karrenbach. *Elastic tensor wavefields*. Ph.D. thesis, Stanford University (1995).
6. N. Gold, S. A. Shapiro and E. Burr. "Modelling of high contrasts in elastic media using a modified finite difference scheme." *67th Annual Internat. Mtg., Soc. Expl. Geophys., Expanded Abstracts*, volume 97, ST 14.6 (1997).

9 Acknowledgments

Being members of the Collaborative Research Center 381 (SFB 381), we wish to thank the German Research Society (DFG) and the Wave Inversion Technology (WIT) Consortium-project for their financial support. We also thank Dr. Martin Karrenbach for providing us with his FD program.

Fast Repeat-Modelling with Born Approximation

Andreas Kirchner[1], Serge A. Shapiro[2], and Peter Hubral[1]

[1] Geophysical Institute, Karlsruhe University, 76187 Karlsruhe, Germany
[2] Fachrichtung Geophysik, Freie Universität Berlin, 12249 Berlin, Germany

1 Introduction

Various branches of science probe the interior of an unknown body with waves. Scientists frequently wish to simulate such experiments to improve understanding of phenomena not directly accessible to observation, or to test experimental configurations before their installation.

Accurate forward modelling of wave propagation with standard techniques, like the Finite Difference method (FD) [1, and references in this article], requires large computational effort. This is particularly true for the propagation of elastic waves in a three-dimensional model representing a solid body, which is required in seismic/seismology, and non-destructive testing of materials. In both disciplines, people often need to examine the change of elastic properties of a small part of a body due to external strain. Such a situation is the in-situ monitoring of strain-induced development of defect zones in certain materials using ultrasonic waves. Seismic monitoring of effects of oil production in a hydrocarbon reservoir is another example.

In these cases the forward modelling has to deal with a very special situation: A lot of simulations of wave propagation are necessary for a set of very similar models. These models represent circumstances at different stages of crack development or oil production. The difference from one model to the next happens only inside the reservoir/defect zone while the biggest part of the model remains the same for all simulations. Standard methods like FD do not make use of this fact. Thus, often much effort is spent without gaining additional information, and a fast and comparably accurate alternative would be desirable.

The above-mentioned problem, originally comes from geophysics. In the following, we will therefore use geophysical terminology to describe our investigations of a new modelling technique called Born Repeat-Modelling (BRM). However, because of the great similarity of defect detection with ultrasound and seismic reservoir monitoring, the proposed method is equally applicable in both fields.

In Section 2, we give a brief description of the procedure of Born Repeat-Modelling. Sections 3 and 4 are devoted to examples. We show a test carried out for a simple model. Although the proportions of this model correspond to geophysical orders of magnitude, the results are also significant for similar models from material science.

2 Born Repeat-Modelling

Born Repeat-Modelling is especially designed for the above-mentioned type of situation. With BRM, only one simulation must be carried out with a standard technique, typically FD. For all further simulations, the change of the measured wavefield is calculated with Born Approximation. This allows to save a considerable amount of computing time. For details about Born approximation see, e.g., [2,3].

The Born-Repeat modelling technique consists of two steps:

Green's function generation: For the original model, i.e. the configuration before the beginning of production (or before application of external stress), one survey (or ultrasonic scan) must be simulated using Finite Differences. In addition, the full wave field in all reservoir points should be recorded. This data set is equivalent to the Green's function for the particular source configuration. It is required to compute the Born Approximation in the following steps.

Repeat-Modelling: It is expected that production modifies the elastic properties inside the reservoir. If appropriate Green's function data are available, we can use Born Approximation to compute the production-induced changes of the seismic or ultrasonic response.

There are no restrictions to the complexity of the overburden, i.e. the constant part of the model. However, for Born Approximation to be valid, the modifications of the reservoir model have to be small. This is usually the case in practice. What the word "small" means in the context of Born approximation, is demonstrated below with an example. In order to receive the complete Green's function data required for the Born Approximation of one seismic trace, we must currently shoot two times. Once from the actual source position, and once at the location of the receiver. The different directivities of the seismic receivers, the geo- or hydrophones, and the seismic source ought to be considered, as well. At first glance, the problem of obtaining the Greens functions seems to make the whole procedure inefficient. But this is not the case: Modelling whole surveys, the Green's function data can be re-used several times for sources and receivers.

Our approach is most useful, if seismograms are required for a large number of different reservoir configurations. Then, it can save a large part of the CPU time, even in 2-D. For the Repeat-Modelling step, the gain in computing time is proportional to the ratio of grid points in the reservoir and the total number of grid points. Therefore, we expect that this effect will be larger in 3-D.

3 Examples

To demonstrate the performance of BRM, we have applied this new technique to a very simple two-dimensional geophysical model (cf. Figure 1). We chose a homogeneous block with compressional wave (P-wave) velocity $v_p = 4000\frac{m}{s}$, shear wave (S-wave) velocity $v_s = 2236\frac{m}{s}$, and density $\rho = 2.0\frac{g}{cm^3}$. Inside a rectangular region near the bottom of the model the elastic parameters were perturbed to simulate the effect of production on a reservoir. The reservoir was chosen $16m$ thick and $360m$ wide. In this simple configuration, it is possible to observe clearly the effect of reservoir perturbations on the seismograms without being confused by waves scattered from the overburden. For the source and receiver locations shown in the figure, we have simulated a few seismic sections with different offsets. The source wavelet was the derivative of a Gaussian with fundamental frequency $35Hz$. A small selection of our results is depicted below in Figures 2-4. Since the effect is usually hard to observe directly in seismic sections, we always consider the difference signal, i.e.,

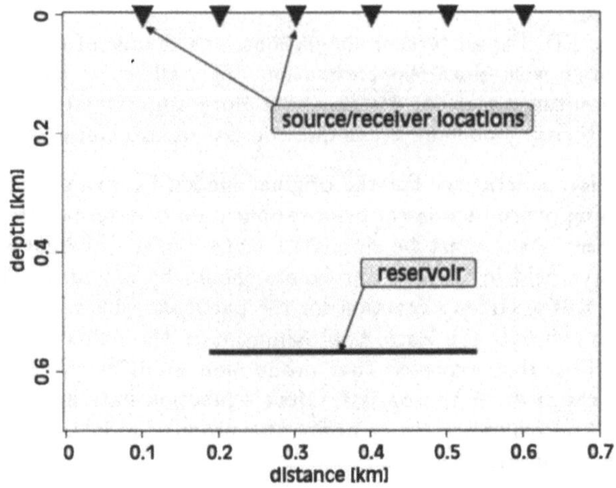

Fig. 1. Simple 2-D earth model as described in the text. The red triangles indicate source and receiver locations.

the change of the seismic response caused by a perturbation of reservoir properties. Figures 2 and 3 offer the possibility to compare results of BRM with corresponding data obtained with FD. In Figure 2, we show the change of displacement at different offsets from the source (common-shot difference section). The seismograms have been computed for the model described above. In all points inside the rectangle, the Lamé parameter λ was increased by 15.3%. This gives rise to a 5% increase of $\lambda + 2\mu \propto v_p^2$. The Lamé parameter $\mu \propto v_s^2$ and the density ρ were kept constant. In this section, we can distinguish two real events: Firstly, there is the reflected P-wave, which arrives at about $t = 0.3s$ at the zero-offset location. The second event (at $t = 0.38s$ in the zero-offset trace) is the P-wave diffracted at the upper right corner of the reservoir. Both events approach each other for larger offsets. Because of symmetry, they arrive at the same time for the largest offset. For larger offsets, at about $t = 0.43s$ another "event" appears in the Born results. However, this is only an artifact caused by numerical inaccuracies.

A common-offset difference section can be seen in Figure 3, i.e., the change of the wave field recorded at constant offset from the source for different source positions. This time, the Lamé parameter μ was increased by 5% with constant λ and ρ. Changing μ implies that the P-wave velocity is changed, as well. Therefore, we expect both, P-wave and S-wave events. Obviously, in Figures 2 and 3, the differently calculated traces coincide quite well for different kinds of perturbations, and for all offsets. Similar results are available for all types of seismic sections, and for perturbations of the density, as well. However, trying to be brief, we had to make a selection. Our approach is also fast. For example, to compute the data for Figure 3, BRM required less than 5% of the CPU time of the corresponding FD calculations. The effort to prepare the Green's function is not included. Of course, it is not completely correct to compare computing times in this way, because FD can generate large numbers of traces at a time, while BRM only produces the trace actually required.

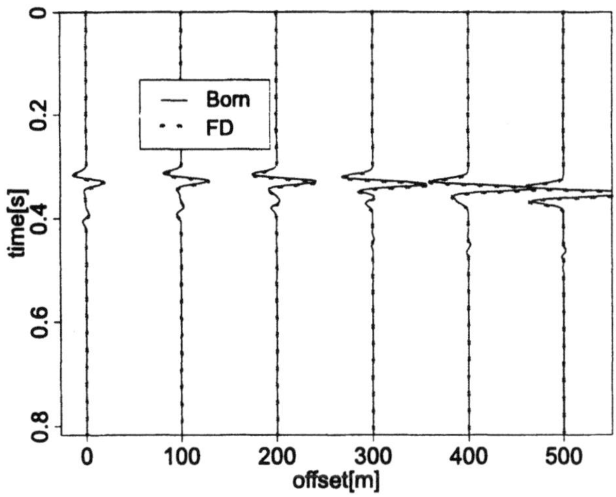

Fig. 2. Common-shot difference section over the model shown in Figure 1. The change of the z component of the seismic response is shown for an increase of the quantity $\lambda + 2\mu$ by 5%. This is equivalent to an increase of the Lamé parameter λ by 15.3%. μ and ρ were kept constant. The leftmost trace is zero-offset.

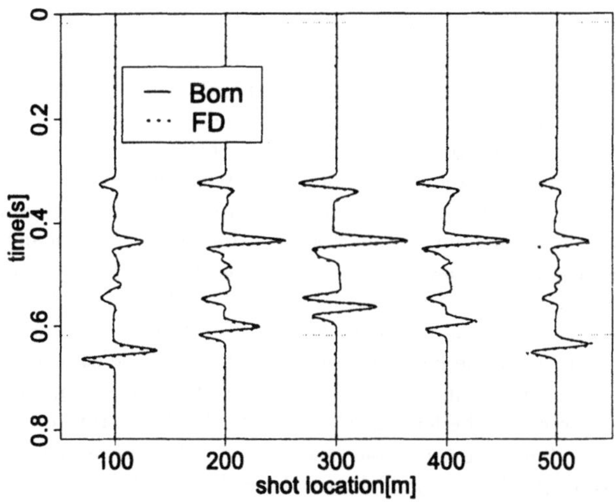

Fig. 3. Common-offset difference section over the model shown in Figure 1. Here, we show the change of the seismic response for an increase of the Lamé parameter μ by 5%.

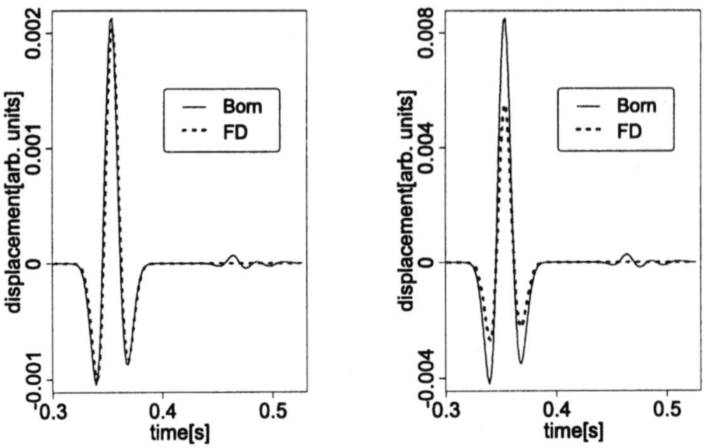

Fig. 4. The change of two seismic traces (offset=500m) for different perturbations of the Lamé parameter λ. The curve on the left hand side corresponds to a perturbation $\delta\lambda = 53\%$. On the right hand side λ was increased by 213%.

4 Validity for large perturbations

We tried to develop a feeling, when a perturbation of the reservoir parameters can be considered small enough for Born Approximation to be valid. Therefore, we have tested very large perturbations of the elastic reservoir properties. Born Approximation treats the contributions to the scattered wave field independently for all reservoir points. Consequently, multiple-scattering effects are neglected. If the frequency and the perturbation are too big, multiple-scattering will start to become important, and the agreement will decrease continuously. Figure 4 shows results for perturbations of the Lamé parameter λ. The traces have been obtained for offset $500m$ (shot at $x = 100m$, and receiver location $x = 600m$). This corresponds to the rightmost trace in Figure 2. We chose $\delta\lambda = 53\%$ for the graph on the right hand side, and $\delta\lambda = 213\%$ for the left hand side. These values are equivalent to 20% and 80% perturbations of the quantity $\lambda + 2\mu$. Apart from the numerical artifacts at $t \approx 0.43s$, the curves on the left match perfectly. In the picture on the right hand side, however, we find a considerable discrepancy. This suggests that the range of validity was exceeded for the model used here. In practical situations, we hardly expect to encounter such a large change of reservoir properties. We rather think that for reservoir models with inner structure even larger perturbations will be feasible, because the important length scales should be smaller.

5 Conclusions

Having compared corresponding synthetic data obtained for a simple model with Born Repeat-Modeling and with Finite Differences (FD), we find that BRM yields excellent results within a wide range of practically relevant perturbations. The most important feature of BRM is the fact that it is much faster than FD. Of course, standard modelling techniques, like FD, are general tools while the range of application of BRM is restricted to special cases. When a large number of simulations are required for very similar models BRM may be applied with great benefit.

Although length scales are different the conditions for certain common types of experiments are very similar in seismic and in non-destructive testing of materials. Therefore, we conclude that the Born Repeat-Modeling technique should be considered a useful tool for the efficient simulation of time-lapse experiments in both branches of science.

References

1. Erik H. Saenger, and Serge A. Shapiro "Modeling of elastic waves in fractured media.", this issue
2. Rhu-Shan Wu "The perturbation method in elastic wave scattering." *PAGEOPH,* Vol. 131, No. 4 (1977), pp. 605-637.
3. J. E. Gubernatis, et al. "Formal aspects of the theory of the scattering of ultrasound by flaws in elastic materials." *Journal of Applied Physics* Vol. 48, No. 7 (1977), pp. 2804-2811.

6 Acknowledgments

Being members of the Collaborative Research Center 381 (SFB 381), we wish to thank the German Research Society (DFG) and the Wave Inversion Technology (WIT) Consortium-project for their financial support. We also thank Dr. Martin Karrenbach for providing us with his FD program.

Simulation and theoretical description of elastic waves in 2D and 3D random media

Tobias Müller[1] and Serge A. Shapiro[2]

[1] Geophysical Institute, Karlsruhe University, 76187 Karlsruhe, Germany
[2] Fachrichtung Geophysik, Freie Universität Berlin, 12249 Berlin, Germany

Abstract. Combining well-known techniques from scattering theory, we construct the Green's function of an initial plane wave propagating in 2D and 3D random media. The Green's function depends on the second order statistics of the medium heterogeneities. We show that averaged logarithmed wavefield attributes obtained in the Rytov-approximation can be used to describe the primary arrivals in single realizations of seismograms. As a consequence we obtain a valuable tool for inversion and modeling of scattering processes in heterogeneous structures. We compare the outcome of finite difference experiments with the theoretically predicted wavefield and find a good agreement: the presented statistical approximations give a smooth version of the primary arrivals.

1 Introduction

In the recent years, the effect of multi-scale heterogeneities in earth and material sciences has been recognized as a vital aspect of the overall behavior of elastic waves. Very much effort has been spent in the understanding of wave propagation in layered media. Especially waves propagating in randomly multi-layered media are subjected to stratigraphic filtering. The physical reason is the multiple scattering by 1D inhomogeneities. In random stationary media, explicit approximations for the transmissivities of obliquely incident P- and SV-plane waves have been found by applying the second-order Rytov approximation to the 1D multiple scattering problem in the frequency domain. This description is known as the generalized O'Doherty-Anstey formalism [2].

However, the earth, the lithosphere and generally composite materials may have a very complex structure. Therefore the concept of 2D or 3D random media can be a more suitable and more general description. The random medium consists of a constant background of a certain medium parameter – the reference medium – and its corresponding fluctuations characterized by a spatial correlation function.

Amplitudes and phases of wavefields fluctuate in random media. Furthermore, averaged wavefields are characterized by scattering attenuation and dispersion; both are important parameters for rock characterization. A suitable description of these wavefield characteristics enables the construction of Green's functions for heterogeneous media. The latter is extremely useful in order to apply any kind of (true-amplitude) imaging or inversion technique.

There is a lack of non-heuristic descriptions of pulse propagation which go beyond the meanfield theory and are valid for seismograms of single realizations of random media. This is exactly the motivation of this study. It is based on two theoretical studies of propagation of a compressional plane wavefield in 2D or 3D random media characterized by isotropic stationary velocity fluctuations (see Shapiro and Kneib (1993) and Shapiro et al.(1996b)). We present a description of the wavefield in a perturbation approximation which enables us to predict the transmitted wavefield around the primary arrivals in a typical single realization of a seismogram.

2 Numerical experiments

The analytical proceeding is confirmed by finite difference simulations of wave propagation in 2-D elastic random media with gaussian, exponential and von Karman correlation functions. Fig. 1 shows a realization of an exponential random medium; Fig 2 displays single traces recorded at several distances in the random medium. Due to the huge computing times our modeling is restricted to 2D random media. However with the extensive use of parallel computing facilities also 3D simulations become possible.

In the present example we simulate a plane wave propagating from the top up to a certain depth (z-direction) in a single random medium realization. The background medium is characterized by a P-wave velocity of 3000 m/s, a S-wave velocity of 1850 m/s and a density of 2.5 g/cm^3. For the modeling we need instead of velocities the stiffness tensor components c_{11} and c_{55} (and density). Only the stiffness tensor component c_{11} exhibits exponentially correlated fluctuations (the correlation length is 40 m, the corresponding standard deviation of the P-wave velocity is 15%). We use a body-force linesource with only a z-component. The wavelet is the second derivative of a Ricker-wavelet with a dominant frequency of about 75 H (this corresponds to a wavelength of 40 m for the P-wave). Note that under these conditions no S-waves will be generated.

In the background of Fig. 3 we see 54 traces (the z-component of the wavefield) recorded on a common travel-distance gather at a depth of 336 m. The distances between geophones along the receiver line is of the order of the correlation length so that statistically correlated measurements are avoided. The thicker black curve denotes the result of convolving the analytically computed Green's function with the input wavelet. It is obvious that the theoretically predicted wavefield represents the simulated wavefield in a somehow averaged form. However, we want to point out that this formalism gives an estimate of the primary wavefield for a given single trace.

Further, we investigate which kind of averaging of the traces is needed in order to compare with the theoretical result. In Fig. 4 we calculate the meanfield (red curve) by simply stacking all traces without any corrections. We observe a big difference between the meanfield amplitude and our wavefield approximation (blue curve). Finally, the green curve denotes the traveltime-corrected section and we see a rather good agreement with our result. We conclude that our approach goes beyond the ordinary meanfield description, where all incoherent parts of the wavefield are removed. Using averaged logarithmed wavefield attributes we give a description of the main part of the transmitted signal.

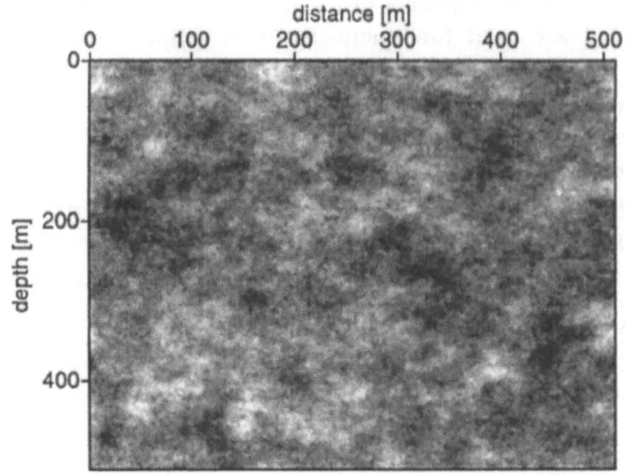

Fig. 1. *Single realization of a 2D random medium. A constant background P-wave velocity is superimposed by exponentially correlated fluctuations. Bright colors denote higher velocities in respect to the mean velocity and vice versa.*

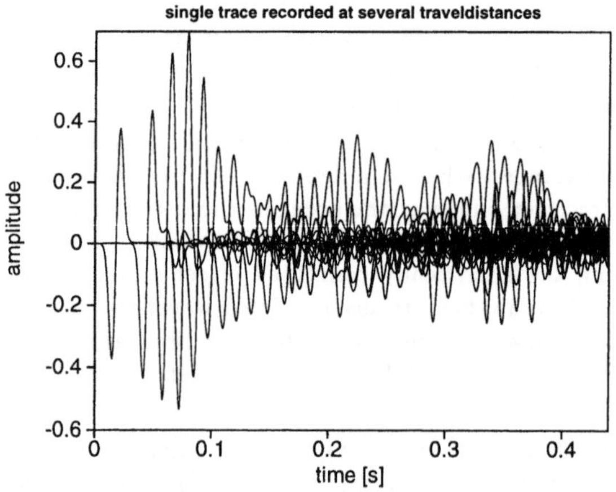

Fig. 2. *Single trace (z-component of the wavefield) recorded at several traveldistances in the random medium (L=0, 16, 56, ..,976 m). Note the random focusing and defocusing of the wavefield.*

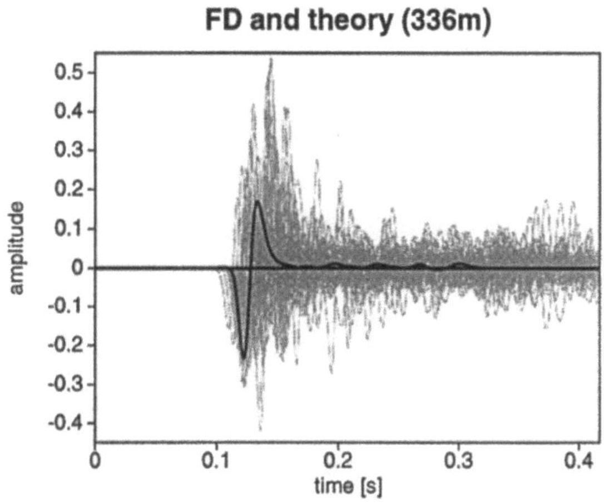

Fig. 3. *Comparison of FD-experiment and statistical wavefield description. In the background are displayed the traces recorded at a depth of 336m. The thicker, black curve denotes the result of convolution of the Green's function with the input wavelet.*

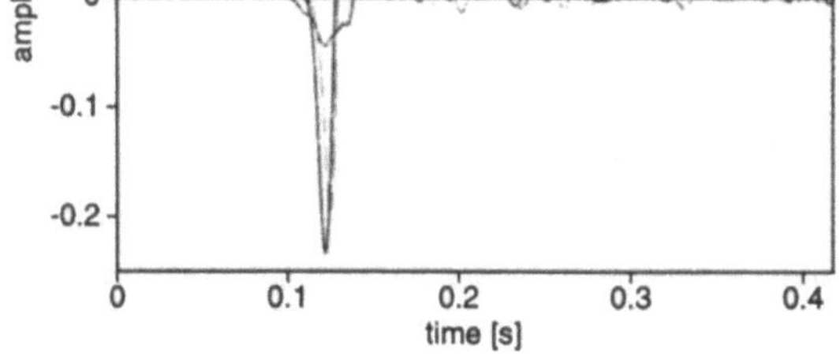

Fig. 4. *For the same FD-experiment we computed the meanfield (red), the traveltime-corrected section (green) and our statistical wavefield description (blue).*

Crack imaging in randomly heterogeneous media

Kai-Uwe Vieth[1], Serge A. Shapiro[2], and Peter Hubral[1]

[1] Geophysical Institute, Karlsruhe University, 76187 Karlsruhe, Germany
[2] Fachrichtung Geophysik, Freie Universität Berlin, 12249 Berlin, Germany

Abstract. *In reservoir characterization or non destructive testing it is necessary to get a better understanding of the effect of fractures on the wave field and to be able to image cracks, respectively. Here, the crack is within a randomly heterogeneous media. The Kirchhoff migration is applied followed by computing the envelope which emphasizes the crack. Additionally, using coherence analysis increases resolution.*

1 Introduction

A reservoir is characterized rather by an ensemble of cracks or fractures than by one single crack. Hence, the extend of the extensive fractured region is of interest and their effect on seismograms which have to be analyzed. At the beginning of our work we concentrate on imaging one crack surrounded by a randomly heterogeneous media. We use the well known Kirchhoff migration and compute afterwards the envelope which improves the image, i.e. it is easier to localize the crack. During the Kirchhoff migration we supplementarily do coherence computations which help us to interpret or verify the result. As we focus on a single crack the work is attractive for non destructive testing where damaged material, e.g. concrete, often has prominent cracks but not necessarily a cluster of it.

2 Generating the model

Before we start producing some synthetic data we have to generate somehow a randomly heterogeneous medium. In the following we describe the parameters of the slightly modified program MEDIUM2D [1] which produces the model needed.

The synthetic model is bounded by a thin layer of air. The width and height of the model is 0.1 m. The ellipses which represent gravel of different sizes make up 50 % of the overall area. The gravel has a P-velocity of 4.03 km/s, a density of 2.59 g/cm^3 and a standard P/S-ratio of $\sqrt{3}$. The background material, i.e. the cement, has a P-velocity of 3.95 km/s, a density of 2.17 g/cm^3 and also a P/S-ratio of $\sqrt{3}$. The little white spots in Figure 1 indicate air inclusions. Their content amounts to 4 %. The area of the crack is about 8 mm^2.

References

1. Shapiro, S. A., and Kneib, G., 1993, Seismic attenuation by scattering: theory and numerical results: Geophys. J. Int., **114**, 37[illegible]–391.
2. Shapiro, S. A., Hubral, P., and Ursin, B., 1996a, Reflectivity/transmissivity for one-dimensional inhomogeneous random elastic media: dynamic-equivalent-medium approach: Geophys. J. Int., **126**, 184–196.
3. Shapiro, S. A., Schwarz, R., and Gold, N., 1996b, The effect of random isotropic inhomogeneities on the phase velocity of seismic waves: Geophys. J. Int., **127**, 783–794.

3 Acknowledgment

The work has been supported in part by the WIT-sponsors and in part by the Deutsche Forschungsgemeinschaft (SFB381).

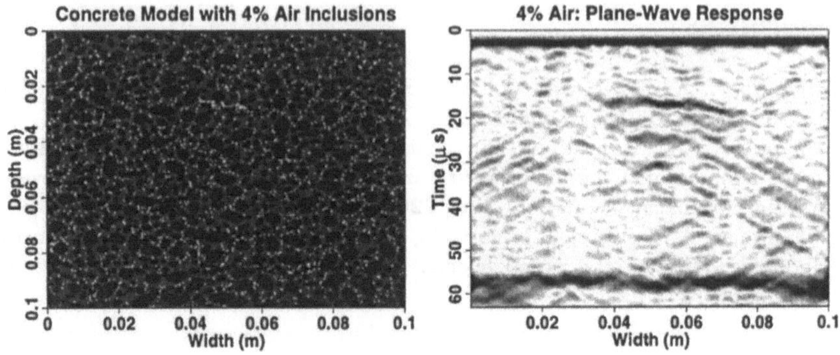

Fig. 1. *On the* **left** *side is a heterogeneous concrete model with a defect, i.e. the crack. The ellipses represent gravel within cement. The* **right** *side shows the plane-wave response calculated with the Finite Difference program* ULTIMOD *using a rotated staggered grid.*

3 Generating and imaging synthetic data

The Finite Difference program ULTIMOD [2] is used to compute the propagation of a plane/wave through the model. In order to handle high contrasts the computations in ULTIMOD are performed on a rotated staggered grid. For more details concerning this modification please refer to [4]. A Ricker1 wavelet (first derivative of the Gaussian distribution curve) is propagated. Its fundamental frequency is 300 kHz, whereas the maximum frequency is 500 kHz. The plane-wave response is depicted on the right in Figure 1. Different observations can be made:

- the seismogram looks wavy, frayed and changes smoothly
- the amplitude of the back-echo is about as strong as the reflection of the crack
- lots of diagonal reflections possessing slopes of opposite sign occur.

As we do not have a layered medium or a different medium where it makes sense to use a velocity function for migration, we have to use an effective velocity. This velocity cannot just be an average P- or S-velocity but also has to take the lengthened ray path due to multiple scattering into account. We pick along the receiver line a few traveltimes of the back-echo in order to calculate the effective velocity of the medium. In Figure 1 it can be seen clearly that the reflection of the bottom of the model varies along the modelwidth because the plane-wave is scattered. Thereafter, this velocity is the migration velocity. Notice the wavefront healing that occurs after the plane-wave passes the crack which is noticeable by the fact that the back-echo is a continuous reflection.

Figure 2 shows the Kirchhoff migration of the plane-wave response on the left. It is quite surprising that the crack is imaged with such a good quality although the unmigrated data do look very complicated. No further processing steps are applied before the migration. We tried to improve the image by suppressing the slanted events in the (τ,p)-domain. This effort did not yield better results.

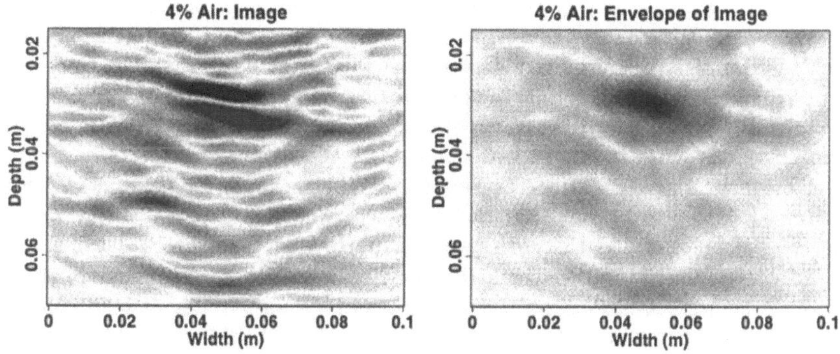

Fig. 2. *The migration of the plane-wave response is on the* **left**. *On the* **right** *is the envelope of the migrated section.*

Another processing step was much more successful. We computed after the migration the envelope or instantaneous amplitude, respectively. Now, the image becomes much more distinct as one can see and compare in Figure 2. Areas of high amplitude which do not correspond to the crack fade out after the envelope has been calculated. Only the reflection of the crack is left over. Unfortunately, these benefits have to be paid by decreasing resolution.

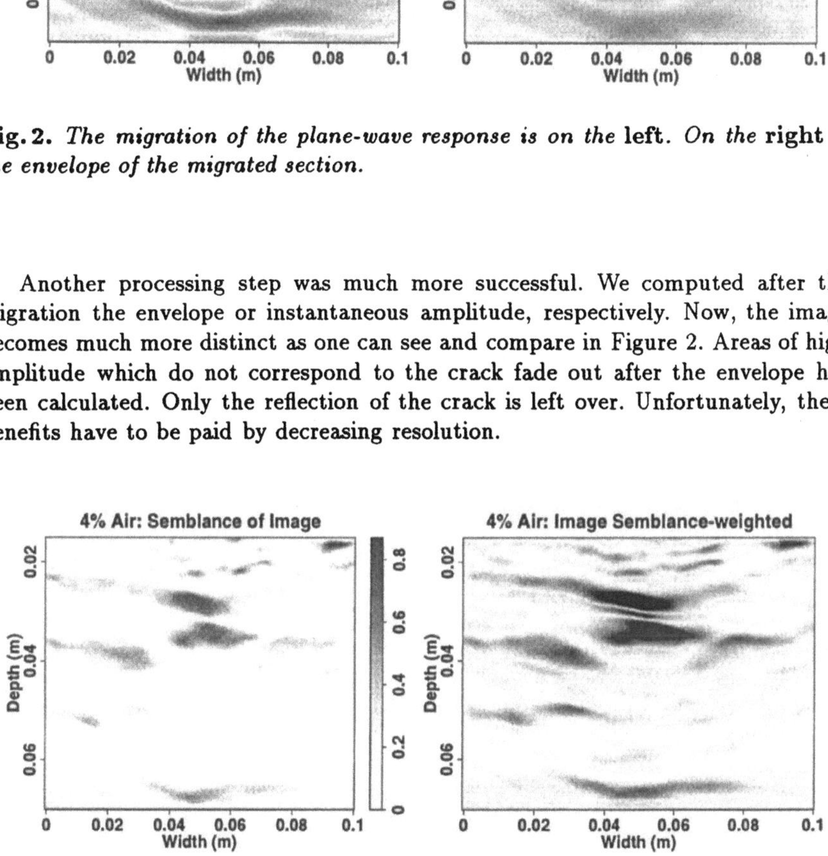

Fig. 3. *On the* **left** *the semblance is depicted which is calculated during the Kirchhoff migration. The* **right** *figure is the result of the migrated seismogram multiplied by the semblance section.*

Using coherence analysis during the migration process we tried to improve the expressiveness of the image. The semblance [3] is calculated along the diffraction parabola while simultaneously summing along it for the migration. The output is shown on the left in Figure 3. The semblance values are in the interval of one and zero. One means maximal coherency, whereas zero is associated with minimal coherency. The reflection of the crack is displayed by high semblance values. On the right side in Figure 3 the output of the multiplication of the migrated section with the semblance is displayed. Comparing this weighted image with the original one of Figure 2 makes the improvement obvious. The reflection of the crack is more pronounced. Nevertheless, the width of the reflection suggests the crack to be nearly twice as big as it is actually in the model.

Obviously, this is not the case for the semblance of the image on the left in Figure 3. To have a closer look at it, Figure 4 depicts a zoom of the crack in the concrete model and the corresponding part of the semblance data. Figure 4 demonstrates nicely that the crack is associated with the top boundary of the first region of high coherence values. The regions of high amplitudes or high coherence values being on top of each other appear due to the input wavelet which has a positive and a negative peak. The extension of the upper coherence region matches the crack perfectly. Thus, the semblance plot yields excellent results concerning detection and resolution.

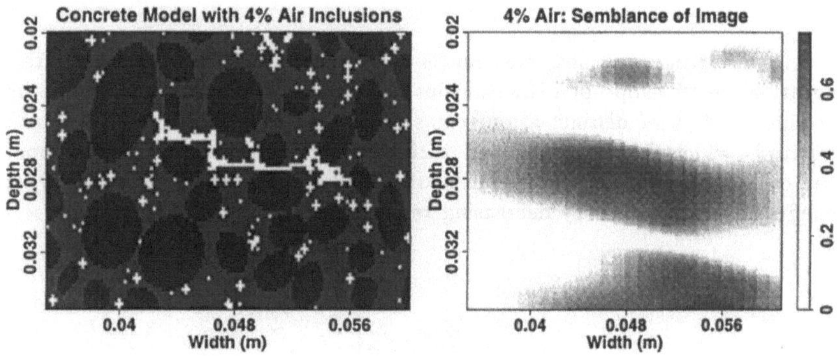

Fig. 4. *The* left *figure shows a zoom of the concrete and the* right *figure is the corresponding zoom of the semblance.*

4 Conclusion

We gained good results in imaging a crack in a randomly heterogeneous medium. It has been shown that the semblance, computed simultaneously, and the envelope, computed after Kirchhoff migration, complement one another, very well. The envelope pronounces the existence of a crack and the semblance provides good resolution in the corresponding area.

References

1. Burr, E., 1996, Theoretische Beschreibung und numerische Simulation elastischer Wellenausbreitung in zufallsverteilten Medien: Master's thesis, Geophysikalisches Institut der Universität Karlsruhe.
2. Karrenbach, M., 1995, Elastic tensor wavefields: Ph.D. thesis, Stanford University.
3. Neidell, N., and Taner, M. T., 1971, Semblance and other coherency measures for multichannel data: Geophysics, **36**, no. 3, 482–497.
4. Saenger, E. H., and Shapiro, S. A., Modeling of high contrasts in elastic media using a modified Finite Difference grid:, Technical report, Wave Inversion Technology Consortium (WIT), 1998.

5 Acknowledgments

We wish to thank the German Research Society (DFG) and the Wave Inversion Technology (WIT) Consortium-project for their financial support.

Development and Astrophysical Applications of a Parallel Smoothed Particle Hydrodynamics Code with MPI

Stefan Kunze, Erik Schnetter, and Roland Speith

Institut für Astronomie und Astrophysik, Abt. Theoretische Astrophysik
Universität Tübingen
Auf der Morgenstelle 10, 72076 Tübingen
E-mail: {kunze,schnetter,speith}@tat.physik.uni-tuebingen.de
Internet: http://www.tat.physik.uni-tuebingen.de/~speith/particle.html

Abstract. Smoothed Particle Hydrodynamics (SPH) is a particle method to simulate compressible fluids. First we present a brief introduction to SPH, where we focus on an approach to treat the physical viscosity. Then we describe in detail the basic principles of our parallel implementation of the SPH method. The efficiency of the code on Cray T3E and IBM SP2 is discussed. In the last part we present a short introduction to accretion disks in interacting binary stars and report on some new results in that field achieved with our code.

1 Smoothed Particle Hydrodynamics

Smoothed Particle Hydrodynamics (SPH) is a grid-free Lagrangian method for solving the hydrodynamic equations numerically. It was introduced in 1977 by Gingold & Monaghan [2] and by Lucy [3]. The method is especially suited for compressible flows involving free boundaries, a commonplace situation in astrophysics. In SPH the matter distribution is divided into small overlapping mass packets, so-called *particles*, which do not exchange matter. The particles move in space following the motion of the fluid while interacting with their neighbor particles.

1.1 Basic Principles

The system of hydrodynamic equations, i.e., the Navier-Stokes equation, the equation of continuity, and the energy equation together with an equation of state, form a system of coupled partial differential equations. Applying the SPH formalism, these are transformed into a set of ordinary differential equations (ODEs) in two steps. These ODEs are subsequently integrated numerically using standard methods. First the field quantities $f(\boldsymbol{r})$ are *smoothed*, i.e., they are replaced by the convolution

$$f(\boldsymbol{r}) \longrightarrow \int f(\boldsymbol{r}')\, W(|\boldsymbol{r} - \boldsymbol{r}'|, h)\, dV' \tag{1}$$

with a differentiable kernel W which is normalized to unity. The smoothing length h is a measure of the (compact) support of the kernel function, leading to

$$f(\boldsymbol{r}) = \int f(\boldsymbol{r}')\, W(|\boldsymbol{r} - \boldsymbol{r}'|, h)\, dV' + O(h^2) \quad . \tag{2}$$

A common choice for the kernel is a cubic spline (see, e. g., Monaghan [4]).

In the second step the above integrals are evaluated at the particle positions $\boldsymbol{r}_i$, and are approximated by sums

$$f_i = f(\boldsymbol{r}_i) \approx \sum_j V_j f(\boldsymbol{r}_j) W(|\boldsymbol{r}_i - \boldsymbol{r}_j|, h) \quad . \tag{3}$$

The quantity V_j describes the volume that is represented by the particle j. Spatial derivatives of field quantities at the particle positions can be approximated by

$$\nabla_i f(\boldsymbol{r}_i) \approx \sum_j V_j f(\boldsymbol{r}_j) \nabla_i W(|\boldsymbol{r}_i - \boldsymbol{r}_j|, h) \quad . \tag{4}$$

Typical particle quantities are the position $\boldsymbol{r}_i$, the velocity $\boldsymbol{v}_i$, the particle mass $m_i = V_i \varrho_i$, where ϱ_i is the mass density, the internal energy e_i, etc.

1.2 SPH Representation of the Hydrodynamic Equations

The particles follow the motion of the fluid ($d\boldsymbol{r}_i/dt = \boldsymbol{v}_i$) and do not exchange matter ($dm_i/dt = 0$). Therefore, the continuity equation is automatically satisfied. With the abbreviation $W_{ij} = W(|\boldsymbol{r}_i - \boldsymbol{r}_j|, h)$ the particle density ϱ_i can be calculated from

$$\varrho_i = \sum_j m_j W_{ij} \quad , \tag{5}$$

which is motivated by evaluating Equation (3) with $f(\boldsymbol{r}) = \varrho(\boldsymbol{r})$.

For example, using Eq. (4), the Euler equation in Lagrangian formulation yields in the SPH formalism

$$\frac{d\boldsymbol{v}_i}{dt} = -\sum_j m_j \frac{p_j + p_i}{\varrho_j \varrho_i} \nabla_i W_{ij} \quad , \tag{6}$$

where we have added the term proportional to p_i in order to anti-symmetrize the inter-particle forces. Other formulations of the SPH equations are also possible. For a more detailed overview of the standard SPH method see, e. g., Monaghan [4].

The advantage of using a kernel with compact support is the finite interaction range of the particles, i. e., only a particle's neighbors within a distance of the smoothing length h contribute to the above SPH sums.

Because SPH is a grid-free method, complicated geometries are relatively easy to model. The local resolution of SPH can be enhanced considerably by allowing for a variable smoothing length. The disadvantage of the method is that neighborhood relationships change with time and have to be re-established every time the particle positions have changed.

1.3 Physical Viscosity

The viscous interactions are implemented according to the viscous stress tensor in the Navier-Stokes equation (Flebbe et al. [1]). This is important, e. g., for the simulation of accretion disks over large periods of time, because their evolution is governed by viscous processes which cannot be treated properly by an artificial viscosity. The reason is that in the hydrodynamical limit, i. e., for a vanishing smoothing length, the SPH equation of motion converges to the Navier-Stokes equation, and the physical viscosity does not vanish in contrast to the artificial viscosity.

One problem with the physical viscosity is the occurrence of second order derivatives of the velocity, and in the SPH formalism there are different possibilities to approximate these terms. It is generally better to treat them as a sequence of two first order derivatives rather than using second order derivatives of the kernel. This leads to the viscous acceleration

$$\left(\frac{dv_\alpha}{dt}\right)_i^{\text{visc}} = \sum_j m_j \left(\frac{\nu_j}{\varrho_i}\left(\sigma_{\alpha\beta}\right)_j + \frac{\nu_i}{\varrho_j}\left(\sigma_{\alpha\beta}\right)_i\right)\frac{\partial W_{ij}}{\partial x_\beta} \tag{7}$$

with the particle form of the shear $\sigma_{\alpha\beta}$

$$\left(\sigma_{\alpha\beta}\right)_i = (V_{\alpha\beta})_i + (V_{\beta\alpha})_i - \tfrac{2}{3}\delta_{\alpha\beta}(V_{\gamma\gamma})_i \quad , \tag{8}$$

where $(V_{\alpha\beta})_i$ is the particle representation of the velocity gradient

$$(V_{\alpha\beta})_i = \left(\frac{\partial v_\alpha}{\partial x_\beta}\right)_i = \sum_k \frac{m_k}{\varrho_k}\left((v_\alpha)_k - (v_\alpha)_i\right)\frac{\partial W_{ik}}{\partial x_\beta} \quad . \tag{9}$$

Here, the Greek indices label spatial coordinates, whereas particles are again labeled by Latin indices.

It can be easily shown that this formulation conserves momentum exactly (Riffert et al. [5]). In addition, the viscous terms of the energy equation can be formulated in such a way that the total energy is conserved exactly.

2 Implementation Details

Depending on the integration time, simulations with currently up to about one million particles can be performed. In order to simulate three-dimensional flows with more than a million particles while keeping the memory requirements and the turnaround time within reasonable limits, we implemented a distributed version of the SPH algorithm. We distribute the problem using the Message Passing Interface (MPI) by assigning a cuboid region of the simulation domain to every processor. Because the particles are not uniformly distributed, it is important that these cuboid regions can have different sizes which are allowed to vary in time.

2.1 Finding Interactions

The naïve and direct approach to evaluate the SPH sums is clearly not reasonable because it leads to a N^2-algorithm where N is the number of particles. Thus, an efficient way to search for all interacting particles is required. We chose to sort the particles into a uniform spatial grid with a cell size of the order of the smoothing length h which is constant in our applications. All possible interaction partners of a particle must reside within nearby grid cells. Another common approach that is especially suited for a variable smoothing length uses tree structures instead of a grid.

In order to calculate the right hand side of the hydrodynamical equations in the SPH formalism one has to evaluate three SPH sums, the first to determine the density at each particle position, the second and the third to calculate first and second order derivatives of the field quantities. Because finding interacting particles takes up a considerable amount of the total computing time, it is reasonable to re-use the interaction information in all three sums.

Previous implementations stored a list of all interacting particle pairs in memory. This list was then traversed three times. Given that there are of the order of 100 interactions per particles it is clear that this approach is not ideal, because this list is then the largest data structure and determines the total memory requirement. A single Cray T3E node with 128 MB of main memory could thus work on no more than about 20,000 particles.

Our implementation never stores the whole list in memory. Instead the set of particles is swept along a certain dimension, determining interacting particle pairs on the fly. As soon as the result of the first sum is known for a particle, the particle's contribution to the next sum is calculated. Particles for which the right hand sides of the dynamical equations are completely known are dropped from the interaction list, thus decreasing the list length. The state of a particle is determined by the position of the particle in the grid as shown in Figure 1.

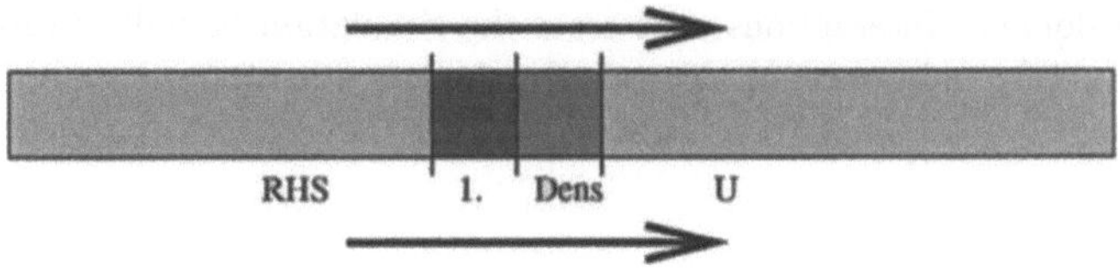

Fig. 1. The particles are swept from left to right. Particles in the region "U" are yet untouched. Only particles from the regions marked "Dens" and "1." are kept in the interaction list, their density and first order derivatives, respectively, are known. Particles in the region marked "RHS" have been completely processed, they have been dropped from the interaction list.

This algorithm drastically reduces the total memory requirement to about one tenth without increasing the overall computing time.

2.2 Boundary Conditions

A very flexible method to formulate boundary conditions in the SPH formalism consists of including ***ghost particles***, i. e., particles that are not integrated in time but are merely a kind of mirror image of other particles. Due to the finite smoothing length h it is never necessary to include a boundary regime with a width larger than h. Periodic boundaries are implemented by copying the ghost particles on one side of the domain from the particles on the other side. For reflecting boundaries the ghost particles are a copy of the particles on the same side, and vector and tensor quantities are additionally reflected at the boundary plane. These ghost particles are inserted prior to each right-hand-side evaluation and dropped afterwards.

There are two optimizations that are important when the number of ghost particles is comparable to the number of real particles, as is the case for simulations with only a few real particles. First, one never calculates any quantities for the ghost particles directly, instead those quantities are copied from the corresponding real particles. Secondly, in the case of periodic boundaries, unnecessary duplicate evaluations of SPH sums are avoided by placing ghost particles only on one side of the domain in each spatial direction (see Figure 2). After these optimizations there is no net computing time spent on the ghost particles except for copy operations.

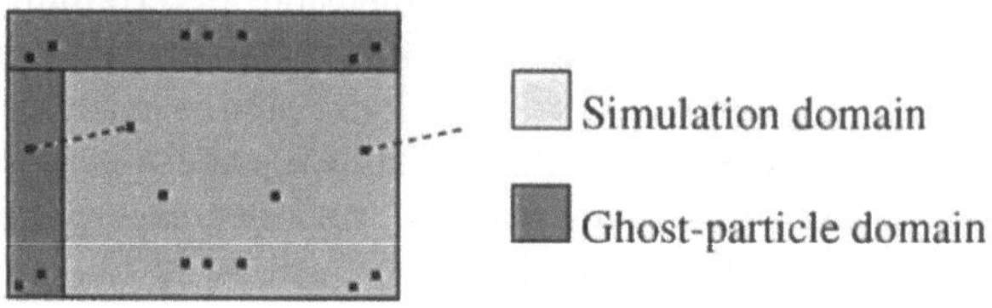

Fig. 2. Periodic boundaries: Ghost particles are placed only on one side of the simulation domain. Interactions that cross the simulation boundary (dotted lines) are accounted for only once, in this case on the left side of the domain.

2.3 Distributed Computing

The MPI library runs a program several times on different processors, there is no uniform view of a single process running distributed. Our aim was to re-use as much as possible from an existing optimized sequential SPH implementation. In our parallel code, each processor is assigned a cuboid region of the simulation domain. The interface to the other regions is treated as a real, physical boundary rather than an algorithmic one.

The resulting (physical) boundary conditions are realized in a way conceptually similar to periodic boundaries but they include communication to neighbor processors. The ghost particles on one processor are mirror images of particles on other processors. Again, as in the case of periodic boundaries, duplicate evaluations of the sums are avoided. Creating the ghost particles and copying values between them and their real counterparts are the only time critical sections of the code that involve communication.

The sweeping algorithm mentioned above (see again Figure 1) takes place on every processor. Copying values from and to ghost particles, i. e., communication between processors, occurs when a particle changes its state. It is possible to delay changing the state of a particle for some time. This delay can be used to hide the time spent on communication. This delay also increases the size of the interaction list and thus acts as a tuning parameter for a size vs. speed tradeoff.

2.4 Load Balancing

Distributing the problem in the way outlined above is relatively inflexible with respect to load balancing. The boundaries between the processors' regions are treated as physical boundaries, and it is difficult to change them dynamically. It is however possible to change them between individual time steps using statistics gathered during the previous time step. This is also the time to remove or insert particles, if necessary.

In order to equally balance the load, a sufficient number of degrees of freedom is required. If $n \times m$ processor regions are placed on a rectangular $(n \times m)$-grid with unequal spacing then there are only $n + m - 2$ degrees of freedom, which is in general less than the $n \times m - 1$ degrees required. To circumvent this problem, a dimensional tree is used instead rather than a rectangular grid. Thus, an individual region can then have more than one neighbor in one direction (see Figure 3).

The domain decomposition is optimized in order to distribute the workload equally among the regions. Before the first time step, the number of particles is used as a measure for the load. After the right hand side has been evaluated, the number of interactions is used instead. Once an acceptable set of domain boundaries has been found, it is eventually fine tuned according to the time a processor spent waiting on other processors. The time spent re-balancing the load between time steps is only an insignificant fraction of the total computing time.

2.5 Efficiency of the Parallel Code

We have tested our implementation on various computer architectures, among these a cluster of workstations (Beowulf), the IBM SP2, and the Cray T3E.

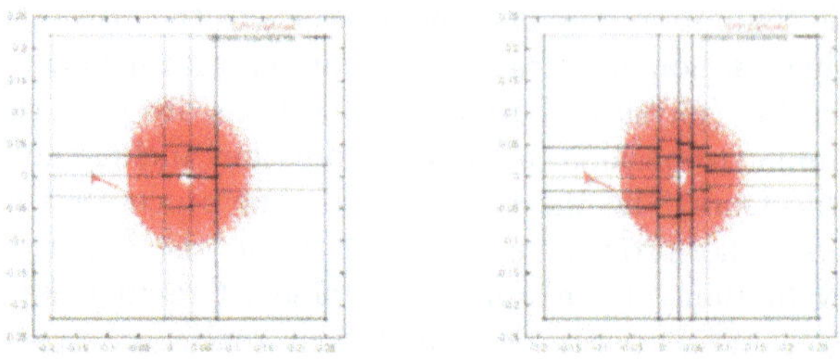

Fig. 3. Sample of our domain decomposition: The simulation domain is cut into slices along the x-axis. These slices are then independently cut along the y-axis, leading to a total of $(n-1)+n\times(m-1)=n\times m-1$ degrees of freedom. The figures show the boundaries of decomposed domains with 15 and with 32 processors, respectively.

Comparison Cray T3E – Pentium2 (300MHz)			
Number of Particles	Number of nodes	runtime Cray	runtime PC
75 000	16	1.77 s	8.08 s
375 000	128	4.16 s	39.8 s

Comparison IBM SP2 – Pentium2 (300MHz)			
Number of Particles	Number of nodes	Runtime IBM SP2	Runtime PC
302 961	16 (wide)	7.902 s	29.7 s
302 961	64 (thin)	1.907 s	29.7 s

Table 1. Comparison of some simulations. The runtime given is the time for one evaluation of the right hand side of the equations on n nodes of the supercomputers vs. 1 CPU on the PC.

A typical moderately sized problem contains 275,000 particles and runs on 64 processors with a total processor usage of more than 92 % on the T3E, as reported by MPI. This we believe to be quite satisfactory.

In Table 1 some comparisons of simulations on Cray TE3 and IBM SP2 vs. standard PC workstations are shown. When comparing CRAY T3E and IBM SP2 one has to keep in mind that a Cray node has only 128 MB memory, whereas the IBM nodes possess 256 MB or 512 MB memory, allowing for larger particle number per node, resulting in less communication costs.

3 Astrophysical Applications: Accretion Disks in Symbiotic Binary Stars

Symbiotic binary stars are so close to each other that their evolution and appearance changes dramatically compared to single stars. Dwarf Novae are a class of variable symbiotic binary stars where mass transfer from one star to the other occurs. The donor is a light main sequence star, the other a more massive, but much smaller White Dwarf (WD). Due to its intrinsic angular momentum the overflowing gas cannot be accreted by the WD right away, instead a thin gaseous disk around the WD forms and the subsequent accretion is governed by viscous processes in the disk. The physics of these accretion disks is far from being well understood. Existing models of long term outburst behavior are essentially 1D and neglect the tidal influence of the donor star. Observationally, the disks show variability on timescales from minutes to decades, occasionally increasing in brightness up to 5 magnitudes. Numerical simulations are very helpful in exploring and validating the theoretical models. Such simulations, especially in 3D, require enormous amounts of grid points – or particles in our case – to achieve the necessary resolution. Since the problem size is so large, and the integration time so long, parallel programs on supercomputers are the only possible way to go.

3.1 The Stream-Disk Overflow in Dwarf Novae

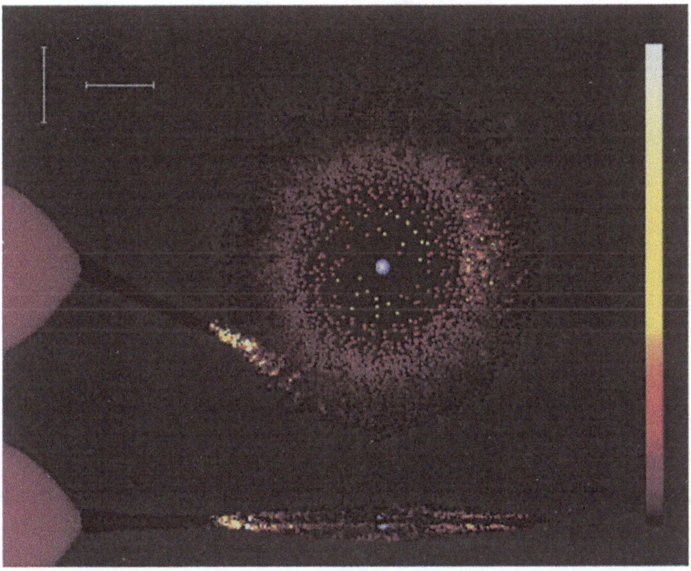

Fig. 4. Pole-on and edge-on view of a simulation of the accretion disk of the Dwarf Nova OY Car. Mass of the White Dwarf: 0.696 $M_\odot$, mass of the donor star: 0.069 $M_\odot$. The scales indicate 0.1 solar radii. The donor star is on the left. Color-coded is the dissipated energy. One can see the bright spot at the impact site of the stream onto the disk, and a secondary bright region right of the White Dwarf, where overflowing material re-enters the disk.

One aspect of Dwarf Nova disks is the impact of the overflowing gas stream onto the rim of the accretion disk. Both flows are highly supersonic and two shock regions form. The shocked gas becomes very hot, a bright spot develops, which sometimes can be brighter than the rest of the disk. The relative heights of the stream and the rim of the disk are unclear. If the stream is thicker than the disk, a substantial portion of the in-falling gas could stream over and under the disk and impact at much smaller radii.

Figure 4 shows a snapshot of the simulation of the accretion disk of the Dwarf Nova OY Carinae. The donor star has 0.069 solar mass, the White Dwarf 0.69 solar mass. Color-coded is the energy release due to viscous dissipation. One can clearly see the bright spot where the streams hits the disk rim. Furthermore, on the far side of the donor star, a secondary bright spot is visible where overflowing stream material finally impacts onto the disk. In this simulation, about 10 to 20% of the stream material can flow over and under the disk.

3.2 Superhumps in AM CVn

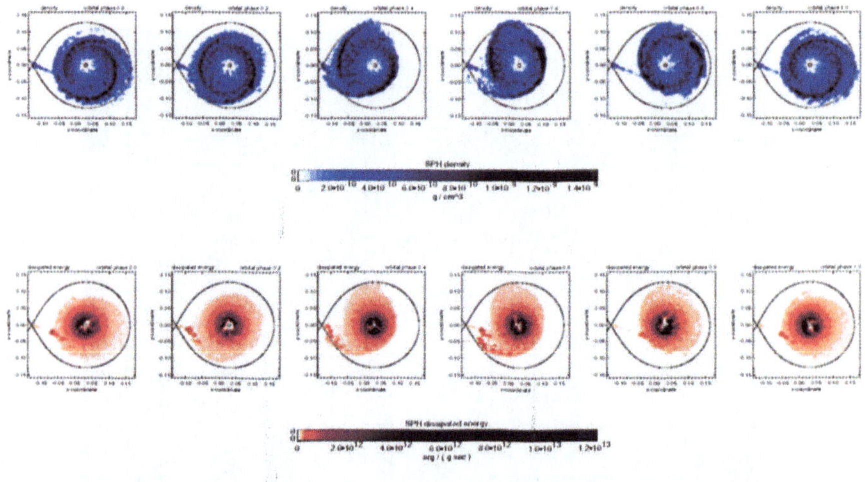

Fig. 5. A series of snapshots of the disk of Am CVn, 0.2 orbital phases apart. Upper panel: density distribution. Lower panel : dissipated energy. The parameters used are: $M_1 = 1$ solar mass, $M_2 = 0.15$ solar masses, mass transfer rate 10^{-10} solar masses per year. A ploytropic equation of state with $\gamma = 1.01$ was used. One can see how the precession of the tidally distorted disk leads periodically to higher dissipation, resulting in superhumps in the light-curve, see Figure 6

AM Canem Venaticorum stars are thought to be the helium counterparts to dwarf novae. So far, there are six systems known. Their common characteristics are hydrogen deficient spectra and photometric variations with very short periods of 1000 to 3000 s. AM CVn stars are believed to consist of two helium white dwarfs, a rather massive primary and a very light, Roche-lobe filling secondary. Roche-lobe overflow feeds an accretion disk around the primary. Tsugawa & Osaki [6] showed that such helium disks undergo thermal instabilities similar to the hydrogen disks in Dwarf Novae. In three AM CVn stars, Dwarf Nova-like outbursts indeed have been observed.

Among the AM CVn stars, AM CVn itself and its near twin EC 15330-1403 have the shortest photometric periods and show no large amplitude outbursts. This, together with a high mass transfer rate, renders it plausible that the disks are thermally stable in a sort of "permanent outburst" state. The hydrogen counterparts of such systems are the so-called permanent superhumpers.

In order to investigate whether AM CVn exhibits superhumps we performed 3D-SPH simulations of the accretion disk. Particles were inserted at the inner Lagrangian point according to a mass transfer rate of $10^{-10} M_{\odot}/\text{yr}$. Already after about 30 orbital periods the disk grew to a point where it was subject to the 3:1 inner Lindblad resonance [8]. Subsequently, the disk became more and more tidally distorted and started to precess rapidly in the frame of reference Co-rotating with the stars, which translates to a slow prograde precession in the observers' frame. Every time the bulk of the disk passes the secondary, the tidal stresses and hence the viscous heating are strongest, giving rise to modulations in the photometric light-curve, the superhumps. A Fourier transform of the obtained light-curve reveals a superhump period excess of 4.4 %. This is in good agreement with the periods given by Warner [7], which differ by 3.8 %.

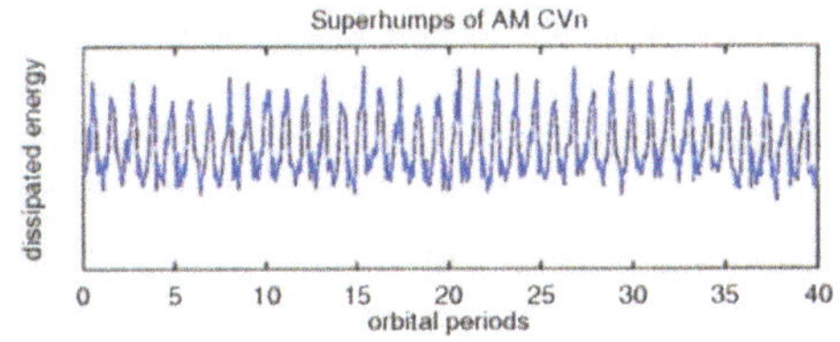

Fig. 6. Shown is the total dissipated energy of the disk over a time span of 40 orbital periods. A Fourier transform of the dissipated energy reveals a superhump period of 4.4 %.

References

1. Flebbe, O., Münzel, S., Herold, H., Riffert, H., Ruder, H. 1994, ApJ **431**, 754
2. Gingold, R. A., Monaghan, J. J. 1977, MNRAS **181**, 375
3. Lucy, L. B. 1977, Astron. J. **82**, 1013
4. Monaghan, J. J., 1992, Ann. Rev. Astron. Astrophys. **30**, 543
5. Riffert, H., Herold, H., Flebbe, O., Ruder, H. 1995, Comp. Phys. Comm. **89**, 1
6. Tsugawa, M., Osaki, Y. 1995, PASJ **49**, 75
7. Warner, B. 1995, Ap&SS **255**, 249
8. Whitehurst, R. 1988, MNRAS **232**, 35

Collisional dynamics around black hole binaries in galactic centres

Marc Hemsendorf

May 3, 1999

Abstract

We present the first results from simulations applying a hybrid "self consistent field" (SCF)[1] and direct Aarseth N–body integrator (NBODY6)[2] which synthesises the advantages of the direct force calculation with the efficiency of the field method. The code is aimed for use on parallel architectures and is therefore applicable for collisional N–body integrations with extraordinarily large particle numbers ($> 10^5$). It opens the perspective to simulate the dynamics of globular clusters with realistic collisional relaxation, as well as stellar systems surrounding a supermassive black hole in galactic nuclei.

1 Introduction

Many nearby galaxies show evidence to harbour massive dark objects in their centres.[3] These observations imply that nearly every larger galaxy should have a massive dark object in its centre, and this implies that the central mass plays a vital role in the galactic evolution. The nature of these central massive dark objects is still unclear, but they are widely believed to be supermassive black holes. Every alternative dense physical object, like supermassive stars or ultra dense relativistic clusters would become unstable and collapse into a black hole.[4] The working model of a black hole sitting in the centre of larger galaxies, can explain nearly all aspects of galactic activity, as in the case for quasars, ultraluminous infrared galaxies, Seyfert galaxies etc. Other explanations like extreme starburst activity fail to explain all observed features and to rule out the creation of a massive black hole as the evolutionary end point.[5] Recent observations of the centre of the Milky Way can constrain the size and mass of the central dark object in so far that it is most likely a black hole with a mass of the order $10^6 M_\odot$.

Under the assumption following from the above that larger galaxies harbour supermassive Black Holes with masses around $10^6 - 10^9 M_\odot$, the question of their formation and their influence on the surrounding galaxy arises. Still the question is not settled whether the black hole form primordial, which means before the first stars, or in the course of galaxy formation.[6] Since galaxies evolve by hierarchical merging of smaller entities, black holes will be involved in these mergers as well. Especially in the young

[1] See Hernquist & Ostriker, (1992), [2].

[2] See Aarseth, (1993), [1].

[3] See Magorrian et al., (1998), [4].

[4] See Rees, (1984), [7].

[5] See Terlevich et al., (1995), [11] and Kormendy & Richstone, 1995, [3].

[6] See Silk & Rees, (1998), [8] and Rees, (1984), [7].

universe these mergers happend frenquently. The subsequent backreaction of the stellar component to the dynamical input of these compact and heavy objects could not be followed by fully collisional N-body simulations. The same is true for studies of the dynamical influence of a central supermassive black hole on its surrounding star cluster. This project is aiming to carry these simulations using a reasonable particle number.

2 Computational Challenges

The enormous complexity of the physics of galactic centres rules out detailed simulations in the moment. Nevertheless, simplified models can as well serve as help for a better understanding of the observed effects. The general starting point for stellar dynamical studies of galactic centres is to place one or more heavy particles into a cluster of light one. Though one can observe immense interaction between the gaseous component and the stellar component in real galactic centres, these interactions are neglected in the N–body models.

In real galactic centres the supermassive black holes resemble up to 0.3% of the bulge's mass. Under the assumption of a mean stellar mass of $1M_{\odot}$ per star the supermassive component is up to 10^9 times heavier. In models of central star clusters the black hole mass lies between 0.01% and 3% of the whole systems mass. The mass difference between the stars the heavy particles is therefore a function of particle number. State of the art cosmological simulations just reach the particle number to accompish a mass contrast of 10^8. Stellar dynamical simulations including realisitic two body relaxation can not be much larger than 2×10^5, which restricts the mass contrast to be about 10^3. Models with realistic two body relaxation have to integrate the orbits by direct force summation, which is a process scaling with $t_{\rm cmp} \propto N^2$. Aarseth type high precision codes applying individual timestep schemes outperform treecodes for this kind of simulations, because the treecode would have to compute to many forces in a direct manner to reach a comparable precision.

As the simulations by Makino et al. show, the mass contrast indeed affects the dynamical behaviour of a black hole binary in a merging cluster.[7] So even these comparatively simple simulations in a highly idealised cluster are computational grand challenge projects. To overcome this problem special purpose hardware or special purpose software is being developed.[8]

To follow the special purpose software approach for reducing the computational effort, we specialise the integrator to nearly spherical stellar dynamical problems. Assuming a spherical stellar component in the system to be investigated, the stars will arrange showing a core halo structure. Since the stars in the halo move essentially in the mean potential of the system and neighbour forces are generally unimportant, a good part of the particles could be integrated much less expensive mean field approximation method, and stars interacting strongly with their neighbouring stars would still be treated with an expensive method for coarse grained potentials. The next chapter will give a brief introduction to the principles of this newly developed hybrid method.

[7] See Makino et al., (1997), [5].

[8] The proposal of *GRAPE* by Sugimoto et al., (1990), [10]. A special purpose software for simulations of galactic centres similar to the one developed in this project was created by Quinlan & Hernquist, (1997), [6].

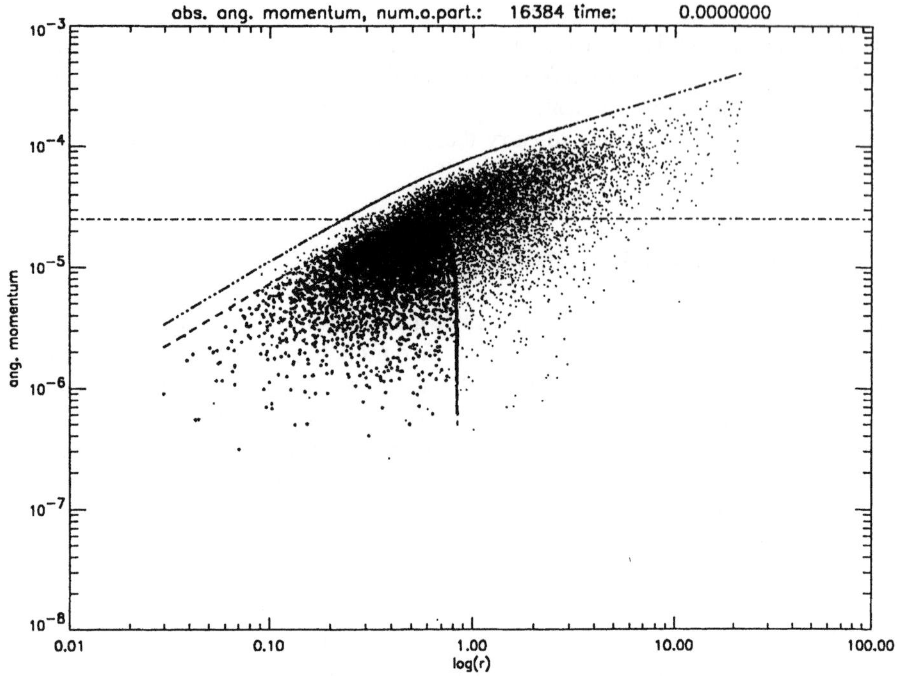

Figure 1: The absolute angular momentum of 16384 particle realisation of a Plummer's model as a function of radius. Exemplaric differences between two selection methods are shown in the case of an angular momentum criterion and an energy criterion. The dot dot dot dashed line at the top shows the angular momentum at which a star would escape the cluster. The dashed line gives the maximal angular momentum a star with a critical energy (E_{crit}) could have. The crosses denote the particles selected by an energy criterion at E_{crit}. The dot dashed line shows the effect of a selection scheme using a critical angular momentum at L_{crit}.

3 A hybrid *NBODY - SCF* code

The idea of this hybrid code is to combine a field method computing the mean potential of the noncollisional parts of the system with a direct force method, which can treat potentials being a superposition of Keplerian potentials. Any particle excange between these two schemes is therefore prone to introducing systematic energy errors. Sufficient criteria for decomposing the stellar system are the integrals of motion, which change for each particle on long timescales. For that it is recommmendable to either the particles energy or angular momentum as the selection criterion.

Figure 1 gives an impression about the effect of selecting particles by either criterium on a Plummer's model. The crosses show the particles which would have been selected by introducing a critical energy E_{crit}. All particles marked with these crosses would then be assigned to *NBODY6* and the rest would be given to the field method *SCF*. (Self consistent field.) Seen from the spatial point of view, the energy criterion selects only a subset of the central particles. In fact, looking at the projection of the

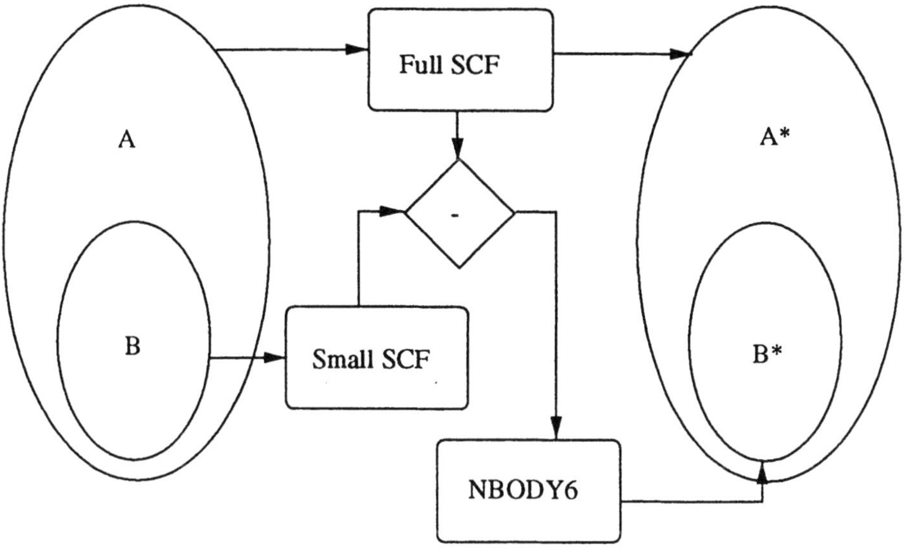

Figure 2: Overview over the particle integration in *EuroStar*.

spatial particle distribution to an arbitrary plane in space, the particle set assigned to the field code has a homogenius central density. The picture changes when the particle set is devided by an angular momentum criterion. In this case, all particles below the dot dashed line in figure 1 would be treated by the direct force integrator, the rest above by the field method. Here clearly is the very central part of the system would be fully assigned to the direct meethod. Nevertheless, for testing purposes simulations using both criteria come to use.

For the fully collsional part of the system *NBODY6++* is chosen, because of its high performance on parallel architectures.[9] The field method chosen for the use with *NBODY6++* called *SCF* is not able to handle potential density pairs having a form not covered by the α-law profiles.[10] *SCF* is restricted to nearly spherical systems, but has the advantage to deliver analytical expressions for the force and the force derivative with respect to time. The second advantage is its nearly ideal scaling on parallel architectures. The following describes the strategies for the orbit integration in the hybrid code ensuring force symmetry between the direct and the field method. The basic idea behind *SCF* is the expansion of the density potential pair into a set of biorthogonal basis functions.[11]

Figure 2 shows which are the major operations necessary to calculate the new particle set A^* within one *SCF* timestep. The particles which are members of the subset B are selected from A by either an energy or angular momentum criterion as stated above.

1. The full *SCF* code computes the new particle set A^* and a new timestep according to the hermite scheme. There is no individual timestep in *SCF*, so that the smallest step for particle is chosen as the systems timestep. Since this work

[9] See Spurzem & Baumgardt, (1998), [9].
[10] See Zhao, (1996), [12].
[11] See Hernquist & Ostriker, (1992), [2].

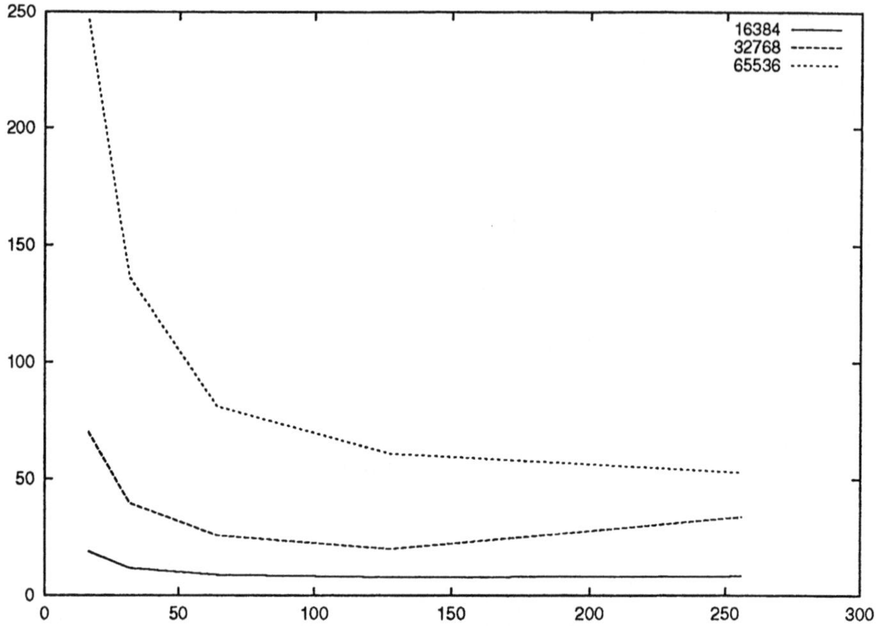

Figure 3: Processing time in seconds as a function of processing elements for one N–body time unit evolving a PLummer's model. The *NBODY6++* part assigned 7284 particles in the 16384 particle run, 14377 particles in the 32768 particle run, and 28627 particles in the 65536 particle run.

can be accomplished with even load balance and data distribution on a parallel computer, this disadvantage against an individual timestep scheme is eased.

2. The small *SCF* program does the same work as the full version on set B, for that it applies the underlying timestep from the full calculation.
3. The difference of the small *SCF* and the full *SCF* potential serves for computing the forces from the living halo. Taking this halo force into account, *NBODY6++* computes new particle positions which are given to the field method as subset B^* for the next timestep.

4 Performance

The parallelisation strategy of *NBODY6++* is to dynamically distribute the work onto the processing elements. This ensures good performance and a good load balance. In the direct method, every particle has its own individual timestep, which are restricted to discrete values obeying $\Delta t = 2^{-n}$. Through that, a group of statistically $n_{\mathrm{gr}} = n_{\mathrm{nb6}}^{2/3}$ particles can be selected at every force calculation step. The field method part distributes load and memory evenly on each node and does not need much communication per timestep. Therefore it scales rather well and takes only approximately 10-20% of the computation time in figure 3. For every particle numbers the code shows an optimal processor number, taking these optimals the scaleup seems to be rather good.

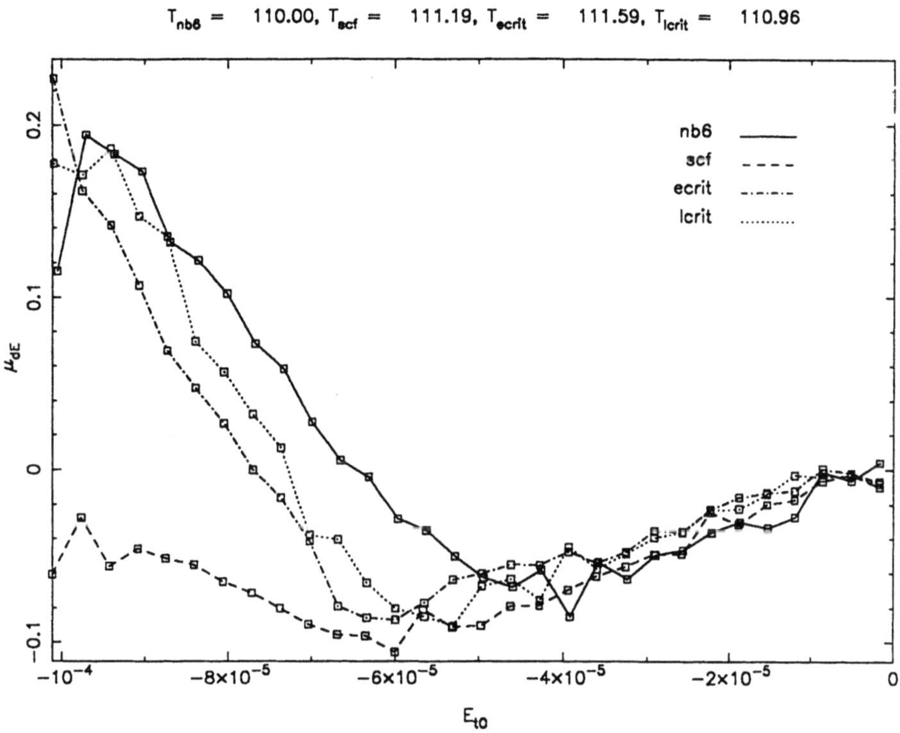

Figure 4: Comparison of the mean energy change in the binned data as a function initial binding energy. The squares are marking the computed data points at the center of each binding energy bin. All runs treat an initially identical dataset of 16384 particles. The hybrid code assigns 4590 particles to the direct method by a critical energy (ecrit) or 7284 by a critical angular momentum (lcrit).

5 Comparison with a purely direct method

To test whether the new hybrid scheme produces reasonable results for the relaxation it was tested against the two methods from which it derives. The simulation using the direct method *NBODY6++* serves as the reference calculation, since the code uses the least assumptions and is fully collisional. The results from collisionless *SCF* method serve as the lower limit for the relaxation processes. To compare how sufficent the selection criteria in the hybrid code are, the Plummer's model is divided by the energy criterion and the angular momentum criterion.

The energy change after ≈ 110 nbody time units (NBU) is compared qualitatively and quantitatively as measure for relaxation. For *SCF* the reference snapshot time is $T_{\rm scf} = 111,19$ nbody time units (NBU), for *NBODY6++* it is $T_{\rm nb6} = 110,00$ NBU, for the hybrid code with energy criterion it is $T_{\rm ecrit} = 111,59$, and for the hybrid code with angular momentum criterion it is $T_{\rm lcrit} = 110,96$.

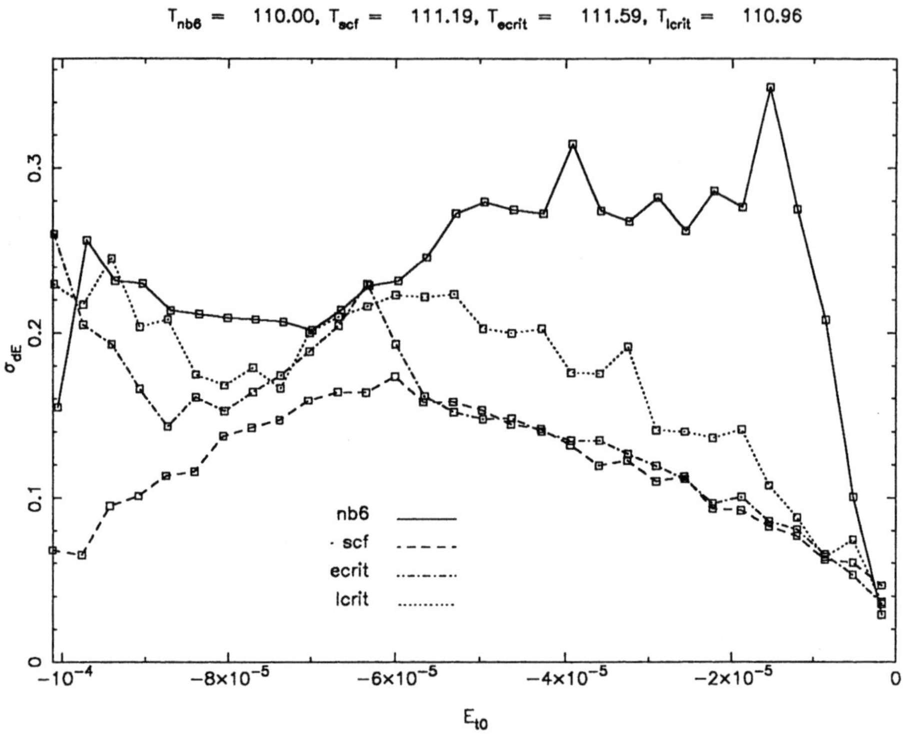

Figure 5: Comparison of the standard deviation in 30 data bins from zero as a function of the initial binding energy of the particles. The squares are marking the computed values at the center of each binding energy bin.

The quantitative comparison in figures 4 and 5 is computed by binning the initial energy distribution of the particles. For highly bound particles, the hybrid scheme can represent the results given by the direct method, when applying the angular momentum criterion (lcrit). Since the plots show the energy change relative to the initial binding energy, the little changes in energy computed with *NBODY6++* show a great effect for particles with low binding energy.

6 First results: Hardening of a black hole binary

Since the phase of thoroughly testing the code could be successfully ended, this section describes the first attemts to study dynamical problems of heavy particles floating in a sea of light ones. The initial conditions are generated by a Plummer's model for the stellar component, which acts as a disturbing force on a black hole binary system. In the undisturbed two body case, the black hole binary is clearly unbound: The two masses would encounter each other on a hyperbolic orbit. The velocity of the massive particles is computed from the following formula:

$$v_{m[12]} = \sqrt{\frac{G(m_1 + m_2)}{a}}. \tag{1}$$

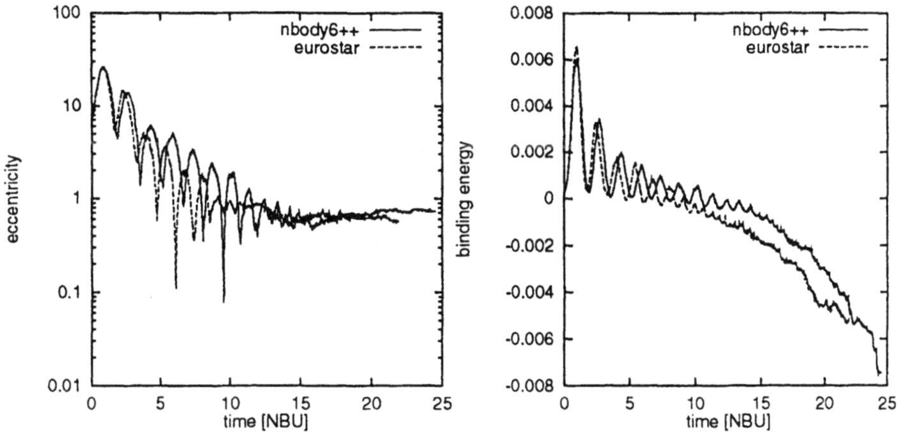

Figure 6: Dynamical friction leads to a sinking and hardening black hole binary in the surrounding Plummer's model. The simulations where carried out using 16382 equal mass particles for the background potential. EuroStar applied the angular momentum criterion assigning 7089 to the collisional part.

Where a denotes the initial radius from collective center of mass of the binary and the Plummer's model and the $m_{[12]}$ denote the masses of the binary components. The resulting velocity $v_{m[12]}$ is resonably lower than the circular velocity of the Plummer's model at radius a. The binary is embedded in a realisation of a Plummer's model with a total mass of unity and a scale radius $R = 3\pi/16$. All units are again in NBODY units, so $G = 1$. The total mass of the system is equal to one, each massive particle resembles approximately 1% of the total mass.

Figure 6 shows the evolution of the eccentricity and the binding energy of the binary for a fully collisional *NBODY6++* and a hybrid code run. Since the particle number is still very low with 16384, the density contrast is approximately 168. The influence of a single 'stellar' particle on the heavy binary is therefore very strong and the reason for the non smooth curves. The right plot in figure 6 shows that the binary becomes bound at around 10 time units for the hybrid scheme and 12 time units for the fully collisional code. After that the evolution takes a parallel path for both schemes. Until the binary becomes bound its is in the dynamical friction regime. Once bound, its relative velocity to the surrounding particles is low enough that two and three body relaxation processes become more important. In the dynamical friction regime the hybrid scheme has significantly less particles to react with the heavy particles in the outer regions. This leads to a significant higher friction force when the heavy particles pass the inner regions of the cluster, so that in total the binary can become bound earlier. Since the relaxational binary hardening is very similar to results from the collisional code, *EuroStar* proves its ability to follow relaxation processes around massive particles.

7 Summary and further work

During the course of this project a new hybrid code was developped for the simulation of collisional dynamics in galactic nuclei. Due to its good scaling behaviour and memory management on parallel architectures it is able to handle extraordinary large particle numbers. The tests show, that the hybrid scheme can reproduce the results from a fully collisional method.

Further work includes detailed studies of the relaxation processes around single massive particles and binary systems. It will furthermore include studies of the backreaction of the stellar system to the heavy particle as a function of total particle number.

Acknowledgements

I would like to thank R. Spurzem and S. Sigurdsson who had the idea for this project. I would like to thank S. Aarseth, D. Heggie, W. Sweatman and G. Hensler for fruitful help and discussion. This project is funded by *Deutsche Forschungsgemeinschaft* (DFG) project Sp 345/9-1 and SFB 349. Technical help and computer resources are provided by *NIC* in Jülich, *HLRS* in Stuttgart, *EPCC* in Edinburgh, *ZIB* in Berlin, University of Heidelberg and University of Kiel.

References

[1] S. J. Aarseth. Direct methods for *N*–body simulations. In G. Contopoulos, N. K. Spyrou, and L. Vlahos, editors, *Galactic Dynamics and N–Body Simulations*, volume 433 of *Lecture Notes in Physics*, pages 365–417, Thessaloniki, 1993. Springer–Verlag.

[2] L. Hernquist and J. P. Ostriker. A self - consistent field method for galactic dynamics. *The Astrophysical Journal*, 386:375–397, 1992.

[3] J. Kormendy and D. Richstone. Inward bound - the search for supermassive black holes in galactic nuclei. *The Annual Review of Astronomy and Astrophysics*, 33:581–624, 1995.

[4] J. Magorrian, S. Tremaine, D. Richstone, R. Bender, G. Bower, A. Dressler, S. M. Faber, K. Gebhardt, R. Green, C. Grillmayr, J. Kormendy, and T. Lauer. The demography of massive dark objects in galaxy centers. *The Astronomical Journal*, 115:2285–2305, 1998.

[5] J. Makino. Merging of galaxies with central black holes II. evolution of the black hole binary and the structure of the core. *The Astrophysical Journal*, 478:58–65, 1997.

[6] G. D. Quinlan and L. Hernquist. The dynamical evolution of massive black hole binaries - II. self-consistent *n*–body integrations. *New Astronomy*, 2:533–554, 1997.

[7] M. J. Rees. Black hole models for active galactic nuclei. *Annual Review of Astronomy and Astrophysics*, 22:471–506, 1984.

[8] J. Silk and M. Rees. Quasars and galaxy formation. *Astronomy and Astrophysics*, 331:L1–L4, 1998.

[9] R. Spurzem and H. Baumgardt. A parallel implementation of an Aarseth *N*–body integrator on the CRAY T3D. ARI–Preprint No. 68, submitted to Monthly Notes of the Royal Astronomical Society, 1997.

[10] D. Sugimoto, Y. Chikada, J. Makino, T. Ito, T. Ebisuzaki, and M. Umemura. A special-purpose computer for gravitational many-body problems. *Nature*, 345(6270):33–35, 1990.

[11] R. Terlevich, G. Tenorio-Tagle, M. Różyczka, J. Franco, and J. Melnick. The starburst model for active galactic nuclei – II. the nature of the lag. *Monthly Notes of the Royal Astronomical Society*, 272:198–212, 1995.

[12] H. Zhao. Analytical models for galactic nuclei. *Monthly Notes of the Royal Astronomical Society*, 278:488–496, 1996.

IMD – A Massively Parallel Molecular Dynamics Package for Classical Simulations in Condensed Matter Physics

Johannes Roth, Jörg Stadler, Marco Brunelli, Dietmar Bunz, Franz Gähler, Jutta Hahn, Martin Hohl, Christof Horn, Jutta Kaiser, Ralf Mikulla, Gunther Schaaf, Joachim Stelzer, and Hans-Rainer Trebin

Institut für Theoretische und Angewandte Physik, Universität Stuttgart, Pfaffenwaldring 57, 70550 Stuttgart, Germany

Abstract. We describe the current development status of IMD (**I**TAP **M**olecular **D**ynamics), a software package for classical molecular dynamics simulations on massively-parallel computers. IMD is a general purpose program which can be used for all kinds of two -and three-dimensional studies in condensed matter physics, in addition to the usual MD features it contains a number of special routines for simulation of mechanical properties of solids, analysis and visualization.

1 Introduction

Molecular dynamics is a method widely used in physics and chemistry to study mechanical stability and thermodynamical properties of all kinds of systems from atoms to galaxies. Starting from the basic equations of motion which are ordinary differential equations with a huge number of variables (the positions and momenta of the particles) one derives a set of difference equations for finite time steps and solves the equations iteratively. In the equilibrium case one extracts thermodynamical averages like temperature, pressure or diffusion constants after the simulation has been finished. But MD simulations are not limited to this possibility: non-equilibrium simulations like heating, cooling, deformation of the sample can also be carried out.

The basic design of IMD is due to Jörg Stadler [1] who wrote most of the code as a contribution to the project A2: "Molecular Dynamics Simulations of Quasicrystals" of the collaborative research center (Sonderforschungsbereich) 382: "Procedures and Algorithms for the Simulation of Physical Processes on Supercomputers". Meanwhile IMD has been enhanced by many new features. The software is currently maintained by a team of programmers. To organize the ongoing development and maintenance we use the software management system CVS. Although we intend to give the most up-to-date description of IMD in this article, we would also like to draw the attention of the reader to the IMD-webpage [2] where the latest news, the source code and the documentation can be found. The design and development of IMD has been described in two articles, called Paper I [1] and Paper II [3]. An additional HLRS report will present some of the results achieved with IMD [4].

1.1 Short overview of IMD 1.3

Interactions and Ensembles. IMD 1.3 allows particles to interact via pair, three-body [4] and EAM[1] potentials. Details of the implementation are given in the second report. Further interactions are the Gay-Berne potentials commonly applied in liquid crystal studies. These interactions are anisotropic modifications of the well-known Lennard-Jones potentials [6]. There is no limit to the number of particle types differing in mass and interaction potential. The pair interaction potentials are read from tables and therefore are not limited to simple analytical expressions. The only exception are Lennard-Jones-potentials where the interaction is computed directly if the system is monatomic. The calculations can be carried out in a number of thermodynamical ensembles [7] like microcanonical (NVE, total energy is constant), canonical (NVT, temperature and volume constant, NPT, temperature and pressure constant, and NσT, temperature and uniaxial stress constant). For NVT a stochastic Andersen thermostat [7] is also available. In general a Nosé-Hoover thermostat [7] is used for NVT, NPT and NσT. In addition to MD simulations it is possible to optimize the potential energy of a sample with the microconvergence method [8]. Mechanical boundary conditions allow the sample to be pulled apart or sheared at constant rates or certain intervals.

Multi-Phase Simulations. In very large simulations it may not be possible to read and write the initial and final data sets. On the other hand it is often necessary to optimize the potential energy of the generated structure with the microconvergence method first, then to equilibrate it with NPT to correct the volume, and finally to carry out an NVT or NVE simulation to collect the data for the thermodynamical averages. With IMD it is possible to carry out such a multi-phase simulation in a single run.

Simulation Setup. Several types of crystal or quasicrystal structures may be generated directly with IMD. In addition there are special setups for fracture simulations to damp out shock waves emitted from a crack [9], and in shock wave simulations to initiate a shock wave. A moving crack generates heat which must be removed from the sample (usually this is done by long range heat conduction). Since the ordinary Nosé-Hoover thermostat mimics an infinitely large thermal conductivity we had to modify the simulation setup: in the central part of the sample a NVE ensemble is simulated, and the Nosé-Hoover thermostat is limited to the outer part of the box.

Online Tools. A common usage of MD simulations is the calculation of materials properties. For this purpose special functions or the collection of histograms of several variables have to be evaluated. This can be done effectively only during simulation. IMD contains routines to compute self-correlation

[1] EAM: **E**mbedded **A**tom **M**ethod, first described by Daw and Baskes [5].

functions which are used to study the dynamical properties of solids. Transport properties like the thermal conductivity for example can also be computed. To monitor atomic jumps of atoms we have implemented flip detectors which allow to keep track of the motion of pre-selected atoms. The flip detectors can be applied only to certain decagonal quasicrystals since the possible jumps depend heavily on the geometry of the simulated structure.

To study the dynamics of a simulation it is helpful to know how the kinetic and potential energy changes. The kinetic energy for example allows to locate the crack tip whereas the potential energy is sensitive to stacking faults. With IMD one can produce color-coded histograms of both energy distributions.

It is also possible to compute displacement fields and energy difference maps if reference states are stored at the beginning of a simulation.

Online Visualization. Large simulations on massively-parallel supercomputers require completely new concepts to deal with the overwhelming flood of data: If there are ten to hundred million particles and the simulation is carried out for several hundred thousand or million steps it is impossible to store the complete particle configurations, even at infrequent time intervals. IMD includes several features to deal with this problem: It is possible to generate pictures of the potential or kinetic energy directly. The program also provides the possibility of online visualization: the simulation on a remote supercomputer is monitored at the same time on a local workstation (even if there is a firewall in between as it is the case at the HLRS).

Metacomputing. The distribution of one application on a cluster of parallel computers is called metacomputing. It is possible to carry out such simulations with IMD.

Implementation Details. The package's main design goals were to create flexible and modular software that reaches a high performance on contemporary computer architectures while it is still as portable as reasonably possible. The target platforms for IMD are RISC workstations and massively-parallel computers with distributed memory. It does not perform well on vector machines since the vectorizable loops are to short. To run IMD efficiently on a distributed-shared architecture would also require some recoding.

IMD is actively supported on DEC and SGI workstations and on PCs running Linux as well as on the Cray T3E. It has been ported to IBM workstations, to the Intel Paragon and to the Hitachi SR2201. IMD is written in C and uses the widely available MPI library for message passing. It has been adapted to metacomputing via the PACX [10] library. For the online visualization a special version of VolRend [11] called VolIMD is available.

Offline Tools. IMD is not only a simple simulation program but a package which contains routines that allow to analyze and visualize the results. This means that there are tools which share the data structures set up in IMD.

The analysis tools usually start from configuration files which contain the positions and velocities of the atoms, their potential and kinetic energies, their type and masses and further properties if required. There is a program which computes the radial distribution functions for each type of atoms from the coordinates. PPM pictures or VRML scenes can also be generated.

Outline. In Paper I the concept of linked cells for simulations of short-range interactions and its the extension to parallel computers has been described in detail. Paper I also deals with communication modes and the performance of IMD in large scale and high speed simulations. In Paper II we have described the concepts of online visualization and metacomputing in detail.

This paper is organized as follows: in the first section we will summarize Paper I and extend it to the concept of dynamical linked cells and many-body interaction. In the second part we will describe the online visualization and the generation of pictures. In the third section we will talk about meta-computing and In the final section we present examples and summarize the applications that have been simulated with IMD.

2 Basic Concepts of Short Range Massively-Parallel MD Simulations

In MD simulations the equations of motion for a system of N particles have to be solved over a time T_{sim}. The integration is done stepwise in small time intervals Δt. At each step the force F_i acting on particle i must be calculated. Since the particle may interact with all other particles, the complexity of the problem grows as $O(N^2)$. Thus the force calculation is the most time consuming part of the simulation. If the forces are short ranged, however, the complexity for the necessary force calculations grows with NN_c, where N_c is the number of atoms in the range of the interactions. This is true for two-body pair potentials as well as for many-body interactions like EAM for example, since the many-body interactions are usually also short ranged.

For short-range interactions the computation of forces is a local problem and lends itself naturally to parallelization by domain decomposition. The simulation box is geometrically divided into cells which represents an easy way to find the neighbors of any particle.

In case of pair interactions the total force F_i acting on particle i is the sum over all forces f_{ij} of the particles j within the interaction range r_c. To calculate F_i we must know the positions r_j of all particles closer than r_c to particle i. To find them we subdivide the simulation box into small cells with a diameter that is at least twice as long as r_c (see Fig. 1). The cells need not be squares or cubes, IMD also permits parallelepipeds. The atoms can easily be assigned to the cells if their coordinates are divided by the length of the simulation box and then multiplied by the number of boxes. Since the diameter of the cells is larger than $2r_c$, all interactions take place within a cell or between neighboring cells. What remains to be done is to set up a linked list of the neighbors of each cell once at the beginning of the simulation. Then the whole computation can be carried out by computing the interaction between atoms in the same cell and between atoms in neighboring cells. Since $f_{ij} = -f_{ij}$ by Newton's law only half the number of neighbors has actually to be taken into account (See Fig. 1). Since the assignment of the atoms to the cells is a linear process the complexity of our simulation has been reduced to $O(N)$.

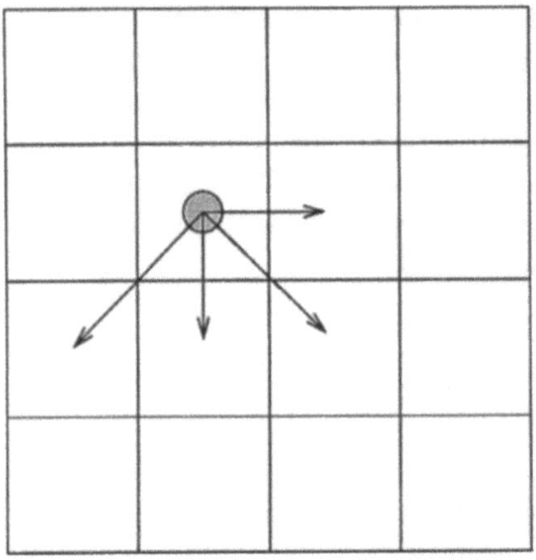

Fig. 1. Decomposition of the simulation box into cells. The arrows indicate the neighbor cells that must be taken into account in the computation of the interaction in the central cell.

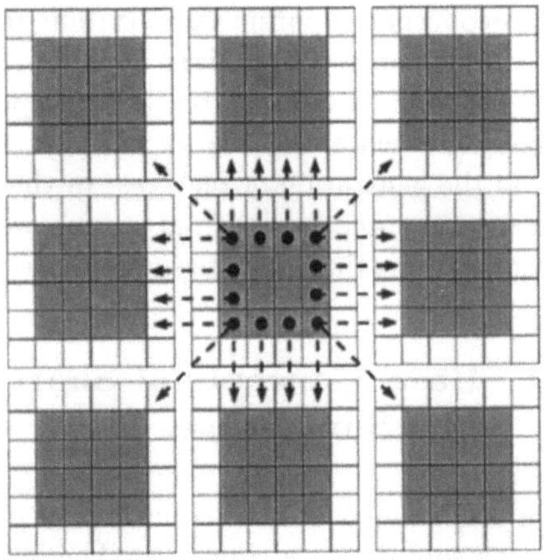

Fig. 2. Assignment of the cells to processors in two dimensions. The white cells are *buffer cells*, the adjacent cells (dotted in the central processor) are the *surface cells*.

In the NVE and NVT simulations the cell structure is set up at the beginning of the simulation and remains unchanged during the whole run. This is not the case in NPT or NσT simulations where the cells must be adjusted to the variable volume. It may happen that the diameter of the cells is becoming too small if the volume shrinks for example during a transition from a gas to a liquid. Or during the opposite transition it may happen that the number of cells is too small and the simulation becomes very inefficient. In NPT and NσT simulations IMD checks if the cells decomposition is appropriate, and if not, it computes a new cell setup.

2.1 Cells on Parallel Computers

The generalization of the cell structure to parallel computers is rather straightforward: IMD assumes that the processors are arranged on a two- or three-dimensional cartesian grid with periodic boundaries, so each processor has 26 neighbors in three dimensions or 8 in two dimensions. Our algorithm subdivides the array of cells into equally sized parts which then are assigned to the processors. Figure 2 shows a two-dimensional example of how the cells are associated with the processors: A 12×12 array of cells (shaded) is distributed onto a 3×3 grid of processors. A layer of empty cells (non-shaded) is added to each processor. These are used as temporary stage for particle data received from neighbors. We call the extra cells *buffer* cells. The outermost layer of non-empty cells at the surface of a processor's part are called *surface* cells.

Particles in non-surface cells interact only with particles on the same processor, whereas those in the surface cells also interact with particles in surface cells on the neighboring processors. To calculate the forces for them, data from the surface cells are copied to the buffer cells of the neighbors, as indicated by the arrows in Fig. 2.

Three communication steps have to be performed at each time step: First, the particle positions have to be exchanged with the neighbor processors for the force calculation, second, the forces accumulated due to Newton's third law by particles in the buffer cells are sent back to the original processor, and third, after the positions have been updated, all atoms that have left or entered a processor's column must be transferred to their new processor.

As indicated in Fig. 2 by the arrows, the data for each surface cell could be communicated cell by cell. This is, however, not optimal since it creates a large number of small messages. IMD first collects the data for a given neighbor into an intermediate data buffer and then translates the whole buffer with only one single MPI call. We have tested the cell by cell and the buffer scheme. Details are given in Paper I.

In IMD periodic boundary conditions are usually assumed in all directions, but it is possible to switch them off and use open or fixed boundaries. If the system is homogeneous and periodic boundary conditions are used, we find that load balancing conditions are fulfilled very well. However, if open or fixed boundaries are present or inhomogeneous parts like cracks or shock fronts we find that the load balancing goes down considerably. The minimum was a computational effectivity of 35%. An improvement of this situation would require the introduction of active load balancing.

3 Online Visualization and Picture Generation

Until recently MD simulations were carried out in the following way: The simulation was run, data were collected and stored for later analysis. With the advent of large massively-parallel supercomputers such a scheme may no longer be feasible since in many cases the amount of data is too large to be stored.

One way to circumvent this problem is to visualize the data online. More advanced schemes even permit to steer the simulation online, for example by changing the applied pressure or temperature, by altering the frequency of picture production or by starting data storage or interrupting the simulation in case of errors.

With IMD online visualization is possible and the basic interfaces for online steering are also given. IMD uses ordinary TCP sockets for communication. This may cause problems if the supercomputer is hidden behind a firewall as it is the case at the HLRS. For this reason IMD works with high port numbers, but it may still be necessary to ask the computing center to open up ports for communication. More details about the interaction of simulation and visualization can be found in Paper II.

Online visualization can be used for interactive jobs as well as for batch jobs. The monitoring may be started at any time during the simulation. Then the visualization client is connected to the simulation server and receives pictures or histograms. If the simulation is running for a long time, one may connect to the server infrequently and check the status of the simulation job.

For the online visualization a program is required which runs on the visualization client and is able to deal with the IMD output. This program is called VolIMD, a modification of VolRend written by R. Niemaier [11]. Although VolIMD is running only on SGI workstations, there exists a version of VolRend called GlutRend which can be used on DEC Alpha workstations and on Linux PCs. All the programs are available at our institute.

3.1 Pictures and Movies

The pictures and histograms produced by the online visualization routines are raw data written in binary format. This has the advantage of being faster and more flexible than writing formatted output. At the beginning of the simulation one has to define an array of bins into which the data are collected. The conversion of the binary data is carried out by the ppm-tools or with the help of VolRend. In two dimensions it is possible to write ppm-files directly from IMD, but then one looses the possibility to manipulate the pictures afterwards.

Picture sequences from the simulation, from VolRend or from the offline tools may be assembled to computer movies using movieconvert from SGI for example or any other movie program. This adds a new dimension to the presentation of the results since the dynamics of a system can be studied directly.

4 Special Requirements for Metacomputing

IMD has been written for massively-parallel supercomputers like the Cray T3E, where the communication between the processors is very fast and and the communication time is short compared to the time required for a simulation step. For metacomputing this is no longer the case. The internal communication time within the T3E is 18 μs, with PACX it is 21 μs. Communication between the supercomputer nodes is at least one order of magnitude slower than within a supercomputer, even if high-speed and high-bandwidth lines are used. The latency time to send a the message between two partitions within the same T3E is 3.5 ms, between Stuttgart and Manchester it is 45 ms, and between Stuttgart and Pittsburgh it is 75 ms! To avoid wrong results the computations have to be stopped until the inter-supercomputer communication has been finished. This is very unsatisfactory and a way to interlace calculation and communication has to be found: If we take a look at the central part of the middle cell in Fig. 2 we find four gray cells without dots. The calculations in these cells can be started at the same time as the communication. Then we may proceed with the surface and boundary cells which have been updated already, thus breaking up the synchronous communication scheme into smaller pieces which can be sent continuously during computation.

5 Applications

Up to now, IMD has been used mainly for simulations of the mechanical stability of quasicrystals. Examples are simulations of cracks, fracture [13–15], shear deformations and dislocations in quasicrystals in two and three dimensions [16], shock waves in crystals and quasicrystals in three dimensions [3,12], interfaces and covalent materials. Details of the latter results are given in a second HLRS report [4]. Upcoming simulations include the study of copper islands on gold under irradiation, the simulation of silicon interfaces and the modeling of milling processes. Here we give a short summary of two applications. Details may be found in the original publications. For a typical run the simulation box contains about 70000 atoms. On the Cray T3E it runs on 64 processors and lasts about 30 minutes. This permits us to run parameter studies at various loads and temperatures.

Crack Simulations. For the crack simulation we use a piece of quasicrystal with fixed boundaries at the bottom and at the top and free boundaries at the other two sides. The material in a notch is removed from the sample, and then the upper and lower boundaries are moved apart by a fixed amount until a crack is emitted from the notch. The breaking of bonds at the crack tip bonds causes the emission of shock waves. To avoid an interaction of the crack with the shock waves reflected at the boundary of the sample, we have generated a stadium around the crack. Waves meeting the stadium are damped out [9]. In a crystal the crack would propagate smoothly. In a quasicrystals the crack is meandering around, thereby avoiding clusters of strongly bound atoms (Fig. 3). The motion of the crack has been studied at various loads: it starts from an intermittent behavior at a small load and gets continuous at higher load. The velocity increases and saturates. At high loads the crack motion gets chaotic since the energy released cannot be transported away fast enough.

Shear Simulations. Further studies are concerned with shear simulations of quasicrystals. In this case the upper border of the cell is shifted parallel to the lower border but in the opposite direction. Dislocations are created which move through the quasicrystal. In contrast to the crystal case the sample is not perfect behind a dislocation but contains a stacking-fault-like phason wall (Fig. 4). On the one hand this weakens the quasicrystal, on the other hand it costs energy. Dislocation pairs are created on the defect wall which modify it in such a way that the energy cost is reduced.

Acknowledgments

The help of Michael Reichenstein (visualization) and the PACX group led by Michael Resch and Edgar Gabriel is gratefully acknowledged.

References

1. Stadler, J., Mikulla, R., Trebin, H.-R.: IMD: A Software package for molecular dynamics studies on parallel computers. Int. J. Mod. Phys. C **8** (1997) 1131-1140
2. www.itap.physik.uni-stuttgart.de/~imd
3. Roth, J.: IMD – A molecular dynamics program and applications. Proceedings of the workshop: MD on Parallel Computers. Eds. Attig, N., Esser, R., to be published by World Scientific, Singapore
4. Hahn, J., Trebin, H.-R.: Molecular dynamics of covalent systems, HLRS report 1999, this volume
5. Daw M.S., Baskes M.T.: Embedded atom method: Derivation and application to impurities, surfaces and other defects in metals. Phys. Rev. B **29** (1984) 6444-7991
6. Gay G.J., Berne B.J.: Modification of the overlap potential to mimic a linear site-site potential. J. Chem. Phys. **74** (1981) 3316-3319
7. Allen M.P., Tildesley D.J.: Computer simulation of liquids. Oxford Science Publications, Oxford (1987)
8. Beeler J.R.: Radiation effects computer simulations. North Holland, Amsterdam (1983) p 27
9. Zhou S.J., Lohmdahl P.S., Holian B.L.: Dynamic crack processes via molecular dynamics. Phys. Rev. Lett. **76** (1996) 2318-2321
10. www.hlrs.de/structure/organisation/par/projects/pacx-mpi/
11. www.uni-stuttgart.de/RUSuser/vis/People/roland/volrend/volrend.html
12. www.hlrs.de/news/events/1998/sc98/sc98_hlrs_booth.html
13. Mikulla R., Stadler J., Krul F., Trebin H.-R.: Gumbsch P.: Crack Propagation in Quasicrystals. Phys. Rev. Lett. **81** (1998) 3163-3166
14. Mikulla R., Stadler J., Gumbsch P., Trebin H.-R.: Molecular dynamics simulations of crack propagation in quasicrystals. Proceedings of the 6th International Conference on Quasicrystals. Eds. Takeuchi S., Fujiwara T., World Scientific, Singapore (1998) 485
15. Mikulla R., Gumbsch P., Trebin H.-R.: Dislocations in quasicrystals and their interaction with cluster-like obstacles, to appear in Phil. Mag. Lett.
16. Schaaf, G.: Numerische Simulationen des dynamischen Verhaltens von Quasikristallen. Diploma Thesis, Stuttgart (1998)

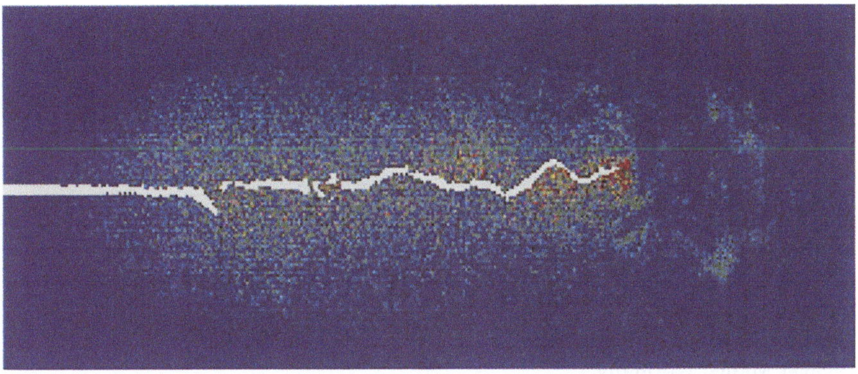

Fig. 3. Crack in a quasicrystal. Blue indicates low energies, green the shock waves emitted from the crack and red the very hot crack tip.

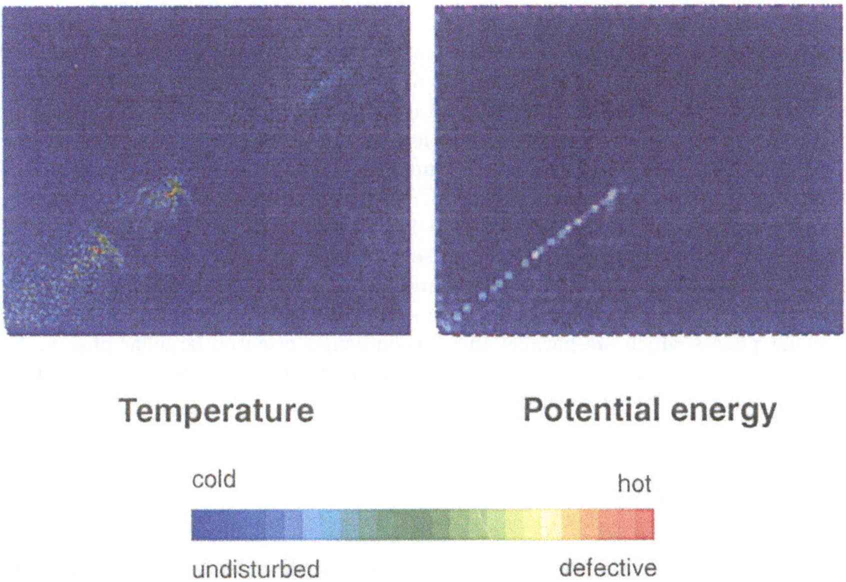

Fig. 4. Shear simulations of quasicrystals. Left side: temperature of the sample. The green parts indicate the position of the dislocation defects. Right side: the green chain of dots marks the location of the stacking fault.

Symmetric diblock copolymers confined into thin films: A Monte Carlo investigation on the CRAY T3E

T. Geisinger, M. Müller, and K. Binder

Institut für Physik
Johannes Gutenberg Universität
D55099 Mainz, Germany

Abstract. We present the results of large scale computer simulations targeted at investigating the phase stability and the structure of symmetric AB diblock copolymers in thin films. The connectivity of the two different monomer species A and B in the diblock copolymer prevents macrophase separation and the molecules assemble into A-rich and B-rich domains on the scale of the molecule's extension. This large length scale of the ordering phenomena makes these polymeric systems a promising candidate for revealing the universal features of self-assembling in amphiphilic molecules. However, the widely spread length and time scales impart protracted long relaxation times to the systems and pose a challenge for computer simulations. Parallel supercomputers like the CRAY T3E are necessary for equilibrating systems with structure on very different length scales. In order to investigate the stability of different phases we quench several systems with identical parameters from the disordered state to a temperature in the intermediate segregation regime and monitor the morphology. Three film thickness are investigated and we find lamellar phases which are parallel and perpendicular oriented lamellar phases. The results are compared to analytical calculations, and the influence of capillary waves on the profiles in the thin film are discussed.

1 Introduction

Joining chemically distinct polymers – A and B – at their ends to form an AB diblock copolymer prevents their phase separation into domains of macroscopic size. In order to reduce the number of energetically unfavorable interactions between distinct blocks in a dense melt of these molecules, the polymers self-assemble into complex spatial morphologies[1]. The symmetry of spatial ordering is selected via a delicate balance between the free energy cost of the internal interfaces and the conformational entropy loss as the molecules stretch to fill space at constant density. The phase diagram in the bulk has been investigated in much detail as a function of the relative length of the blocks f and the incompatibility between the different moieties χN[1–3].

The morphologies found in copolymer melts resemble the spatially structured phases of other amphiphilic systems and biological systems (e.g., lipid

/ water mixtures). Polymeric systems are particularly convenient for investigating mechanisms of self-assembly, because they are describable by only a small number of parameters, namely the fraction f of A monomers in the diblock copolymer, the end-to-end distance of the molecules R_e, and the incompatibility χN. The Flory-Huggins parameter χ denotes the repulsion between monomers of different species and N is the number of monomers per molecules. The extension of the molecules sets the length scale of the spatial structure, and it is much larger than the extension of the monomers. This difference in length scales, between the monomer extension and the scale of the morphology of the A and B rich domains, gives rise to extremely long relaxation times. Hence, it is very difficult in experiments[4–7] and simulations[8] to achieve equilibrium. Investigating these problems with Monte Carlo simulations calls for powerful massively parallel supercomputers.

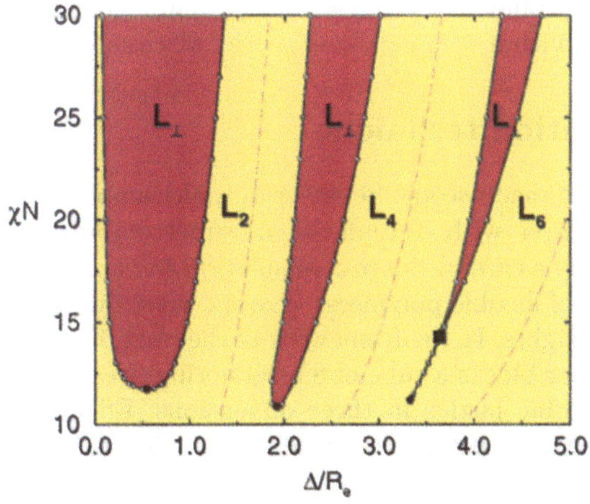

Fig. 1. Phase diagram of a symmetric diblock copolymer confined into a thin film as a function of the film thickness Δ (measured in units of the end-to-end distance R_e) and the incompatibility χN. Both surfaces attract the A component of the diblock. L_2, L_4, and L_6 denote parallel lamellar phases in which 2,4 or 6 AB interfaces run parallel to the confining walls, respectively. $L_\perp$ denotes the perpendicular oriented morphology.

Confining these complex fluid into thin film has attracted abiding interest. From the theoretical point of view, the confinement of complex fluids might lead to pronounced shifts of the phase boundaries. For spatially structured phases the confinement leads additionally to transitions between phases with identical symmetry but different orientation with respect to the confining walls. Moreover, the surfaces alter the structure of the fluid in its vicinity and lead to an enrichment of one component. From a practical point of view, confining and careful tuning of the interactions between the surfaces and complex fluid, one might control the orientation of the spatial structure over rather large length scales. This is important to many technical applications, e.g., structures with perpendicular orientation to the substrate/wall have been successfully employed as a template for lateral structure on the nanometer scale[9,10].

In this paper we investigate the phase stability and the structure of diblock copolymers via large scale computer simulations in the framework of a coarse grained lattice model, and we compare our results to the predictions of the self-consistent field theory[11]. The phase diagram in the self-consistent field theory is presented in Fig.1. At intermediate segregation ($\chi N = 30$) we find a sequence of lamellar phases with parallel and perpendicular orientation with respect to the confining surfaces of the film. At lower segregation the perpendicular oriented morphology ends in a critical point for small film thicknesses, or we encounter a triple point. At the latter point three phases coexist: two parallel oriented lamellar phases with a different number of interfaces and a perpendicular one. At even lower segregation, there are direct transitions between parallel lamellar phases whose number of interfaces differ by two.

Our report is arranged as follows: In the next section we describe the computational model and technique. Then we present our results on the phase stability and discuss composition profiles and profiles of individual segements across the film. The paper closes with a summary and a brief discussion.

2 Model and the simulation technique

The accurate determination of the macroscopic behavior of multicomponent polymeric systems is not feasible even with state-of-the-art supercomputers. Yet, there is ample evidence that by a careful choice of simulation and analysis techniques, coarse grained model of flexible polymers – like the bond fluctuation model[12] – provide useful insights. In the framework of the bond fluctuation model each effective monomer blocks a cube of 8 neighboring sites from further occupancy on a simple cubic lattice in three dimensions. Effective monomers are connected by effective bond vectors of length $2, \sqrt{5}, \sqrt{6}, 3$, or $\sqrt{10}$ in units of the lattice spacing. Each effective bond represents a group of $n \approx 3-5$ subsequent $C-C$-bonds along the backbone of the chain[13]. Hence, the chain length $N = 32$ employed in the present simulations corresponds to a degree of polymerization of $100 - 160$ in a real polymer. For a number fraction $\rho = 1/16$ of occupied sites, properties of a concentrated solution or dense melt are well reproduced. Binary interactions between monomers are catered for by a short ranged square well potential $-\epsilon_{AA} = -\epsilon_{BB} = \epsilon_{AB} = 1/k_BT$ which is extended up to a distance $\sqrt{6}$. This interaction range comprises 54 sites which contribute to the first peak of the monomer-monomer density pair correlation function. The statistical segment length b in the relation for the radius of gyration $R_g = b\sqrt{N/6}$ is $b = 3.05$ ($R_g \approx 7$ for $N = 32$). The Flory-Huggins parameter is $\chi = 2z_{\text{eff}}\epsilon$ where $z_{\text{eff}} \approx 2.65$ denotes the effective coordination number in the bulk[14]. The statistical segment length and the Flory-Huggins parameter are used to map the computational model onto the Gaussian chain model with δ functional repulsion between unlike species.

The program code is written in C and communications are implemented via `shmem` routines. Details of the programming techniques can be found in reference[15]. The Monte Carlo simulation comprises local moves, slithering snake-like motions and a swap of the identity of A and B monomers along the molecules. This allows an efficient relaxation. Each Monte Carlo steps comprises 3 slithering snake like movements per chain, 1 local hopping attempt of each monomer, and 1 $A \rightleftharpoons B$ flip per diblock. Every 12500 Monte Carlo steps a configuration was stored for further analysis.

Since we are interested in studying the stability of different morphologies, we do not impose an orientation on the system. Rather we quench several systems with identical parameters from the disordered state $\chi = 0$ to the ordered state at $\chi N = 30$, and monitor the morphologies which occur in several independent runs with identical parameters. We consider three different system sizes with geometry $30 \times 96 \times 96$, $46 \times 93 \times 93$ and $56 \times 96 \times 96$ in units of the lattice spacing. Hence, our largest system comprises more than 500,000 grid points. Periodic boundary conditions are applied in the lateral directions yz, while there are two impenetrable walls at $x = 0$ and $x = \Delta_0 + 1$. Each A monomer less than 2 lattice spacings away for the walls reduces the energy by an amount $\epsilon_w = 0.1 k_B T$, whereas a B monomer in this region increases the energy by the same amount. According to the self-consistent field calculations these system geometries correspond to the L_2, the $L_\perp$, and the L_4 phase respectively. The lateral extension in the $L_\perp$ has been chosen compatible with the lamellar spacing in the self-consistent field theory. For each of these geometries we have simulated at leat 4 systems. We do not find transitions from one morphology to a different one during the course of a single simulation run which exceed for each system at least $3.8 \cdot 10^6$ Monte Carlo steps. Hence, we cannot rule out metastability effects completely. However, the simulation time is long enough that the observed structures are free of defects (on the scale of the simulation cell) and the composition profiles of systems which assembled into the same morphology agree.

Fig. 2. Snapshot of a configuration for $\chi N = 30$ and film thickness $\Delta_0 = 30$. The system has assembled into the L_2 phase, i.e., there are two AB interfaces running parallel to the surfaces. A monomers are shaded darkly, B monomers are bright. The surfaces on the top and the bottom attract the A component.

3 Results

For film thickness $\Delta = 30$ all four systems have assembled into the L_2 phase in accord with the predictions of the self-consistent field theory. In the snapshots the A component corresponds to the darker species. Half an A lamella is located at each wall. However, one also observes rather strong fluctuations which allow the B component to protude up to the wall.

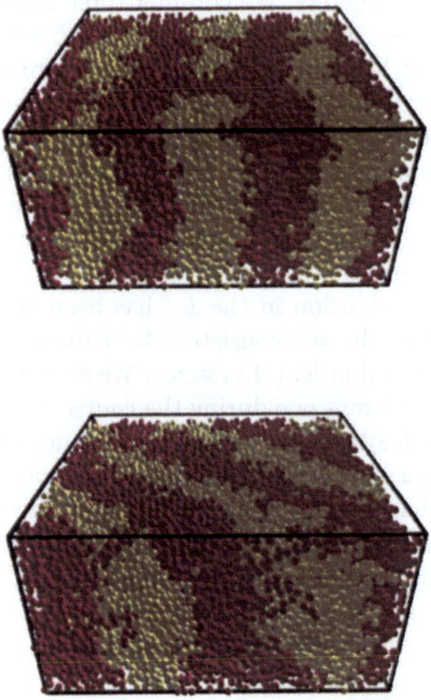

Fig. 3. Snapshots of configurations for $\chi N = 30$ and film thickness $\Delta_0 = 46$. The systems have assembled into the perpendicular oriented lamellar phase ($L_\perp$). The repeat distance in (**a**) is close to the value predicted by the SCF theory, while it exceeds the SCF value by 6% in panel (**b**).

The film thickness $\Delta_0 = 46$ corresponds to roughly 1.5 of the period of the lamella in the bulk. Hence, the competition between the intrinsic length scale and the film thickness leads to a perpendicular orientation of the lamellae. In this orientation the system does not benefit much from the surface interactions, because the surface is covered by both components to roughly the same amount. However, the system can minimize its free energy by optimizing the periodicity of the morphology perpendicular to the surfaces. In agreement with the self-consistent field calculations all eight systems have chosen the perpendicular orientation, however, two distinct periodicities occur (cf. Fig.3(**a**) and (**b**)).

In two cases we find a morphology with a repeat distance $D = 1.827R_e$, which agrees with the result of the self-consistent field theory $D_{\mathrm{SCF}} = 1.822R_e$. In six systems the morphology makes an angle with the axis of the simulation cell and the repeat distance is 6% larger. In the framework of the self-consistent field calculations such a deviation from the prefered repeat distance increases the free energy by 0.01 k_BT per molecules or 8 k_BT for the whole system. Hence, the probability that a thermal fluctuation leads to a deviation from the mean field theory is of the order $e^{-8} \approx 3 \cdot 10^{-4}$. The difference might be traced back to the fact that our self-consistent field calculations use the chain dimension in the athermal state as input. However, the chains stretch already in the disordered state to avoid the energetically unfavorable contacts between the different blocks and this might cause a temperature dependence of the statistical segment length b, which is not captured by the self-consistent field calculations.

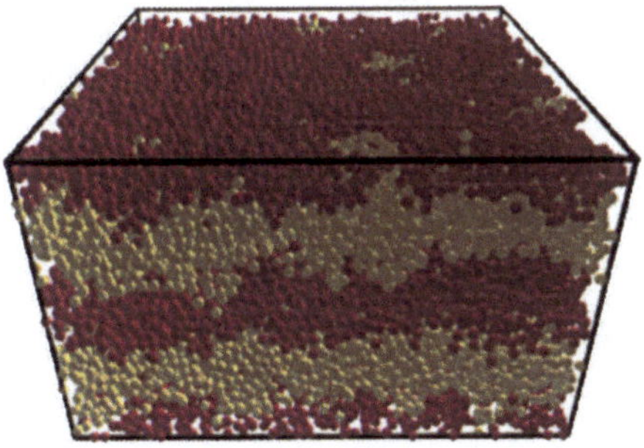

Fig. 4. Snapshot of a configuration for $\chi N = 30$ and film thickness $\Delta_0 = 56$. The figure show the L_4 phase, in which 4 AB interfaces run parallel to the surfaces.

Three out of four films of thickness $\Delta_0 = 56$ assemble into the parallel oriented L_4 phase as predicted by the self-consistent field theory. A typical snapshot is shown in Fig.4. One systems, however, chose the perpendicular oriented state. Besides this one system, the dependence of the morphology on the film thickness in the Monte Carlo simulations is in qualitative agreement with the self-consistent field calculations[11,16,17].

The results of the Monte Carlo simulations have been compared in more detail with the self-consistent field calculations: In the perpendicular oriented phase the AB interface bends in the vicinity of the surface, such that the component with the lower surface free energy covers a larger area. Since the surface interactions are small, the prefered component does not wet the surface and the AB interface makes a finite contact angle with the surface. The distortion of the interface in the vicinity of the surface interferes with the distortions which originates from the opposite side of the film and leads to a slight oscillation of the interface across the entire film thickness. This occurs in the Monte Carlo simulations and the self-consistent field calculations. However, there are also differences between the self-consistent field calculations and the MC simulations: In the SCF calculations the monomer density decays smoothly to zero at the surface, while there are pronounced packing effects in the MC simulations. In both cases the interfacial width of the AB interface increases at the surface in the perpendicular phase, because the effective incompatibility is reduced in the vicinity of the surface. This gives rise to a negative line tension and tends to stabilize the $L\perp$ phase. In the SCF calculations the effect stems from the reduction of the density at the surface; in the Monte Carlo simulations it is caused by the finite range of the monomer–monomer interactions ("missing neighbor effect").

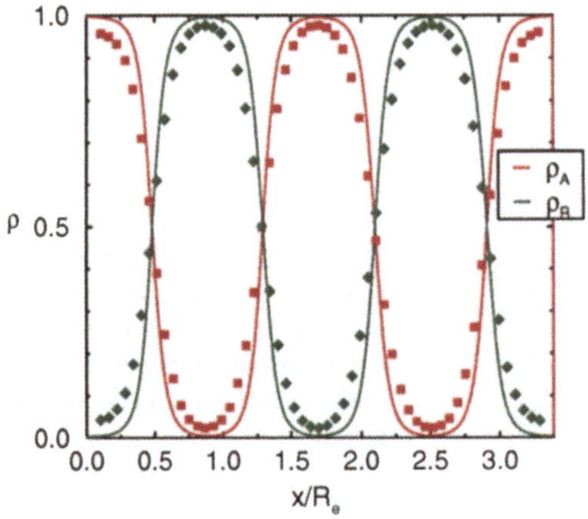

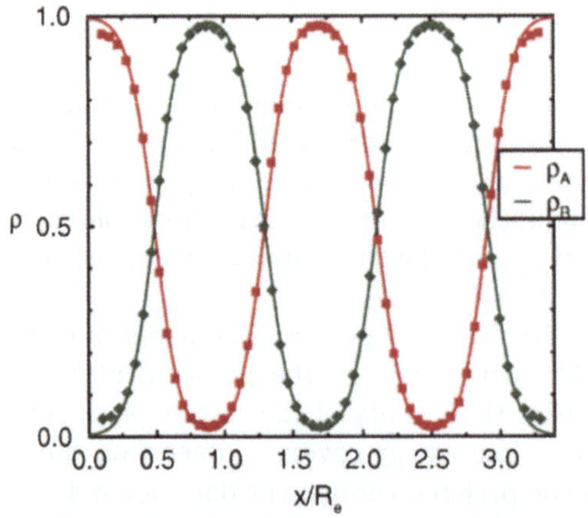

Fig. 5. (**a**) Comparison between the composition profiles in the MC simulations (symbols) and the SCF calculations. (**b**) The same as (**a**) but the SCF results have been convoluted with a Gaussian ($s = 0.12R_e$) to take account of interfacial fluctuations.

There is another important difference between both methods: The AB interfaces in the lamellar phase are assumed to be ideally flat in the SCF calculations while the snapshots Fig.2-4 show long wavelength fluctuations of the local interfacial position (capillary waves). They lead to a broadening of the interfacial profiles measured in experiments[18,19] and simulations[20,21]. The additional broadening of the laterally averaged composition profiles in the simulations is clearly observed in Fig.5(**a**) for the L_4 phase. Comparing

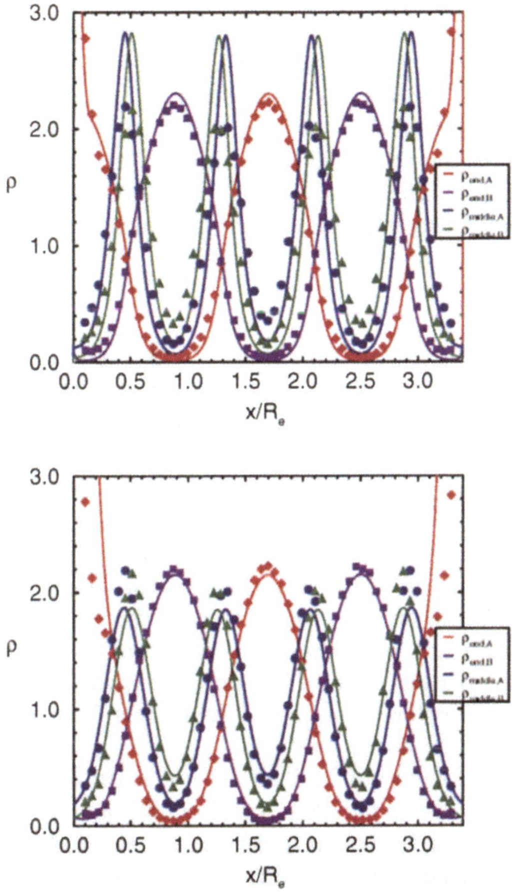

Fig. 6. (**a**) Comparison between the profiles of the end segments and the middle segments (monomer number 16 and 17) in the MC simulations (symbols) and the SCF calculations. (**b**) The same as (**a**) but the SCF results have been convoluted with a Gaussian ($s = 0.12R_e$) to take account of interfacial fluctuations. Note that this is the same value than used in the previous figure.

the MC results to the SCF calculations due account of these interfacial fluctuations has to be taken. Lacking a simple analytical prediction for the fluctuation spectrum of a strongly interaction stack of interfaces, we convolute the SCF profiles with a Gaussian of width s. This mimics the effect of the fluctuations of the local interfacial positions. The strength s is used as a free parameter, which is adjusted as to match the interfacial width in the convoluted SCF profiles and the MC results. This procedure results in excellent quantitative agreement between MC simulations and SCF calculation Fig.5(**b**). Using the same value of s we obtain almost quantitative agreement for other profiles (e.g., the distributions of the end-segments or the joining pints between the two distinct blocks). The end segments are located in the middle of the lamellae, while the joining point are centered around the interfaces. In accord with experimental observations[22], the spatial distribution of the joining points is narrower than the distribution of the end segments (cf. Fig.6).

This procedure results in $s/R_e = 0.1, 0.12$ and 0.127 for the L_2 phase, the L_4 phase and the perpendicular alinged $L_\perp$ phase, respectively. The increase of the strength of fluctuations is compatible with intuiton that the confined geometry reduces the interfacial fluctuations more in the L_2 phase tha in the L_4 phase or the $L_\perp$ phase. A detailed investigation of the interfacial fluctuations in the system reveals that the Fourier spectrum of interfacial fluctuations deviates strongly from the analogous spectrum of a free unconfined homopolymer interface. The fluctuation are strongly suppress by the mutual interactions between neighboring lamellae and the confining surfaces.

4 Summary and Discussion

We have presented large scale MC simulations for symmetric diblock copolymers confined into a thin film at intermediate segregation ($\chi N = 30$). Depending on the film thickness, we find lamellar phases with two or four AB interfaces running parallel to the walls or perpendicular aligned lamellae. The sequence of phases observed in the simulations is in accord with SCF calculations. Taking due account of interfacial fluctuations we find also good agreement for the detailed composition profiles and profiles of individual segments. This is rather surprising, because the interactions on the monomer scale $\chi = 30/32$ are not small, and the smallest length scale of the morphology, namely the width of the interface between A and B domains, is not much larger than the microscopic length scale of the polymer architecture (e.g., the persistence length b)i. Therefore pronounced deviations from the Gaussian chain model might have been anticipated[23].

Though we find clear evidence for interfacial fluctuations, their Fourier sepctrum is not describable with a simple capillary wave Hamiltonian using the interfacial tension between A and B homopolymers at the same segregation χ. In the simulations interfacial fluctuations are strongly suppressed compared to unconfined interfaces in homopolymer blends. Both the confinement into a film and the mutual interactions between neighboring interfaces strongly modify the Fourier spectrum of fluctuations of the local interfacial position.

Acknowledgments

We acknowledge support from the Deutsche Forschungsgemeinschaft (DFG) grants No Bi 314/17. Generous access to the CRAY-T3E at the HLR Stuttgart is acknowledged. It is a great pleasure to thank P.K. Janert, F. Schmid and M.W. Matsen for helpful discussions.

References

1. L. Leibler, Macromolecules **13**, 1602 (1980).
2. A.N. Semenov, Soviet Physics JETP **61**, 733 (1985).
3. M.W. Matsen and M. Schick, Phys.Rev.Lett. **74**, 4225 (1995).
4. P. Lambooy, T.P. Russell, G.J. Kellog, A.M. Mayes, P.D. Gallagher, and S.K. Satija, Phys.Rev.Lett. **72**, 2899 (1994).
5. G.J. Kellogg, D.G. Walton, A.M. Mayes, P.Lambooy, T.P. Russell, P.D. Gallagher, and S.K. Satija, Phys.Rev.Lett. **76**, 2503 (1996).
6. M. Koneripalli, R. Levicky, F.S.Bates, J. Ankner, H. Kaiser, and S.K. Satija, Langmuir **26**, 6681 (1996).
7. N. Koneripalli, F.S. Bates and G.H. Fredrickson, Phys.Rev.Lett. **81**, 1861 (1998).
8. M. Kikuchi and K. Binder, J.Chem.Phys. **101**, 3367 (1994); Europhys.Lett. **21**, 427 (1993).
9. T.P Russell, Current Opinion in Colloid & Interfacial Science **1**, 107 (1996).
10. M. Park, C. Harrison, P.M. Chaikin, R.A. Registar, and D.H. Adamson, Science **276**, 1401 (1997).
11. M.W. Matsen, J.Chem.Phys. **106**, 7781 (1997).
12. I. Carmesin and K. Kremer, J.Phys. (France) **51**, 915 (1990); H.-P. Deutsch and K. Binder, J.Chem.Phys. **94**, 2294 (1991).
13. K. Binder, Adv.Pol.Sci **138**,1 (1999).
14. M. Müller and K. Binder, Macromolecules **28**, 1825 (1995).
15. A. Werner, F.Schmid, M. Müller, and K.Binder, in High Performance Computing in Science and Engineering, p.176, (1998), E.Kramer and W. Jäger (Eds), Springer
16. G.T. Pickett and A.C. Balazs, Macromolecules **30**, 3097 (1997).
17. T. Geisinger, M. Müller, and K. Binder, (to be submitted to J.Chem.Phys.)
18. K.R Shull, A.M. Mayes, and T.P. Russell, Macromolecules **26**, 3929 (1993).
19. N. Koneripalli, R. Levicky, F.S. Bates, M.W. Matsen, S.K. Satija, J. Ankner, and H. Kaiser, Macromolecules **31**, 3498 (1998).
20. A. Werner, F. Schmid, Müller, and K. Binder, J.Chem.Phys. **107** 8175 (1997).
21. M. Müller and A. Werner, J.Chem.Phys. **107**, 10764 (1997).
22. A.M. Mayes, R.D. Johnson, T.P. Russell, S.D. Smith, S.K. Satija, and C.F. Majkrzak, Macromolecules **26**, 1047 (1993).
23. S.K. Nath, J.D. McCoy, J.G. Curro, and R.S. Saunders, J.Chem.Phys. **106**, 1950 (1997). J.D. McCoy, S.K. Nath, J.G. Curro, and R.S. Saunders, J.Chem.Phys. **108**, 3023 (1998).

Molecular Dynamics of Covalent Crystals

J. Hahn and H.-R. Trebin

Institut für Theoretische und Angewandte Physik, Universität Stuttgart, D-70550 Stuttgart, Germany

Abstract. A molecular mechanics-like two and three body potential is implemented in the ITAP molecular dynamics program *IMD* [1]. First results on the simulation of covalent crystals are presented.

1 Introduction

Many predominantly covalent bonded crystals are composed of atoms which are simultaneously bonded to a given number of other atoms. The covalent character of the nearest atom-atom interactions gives rise to a particular crystal symmetry. Atoms with sp^3 hybridization like carbon and silicon exhibit a tetrahedral configuration with four nearest neighbours, while sp^2 hybrids of carbon and nitrogen may form trigonal planar arrangements with three atoms bonded to the center atom. For example, the tetrahedral coordination is found in diamond, cubic close packed silicon, and wurtzite or zinc blende structure of silicon carbide.

It is of increasing interest to material science to investigate the microscopic structure of such covalent systems, including a variety of crystal defects like vacancies, interfaces and grain boundaries. Among others, computer simulations of covalent crystals based on atomistic models might help to gain insight into the correlations of microscopic and macroscopic properties. However, in order to calculate reliable thermodynamical and mechanical properties by molecular dynamic methods adequate large systems should be explored. In turn, efficient software codes in combination with fast computer equipment are required for large scale simulations. We use the *IMD* package, which is a molecular dynamics software designed for risc workstations as well as for massively parallel computer architectures [1,2].

We are interested in the simulation of the atomistic structure of covalent materials, especially of interfaces . At first, a potential function based on a molecular mechanics approach is developed and the code for it is implemented in the *IMD* program. In the present report, the basic features of the model and first molecular dynamics results on the bulk structures of several tetrahedral and model systems are presented.

[1] See HLRS project: ITAP Molecular Dynamics (IMD)

2 Basic concept on the simulation of covalent crystals

2.1 Covalent crystals

The bond strengths and mutual bond orientations are the most typical atomistic features of covalent materials, like for example diamond, silicon carbide and silicon nitride. In order to describe covalent bonds a pair interaction potential is applied. The potential depth is related to the dissociation energy and the equilibrium distance characterizes the bond length. It is interesting to note, when comparing the pair interactions of the van der Waals type and of the covalent bond the latter is about ten to hundred times stronger than the former. Therefore, the van der Waals interaction may be neglected in a first approximation. (This assumption is not valid for molecular crystals.)

Although the strength and the form of the pair interaction significantly influences the material characteristics it cannot describe the anisotropy of the material. Therefore, another potential term must be taken into account. Essentially, the hybridization of the atom gives rise to preferred bond directions. The bonds i-j and i-k at atom i form the so-called valence angle. Depending of the hybridization of atom i typical values of the valence angle appear. In case of silicon and carbon, each with sp^3 hybridization, the angle is 109.47°, while for sp^2 carbon the angle is 120°. Any deviation from the equilibrium angle costs energy and should be described by an adequate potential which depends on the positions of the atoms involved, i.e. a three body function must be defined.

Higher order terms like torsion, inversion or cross functions are of less importance to the bulk properties [3]. Therefore, it is a well-established approach in atomistic simulations of covalent crystals that the pair interaction term and the three body potential may well characterize the structure and energetics of the system under investigation [4].

2.2 The model

In order to simulate covalently bonded systems a suitable force field ansatz must be chosen. The structure and energetics of the systems under investigation are essentially determined by the functional form and the parameters of the force field applied [5]. In the present study the potential used is based to some extend on the universal force field approach [6]. The force field parameters are estimated by general rules from a set of fundamental (atomic) parameters. The atomic parameters are based only on the element, its hybridization and its connectivity [6].

In the following, we focus on a force field apporach which is capable to describe tetrahedral systems. However, the application of the model on trigonal planar and other systems is straightforward [6].

First, the analytical potential functions are discussed. The total potential energy is

$$E_{POT} = \sum_{i,j} V_{ij} + \sum_{i,j,k} V_{ijk}. \tag{1}$$

The pair interaction V_{ij} describes the bond between two atoms, while the three body potential V_{ijk} characterizes the direction of the bonds of an atom. The pair interaction of covalent bonds is written in the usual form of a Morse potential while for the valence angle function a fourier expansion is chosen

$$V_{ij} = D_{ij}(e^{-\alpha_{ij}(r_{ij}-r^o_{ij})} - 1)^2 \tag{2}$$

$$V_{ijk} = k_{ijk} \sum_{n=p}^{m} C_n \cos(n\Theta_{ijk}) f(r_{ij}) f(r_{ik}). \tag{3}$$

The distance r_{ij} and the angle Θ_{ijk} are the actual values calculated during the simulation.

The parameters involved are given as follows. The equilibrium distance r^o_{ij} is assumed to be the sum of the atom type covalent radii: $r^o_{ij} = r_i + r_j$. In case of C and Si these radii are 0.77 and 1.17 Å, respectively. The dissociation energy D_{ij} is the geometric mean of D_i and D_j (C: 101, Si: 80 kcal/mol), and α_{ij} is calculated from D_{ij} and the force constant k_{ij}

$$\alpha_{ij} = \sqrt{k_{ij}/2D_{ij}}. \tag{4}$$

Considering the fourier term, the parameters p and m and in turn the expansion coefficients C_n depend on the hybridization of the atoms. In case of sp^3 hybrids like carbon and silicon the parameters are $p = 0$ and $m = 2$, and the coefficients C_n are given by

$$C_2 = 1/(4\sin^2\Theta^o) \quad , \quad C_1 = -4C_2\cos\Theta^o \quad , \quad C_0 = C_2(2\cos^2\Theta^o + 1), \tag{5}$$

with the equilibrium angle $\Theta^o = 109.47°$.

The function $f(r)$ is defined such that it is equal to one in case of the equilibrium distance r^o (1-2 interaction) and zero in case of the 1-3 equilibrium distance. In-between a smooth function is chosen, like for example a normalized cosine or error function.

The force constants k_{ij} and k_{ijk} are given by a generalization of Badger's rule [7]. The procedure to calculate the force constants is described in detail in reference [6]. It is noteworthy, that again solely atom based parameters are used. The force constants k_{ij} are 695, 462 and 275 kcal/mol/Å^2 for C-C, C-Si, and Si-Si, respectively. However, in case of silicon the force constant k_{ij} leads to unstable bulk structures because of quite strong 1-3 interactions. Therefore, α is set equal to five (this approximation will be refined soon). In case of k_{ijk} the force constants of atom i are, at present, assumed to be independent of the other atoms involved: 215 and 84 kcal/mol/deg^2 for C and Si, respectively.

2.3 Upgrading IMD

Using a force field approach in conjunction with well-known molecular mechanics and molecular dynamics methods enables one to analyse the phase space of the atomistic system under investigation [8]. At ITAP the *IMD* software package is developed for molecular dynamics simulation of large scale systems [1]. A detailed description of the actual procedures available in *IMD* is presented elsewhere [1,2]. In the following, a short description on the new modul is given in which the three body interaction energy is calculated.

So far, *IMD* considers pair interactions only. Now, the code is extended in order to take preferred bond orientations into account. This requires merely another loop over the particles. The calculation is performed such that first the pair interaction of the particles in one cell [2] is computed, meanwhile, the coordinates of the nearest neighbours are saved. Then, while still in the same cell, the neighbour list is used in a second loop to determine the valence angle energy. Fortunately, the three body term could be implemented in *IMD* in a straightforward way. The valence angle function is calculated analytically, while the pair functions (here: V_{ij} and $f(r_{ij})$) are tabulated in the *IMD* program as usual: for each type of atom-atom interaction the potential is listed as a function of the distance.

3 Recent Results

Comparing the computational time of the two and three body interaction code and of the 'classical' pair interaction program one finds, that the new code is less than about two times slower than the classical one. This is expected, because of the additional loop over the particles and of the analytical calculation of the three body force. Due to the implementation strategy the MPI timing remains similar to runs with the classical code. However, the memory access for the neighbour list update slightly slows down the efficiency.

In the following, first results on the simulation of covalent and model systems are presented.

Covalent crystals

The program, and in turn the force field, is tested by investigating the stability of given bulk structures. Therefore, the geometries of diamond, cubic close packed silicon and silicon carbide in zinc blende structure are calculated. Up to $N = 262144$ atoms and 64 cpus are used by molecular dynamics simulations. The microconvergence method and microcanonical ensemble as well as canonical ensemble (NVT) techniques are applied to equilibrate the systems. The calculations are performed using periodic boundary conditions as well as open boundaries. The calculations show that the system under investigation remains stable and the calculated geometries are independent on the simulation method. The results are in good agreement with experimental data (Table 1).

		UFF	IMD	exp
C	C-C	1.57	1.55	1.54
	-C-	109	109	109
Si	Si-Si	2.23	2.33	2.35
	-Si-	110	109	109
SiC	Si-C	1.89	1.86 *	1.87
	-Si-	110	109	109
	-C-	110	109	109

Table 1. Geometries of some selected tetrahedral crystals (Distances in Å and angles in deg). The results of the *IMD* simulations are compared with calculations using the universal force field (UFF) [6] and with experimental datas [9]. (* The covalent radius of Si is chosen $r_{Si} = 1.10$ Å.)

In another test, the diamond structure is investigated again. However, the initital configuration is modified significantly. The atoms are displaced randomly: $\boldsymbol{x} \to \boldsymbol{x} + \boldsymbol{\lambda}$, with $\boldsymbol{\lambda}$ a three dimensional random number $\lambda \in [-0.15, 0.15]$, leading to a maximum displacement of $|\boldsymbol{\lambda}| = 0.5$ Å. After the system is equilibrated by molecular dynamics simulations the diamond structure is recreated.

Covalent interface

The (100) interface of SiC/CSi is analysed by molecular dynamics simulations using the microcanonical ensemble technique. The initial structure is generated as follows. First, the SiC zinc blende structure is chosen. Then, in one half of the simulation box the particle types are interchanged (Si $\to$ C; C $\to$ Si) leading to an SiC/CSi interface. Periodic boundary conditions as well as open boundaries are applied. The number of particles in the box is $N = 162432$, 10^4 steps and 64 processors are chosen. The molecular dynamics run took 59 cpu hours.

The simulations show that the crystal remains stable during the molcular dynamics run (Figure 1). However, the carbon atoms in the interface region are displaced from the initial positions to form strong C-C bonds of optimum length. This way, defects appear in the interface layer.

Model system

In order to investigate the influence of the bond strength on the material characteristics a two dimensional binary model system is considered. The total potential energy is given by the Lennard-Jones potential. Here, the

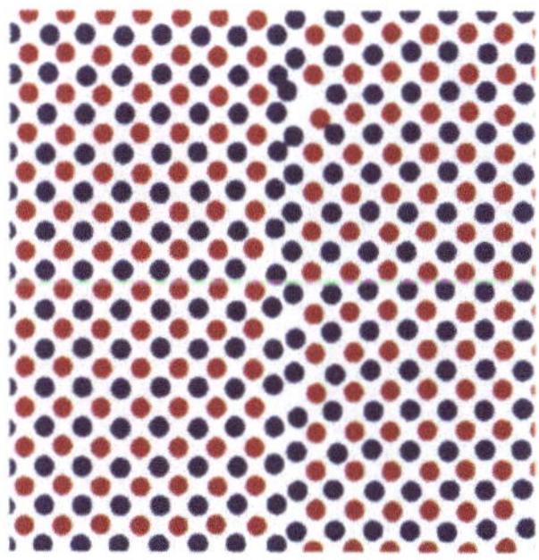

Fig. 1. Detail of the SiC/CSi (100) interface (C: blue; Si: red).

three body interaction is not applied, because of the primary focus on the strength of the pair interaction.

$$E_{POT} = \sum_{ij} \epsilon_{ij}[(r^o_{ij}/r_{ij})^{12} - 2(r^o_{ij}/r_{ij})^6]. \qquad (6)$$

The equilibrium distance r^o_{ij} is set to unity for each particle type, A and B. On the contrary, the potential depths ϵ_{ij} are varied: The interaction strength B-B is chosen twice the one for A-A, and the unlike bond strength A-B is given by the arithmetic mean $(A+B)/2$. The initial structure represents an interface of two blocks, each of particles with similar type. Periodic boundary conditions are applied and a temperature significantly below the melting point is chosen. Then, the simulation box is elongated in x direction step by step. At each step a molecular dynamics run is performed using the Anderson thermostat to equilibrate the system. Usually, $46 * 10^3$ particles are used on 64 processors. In case of the strong interaction simulation the equilibration of one elongation step took about 28 hours of cpu time on T3E.

As expected, in case of weak interactions ($\epsilon_{AA} = 1$) the elongation of the particle box does not lead to any significant change of the crystal structure; this is valid until the system melts. However, for increasing interaction strength the material gets harder. In the latter case, the elongation of the box eventually results in a crack in the block of particles with stronger pair interaction (Figure 2).

4 Outlook

Our aim is to apply the *IMD* program on problems concerning the atomistic structure of precursor derived covalent ceramics, like silicon carbide and silicon nitride [10]. In particular, the technological production process yields covalent interfaces, too. For example, SiC/SiC and SiC/Si_3N_4 interfaces occur. The material behavior of the ceramics depends strongly on the interfaces

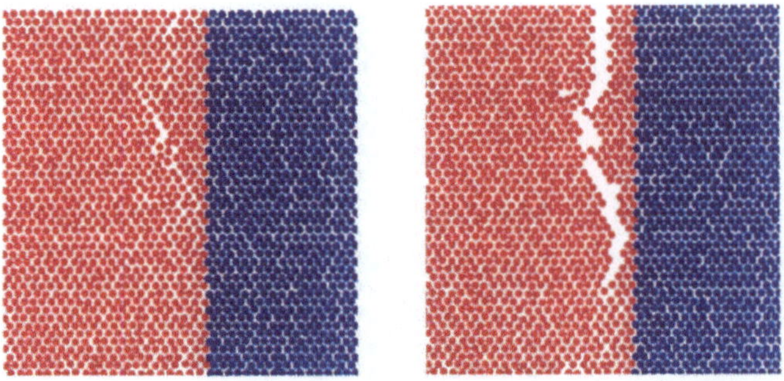

Fig. 2. Detail of the two dimensional interface of a binary system (A: blue; B: red; $\epsilon_{AA} = 10$). **Left:** Snapshot at the beginning of the fracture (3.25% elongation). **Right:** Snapshot of the crack (3.375% elongation). Interestingly, the fracture occurs inside the harder material, while the softer material as well as the interface remain undestroyed.

and, therefore, the knowlegde of the atomistic structure is of special interest to material science.

The HLRS project started late 1998. So far, first steps in establishing the model and transfering the program code to massively parallel computers have been performed. The next steps will be to refine the pair potential approach and the memory allocation procedure for the nearest neighbour list. After that, the covalent crystals and interfaces are investigated in detail.

Acknowledgement

The support of the Deutsche Forschungsgesellschaft (GRK 285) is gratefully acknowledged. The authors thank F. Gähler and J. Roth and all other member of the *IMD*team for helpful discussions during the development of the program code.

References

1. Stadler, J., Mikulla, R., Trebin, H.-R.: Int. J. Mod. Phys. C **8** (1997) 1131
2. Roth, J., Stadler, J., Brunelli, M., Bunz, D., Gähler, F., Hahn, J., Hohl, M., Horn, C., Kaiser, J., Mikulla, R., Schaaf, G., Stelzer, J., Trebin, H.-R.: *IMD - A Massively Parallel Molecular Dynamics Package for Classical Simulations in Condensed Matter Physics* HLRS Project Report (1999)
3. Carlson, A. E.: *Beyond Pair Potentials in Elemental Transition Metals and Semiconductors* in Solid State Physics, Advances in Research and Application by Ehrenreich, H., Seitz, F., Turnbull, D., Eds., **43** p.1 Academic, New York (1990)
4. Stillinger, F. H., Weber, T. A.: Phys. Rev. B **31** (1985) 5262

5. Pertsin, A. J., Kitaigorodsky, A. I.: *The Atom-Atom Potential Method* in Springer Series in Chemical Physics **43** (1987)
6. Rappe, A. K., Casewit, C. J., Colwell, K. S., Goddard III, W. A., Skiff, W. M.: J. Am. Chem. Soc. **114** (1992) 10024; C. J. Casewit, K. S. Colwell, and A. K. Rappe, J. Am. Chem. Soc. **114** (1992) 10035 and 10046
7. Badger, R. M.: J. Chem. Phys. **2** (1934) 2128
8. Allen, M. P., Tildesley, D. J.: *Computer Simulation of Liquids* Oxford Science Publications, Clarendon Press (1992); Frenkel, D., Smit, B.: *Understanding Molecular Simulations* Academic Press (1996)
9. CRC Handbook of Chemistry and Physics, 78th Ed. (1997-1998)
10. Bill, J., Seitz, J., Thurn, G., Dürr, J., Canel, J., Janos, B. Z., Jalowiecki, A., Sauter, D., Schempp, S., Lamparter, H., Mayer, J., Aldinger, F.: Phys. Stat. Sol. A **166** (1998) 269

Simulation of random copolymers at selective interfaces and of cross-linked polymer blends

Jens-Uwe Sommer, Stefan Lay, and Alexander Blumen

Theoretische Polymerphysik, Universität Freiburg, Hermann-Herder-Str. 3, 79104 Freiburg

Summary. We have investigated two complex statistical systems where the interfacial properties of a two-phase polymer system can be controlled. Using the bond fluctuation model (BFM) we have first studied the adsorption behavior of random copolymer chains at a single selective interface separating two phases. We found a new regime where the adsorbed chains extend in a brush-like fashion perpendicular to the interface as predicted from scaling arguments. We have extended our simulation model to investigate the influence of cross-linking on the phase behavior of polymer blends consisting of A- and B- chains. Here very large lattices are necessary to avoid finite size effects due to the network superstructure on large scale. A large cubic lattice consisting of $400 \times 400 \times 400$ sites was sliced into 64 parts to allow for parallelization. Here we equilibrated 40 000 chains of length $N = 100$ before starting cross-linking under conditions without AB-interactions with different cross-link densities. Investigating the network topology we found an effective chemical dimension larger than three in agreement with previous investigations on smaller systems. This differs strongly from a simple homogeneous structure such as a diamond topology. Switching on the AB-interaction between the different chains within the network results in the formation of A- and B- enriched components, i.e. to microphase separation. The sizes of these regions are much bigger than expected from a homogeneous network approach. The structure factor $S(k)$ shows at intermediate wave numbers k a k^{-4}-behavior, in contrast to the k^{-2}-behavior expected from simplified models.

1. Introduction

The simplest method to design and improve material properties of polymers is "polymer metallurgy": Blending different chemical chains together results in new material properties, which are frequently determined by a superposition of the properties of the pure components. In this way, for instance, the glass transition temperature and the fracture toughness can be tuned. However, in contrast to most low molecular materials such as metals, long polymer chains are highly incompatible. In order to increase the technically important miscibility of polymers one can consider two possibilities. First, the interface tension between the both phases can be reduced using *compatibilizer* chains which adsorb at the interfaces (polymer "surfactants"). Besides regular blockcopolymers, which are expensive to synthesize, random copolymers can be used. Second, cross-linking the blend with additional chemical bonds also prevents the demixing on larger scales. Unfortunately both possibilities lead to highly complex statistical properties. It is the goal of the presented simulations to elucidate the basic properties of both cases.

In general, dense polymer systems consisting of two or more chemical species (connected along a single chain or via a network structure) show a rich physical behavior resulting from competing interactions. In most cases the intrinsic disorder which is inherent in such systems makes a rigorous analytical treatment difficult or even impossible. Each polymer chain interacts with many other chains at once, summing up relatively small contributions of the individual interaction events. Many constraints (say chain connectivity) considerably restrict the phase space accessible and create novel dynamical effects, such as reptation. The macroscopic (material) properties are averages of the microscopic ones over length scales ranging from single monomers up to hundreds of chains. Only very intuitive and simplifying models are able to describe analytically such systems but they do not always rely on proven assumptions.

In such a situation computer simulations are of great importance, either to test simple theoretical ideas such as scaling approaches, or to select out the most interesting universal properties and, by recreating special situations which are not accessible experimentally, give the possibility to distinguish between different model assumptions. While simple, rather homogeneous polymer systems such as homopolymer solutions can be simulated on modern home computers and workstations, structured polymers such as networks require much larger scales to obtain reliable information about their statistical behavior. In previous works [1–6] we already encountered systems whose heterogeneities extended over the whole length the box size of the simulation, i.e. over more than 50 000 monomers. Only massive parallelization including high speed communication between the processing elements (PE) makes it possible to simulate for the first time polymer networks which are large enough to show the characteristics of true macroscopic systems. This allows to check intuitive models supposed to work in such a complex situation.

2. Simulation of random copolymers near selective interfaces

The behavior of single chains composed of a random succession of two species A and B (random copolymers) near interfaces which separate two half spaces preferred by only one species each (selective interfaces) has been investigated experimentally [7, 8], theoretically [9–12] as well as using computer simulations [13–16]. However, even in the simplest situation, where the chains are treated as ideal (disregarding any self-interactions), rigorous analytical treatments are difficult [9]. Recently, new results could be obtained [11, 12], but only in an approximate way. However, in practice the situation is even more complex: The chains are always present in finite concentration so that interaction effects among the adsorbed chains are always important. It is well known from the theory of homopolymers that the adsorption isotherm, i.e.

the graph of the interface concentration versus the bulk concentration at a given temperature is an extremely rapidly increasing function for very small bulk concentrations [17]. Therefore for most adsorbing interfaces even very dilute bulk solutions lead to a rather complete coverage of the interface. At high interface coverage the excluded volume interactions are dominant. As a consequence the chains are forced to form large loops into the bulk. Finally the surface will be completely occupied by chains and the excluded volume interactions prevent further adsorption. The adsorption isotherm saturates at a plateau value. In this state the adsorbed chains form loops in the direction perpendicular to the interface up to lengths comparable to the unperturbed chain extension [17, 18].

However, for random copolymers at selective interfaces the situation is different. Here the monomers do not gain energy if they are in contact with the interface but only if they lie in their preferred solvent. Therefore the free energy decreases with increasing concentration by having segments stretching perpendicular to the interface. This allows for more chains to be polarized at the interface [19]. As a consequence two effects should be observable. First the adsorption isotherm does not saturate to a plateau value but continues to increase with the concentration (but at a smaller rate). Such an *over-saturation* accounts for the stretching effect discussed above. Second, as a direct consequence, a stretching of the adsorbed chains in the direction perpendicular to the interface beyond the isotropic chain behavior must occur. It was the task of this part of our simulations to investigate this behavior.

Our Monte Carlo simulations for copolymers were performed using the bond–fluctuation method (BFM) [24]. The BFM is a lattice algorithm where each monomer is represented by a lattice cell. Thus on a three–dimensional simple cubic lattice each monomer occupies eight neighboring lattice sites. The length of a bond connecting two neighboring monomers fluctuates between 2 and $\sqrt{10}$ lattice spacings. In this way, lattice artifacts as they occur in simpler models are avoided and the phase space is more effectively sampled. Self–avoidance (the excluded volume interaction) is satisfied by not allowing any two monomers to have any lattice sites in common. To avoid bond-crossing the allowed bonds are restricted to a set of 108 vectors. Here we have studied the behavior of n random copolymers, each of length N, all placed in a box of size $L \times L \times H$, with periodic boundary conditions in the x– and y–directions and two impenetrable surfaces at $z = 0$ and $z = H$. The copolymers are random; for each of them separately the N-monomer sequence consists of randomly chosen A- and B-monomers. The assignment of the A's and B's along each chain is uncorrelated, each species being chosen with a *a priori* probability of 1/2. Therefore, on each chain the number of A and B monomers differs because of the statistical fluctuations associated with the assignment procedure.

For adsorption we assume a symmetrical situation: the interaction energy of each monomer with its unfavorable solvent is $\chi k_B T$ and with its favorable solvent is zero. The interface is located at $z_0 = (H+1)/2$. In the Monte Carlo algorithm the chains move by position changes of their monomers, which attempt nearest neighbor steps on the underlying cubic lattice. A move is taken into consideration only if it satisfies the requirements of self–avoidance and of non–breaking of bonds. Furthermore, energetically unfavorable moves are statistically permitted according to the usual Metropolis prescription involving the Boltzmann factor.

We have investigated the density dependence of the adsorption properties by changing n, the number of polymers in the box. Here we take $L = 50$ and $H = 100$ and focus on the results obtained using copolymers of length $N = 64$ with a monomer–solvent interaction parameter of $\chi = 3.15$. In the single chain case ($n = 1$) these parameters let the system be located in the well adsorbed scaling regime [14, 15]. In order to improve the statistics we average over results from 8 parallel runs.

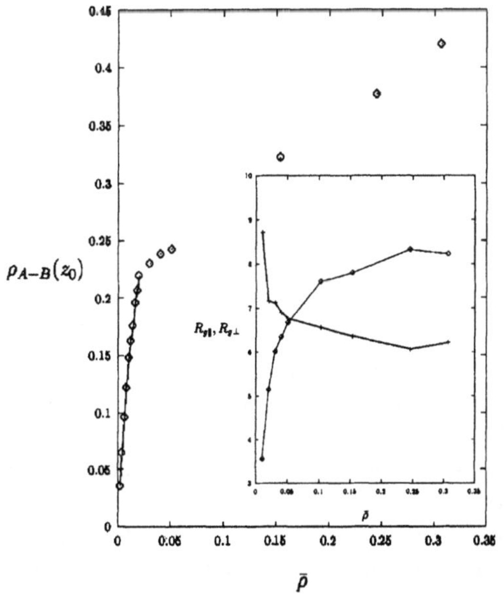

Fig. 2.1. Adsorption isotherm of random copolymers at a symmetric selective interface. Inset: Extension of adsorbed chains in the directions perpendicular ◇ (from lower left to upper right) and parallel + to the interface.

Two important results are displayed in Fig.2.1; for more details see Ref. [25]. The adsorption isotherm is plotted in Fig.2.1. As expected, for very small bulk concentrations an extremely steep increase is observed. However, in contrast to the behavior of homopolymers the isotherm does not saturate into a plateau value. Instead we obtain a further linear increase in the interface concentration. This indicates clearly a second adsorption regime which can be associated to the expected stretching of the adsorbed chains in the direction perpendicular to the interface. In fact this behavior can be verified directly in the inset of Fig.2.1. Here the extension of the adsorbed chains both in the direction parallel and perpendicular to the interface is plotted versus $\bar{\rho}$, the overall density of monomers in the box. At about $\bar{\rho} = 0.05$ the chains become stretched perpendicular to the interface.

To conclude, for random copolymers at selective interfaces we have found a new adsorption regime; here the geometry of adsorbed chains changes from flat, pancake-like shapes into brush-like assemblies of stretched loops. The possibility of such over-saturation effects of random copolymers adsorbed at interfaces might be interesting in view of their use as compatibilizers in polymer blends.

3. Simulation of cross-linked polymer blends

We studied the behavior of polymer blends consisting of A and B chains cross-linked by additional chemical bonds. Both the cross-linking process as well as the resulting structure and microphase separation were considered. Since we knew from previous simulations of polymer networks [1–6] that the cross-link topology is inhomogeneous on very large scales, a simulation volume of $400 \times 400 \times 400$ covering 40 000 chains of length 100, i.e. $4 * 10^6$ monomers was chosen. Here a lattice parallelization algorithm was developed in order to cut the lattice into 64 sublattices. As for random copolymers, analytic calculations of randomly cross-linked blends or inter-penetrating networks are extremely difficult because of the frozen randomness and competing interactions. Qualitatively, the cross-links prevent a demixing of the A and B chains, resulting in a so called microphase separation, i.e. the presence of finite A- and B-rich regions. In scattering experiments this results in a peak of the structure factor at finite wave numbers k^*. This value corresponds to the mean extension of the microphase patterns. A first attempt to explain this behavior was made by de Gennes using an intriguing correspondence to the polarization of a dielectric medium [20]. Based on a simple homogeneous network model the predicted characteristic length scale of the microphase separation is of the order of the radius of gyration of the networks strands (parts of the precursor chains between two cross-links). On the other hand experiments have shown the appearance of much larger scales [21]. However, the experimental situation is not completely clear since polymer blends are usually strongly incompatible as discussed in the beginning. As a consequence it is almost impossible to make a clear-cut separation between the demixing process before and during cross-linking, and the result of the microphase separation in thermodynamic equilibrium as assumed in Ref. [20], i.e., it cannot be decided whether the resulting structures emerges due to a competition between demixing and cross-linking *kinetics* or due to a thermodynamic balance between repulsive AB-interactions and the local deformation of the network structure.

Here computer simulation are a unique tool, by means of which the AB-interaction can be completely switched off before and during the cross-linking process. Hence a clear separation of kinetic and thermodynamic effects is possible. Our results show, however, that also in this well defined situation demixing takes place on much larger scales compared to the size of the network strands (usually one has to take into account the even smaller size between two elastic active entanglements). This concurs with our results regarding the heterogeneity of the cross-linked topology on large scales which should be considered as a general aspect of polymer networks [22, 23]. As a consequence our simulations lead to the following interpretation of the experimentally observed effects: Large scales in the demixing of cross-linked blends can occur also if kinetic effects are neglected but kinetic effects can still additionally enlarge the characteristic scales in the microphase separated structure.

Extending previous work on cross-linked polymers [1, 6] we simulated cross-linking in polymer melts (consisting of two species of polymer chains A and B) and the phase behavior of the resulting network structure. The chains are equivalent except for their mutual repulsion χ. During the cross-linking process we switch off the AB-interactions, i.e. we treat the system as a homopolymer melt. Exactly this procedure allows us to distinguish between kinetic and thermodynamic demixing effects.

The BFM is used as described in section 2.. Throughout all our simulation runs we have chosen the following parameters: Our lattice consists of $400 \times 400 \times 400$ sites. We generated 40 000 chains of length 100 simultaneously, leading to a lattice occupation density of 0.5, which corresponds to a dense melt in the framework of the BFM. One half of the chains was labeled as A, the other half as B. The melt was equilibrated during $t \approx 10^6$ MCS .

The requirement of memory and computing time is very high for such a large system, which (as discussed above) is absolutely necessary in the present situation. Thus a parallelization of the algorithm is imperative. We have distributed our lattice on 64 processing elements (PE's) on the CRAY-T3E by cutting it three times in each direction. Thus each PE is working on lattice consisting of $100 \times 100 \times 100$ sites. To consider bond- and excluded volume restrictions and interaction energies at the interfaces between the PE's each PE also has to possess information of a layer of neighboring lattice sites. In our case this layer has a thickness of five lattice units. We move the polymer segments in the following way: First we choose randomly one of 125 sublattices of our system, each of them consisting of sites having a distance of 5 from its nearest neighbors, necessary to separate excluded volume, bond length and interaction effects. Segments occupying the sites of this sublattice can now be moved independently from each other. Eventually, we update the configuration of the border layers. Thus we achieve $80*80*80*0.5 = 256\,000$ attempted monomer moves in the whole system between two communication processes. For the communication between the PE's we use the Message Passing Interface (MPI). The sum of MPI time per sum of application CPU time is approximately 10%. We achieve in average $3*10^5$ attempted monomer moves per PE and second, respectively. This performance is obtained for the system with full network constraints and AB-interactions. During a typical job on the CRAY-T3E which lasts three hours 50 000 MCS can be calculated. After an interval of 5000 MCS we store the positions of all the $4 * 10^6$ monomers into a file (having a size of 4.8 MB). The data analysis is performed afterwards on our local system.

After equilibrating the polymer melt we cross-link the chains stochastically. First each chain segment is radicalized with a given probability, i.e. it gets the ability to form a crosslink. Then the melt is moved according to the BFM. If a radical hits a non-radical during the course of its motion a new bond is created with a certain probability. When all radicals have reacted a network is formed with a given probability P_{net} that a segment is part of a crosslink. We studied the resulting network topology for different crosslink

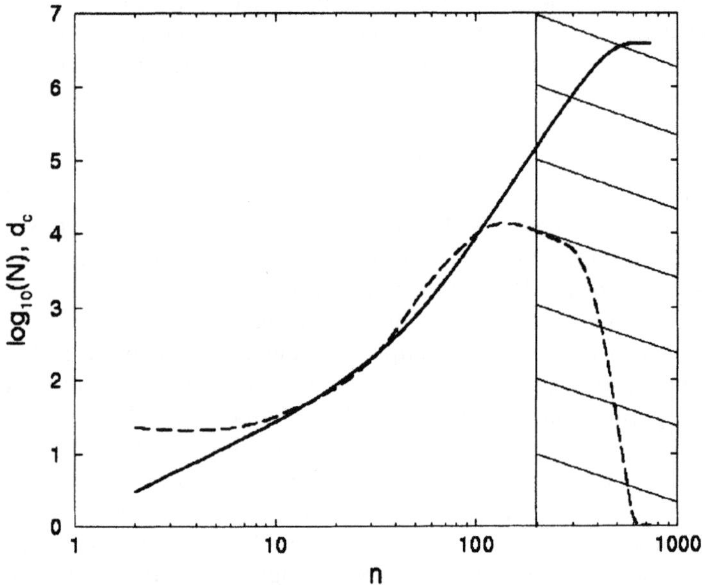

Fig. 3.1. The full line shows the average number $N(n)$ of monomers within the chemical distance n from a randomly chosen starting point on the structure ($P_{net} = 0.07$). The slope of the graph is shown as a dashed line and can be interpreted as an effective chemical dimension d_c. For $n > 200$ the boundaries of the system are reached.

densities P_{net}. First, we counted the number of segments $N(n)$ which can be reached from a given starting point in at most n steps along the network structure. From the slope of the graph of $N(n)$ in a log-log-plot we derive an effective chemical dimension d_c. By plotting the radius of gyration $R_{gyr}(N)$ of the set of segments which we reach in at most n steps double-logarithmically against its mass N we determined a second fractal dimension d_o, which characterizes the embedding of the network clusters into the three-dimensional space. An example for the chemical dimension is shown in Fig.3.1 for $P_{net} = 0.07$ (average strand length 10.8 segments).

Both fractal dimensions reveal large deviations of the network structure from a regular topology. For both dimensions the value 3 is first reached for network clusters consisting of many strands. This means that on this scale the space is only partly filled by monomers within a given chemical distance from the starting monomer. Afterwards there is a large crossover-region where the dimensions first increase to values of about four, before they decay again. Only at an extremely high crosslink densities of 40 net points per chain a saturation of d_o to three dimensions within our system size can be observed.

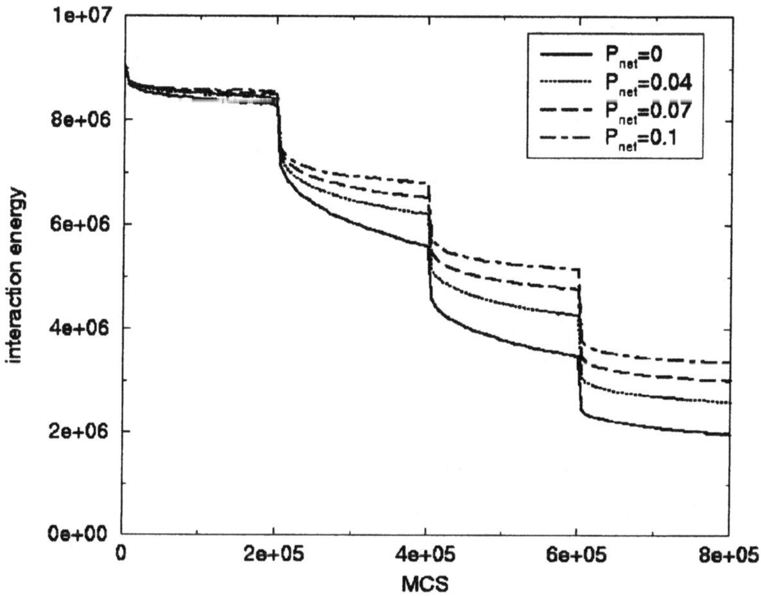

Fig. 3.2. Relaxation of the total AB-interaction energy. For details see text.

When we now switch on the repulsive AB-interaction microphase separation sets in. In Fig.3.2 the total interaction energy for the uncross-linked blend and for three different P_{net} is shown. The interaction parameter χ is increased progressively after each $2 * 10^5$ MCS from 0.02 over 0.1 and 0.2 to 0.4. Fig.3.3 shows a cross-section of the system for $P_{net} = 0.07$, $\chi = 0.4$ at $t = 1.2 * 10^6$ MCS. Furthermore we observe the development of a peak in the corresponding AB-structure factors $S(k)$ at a value of the wave vector k^* which is approximately four to five times smaller than predicted by de Gennes assuming a simple regular network structure. In the inset of Fig.3.4 we show the development of the peak for $P_{net} = 0.07$ for $\chi = 0.4$ from $t = 6 * 10^5$ MCS to $t = 1.2 * 10^6$ MCS.

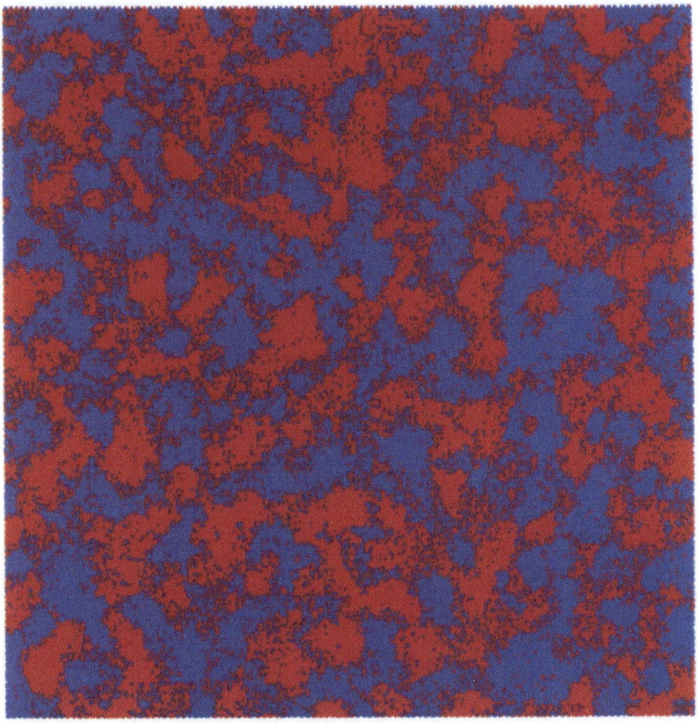

|—| **Average initial chain length.**

H **Average strand-length between net-points.**

Fig. 3.3. Cross-section of the system for $P_{net} = 0.07$, $\chi = 0.4$ at $t = 1.2 * 10^6$ MCS.

Assuming exponential relaxation behavior for each wave vector k we extrapolate the structure factor to its equilibrium value, which is depicted in the inset of Fig.3.4 as a dot-dashed line. In Fig.3.4 we show the same curve in double logarithmic scale and, in addition, fits with three different theoretically predicted functions. The full line is the fit to the structure factor put forward by de Gennes in [20],

$$S(k)^{-1} \sim Ak^2 + B + \frac{C}{k^2}. \tag{3.1}$$

At intermediate wave vectors it predicts a decay of $S(k)$ with k^{-2}, whereas we find a k^{-x}-behavior with an exponent $x \approx 4$. Benmouna et al. [26] generalized the analogy of de Gennes to a dielectric medium to include the concept of screening. They found a structure factor

$$S(k)^{-1} \sim Ak^2 + B + \frac{C}{k^2 + \kappa^2}. \tag{3.2}$$

The fit to this function is shown in Fig.3.4 as the long-dashed line. Alig et.

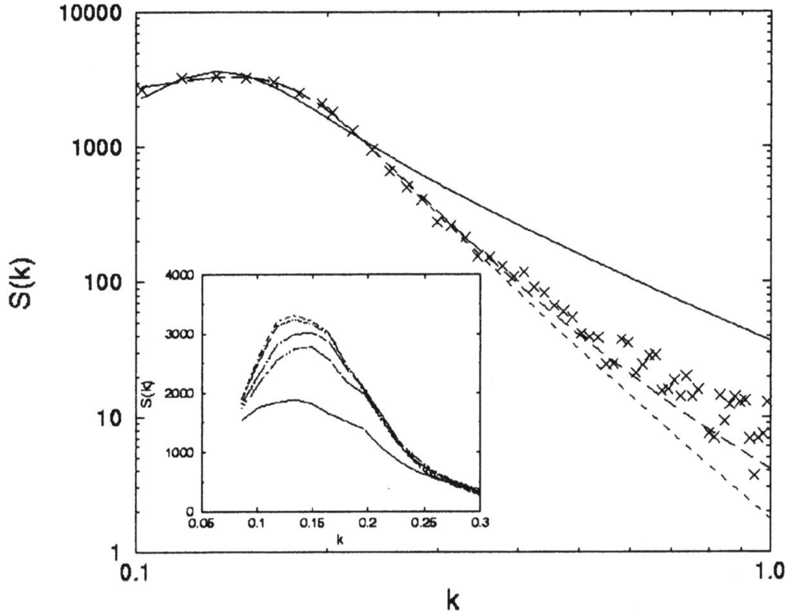

Fig. 3.4. Fits of Eqs. (3.1) (full line), (3.2) (long-dashes) and (3.3) (short dashes) to the structure factor obtained for $P_{net} = 0.07$ and $\chi = 0.4$. Inset: Time development of the structure factor. The dot-dashed line shows the extrapolated equilibrium values.

al [27] considered the simultaneous cross-linking and spinodal demixing of polymer blends. They found a resulting structure factor which has basically the following k−dependence:

$$S(k)^{-1} \sim k^2(Ak^2 + B + \frac{C}{k^2}). \tag{3.3}$$

Interestingly this function (dashed line in Fig.3.4) also fits our data much better than the result of de Gennes . In general our data is fitted well at the peak by Eq.3.2 and by Eq.3.3, whereas at intermediate k-values of k the data points lie between these fits.

4. Conclusions

We have applied massive parallelization of the bond fluctuation model on a T3E system. This made it possible for the first time to simulate heterogeneous polymer networks on scales which can reflect the macroscopic behavior of these systems properly. The performance on 64 PE's including all interaction effects is approximately 20 millions attempted monomer moves per second, which is more than 10 times higher compared to previous highly optimized algorithms on vector computers. The high speed communication hardware of T3E systems combined with optimized lattice update strategies are responsible for such results.

Simulating the demixing process in a cross-linked polymer blend consisting of A and B chains which repel each other has revealed non-trivial structural properties of the network connectivity as well as strong deviations from a simplified physical model proposed by de Gennes. In contrast to experiments, which generally show similar effects, we could make a clear-cut separation between kinetic and thermodynamic effects by switching on the AB-interaction after the cross-linking process is completed. We have obtained much larger scales in the microphase separated structure than expected using complete homogeneous conditions for the network topology. Composition fluctuations in the interaction-free melt combined with strong network heterogeneities are responsible for this effect.

In the first part we have investigated a simpler system reflecting the properties of random copolymer chains at AB-interfaces, as occurs when compatibilizing blends of different polymer species. In the absence of analytical solutions for the practically important many chain systems we have tested simple scaling arguments for this regime. Here we found agreement of our simulations with these ideas. In particular a over-saturated adsorption regime formed of brush-like stretched loops from the adsorbed chains has been obtained.

The presented results demonstrate that highly complex polymer systems, which are both of practical relevance and of theoretical interest, can be successfully simulated using massive parallel supercomputers.

Acknowledgement. We acknowledge support by the DFG through SFB 428 and Graduiertenkolleg and by the Fonds der Chemischen Industrie. We are grateful to the Höchstleistungsrechenzentrum Stuttgart (HLRS) for a grant of computer time on CRAY-T3E.

References

1. J.-U. Sommer and H. Trautenberg, Directory of CRAY Sponsored University Research & Development Grants , 227 (1995).
2. J.-U. Sommer, T. Vilgis, and G. Heinrich, J. Chem. Phys. **100**, 9181 (1994).

3. J.-U. Sommer, Macrom. Symp. **81**, 139 (1994).
4. H. Trautenberg, J.-U. Sommer, and D. Göritz, J. Chem. Soc. Faraday Trans. **91**, 2649 (1995).
5. H. L. Trautenberg, J.-U. Sommer, and D. Göritz, Macrom. Symp. **81**, 153 (1994).
6. S. Lay, J.-U. Sommer, and A. Blumen, J. Chem. Phys. **110**, 12173, (1999)
7. H. Brown, V. Deline, and P. Green, Nature **341**, 221 (1989).
8. C.-A. Dai et al., Phys.Rev.Lett. **73**, 2472 (1994).
9. T. Garel, D. Huse, S. Leibler, and H. Orland, Europhys.Lett. **8**, 9 (1989).
10. J.-U. Sommer and M. Daoud, Europhys. Lett. **32**, 407 (1995).
11. S. Stepanow, J.-U. Sommer, and I. Y. Erukhimovich, Phys. Rev. Lett. **81**, 4412 (1998).
12. V. Ganesan and H. Brenner, Europhys. Lett. **46**, 43 (1999).
13. C. Yeung, A. Balazs, and D. Jasnow, Macromolecules **25**, 1357 (1992).
14. J.-U. Sommer, G. Peng, and A. Blumen, J. Phys. II France **6**, 1061 (1996).
15. G. Peng, J.-U. Sommer, and A. Blumen, Phys.Rev **E 53**, 5509 (1996).
16. J.-U. Sommer, G. Peng, and A. Blumen, J. Chem. Phys. **105**, 8376 (1996).
17. E. Bouchaud and M. Daoud, J. Physique **48**, 1991 (1987).
18. P. G. de Gennes, Macromolecules **14**, 1637 (1981)
19. E. Leclerc, M. Daoud, and R. Douillard, Nuovo Cimento **16 D**, 641 (1994).
20. P. G. de Gennes, J. Phys. Lett. France **40**, 69 (1979)
21. R. M. Briber and B. J. Bauer, Macromolecules **21**, 3296 (1988)
22. J. Bastide, L. Leibler, and J. Prost, Macromolecules **23**, 1821 (1990).
23. J. Bastide and S. J. Candau, in:*Physical Properties of Polymeric Gels*, edited by J. P. Cohen Addad, John Wiley and Sons, N.Y. 1996, p.143.
24. I. Carmesin and K. Kremer, Macromolecules **21**, 2819 (1988).
25. G. Peng, J.-U. Sommer, and A. Blumen, Europ. Phys. J. B **8**, 73 (1999).
26. M. Benmouna, T. A. Vilgis, M. Daoud, and M. Benhamou, Macromolecules **27**, 1172 (1994)
27. I. Alig, M. Junker, M. Schulz, and H. L. Frisch, Phys. Rev. B **53**, 11481 (1996)

Towards the Limits of present-day Supercomputers: Exact Diagonalization of Strongly Correlated Electron-Phonon Systems

Gerhard Wellein[1] and Holger Fehske[2]

[1] Regionales Rechenzentrum Erlangen, Universität Erlangen, D-91058 Erlangen
[2] Physikalisches Institut, Universität Bayreuth, D-95440 Bayreuth

Abstract. The problem of describing lattice dynamical effects in electronically low-dimensional highly correlated systems, such as, e.g., the quasi-1D charge-density wave and spin-Peierls materials or the quasi-2D high-T_c superconductors, has been a challenge to solid state physicists for a long time. At present, perhaps the only realible results come from purely numerical finite-lattice calculations. Employing the Lanczos algorithm in combination with a kernel polynomial moment expansion and the maximum entropy method on massive parallel computers, we show a way of calculating ground-state and spectral properties for models of electrons or spins strongly interacting with quantum phonons.

1 Introduction

There has been renewed interest in the analysis of strongly coupled electron-phonon (EP) systems since it was recognized that intrinsic polaron-like lattice distortions are a common feature of several important classes of perovskites such as the high-T_c cuprates [1], the non-metallic nickelates [2], and the colossal magnetoresistance manganites [3]. In these compounds, e.g., in $La_{2-x}Sr_x[Cu/Ni]O_4$ or $La_{1-x}Ca_xMnO_3$, the charge carriers susceptible to lattice polaron formation mediate the antiferromagnetic (AFM) or ferromagnetic (FM) interactions between the Cu^{2+}, Ni^{2+} ($S = 1/2, 1$) or Mn^{3+} ($S = 3/2$) ions (core spins), suggesting that the local lattice or Jahn-Teller distortions play also a significant role in determining the electronic and magnetic properties [4].

To discuss the interplay of electronic and lattice degrees of freedom in such systems, exhibiting besides a substantial EP coupling strong Coulomb correlations, the investigation of generic microscopic models is of great importance. The probably simplest prototype model Hamiltonian with respect to a strong spin-charge-lattice coupling is the two-dimensional (2D) t-J model appended by an additional Holstein-like interaction term with a dispersionless (optical) phonon branch:

$$\mathcal{H} = -t \sum_{\langle ij \rangle \sigma} \left(\tilde{c}_{i\sigma}^{\dagger} \tilde{c}_{j\sigma} + \text{H.c.} \right) + J \sum_{\langle ij \rangle} \left(\boldsymbol{S}_i \boldsymbol{S}_j - \frac{1}{4} \tilde{n}_i \tilde{n}_j \right)$$
$$- \sqrt{\varepsilon_p \hbar \omega_0} \sum_i \left(b_i^{\dagger} + b_i \right) \tilde{h}_i + \hbar \omega_0 \sum_i \left(b_i^{\dagger} b_i + \frac{1}{2} \right). \quad (1)$$

The Holstein t-J Hamiltonian (1) acts in a projected Hilbert space without doubly occupied sites, where $\tilde{c}_{i\sigma}^{(\dagger)} = c_{i\sigma}^{(\dagger)}(1 - \tilde{n}_{i,-\sigma})$ is a spin-σ electron annihilation (creation) operator on Wannier site i, $\tilde{n}_i = \sum_\sigma \tilde{c}_{i\sigma}^{\dagger} \tilde{c}_{i\sigma}$, and $\boldsymbol{S}_i = \sum_{\sigma,\sigma'} \tilde{c}_{i\sigma}^{\dagger} \boldsymbol{\tau}_{\sigma\sigma'} \tilde{c}_{i\sigma'}$. Within an effective single-band description, the parameters t and J measure the transfer amplitude and the antiferromagnetic exchange interaction between nearest-neighbour sites on a square lattice, where $t > J$ corresponds to the situation in the cuprates. For example, $J/t \simeq 0.4$ with $t \simeq 0.3$ eV is commonly used to model the $La_{2-x}Sr_xCuO_4$ system. In (1), we have chosen the coupling to the hole, $\tilde{h}_i = 1 - \tilde{n}_i$ is the local density operator of the spinless hole, as the dominant source of the EP coupling, because in the t-J model the unoccupied site corresponds to the Zhang-Rice singlet (formed from Cu $3d_{x^2-y^2}$ and O $2p_{x,y}$ hole orbits) for which the coupling should be much stronger than for the occupied (Cu^{2+}) site. The hole-phonon coupling constant is denoted by ε_p, and $\hbar\omega_0$ is the bare phonon frequency (below $\hbar = 1$ and all energies are measured in units of t). Then $q_i = (b_i^{\dagger} + b_i)/\sqrt{2M\omega_0}$ may be thought of representing an internal vibrational degree of freedom of the lattice site i ($b_i^{(\dagger)}$ annihilates (creates) an optical phonon). Indeed for $La_2[Cu,Ni]O_4$ the oxygen vibrational modes have a small dispersion in the Brillouin zone, i.e., they are revealed to be very localized in real space and can be considered as independent.

From a theoretical point of view, the main problem for a proper analytical treatment of this model is that the coherent motion of charge carriers, heavily dressed by interaction with magnons, takes place on a strongly reduced energy scale being almost comparable to the relevant phonon frequencies. As a result, e.g. the standard strong-coupling Migdal-Eliashberg approach based on the adiabatic Migdal theorem might break down, and it has been argued that non-adiabatic vertex corrections are responsible for the enhancement of T_c. Furthermore, as a consequence of the pre-existing magnetic "self-localization" of the charge carriers, polaronic effects play an important role even at rather moderate EP coupling strengths. This makes a weak-coupling approach inadequate.

The lack of well-controlled analytical techniques in treating this complicated many-body problem has stimulated the development of purely numerical methods, such as quantum Monte Carlo simulations, density matrix renormalization group calculations and exact diagonalizations (ED).

2 Numerical Algorithms

At yet exact finite-lattice diagonalization is the only method which enables an extensive and reliable study of microscopic electron-phonon models in a wide range of exchange interactions, EP couplings and phonon frequencies for all dimensions. Motivated by this situation we use the standard Lanczos algorithm [5] to compute ground states and low lying eigenstates of the corresponding Hamilton matrices (HM). Besides we are able to calculate excitation spectras, using a maximum entropy method (MEM) which was recently developed by Silver [6]. The cpu-time and memory requirements of both algorithms are determined by the matrix vector multiplication (MVM). Although the HM are extremly sparse, even their non-zero entries cannot be stored because of the large dimensions of the Hilbert spaces ($D > 10^9$). We have developed a new, parallel and very efficient algorithm which performs the calculation of the non-zero elements and the MVM in one step. Using FORTRAN and C in combination with the MPI communication subroutines the PARPHON code can be used on distributed and shared memory machines. In the following we sketch the generation of an appropriate "parallel" basis set, the parallel implementation of the MVM, some performance results for CRAY T3E and IBM SP supercomputers and the basics of the MEM. For more detailed information we refer to references [7,12].

2.1 Basis Implementation

The microscopic EP models inhibit natural parallelism because the corresponding Hilbert space can be constructed as the tensorial product space of electrons and phonons. Denoting $|u\rangle_{el}$ [$|v\rangle_{ph}$] the basis states of the electronic [phononic] subspace with dimension $D_{el} = \binom{N}{N_\sigma}\binom{N-N_\sigma}{N_{-\sigma}}$ [$D_{ph} = \infty$], the complete basis set can be written as:

$$\{|\Phi_{uv}\rangle = |u\rangle_{el} \otimes |v\rangle_{ph}\} \; . \tag{2}$$

Since the Hamiltonian commutes with the electron number operator $\mathcal{N}_{el} = \sum_{i=1}^{N}(\tilde{n}_{i,\uparrow} + \tilde{n}_{i,\downarrow})$ and the z-component of the total spin $\mathcal{S}^z = \frac{1}{2}\sum_{i=1}^{N}(\tilde{n}_{i,\uparrow} - \tilde{n}_{i,\downarrow})$, the basis (2) has been constructed for fixed $N_h = N - N_{el}$ and $S^z = S^z_{min}$. In contrast to the electronic subspace the Hilbert space associated to the phonons is infinite even for a finite system. Therefore we restrict ourselves to a finite subspace (with dimension $D_{ph}^M = \frac{(M+N)!}{M!N!}$), that retains only phononic basis states with at most M phonons [8,9].
To further reduce the dimension of the total Hilbert space ($D_{tot}^M = D_{el} \times D_{ph}^M$), we can exploit the space group symmetries [translations (G_T) and point group operations (G_L)] and the spin-flip invariance [(G_S); $S^z = 0$–subspace only].

Restricting ourselves to the one–dimensional non–equivalent irreducible representations of the group $G(\boldsymbol{K}) = G_T \times G_L(\boldsymbol{K}) \times G_S$, we can use the projection operator

$$\mathcal{P}_{\boldsymbol{K},rs} = \frac{1}{g(\boldsymbol{K})} \sum_{\mathcal{G}\in G(\boldsymbol{K})} \chi^{(\mathcal{G})}_{\boldsymbol{K},rs}\, \mathcal{G} \tag{3}$$

in order to generate a new symmetrized basis set: $\{|\Phi_{uv}\rangle\} \xrightarrow{\mathcal{P}} \{|\tilde{\Phi}_{\tilde{w}}\rangle\}$. In (3), $\mathcal{G}$ denotes the $g(\boldsymbol{K})$ elements of the group $G(\boldsymbol{K})$ and $\chi^{(\mathcal{G})}_{\boldsymbol{K},rs}$ is the (complex) character of $\mathcal{G}$ in the $[\boldsymbol{K}, rs]$–representation, where $\boldsymbol{K}$ refers to one of the N allowed wave vectors in the first Brillouin zone, r labels the irreducible representations of the little group of $\boldsymbol{K}$, $G_L(\boldsymbol{K})$, and s parameterizes G_S. In general the new basis set $[\{|\tilde{\Phi}_{\tilde{w}}\rangle\}]$ does not preserve the inherent parallelism, because the projection operator acts both on electronic and phononic basis elements. However using the procedure shown in Fig. 1 we are able to construct a "parallel" symmetrized basis set $\{|\tilde{\Phi}_{\tilde{u}v}\rangle\}$. Starting from the complete electronic basis set, we construct a symmetrized electronic basis set with reduced dimension $\tilde{D}^{g(\boldsymbol{K})}_{el}$ $[\tilde{D}^{g(\boldsymbol{K})}_{el} \sim D_{el}/g(\boldsymbol{K})]$. In the second step a new subset of (2) is defined as the product of the electronic representatives $[\{|\tilde{u}\rangle_{el}\}]$ and the complete phononic basis elements. Finally, using the projection operator, a "parallel" symmetrized basis set can easily be constructed. The factors $N^{[\boldsymbol{K}rs]}_{\tilde{u}v}$ are used to ensure both normalization and linear independence of the new basis set $\{|\tilde{\Phi}_{\tilde{u}v}\rangle\}$. Fixing $[\boldsymbol{K}, rs]$ the resulting Hilbert space has a total dimension $\tilde{D}^M_{tot} = \tilde{D}^{g(\boldsymbol{K})}_{el} \times D^M_{ph}$ $[\sim D^M_{tot}/g(\boldsymbol{K})]$, and a general state of the Holstein t-J model (HtJM) is represented as

$$|\Psi_{\boldsymbol{K},rs}\rangle = \sum_{\tilde{u}=1}^{\tilde{D}^{g(\boldsymbol{K})}_{el}} \sum_{v=1}^{D^M_{ph}} \tilde{c}_{\tilde{u}v}\, |\tilde{\Phi}_{\tilde{u}v}\rangle\,. \tag{4}$$

Clearly, in general this procedure generates complex matrix elements and may require some memory overhead, since we also consider basis states with $N^{[\boldsymbol{K}rs]}_{\tilde{u}v} = 0$ which are not relevant to the calculations. These minor "costs" are by far compensated by the fact, that the matrix-dimension is reduced by a factor of $g(\boldsymbol{K})$ $[\approx 10-100]$, preserving the inherent parallelism of the model.

2.2 Coding of the Matrix-Vector-Multiplication

Since the reduced electronic dimension is in general much smaller than the phononic dimension ($\tilde{D}^{g(\boldsymbol{K})}_{el} \ll D^M_{ph}$) the MVM is parallelized by the electronic elements and the coefficients of $|\Psi_{\boldsymbol{K},rs}\rangle$ will be distributed through the processor by the first sum in (4). For the sake of simplicity we discuss the case, where the number of processors (N_{pro}) is equal to the reduced electronic dimension $\tilde{D}_{el}^{g(K)}$. In other words for fixed $\tilde{u}$ all coefficients of the general state (4) $\{\tilde{c}_{\tilde{u}v}\,; v = 1, \ldots, D_{ph}\}$ reside in the memory of processor $\mathbf{P}_{\tilde{u}}$ (cf. Fig. 2). The parallel implementation of the MVM

$$\boldsymbol{A} \leftarrow \boldsymbol{A} + \mathbf{H} \times \boldsymbol{B} \tag{5}$$

is discussed in Fig. 3. The processor $\mathbf{P}_{\tilde{u}}$ computes the non-zero elements of the sparse $(D_{ph}^{M} \times D_{ph}^{M})$ submatrices $\mathbf{h}_{\tilde{u},1}, \ldots, \mathbf{h}_{\tilde{u},\tilde{D}_{el}^{g(K)}}$, defined as

$$\mathbf{h}_{\tilde{u},\tilde{u}'} = \Big(\langle \tilde{\Phi}_{\tilde{u}v} \,|\, \mathcal{H} \,|\, N_{\tilde{u}'v'}^{[Krs]} [|\tilde{u}'\rangle_{el} \otimes |v'\rangle_{ph}] \Big)_{v,v'=1,\ldots,D_{ph}^{M}} . \tag{6}$$

As easily can be seen from Fig. 3 communication between two processors $\mathbf{P}_{\tilde{u}}$ and $\mathbf{P}_{\tilde{u}'}$ is required if the submatrix $\mathbf{h}_{\tilde{u},\tilde{u}'}$ contains nonvanishing entries $(\mathbf{h}_{\tilde{u},\tilde{u}'} \neq \mathbf{0})$. Since third and fourth term in (1) generate non-zero entries only in $\mathbf{h}_{\tilde{u},\tilde{u}}$, the communication pattern is determined by the electronic subsystem and can be stored before starting the first MVM. Besides the simplicity of the communication pattern we want to emphasize two other features of the above implementation, which are important to avoid data coherence problems: First, non local memory access only consists of a read operation on vector $\mathbf{B}$, which is not changed during the MVM. Second, the write access is restricted to the local elements of $\mathbf{A}$.

2.3 Perfomance analysis of MVM

The performance of the MVM is dominated by the inter-processor communication and the calculation of the non-zero elements of the HM. Since the electronic dimension is several orders of magnitude smaller than the Hilbert space dimension all information regarding the electronic part of the Hamilton-operator can be stored in the local memory of each processor. Using a (M+1)-coding of the phononic basis elements enables also a local and memory saving calculation of the effect of the phonon operators $(b_i, b_i^{\dagger}; i = 1, \ldots, N)$.
In Table 1 we present problem sizes and times per MVM step for typical production jobs on the CRAY T3E of the HLRS. The number of MVM performed per job ranges from 70 to 400 depending on the model parameters. Obviously, the available memory limits the accessible model parameters in our calculations. For example, one vector of dimension $\tilde{D}_{tot}^{M} \sim 1.6 \times 10^{9}$ - stored with 8-Bytes per element - requires 13 GBytes or 20% of the total available memory. Please note, the storage of the corresponding matrix - containing approximately 30 non-zero entries per column - would require more than 250 GBytes!
Next we want to discuss the parallel features of the PARPHON code, which is based on MPI communication subroutines to ensure a broad portability. The use of non blocking MPI_ISEND and MPI_IRECV calls permits an overlapping of communication and calculation in the MVM. Considering the NUMA architecture of CRAY MPP systems we have replaced these calls by

SHMEM_GET calls from the CRAY shmem library. Fig. 4 presents a scalability analysis for CRAY T3E/T3D and IBM SP supercomputers. We display the time per MVM step (T) and the corresponding parallel speed up S_{par}, defined as

$$S_{par}(N_{pro}) = \frac{\text{Time per MVM on 1 processor}}{\text{Time per MVM on } N_{pro} \text{ processors}} . \tag{7}$$

A comparison of the timing results for the MPI implementations in Fig. 4 (a) clearly reflects the different single processor peak performance of the CRAY T3E and IBM SP nodes. Moreover, the use of the shmem call increases the gap dramatically and even enables the CRAY T3D to pass the IBM system at intermediate processor numbers. Another striking feature of the shmem version is the improvement of the parallel efficiency

$$E_{par}(N_{pro}) = \frac{S_{par}(N_{pro})}{N_{pro}} \tag{8}$$

as against the MPI code [cf. Fig. 4(b)]. For the parameter set used in Fig. 4 we get $E_{par} \approx 0.95$ and $E_{par} \approx 0.5$ for the shmem and MPI implementation, respectively. Note that the high efficiency does not hold for a few parameter sets and the scalability is limited by $N_{pro} \leq \tilde{D}_{el}^{g(\boldsymbol{K})}$. Nevertheless the above results clearly demonstrate the effecient use of the CRAY T3E supercomputer within the project PARPHON.

2.4 MEM Algorithm for Estimating Spectral Functions

The numerical calculation of spectral functions,

$$\begin{aligned} A^{\mathcal{O}}(\omega) &= - \lim_{\varepsilon \to 0^+} \frac{1}{\pi} \Im m \left[\langle \Psi_0 | \mathbf{O}^\dagger \frac{1}{\omega - \mathbf{H} + E_0 + i\varepsilon} \mathbf{O} | \Psi_0 \rangle \right] \\ &= \sum_{n=0}^{\tilde{D}_{tot}^M - 1} |\langle \Psi_n | \mathbf{O}^\dagger | \Psi_0 \rangle|^2 \delta[\omega - (E_n - E_0)] , \end{aligned} \tag{9}$$

where $\mathbf{O}$ is the matrix representation of any operator $\mathcal{O}$, involves the large sparse HM $\mathbf{H}$. Therefore direct methods - which scale in cpu time and memory requirements as $(\tilde{D}_{tot}^M)^3$ and $(\tilde{D}_{tot}^M)^2$, respectively - cannot be applied. We have implemented a linear scaling algorithm the maximum entropy method [6] (MEM) instead. The MEM computes a direct approximation of $A^{\mathcal{O}}(\omega)$ and was found both to be very efficient and numerical stable for high-energy resolution applications such as the lattice polaron formation. [12] The basic idea of Chebyshev recursion techniques such as MEM is an expansion of the δ-function contained in $A^{\mathcal{O}}(\omega)$ in a series of Chebyshev polynomials $T_m(x)$,

$$A^{\mathcal{O}}(x) = \frac{1}{\pi\sqrt{1-x^2}} \left(\mu_0^{\mathcal{O}} + 2 \sum_{m=1}^{\infty} \mu_m^{\mathcal{O}} T_m(x) \right) , \tag{10}$$

with the coefficients (moments)

$$\mu_m^{\mathcal{O}} = \int_{-1}^{1} dx\, T_m(x) A^{\mathcal{O}}(x) = \langle \mathbf{O}^\dagger \Psi_0 | T_m(\mathbf{X}) | \mathbf{O}^\dagger \Psi_0 \rangle , \tag{11}$$

$x = (\omega - b)/a$, $a = (E_{max} - E_{min})/2(1-\epsilon)$, and $b = (E_{max} + E_{min})/2$. Since the Chebyshev polynomials are defined on the interval $[-1, 1]$, a re–scaled Hamiltonian $\mathbf{X} = (\mathbf{H} - b)/a$ has to be considered, with the eigenvalue interval $[-(1-\epsilon), 1-\epsilon]$. A small constant ϵ is introduced in order to avoid convergence problems at the endpoints of the interval (a typical choice is $\epsilon \sim 0.01$). Using the recurrence relations for Chebyshev polynomials the moments $\mu_m^{\mathcal{O}}$ can be computed by iterative MVM (two moments per MVM step). However, the truncation of the series at a finite number of moments $\tilde{M}$ produces the well known Gibbs oscillations. Within the kernel polynomial method [10] (KPM) - which provides a linear Chebyshev approximation to the spectrum ($\tilde{A}^{\mathcal{O}}(x) \simeq A^{\mathcal{O}}(x)$) - these oscillations are minimized by multiplying the moments by appropriate damping factors [11].

The MEM is a *non–linear* optimization procedure for calculating from $\hat{M}$ known moments a representation of a (reconstructed) spectral function, $\hat{A}^{\mathcal{O}}(\phi) = \hat{A}^{\mathcal{O}}(x)\sin\phi$ with $x = \cos\phi$, that corresponds to a Chebyshev series with $\hat{M}^{eff} = F \cdot \hat{M}$ summands. The $\hat{M}^{eff}$ moments

$$\hat{\mu}_m^{\mathcal{O}} = \int_0^{\pi} d\phi \cos(m\phi) \hat{A}^{\mathcal{O}}(\phi) \tag{12}$$

are determined by maximizing the information theoretical relative entropy [6]

$$\mathcal{S} = \int_0^{\phi} d\phi \left[\hat{A}^{\mathcal{O}}(\phi) - \hat{A}_0^{\mathcal{O}}(\phi) - \hat{A}^{\mathcal{O}}(\phi) \ln \left(\frac{\hat{A}^{\mathcal{O}}(\phi)}{\hat{A}_0^{\mathcal{O}}(\phi)} \right) \right] \tag{13}$$

subject to the constraint that the first $\hat{M}$ moments of $\hat{A}^{\mathcal{O}}(\phi)$ are equal to the $\hat{M}$ known moments. As the default model $\hat{A}_0^{\mathcal{O}}(\phi)$ for $\hat{A}^{\mathcal{O}}(\phi)$ the KPM result $\hat{A}_0^{\mathcal{O}}(x) = \tilde{A}^{\mathcal{O}}(x)$ can be choosen. The values of F and $\hat{M}$ are typically $8 \ldots 32$ and $128 \ldots 256$, respectively. Of course, the calculation of the Chebyshev moments is the time and memory consuming step within the MEM and is performed on the CRAY T3E. The reconstruction of the spectrum using $\hat{M}$ moments can easily be done on a PC or workstation.

Finally we would like to demonstrate in Fig. 5 the accuracy of the MEM for a small system ($\tilde{D}_{tot}^{M} = 126$). Besides the true spectral function we have also depicted the KPM results. Clearly, the convergence to the exact data is much better applying the MEM than the KPM. It must be emphasized, that the number of moments and also the number of MVM involved in the calculation of the spectrum is four times larger for the KPM than for the MEM! Moreover the KPM spectrum still contains not fully damped Gibbs oscillation, whereas the peaks of the MEM reproduces the exact peaks with a high accuracy.

3 Polaronic Effects in Strongly Correlated Fermion Systems - Discussion of Selected Results

As an highly non-trivial application of the numerical scheme outlined so far, in this section we would like to discuss exemplarily polaronic features of dopant-induced charge carriers within the framework of the 2D HtJM. [12]

3.1 Single-particle spectral properties

In the analysis of the HtJM we start with the study of just a single dynamic hole. Increasing the EP coupling in the adiabatic regime, we notice a continuous but rather sharp crossover from a nearly-free polaron, described by an effective transfer amplitude that is only weakly reduced from its value in the pure t-J model, to a less mobile (small-size) adiabatic Holstein hole-polaron.

To elucidate the difference between the nearly-free and small polaron limits and to demonstrate the formation of a hole-polaron band at large ε_p, in Fig. 6 we present the results for the $\boldsymbol{K}$-resolved spectral function

$$A_{\boldsymbol{K}}(\omega) = \sum_{n,\sigma} \left| \langle \Psi_{n,\boldsymbol{K}}^{(N-1)} | \tilde{c}_{\boldsymbol{Q}-\boldsymbol{K},\sigma} | \Psi_{0,\boldsymbol{Q}}^{(N)} \rangle \right|^2 \delta\left[\omega - (E_{n,\boldsymbol{K}}^{(N-1)} - E_{0,\boldsymbol{Q}}^{(N)})\right] , \tag{14}$$

at the allowed values of the momentum $\boldsymbol{K}$ of a ten-site square lattice. In order to visualize the intensities (spectral weights) connected with the various peaks (excitations) in each $\boldsymbol{K}$-sector we also have shown the integrated density of states

$$N(\omega) = \int_{-\infty}^{\omega} d\omega' \frac{1}{N} \sum_{\boldsymbol{K}} A_{\boldsymbol{K}}(\omega') . \tag{15}$$

Of course in the very weak-coupling regime the spectral function is barely changed from that of the pure t-J model, which is characterized by a "quasiparticle" (QP) peak at the bottom of the spectrum followed by a lump of spectral weight with some internal structure, whereby the QP pole directly corresponds to the coherent single-hole ground state with momentum $(3\pi/5, \pi/5)$ and contains a substantial amount of spectral weight. In the weak-coupling regime the mass renormalization of the coherent QP band due to the hole-phonon coupling is generally small compared with that arising from the hole-spin interactions (magnetic polaron regime). The new structures, nevertheless observed in the $A_{\boldsymbol{K}}$ spectra shown Fig. 6 (b), correspond to predominantly "phononic" side bands separated from the particle-spin excitations by multiples of the bare phonon frequency ω_0. However, these phonon resonances have less and less "electronic" spectral weight the more phonons are involved. This is because $A_{\boldsymbol{K}}(\omega)$ measures the overlap of these excited states with the state obtained by creating a hole in the *zero-phonon* Heisenberg ground state.

With increasing ε_p the lowest peaks in each A_K start to separate from the rest of the spectrum. These states become very close in energy and finally a narrow well-separated lattice hole-polaron band evolves in the strong-coupling case ($\varepsilon_p = 4.0$). Since the gap to the next energy "band" is of the order of ω_0 (see panel (b)), these excitations will be triggered by an one-phonon absorption process. The transition to the lattice hole-polaron state is accompanied by a strong increase in the on-site hole-phonon correlations, indicating that the lattice polaron QP comprising a "quasi-localized" hole and the phonon cloud is mainly confined to a single lattice site (small-size lattice hole polaron). Compared to the non-interacting single-electron (Holstein model) [13] or spinless fermion (Holstein-t model) [14] problems, the critical EP coupling strength for lattice polaron formation is considerably reduced due to pre-localization effects of the hole in the antiferromagnetic spin background.

3.2 Optical conductivity

Since polaronic effects might be responsible for some of the puzzling thermodynamic and transport properties of the cuprates, nickelates and manganites, increasing effort has been aimed to test the polaronic nature of charge carriers by means of optical experiments. One of the physical quantities which contains valuable information about the low-energy excitations in interacting EP systems is the optical conductivity, $\sigma(\omega)$, usually determined from reflectivity measurements. Here the anomalous broad optical absorption detected for the $La_{2-x}Sr_x[Cu, Ni]O_4$ compounds in several (normal-state) infrared experiments has attracted a lot of attention. In particular for the nickelate system, where the existence of self-trapped charge carriers is well established, the mid-infrared absorption band has been interpreted successfully in terms of photon-assisted hopping of small-size lattice polarons [15]. On the other hand, from a comparison of the optical absorption of several perovskites ($La_{1.67}Sr_{0.33}NiO_4$, $Sr_{1.5}La_{0.5}MnO_4$, and $Nd_{1.96}Ce_{0.04}CuO_4$), it has been argued that in the low-doping cuprates the lattice polarons appear to be rather large.

Using standard linear response theory, the Kubo formula for the real (absorbent) part of the frequency-dependent optical conductivity gives two physically distinct contributions:

$$\Re e\, \sigma_{xx}(\omega) = \mathcal{D}\delta(\omega) + \sigma_{xx}^{reg}(\omega)\,, \qquad \omega \geq 0\,. \tag{16}$$

The first so-called Drude term at $\omega = 0$ is due to the free acceleration of the charge carriers by the electric field and the second term, frequently called the "regular term", is due to finite frequency dissipative optical transitions to excited quasiparticle states. More explicitly, the regular part can be written in spectral representation at $T = 0$ as

$$\sigma_{xx}^{reg}(\omega) = \frac{e^2\pi}{N} \sum_{n\neq 0} \frac{\left|\langle \Psi_{n,K}^{(N-1)} | j_x^{(p)} | \Psi_{0,K}^{(N-1)} \rangle\right|^2}{E_{n,K}^{(N-1)} - E_{0,K}^{(N-1)}} \, \delta\left[\omega - (E_{n,K}^{(N-1)} - E_{0,K}^{(N-1)})\right], \tag{17}$$

where the summation is taken over the complete set of eigenstates with excitation energies $\omega = [E_{n,K}^{(N-1)} - E_{0,K}^{(N-1)}]$ in the one-hole ($N-1$ electron) subspace. For the HtJM the (paramagnetic) current density operator $j_x^{(p)}$ has the form

$$j_x^{(p)} = it \sum_{i\sigma} (\tilde{c}_{i,\sigma}^{\dagger} \tilde{c}_{i+x,\sigma} - \tilde{c}_{i+x,\sigma}^{\dagger} \tilde{c}_{i,\sigma}) \,. \tag{18}$$

Actually in (17) an optical transition can take place only within the $\boldsymbol{K}$-sector of the ground state.

Let us now make contact with the experimentally observed characteristics of the mid-infrared spectra in the doped perovskites. Needless to say, that it is out of the scope of our ten-site cluster diagonalization study to give a quantitative correct theoretical description of the complex optical properties of particular copper and nickel oxides. But we would like to stress that the 2D HtJM seems to contain the key ingredients to reproduce, at least qualitatively, the principal features of the optical absorption spectra of these compounds. This can be seen by comparing the results shown in Figs. 7 (a) and (b), which correspond to the weak and strong EP coupling situations realized in the cuprate ($La_{2-x}Sr_xCuO_4$) and nickelate ($La_{2-x}Sr_xNiO_4$) systems, respectively. The "effective" EP interaction ε_p/t in the nickelates is estimated to be about one order of magnitude larger than in the cuprates simply because of the much smaller transfer amplitude ($t \simeq 0.08$ eV [15]). According to the internal structure of the low-spin state, the hopping transport of spin-1/2 composite holes in a spin-1 background is rather complex; implying, within an effective single-band description, a strong reduction of the transition matrix elements. A striking feature of the absorption spectra in the cuprate superconductors is the presence of a broad mid-infrared (MIR) band, centered at about 0.5 eV in lightly doped $La_{2-x}Sr_xCuO_4$ (which, using $t \sim 0.3$ eV, means that $\omega \sim 1.5$). Such a strong MIR absorption is clearly observed in Fig. 7 (b), which refers to the weak EP coupling case. Since it also appears in the pure t-J model, this MIR band seems to be caused by the spin fluctuations around the charge carrier. Obviously, it is quite difficult to distinguish the spectral weight, produced by the dressing of the hole due to the "bag" of reduced antiferromagnetism in its neighbourhood, from other (e.g. hole-phonon coupling) processes that may contribute to the MIR band observed experimentally. The results presented for the HtJM in Fig. 6 (b) support the claims, however, that the MIR band in the cuprates has a mainly "electronic" origin, i.e., the lattice polaron effects are rather weak. The opposite is true for their isostructural counterpart, the nickelate system, where the MIR absorption band has been ascribed by many investigators to "polaronic" origin [15,16]. Within the HtJM such a situation can be modeled by the parameter set used in Fig. 6 (c). If we fix the energy scale by $t \sim 0.08$ eV (which is the estimate for the nearest neighbour transfer integral in $La_{2-x}Sr_xNiO_4$ given in Ref. [15]), the maximum in the optical absorption is again located at about 0.5 eV. The whole spectrum clearly shows lattice polaron characteristics, where it seems that the lattice hole polarons are of small-to-intermediate size. Most notably, we are able to reproduce the experimentally observed asymmetry in the shape of the spectrum, in particular the very gradual decay of $\sigma^{reg}(\omega)$ at high energies. It is worth mentioning that this behaviour cannot be obtained from a simple fit to the analytical expressions derived for the small polaron hopping conductivity. Exploiting the f-sum rule we found that there are almost no contributions from band-like carriers in agreement with the experimental findings [15,16].

4 Summary

We have investigated the spectral properties of the two-dimensional Holstein t-J model as a generic model for studying polaronic effects in systems with strong Coulomb correlations. The use of purely numerical techniques allows to treat the electron and phonon degrees of freedom on an equal footing. In the numerical work the full dynamics of all the phonon modes was taken into account on finite lattices with up to 16 sites. The efficiency and accuracy of the employed phonon Hilbert space truncation, Lanczos diagonalization, Chebyshev recursion and maximum entropy methods have been demonstrated in detail.

Finally we would like to stress that the numerical techniques presented in this work, in the meantime have been applied sucessfully to various issues, e.g., to the study of self-trapping phenomenon of electrons and excitions [17], to the investigation of polaronic superlattices [18] or to the discussion of the effects of quantum lattice fluctuations on the Peierls [19] and spin-Peierls [20–22] transition.

Acknowledgements

This work was performed under the auspices of Deutsche Forschungsgemeinschaft, SFB 279. Special thanks go to the ZI Berlin, HLRZ Jülich, SSC Karlsruhe, LRZ München and HLR Stuttgart for the generous granting of their parallel computer facilities. Concerning the computational work we have benefited from a fruitful cooperation with A. Basermann, B. Bäuml, H. Röder, R. N. Silver, V. Waas and A. Weiße.

References

1. Salje, E. K. H., Alexandrov, A. S., Liang, W. Y.: *Polarons and Bipolarons in High Temperature Superconductors and Related Materials*, Cambridge Univ. Press, (Cambridge 1995).
2. Chen, C. H., Cheong, S.-W., Cooper, A. S.: Phys. Rev. Lett. **71**, 2461 (1993).
3. Millis, A. J., Littlewood, P. B., Shraiman, B. I.: Phys. Rev. Lett. **74**, 5144 (1995).
4. Zhao, G., Conder, K., Keller, H., Müller, K. A.: Nature **381**, 676 (1996).
5. Cullum, J. K., Willoughby, R. A.: *Lanczos Algorithms for Large Symmetric Eigenvalue Computations*, volume I & II, Birkhäuser, (Boston 1985).
6. Silver, R. N., Röder, H.: cond-mat9703229 (1997).
7. Wellein, G.: *Gitterinstabilitäten in stark korrelierten Elektron-Phonon Systemen*, Thesis, University Bayreuth (1998).
8. Inada, Y., Ishii, C.: J. Phys. Soc. Jpn. **59**, 612 (1990).
9. Wellein, G., Röder, H., Fehske, H.: Phys. Rev. B **53**, 9666 (1996).
10. Silver, R. N., Röder, H.: Int. J. Mod. Phys. C **5**, 935 (1994).
11. Silver, R. N., Röder, H., Voter, A. F., Kress, J. D.: Int. J. Mod. Phys. C **124**, 115 (1996).
12. Bäuml, B., Wellein, G., Fehske, H.: Phys. Rev. B **58**, 3663 (1998).
13. Fehske, H., Loos, J., Wellein, G.: Z. Phys. B **104**, 619 (1997).
14. Fehske, H., Röder, H., Wellein, G., Mistriotis, A.: Phys. Rev. B **51**, 16582 (1995).
15. Bi, X.-X., Eklund, P. C.: Phys. Rev. Lett. **70**, 2625 (1993).
16. Calvani, P. *et al.*: J. Supercond. **10**, 293 (1997).
17. Wellein, G., Fehske, H.: Phys. Rev. B **58**, 6208 (1998).
18. Wellein, G., Fehske, H., Büttner, H., Bishop, A. R.: Physica C **282-287**, 1827 (1997).
19. Weiße, A., Fehske, H.: Phys. Rev. B **58**, 13526 (1998).
20. Büchner, B., Fehske, H., Kampf, A. P., Wellein, G.: Physica B **259-261**, 956 (1999).
21. Weiße, A., Wellein, G., Fehske, H.: Phys. Rev. B submitted (1999).
22. Wellein, G., Fehske, H., Kampf, A. P.: Phys. Rev. Lett **81**, 3956 (1998).

Fig. 1. Construction of a 'parallel' symmetrized basis set:

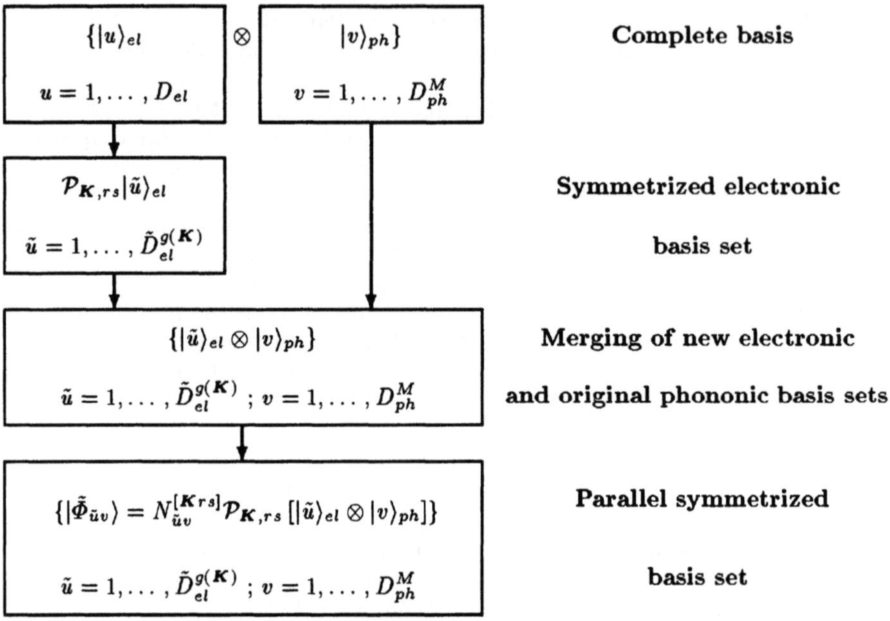

$$
\begin{array}{r}
(\tilde{c}_{\tilde{u}=1,v=1}, \ldots, \tilde{c}_{\tilde{u}=1,v=D_{ph}^M}) \longrightarrow \mathbf{P}_1 \\
(\tilde{c}_{\tilde{u}=2,v=1}, \ldots, \tilde{c}_{\tilde{u}=2,v=D_{ph}^M}) \longrightarrow \mathbf{P}_2 \\
\vdots \\
\vdots \\
(\tilde{c}_{\tilde{u}=\tilde{D}_{el}^{g(\boldsymbol{K})},v=1}, \ldots, \tilde{c}_{\tilde{u}=\tilde{D}_{el}^{g(\boldsymbol{K})},v=D_{ph}^M}) \longrightarrow \mathbf{P}_{\tilde{D}_{el}^{g(\boldsymbol{K})}}
\end{array}
$$

Fig. 2. Parallel distribution of the coefficients of a general state (4) over $N_{pro} = \tilde{D}_{el}^{g(\boldsymbol{K})}$ processors $(\mathbf{P}_1, \ldots, \mathbf{P}_{\tilde{D}_{el}^{g(\boldsymbol{K})}})$.

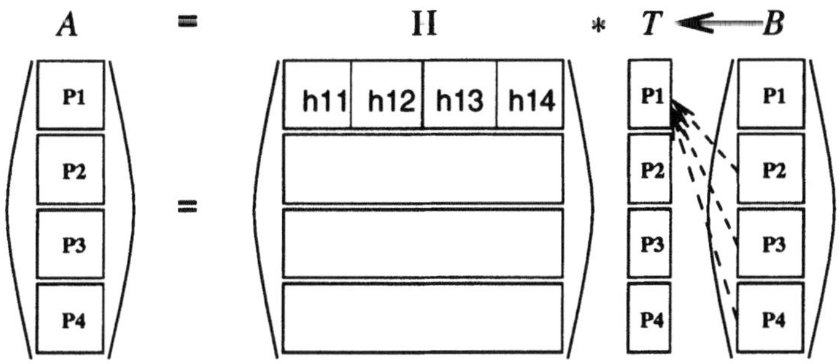

Fig. 3. Matrix vector multiplication with $N_{pro} = \tilde{D}_{el} = 4$ processors $(\mathbf{P}_1, \dots, \mathbf{P}_4)$. In the case of $\mathbf{P}_1$ both the submatrices to be calculated and the corresponding interprocessor communication (arrows) are shown explicitly. $\mathbf{P}_1$ recieves the non local parts of vector $\boldsymbol{B}$ in an temporary array $\mathbf{T}$.

N	N_h	M	D_{el}	$\tilde{D}_{el}^{g(\boldsymbol{K})}$	D_{ph}^M	$\tilde{D}_{tot}^M$	MVM [seconds]	N_{pro}
10	1	8	1260	126	43758	5.5×10^6	4.7	21
10	1	14	1260	126	3268760	4.1×10^8	56	126
10	2	14	3150	52	3268760	1.7×10^8	233	52
16	1	7	102960	6435	245157	1.6×10^9	59	512

Table 1. Time per matrix vector multiplication in seconds, using the CRAY specific version of PARPHON on the CRAY T3E of the HLRS. Results are shown for different matrix-dimensions $(\tilde{D}_{tot}^M)$ and model parameters (N, N_h, M).

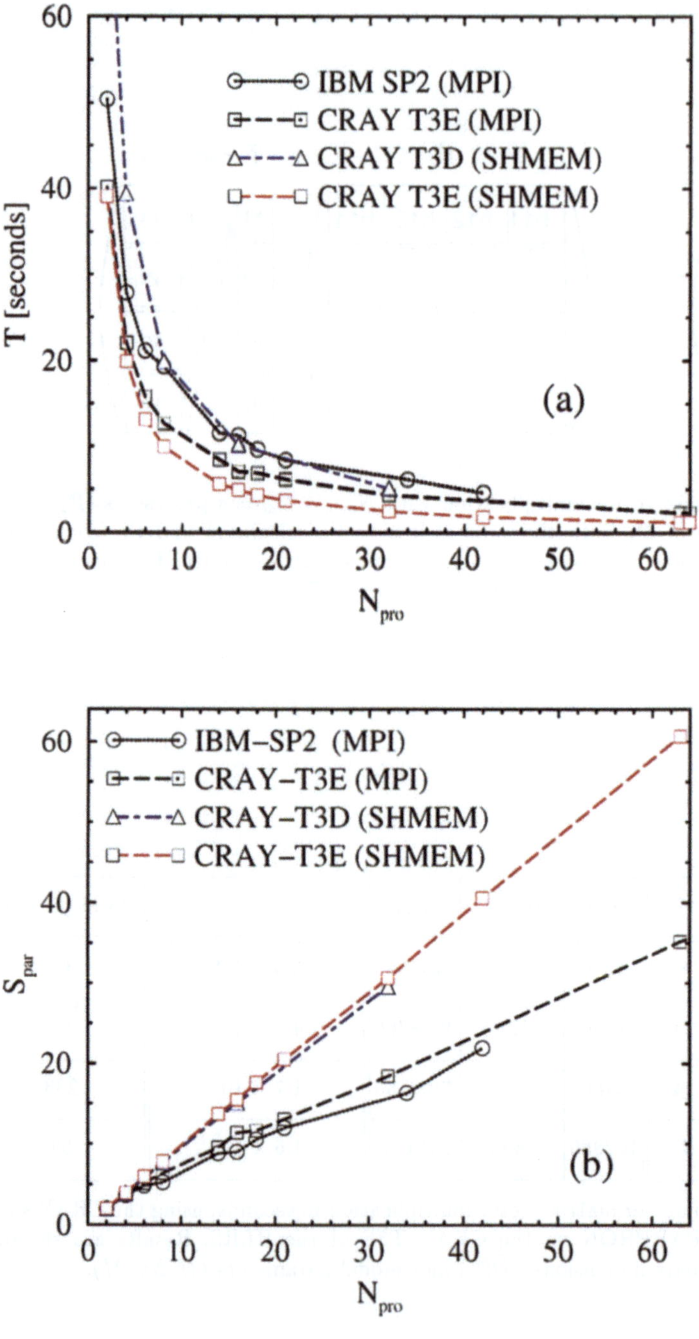

Fig. 4. Time per MVM (a) and parallel speed up S_{par} (b) as a function of the number of processors on CRAY T3E/T3D and IBM SP supercomputers. Results are presented for one set of model parameters ($\tilde{D}^{M}_{tot} = 5.5 \times 10^6$; cf. table 1) and both MPI and CRAY specific shmem implementation. (Single processor peak performance: IBM SP2: 267 MFLOPS; CRAY T3D: 150 MFLOPS; CRAY T3E: 600 MFLOPS)

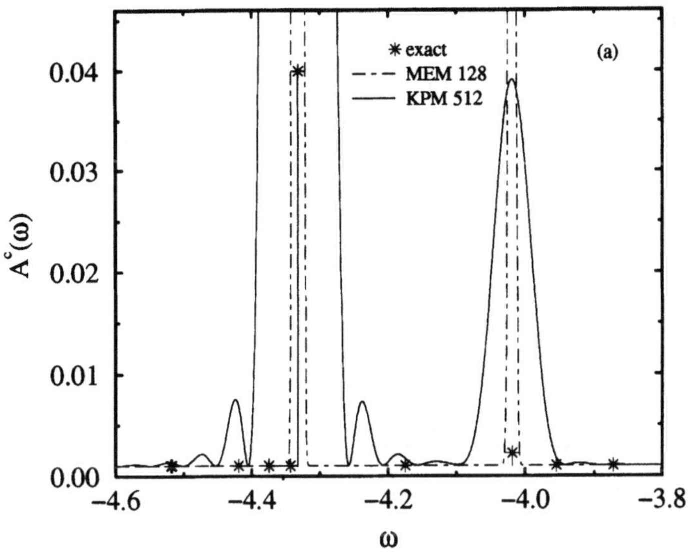

Fig. 5. Low energy region of the excitation spectrum of a single hole in the 2D HtJM. KPM and MEM reconstructed spectral functions have been calculated using 512 and 128 moments, respectively. For comparison the exact peak positions and spectral weights are shown (stars).

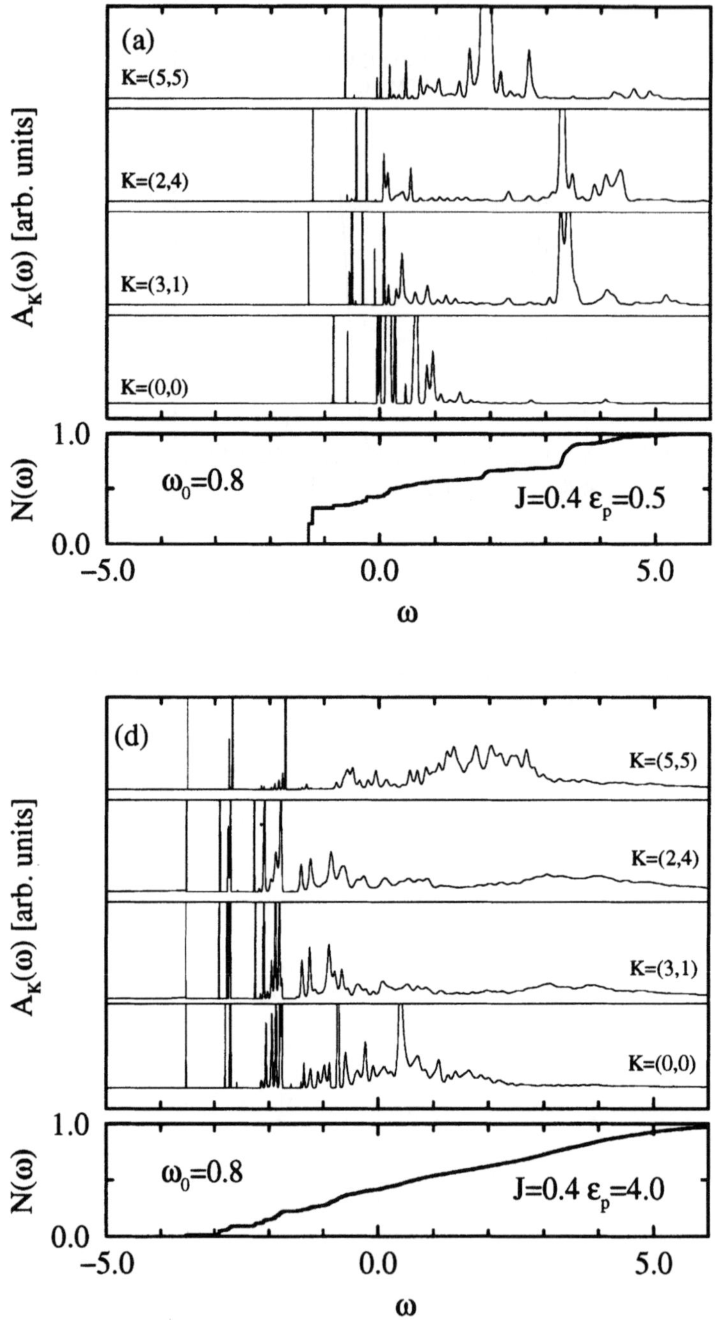

Fig. 6. Single–hole spectral function $A_{\boldsymbol{K}}(\omega)$ and integrated spectral weight $N(\omega)$ for the 2D HtJM at $\omega_0 = 0.8$ for different values of ε_p and J. Note that the lowest peak in (b) contains contributions from each $\boldsymbol{K}$-sector.

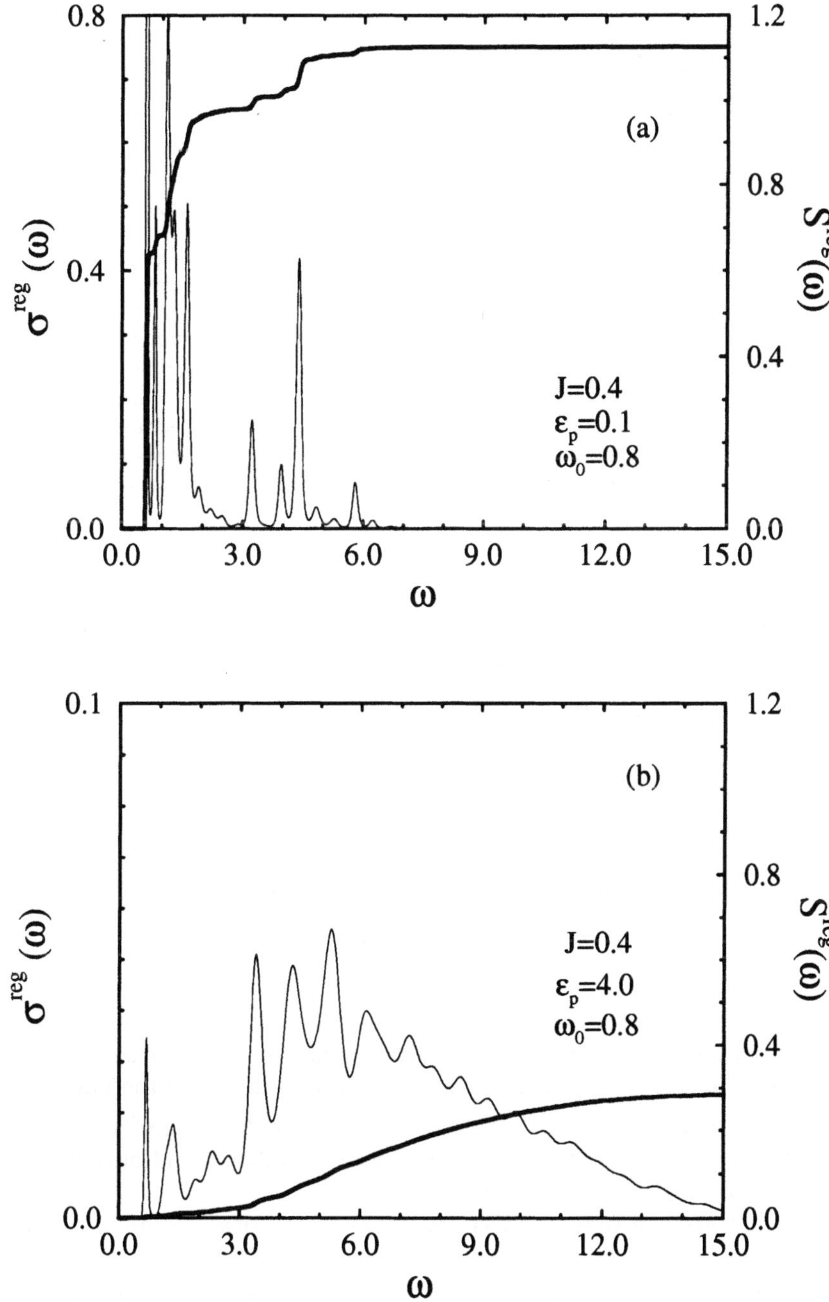

Fig. 7. Optical conductivity in the 2D (Holstein) t-J model for $J = 0.4$. $\sigma^{reg}(\omega)$ and $S^{reg}(\omega)$ are obtained for the ten-site lattice with 15 phonons. The single-hole ground state has momentum $\boldsymbol{K} = (3\pi/5, \pi/5)$.

The Metal-Insulator Transition in the Hubbard Model

M. G. Zacher, A. Dorneich, C. Gröber, R. Eder, and W. Hanke

Institut für Theoretische Physik, Universität Würzburg, D-97074 Würzburg

Abstract. The Mott–Hubbard metal–insulator transition is studied in the two–dimensional Hubbard Model with and without next–nearest neighbor hopping at half–filling by a combination of Quantum–Monte–Carlo and exact diagonalization techniques. In the case without next–nearest neighbor hopping, antiferromagnetic correlations are suppressed by large fluctuations due to a relatively high temperature. The single particle spectral function and the spin– and charge–excitations of the metallic state below the critical Hubbard interaction U_c, where the transtion occurs, are similar to the tight–binding result for $U = 0$. Above U_c, a non–isotropic gap in the spectral function opens, and flat sidebands appear in the spectrum. The gap formation is accompanied by pronounced spin– and charge–modes. In the case including next–nearest neighbor hopping, both finite–temperature Quantum–Monte–Carlo and exact diagonalization ($T = 0$) results yield a metal–insulator transition at about the same critical interaction U_c. However, whereas the finite–temperature transition is again of the Mott–Hubbard type, the $T = 0$ transition is of Mott-Heisenberg type induced by long–range antiferromagnetic correlations.

1 The Mott-Hubbard Metal–Insulator Transition

In this article we study the half-filled Hubbard model

$$H = U \sum_i n_{i,\uparrow} n_{i,\downarrow} - t \sum_{\langle i,j \rangle, \sigma} (c^\dagger_{i,\sigma} c_{j,\sigma} + \text{h.c.}) \ , \tag{1}$$

where U and t are positive interaction constants, i and j lattice site indexes and $c^\dagger_{i,\sigma}$, $c_{i,\sigma}$ and $n_{i,\sigma}$ quantum mechanical operators: $c^\dagger_{i,\sigma}$ creates an electron with spin σ on site i, $c_{i,\sigma}$ annihilates one, and $n_{i,\sigma}$ measures the number of spin-σ electrons on site i. (Thus $c^\dagger_{i,\sigma} c_{j,\sigma}$ models the hopping of an electron from site j to site i, and $n_{i,\uparrow} n_{i,\downarrow}$ has a nonzero value if and only if site i harbors both an $\uparrow$ and a $\downarrow$ electron).

In the half-filled (one electron per site) case we observe the interesting competition between kinetic energy and Coulomb repulsion.

In the limit of a strong on-site Coulomb repulsion ($U \gg t$) the hopping process of a single electron costs the system the large energy U. Even with further hopping processes the electron has no means to regain this initally paid energy (this is indicated in Fig. 1). Therefore the electrons stay localized and the system is in an insulating state. This is an example of a Mott-Hubbard insulator because only the pure electron-electron interactions are responsible for the insulating behavior.

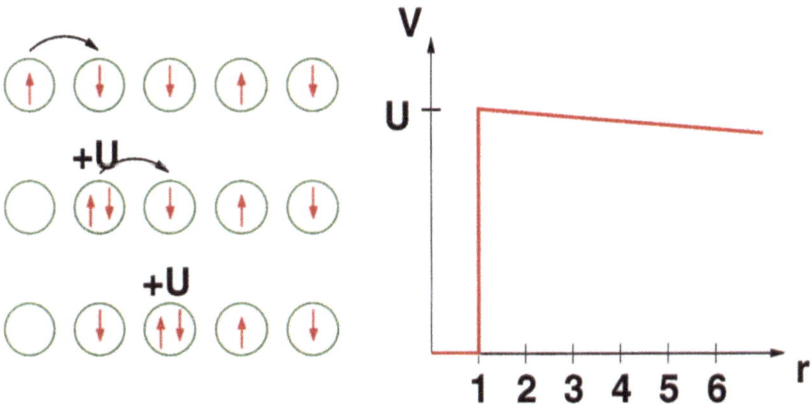

Fig. 1. The effective potential for a single electron in the halffilled Hubbard model in the limit of strong Coulomb interaction ($U \gg t$). Even with a strong electric field (indicated by the with distance decreasing potential) the electrons have no means to regain the Coulomb repulsion they had to overcome in the first hopping process. Therefore the system is insulating.

In the other limit, with $U = 0$ the Hubbard model reduces to the simple tight binding model

$$H = -t \sum_{\langle \boldsymbol{i},\boldsymbol{j} \rangle,\sigma} c^{\dagger}_{\boldsymbol{i},\sigma} c_{\boldsymbol{j},\sigma} + \text{h.c.} = \sum_{\boldsymbol{k},\sigma} \varepsilon(\boldsymbol{k}) c^{\dagger}_{\boldsymbol{k},\sigma} c_{\boldsymbol{k},\sigma} \tag{2}$$

with the free band structure $\varepsilon(\boldsymbol{k})$ which is plotted in Fig. 2. At half-filling all the (single-particle) states below $\omega = 0$ are occupied but gapless charge excitations are possible. The system is metallic.

The interesting question is now whether there is a critical interaction strength U_c where the system changes its state from metallic to insulating and a metal-insulator transition (MIT) occurs.

However, with dimensions greater than one and on square lattices, the MIT is masked by long range antiferromagnetic order: The electron spins align antiferromagnetically, which effectively doubles the unit cell of the lattice and therefore halves the first Billouin zone in momentum space (see Fig. 3). The free band from Fig. 2 becomes reflected at the new zone boundary (the bands originally outside the magnetic Brillouin zone are simply folded inside) and at the magnetic zone boundary the Bragg reflection trivially opens a gap which drives the system to an insulator [1] (see Fig. 4). We can call this kind of insulator caused by long range order a Mott-Heisenberg insulator.

Many calculations [2] show that the Mott-Heisenberg insulator is the true groundstate of the Hubbard model on a two–dimensional square lattice already for infinitesimally small $U > 0$. In three dimensions we expect (although no numerical studies have been performed so far) the long-range or-

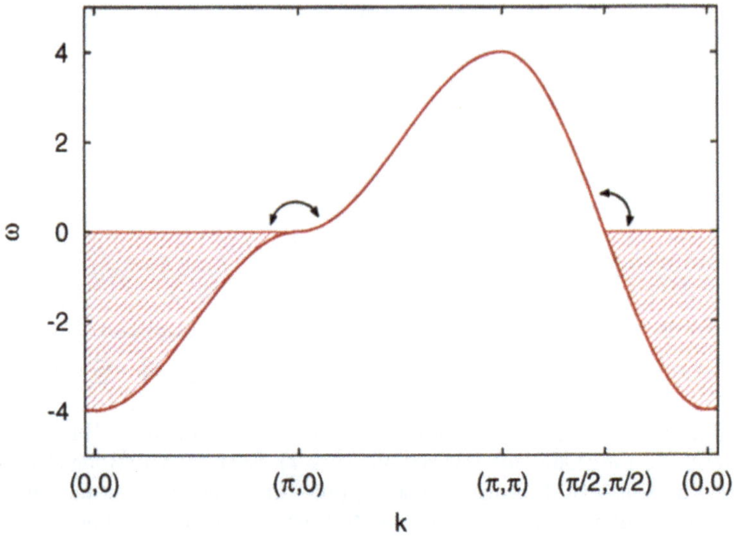

Fig. 2. With $U = 0$ the Hubbard model reduces to the non-interacting tight-binding model, whose band structure is shown here for a two-dimensional square lattice. At half-filling all the states below $\omega = 0$ are occupied, but gapless charge excitations are possible (indicated by the double-arrows). The system is an ideal metal.

Fig. 3. The long range antiferromagnetic order effectively doubles the unit cell of the lattice in real space (left) and therefore halves the Brillouin zone in momentum space (right).

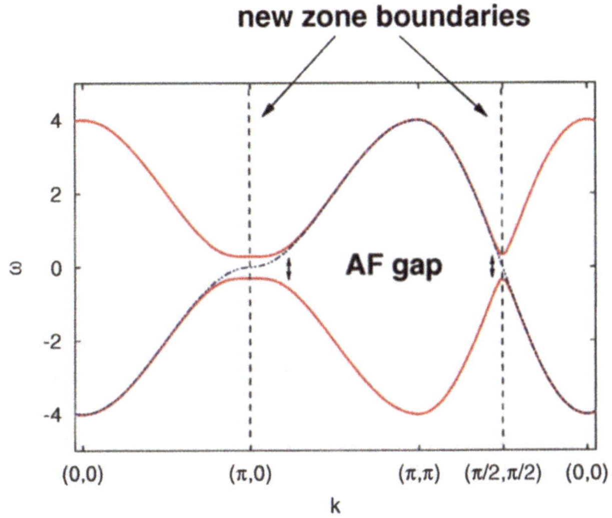

Fig. 4. The energy bands outside the magnetic Brillouin zone are folded inside. The reflected bands at the zone boundary ($(\pi, 0)$ and $(\pi/2, \pi/2)$ here) trivially yield an energy gap. At half-filling all the states below $\omega = 0$ are completely filled and no gapless charge excitations are possible. The system is insulating (Mott-Heisenberg insulator).

der and the insulating behavior already at finite temperature below the Néel temperature T_N also for infinitesimally small $U > 0$ [3,4].

Studying the Mott-Hubbard metal-insulator transition now requires extra care for the antiferromagnetic correlations (and the resulting trivial gap) not to interfere. Unfortunately this is also true for finite temperatures in two dimensions where one is (due to the Mermin-Wagner Theorem [3]) safe from long range order and would expect only short range antiferromagnetic behavior, no unit cell doubling and therefore no gap. However, while studying finite clusters (for example with the QMC algorithm), the finite system only "sees" the antiferromagnet as soon as the spin-spin correlation length is of the order of the lattice size.

2 MIT in Infinite Dimensions

Although it seems like an unrealistic approach to study the Hubbard model in infinite dimensions, calculations show that for properties like the density of states (DOS) or the quasiparticle weight Z the infinite–dimensional Hubbard model is very close to the three–dimensional case and similar to the two–dimensional case [5]. The great simplification of this approach is that n-electron irreducible n-vertices become diagonal in real space [6–8] which (for translationally invariant systems) yields a purely local self energy. However, the self energy

$$\begin{aligned} \Sigma(\boldsymbol{k},\omega) &\stackrel{d\to\infty}{\longrightarrow} \Sigma(\omega) \\ \Sigma(\boldsymbol{r}_1,\boldsymbol{r}_2;\omega) &\stackrel{d\to\infty}{\longrightarrow} \Sigma(\boldsymbol{r}_1,\boldsymbol{r}_2;\omega)\delta_{\boldsymbol{r}_1,\boldsymbol{r}_2} \end{aligned} \tag{3}$$

keeps its nontrivial ω-dependence and therefore its full dynamics.

Since, according to (3), the self energy is diagonal in real space, one can in principle construct an effective single-site model that is equivalent to the infinite–dimensional Hubbard model. In analogy to the mean field solution of spin systems where the single spin couples to a *static* mean field describing the effect of the neighboring spins (Curie-Weiß field) one calls the infinite–dimensional formulation of the Hubbard model "dynamical mean field theory" (DMFT) because the problem reduces to electrons on a single site coupling to a *dynamical* mean field described by the ω-dependence of the self energy. Further analogy comes from the fact that also the mean-field description of spin systems becomes exact in infinite dimensions.

The effective single-site formulation of the infinite–dimensional Hubbard model is the so called single-impurity Anderson model with

$$\begin{aligned} H_{\text{And}} = &\sum_\sigma \int \mathrm{d}\varepsilon \varepsilon \psi^\dagger_{\varepsilon,\sigma}\psi_{\varepsilon,\sigma} + E_f \sum_\sigma f^\dagger_\sigma f_\sigma + U f^\dagger_\uparrow f_\uparrow f^\dagger_\downarrow f_\downarrow \\ &+\sum_\sigma \left[\int \mathrm{d}\varepsilon \sqrt{F_{\sigma,0}(\varepsilon)}V(\varepsilon)\psi^\dagger_{\varepsilon.\sigma} f_\sigma + \text{h.c.}\right] \end{aligned} \tag{4}$$

where the impurity electrons f_σ couple to a bath of conduction electrons $\psi_{\varepsilon,\sigma}$ via the hybridization energy $\sqrt{F_{\sigma,0}(\varepsilon)}V(\varepsilon)$. In order to correctly describe the infinite–dimensional Hubbard model, the hybridization energy has to be properly adjusted. This can be performed by a set of self-consistency equations [3,9,10].

However, even the single site impurity model is too complicated to be solved exactly. One has to rely on approximative or numerical treatments. Unfortunately, the different approximation schemes respectively numerical approaches yield two radically different scenarios for the Mott-Hubbard metal-insulator transition [1]:

The so called "iterated perturbation theory" [11] uses the second order in U/t perturbation solution of the single impurity problem in the self consistency cycle. It has been shown that the perturbation treatment works indeed surprisingly well for the single impurity model itself [12,13], but using the approximated solution in the self consitency loop is not well controlled. When studying the metal-insulator transition the authors start the self consistency

[1] In the $d = \infty$ treatment one can choose whether the strong-coupling groundstate is of the Mott-Heisenberg or Mott-Hubbard type by explicitely including/excluding the perfect nesting condition [3].

loop with either the metallic or the insulating Green's function as the "trial" Green's function and then see if it yields a self-consistent solution.

The interesting result is that there is a finite interval in $U \in [U_{c1}, U_{c2}]$ where both the insulator and the metal have a solution; in this case an energy comparison gives the true groundstate and they find that in the whole interval the metal "wins". This has the consequence that the metal-insulator transition becomes discontinuous because the insulating gap Δ evolves continuously from $\Delta = 0$ at $U = U_{c1}$ and already has a finite value at U_{c2}. Since the true groundstate for $U < U_{c2}$ is the metal the system jumps to an insulating state with a preformed finite gap at U_{c2} (see Fig. 5).

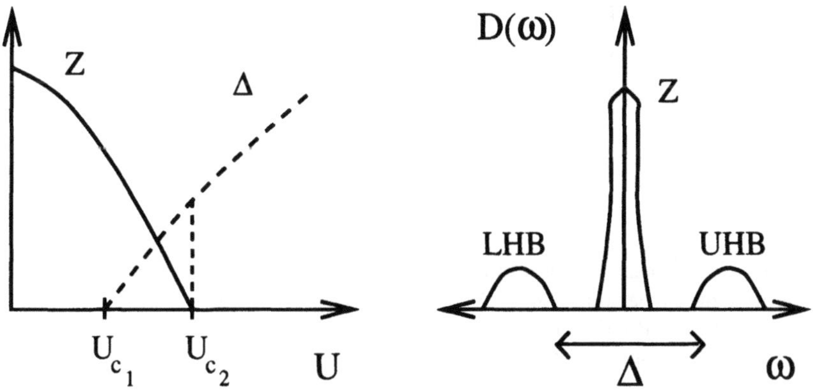

Fig. 5. Iterated perturbation theory results for the infinite–dimensional Hubbard model: The metallic state has a solution for all $U < U_{c2}$ and the insulator for all $U > U_{c1}$. Both the quasiparticle weight Z (characterizing the Fermi-liquid) and the single-particle gap Δ (characterizing the insulator) go to zero continuously in their their respective region. Since the metallic state always has a lower energy, the system jumps to an insulator with a preformed gap $\Delta(U_{c2})$ at $U = U_{c2}$. The evolution of the density of states would look like the right part of the figure: A preformed gap separates the precursors of the lower and upper Hubbard bands; the height and the width of the quasiparticle peak at $\omega = 0$ go to zero continuously for $U < U_{c2}$.

Including temperature in the calculation yields the phase diagram in figure (6). For $T = 0$ the metal-insulator transition occurs at $U = U_{c2}$ and for $T > 0$ a first order phase transition separates the insulating from the metallic region below a critical temperature T_c. Above T_c a crossover behavior is expected. The numerical result in [11] is

$$\begin{aligned} U_{c1} &\approx 1.3W \\ U_{c2} &\approx 1.7W \end{aligned} \tag{5}$$

with W being the bandwidth of the free model ($W = 4dt$ for a d–dimensional "cubic" lattice).

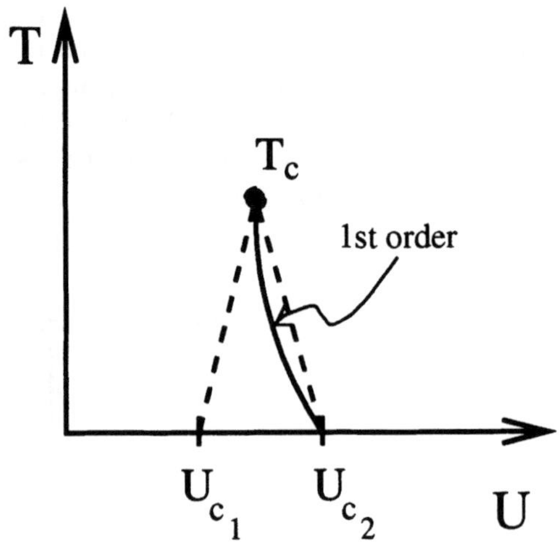

Fig. 6. Phase diagram of the infinite–dimensional Hubbard model as obtained by the iterated perturbation theory: At $T = 0$ the metal-insulator transition occurs at $U = U_{c2}$. For finite $T < T_c$ a first order phase transition line is separating both states. Above T_c one expects a crossover behavior.

However, as interesting as the above result is, most other approaches [3,14,15] to this problem give a radically different and much simpler scenario: Here the quasiparticle weight Z vanishes at the same critical interaction U_c where the insulating gap opens (see Fig. 7).

For finite temperature there is no sharp phase transition in a thermodynamic sense but a simple crossover behavior as illustrated in Fig. 8. Also the density of states looks less pathological: In the metallic region the quasiparticle peak is connected to the precursors of the lower and upper Hubbard band with its weight continuously vanishing when the system is approaching the critical interaction strength U_c. At U_c a gap is opening also continuously starting at $\omega = 0$ (right part of Fig. 7).

The finite temperature results were obtained by a Quantum-Monte Carlo method [15] and the zero temperature results by the random dispersion approximation (RDA) where exact diagonalization in momentum space is used combined with a random dispersion $\varepsilon(\boldsymbol{k})$ [14]. However, there is some disagreement about the value of the critical interaction strength U_c. By extrapolating to $T = 0$ the QMC method finds $U_c \approx 1.47W$ whereas Noack and Gebhard's RDA method finds $U_c \approx W$ (with W again being the bandwidth of the interaction-free model).

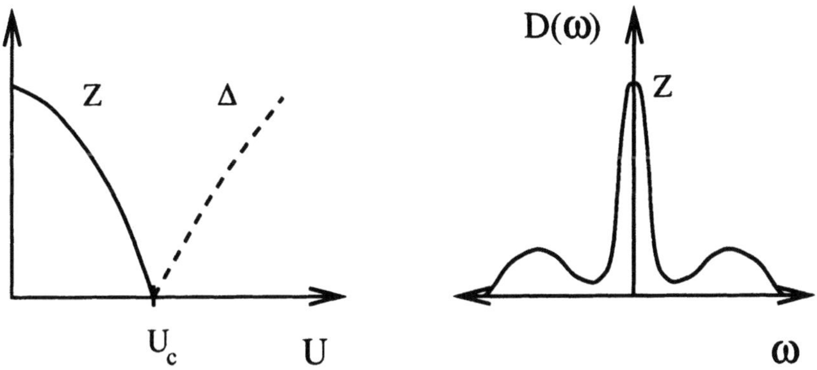

Fig. 7. Continuous scenario for the metal insulator transition of the infinite–dimensional Hubbard model: At the critical interaction strength $U = U_c$ both the quasiparticle weight Z and the insulating gap Δ vanish. The density of states in the metallic region is connected.

3 MIT in Two Dimensions

The still ongoing dispute about the nature of the Mott-Hubbard metal-insulator transition of the infinite–dimensional Hubbard model described in the previous section motivated us to study this phenomenon with our means (i.e. Quantum Monte Carlo and exact diagonalization) in two dimensions [16,17].

3.1 Nearest Neighbor Hopping

Since the Quantum Monte Carlo method is limited to finite clusters we face the problem described in section 1: Although long range order and with it antiferromagnetism is forbidden for two dimensions and finite temperature by the Mermin-Wagner theorem, our finite system already "sees" the antiferromagnet for low temperatures as soon as the spin-spin correlation length becomes larger than the lattice extension in any direction. Therefore great care is needed to not confuse a Mott-Hubbard insulator with the precursors of the Mott-Heisenberg antiferromagnetic groundstate. Fortunately there are some diagnostic tools to exclude such a possibility: First of all Fig. 9 shows the staggered real space spin-spin correlation function

$$\chi_{sz}(\boldsymbol{r}) = \frac{1}{N} \sum_{i} \langle \mathbf{S}_i \mathbf{S}_{i+\boldsymbol{r}} \rangle \, e^{i\boldsymbol{Q}\boldsymbol{r}} \tag{6}$$

as a function of the distance $|\boldsymbol{r}|$ for inverse temperature $\beta t = 3$ on an 8×8 lattice. The correlation is vanishing by two orders of magnitude over the lattice extension and we estimate the spin-spin correlation length to be about

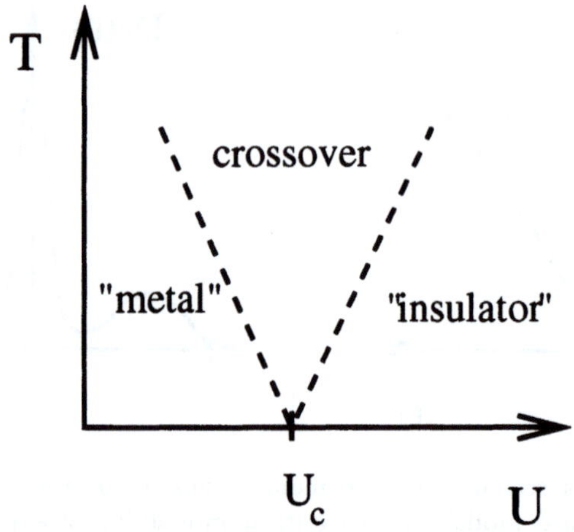

Fig. 8. Phase diagram of the infinite–dimensional Hubbard model as obtained by for example the Quantum-Monte Carlo method supporting a continuous scenario: At $T = 0$ the metal-insulator transition occurs at $U = U_c$. For finite temperatures no phase transition occurs but a simple crossover behavior.

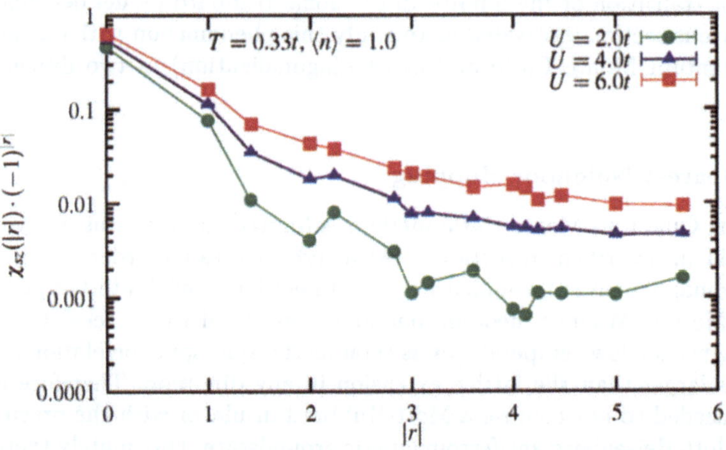

Fig. 9. Staggered spin-spin correlation as a function of distance $|\mathbf{r}|$ plotted on a semi-logarithmic scale. QMC results on an 8×8 lattice with inverse temperature $\beta t = 3$ at half-filling using several values of interaction strength U. We observe that the correlation is vanishing by two orders of magnitude over the lattice extension and the correlation length is estimated to be ≈ 1.5 lattice spacings.

≈ 1.5 lattice units. Therefore we have faith in the antiferromagnetic correlations not interfering with our study. Further evidence for the exclusion of antiferromagnetism will be provided below.

To find the metal-insulator transition point we study the single particle spectral function $A(\boldsymbol{k}, \omega)$ where gaps result in insulating behavior. For low interaction strength U there is hardly a deviation from the metallic tight-binding model (Fig. 10). Only some broadening of the peaks indicates additional scattering channels due to the interaction U.

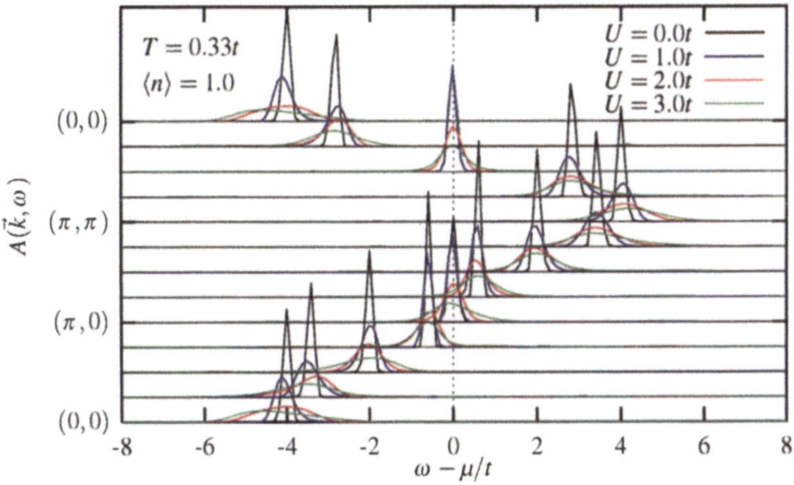

Fig. 10. Single particle spectral function $A(\boldsymbol{k}, \omega)$ for small interaction strength $U/t = 0, 1, 2, 3$. QMC results for an 8×8 system with $\beta t = 3$. There is hardly a deviation from the noninteracting model; only some peak-broadening due to additional scattering is observed.

We proceed to higher interactions (Fig. 11) where some pronounced changes are appearing in the spectrum: For $U = 4t$ the peak on the fermi-surface at $k = (\pi, 0)$ shows a tendency to split; for $U = 5t$ the peak at $k = (\pi/2, \pi/2)$ is split and the one at $k = (\pi, 0)$ is split and separated and for $U = 6t$ the spectral function is clearly gapped over the whole Brillouin zone. Furthermore we observe the formation of additional sidebands at about $\omega \approx 4t$ and $\omega \approx t$ which become more pronounced for larger interactions. Clearly we see a metal-insulator transition at about $U_c \approx 4t$.

As already stressed above, it is very important to ensure that the insulating state is of the Mott-Hubbard type and not a Mott-Heisenberg state with dominating antiferromagnetic correlations driving our system to a "trivial" insulator (see the discussion in section 1). In our opinion the spin-spin correlation length of about ≈ 1.5 lattice sites is indication enough that we

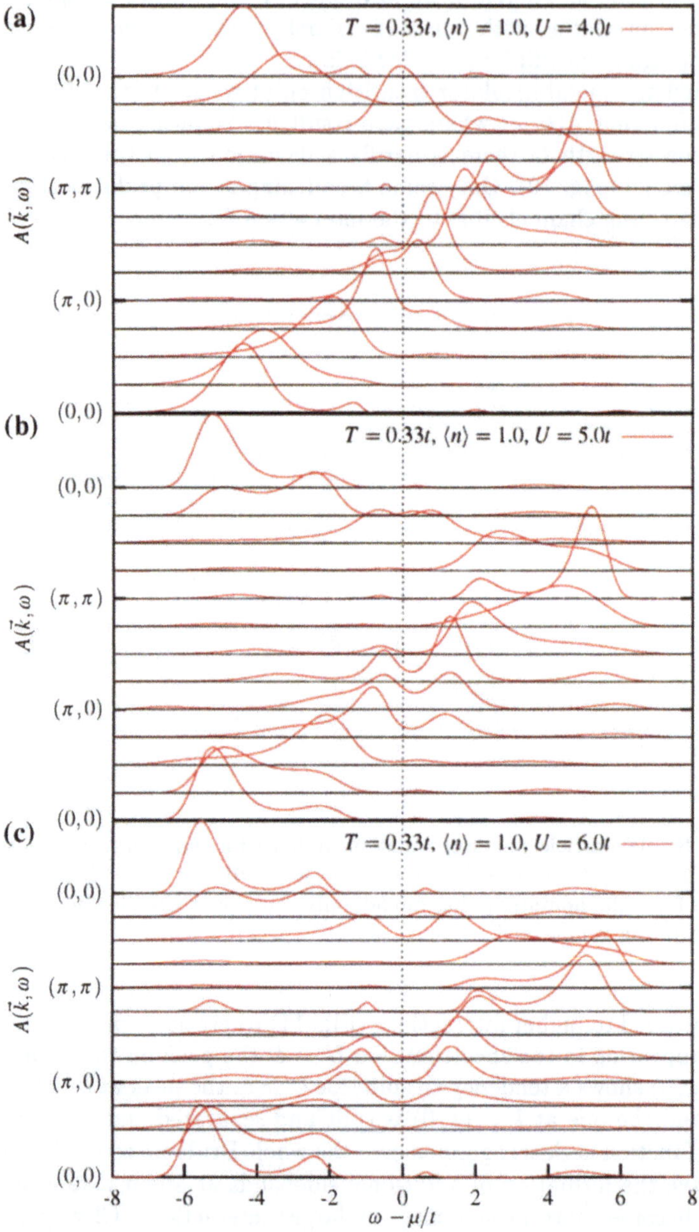

Fig. 11. Single particle spectral function $A(\boldsymbol{k},\omega)$ for interaction strength $U = 4t(a), 5t(b), 6t(c)$. QMC results for an 8×8 system with $\beta t = 3$.

do not observe the "trivial" case but there is further evidence: Since antiferromagnetic long range order is prohibited for the infinite system a Mott-Heisenberg state on a finite cluster obviously has to suffer from large finite size effects. Figure 12 shows the spectral function at the important k-points $k = (\pi/2, \pi/2)$ and $k = (\pi, 0)$ (on the fermi surface of the noninteracting model) for $U = 5t$ for different lattice sizes. The splitting is well pronounced for all lattice sizes and the peak positions are identical and not suffering from finite size effects. This excludes the possibility that we are seeing the precursors of the Mott-Heisenberg antiferromagnetic groundstate[2].

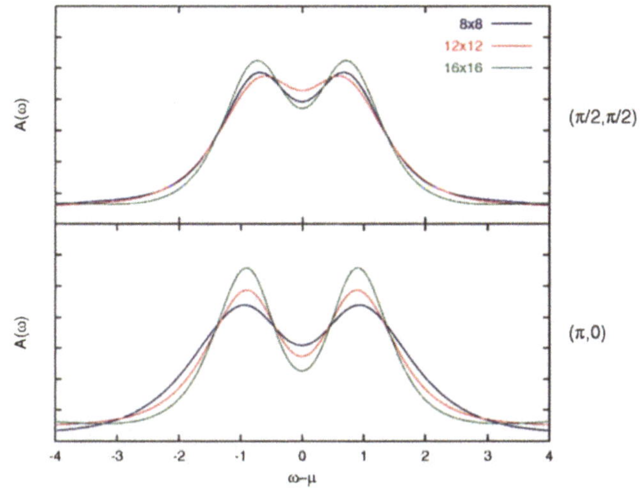

Fig. 12. Finite size behavior of the spectral function at the k-points $k = (\pi/2, \pi/2)$ and $k = (\pi, 0)$: QMC results are shown for $U = 5t$, $\beta t = 3$ and 8×8, 12×12 and 16×16 lattices. Particle-hole symmetry implies that the spectrum at these points is symmetrical which has been used as additional input for the Maximum Entropy technique.

Additional proof for the absence of a Mott-Heisenberg insulator is provided by the absence of any reflected bands in Fig. 13 for the even stronger interaction $U = 8t$.

As discussed above, antiferromagnetism yields a doubling of the unit cell and a halving of the Brillouin zone. This would result in reflected bands in

[2] There is a noticable difference in lineshape. This most certainly stems from the statistical errors of the Monte-Carlo results and the resulting limited resolution of the Maximum-Entropy method. Generally, Maximum-Entropy is not able to answer lineshape questions unless one uses more sophisticated schemes like for example comparing the results of several independent Monte-Carlo runs each with negligible errors.

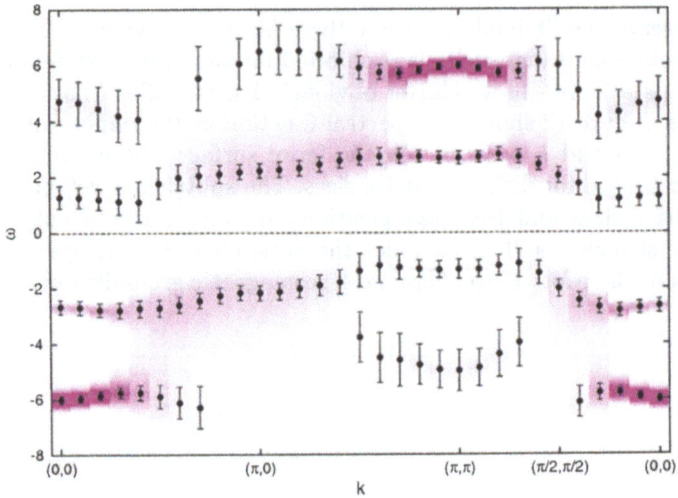

Fig. 13. Density plot of the single particle spectral function for a 20×20 lattice with $\beta t = 3$ and $U = 8t$ as obtained by QMC. The absence of reflected bands in the vicinity of $k = (\pi, 0)$ and $k = (\pi/2, \pi/2)$ (compare with Fig. 4) shows that the system is not in an antiferromagnetic state but in a Mott-Hubbard insulating state. One nicely observes the dispersive four-band structure also.

the vicinity of the magnetic zone boundary at for example $k = (\pi, 0)$ and $k = (\pi/2, \pi/2)$ (see Fig. 4). There is absolutely no hint of this behavior visible in Fig. 13.

We proceed to a discussion of the two-particle excitations: Fig. 14 shows the dynamical spin- and charge-response functions for different interaction strenght U. For interaction strengths well below the transition ($U = 2t$) the spin- and charge-response functions do not exhibit any spectacular features. At about the transition point ($U = 4t$) we see the formation of coherent, dispersive modes in both channels. The spin mode shows some similarity to a spin wave safe for the weight distribution (in a spin wave virtually all the weight is situated at $k = (\pi, \pi)$ [18,19]). For an interaction above the transition ($U = 6t$) the two modes become more coherent and the insulating gap is clearly visible in the charge response.

3.2 Second Nearest Neighbor Hopping

If a diagonal second nearest neighbor hopping term t' is included in the two–dimensional Hubbard model, the long range ordered antiferromagnetic state can be suppressed even for zero temperature. In this case exact diagonalization studies are of use to study the metal-insulator transition. Unfortunately, the Quantum Monte Carlo method cannot be used for low temperatures in

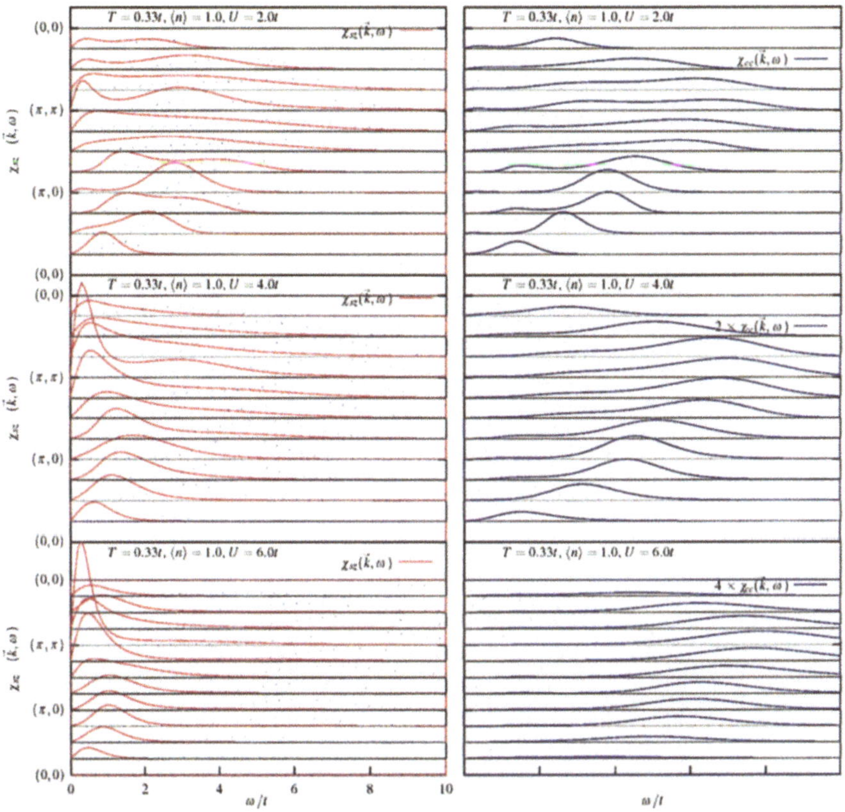

Fig. 14. Dynamical response functions for $U/t = 2, 4, 6$ (from top to bottom), an 8×8 lattice and $\beta t = 3$ as obtained by QMC. Left: spin response, right: charge response.

this case since the second nearest neighbor hopping destroys the particle-hole symmetry and consequently the calculations suffer from the *minus-sign* problem.

Nevertheless we start with the Quantum Monte Carlo treatment of the problem using $t' = 0.5\,t$ and a relatively high temperature $\beta t = 3$ where the *minus-sign* problem is not so severe and increase the interaction strength U analog to the treatment described in the previous section. Figure 15 shows the results for an 8×8 lattice: Similar to the case described above where hopping was restricted to nearest neighbors, the dispersion of the single particle spectral function for small interactions ($U = 3\,t$) closely follows the noninteracting metallic dispersion. For large interactions a gap is opening with the critical interaction being somewhere between $U = 6\,t$ and $U = 7\,t$ here.

As expected, a Mott-Heisenberg state can be excluded for the t' case as well: Fig. 16 shows the staggered spin-spin correlation function ($U = 8\,t$) for

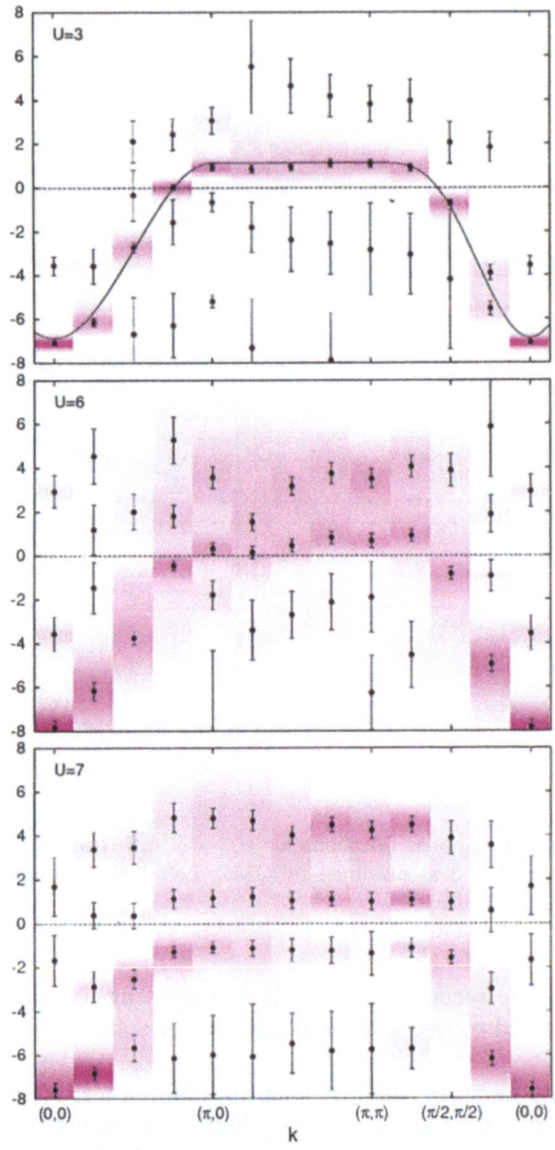

Fig. 15. Density plot of the single particle spectral function with nearest neighbor hopping t' for an 8×8 lattice with $\beta t = 3$, $U/t = 3, 6, 7$ (from top to bottom) and $t' = 0.5\,t$ as obtained by QMC. For the small interaction $U = 3\,t$ the dispersion follows the free dispersion (black line). Between $U = 7\,t$ and $U = 8\,t$ a metal-insulator transition is happening. The insulating state is of the Mott-Hubbard type due to the absence of reflected structures.

Matsubara frequency $i\omega = 0$

$$\langle S_i S_{i+r} \rangle (i\omega = 0) \sim \frac{1}{N} \sum_{r'} \int_0^\beta \mathrm{d}\tau \langle S^z_{r'+r}(\tau) S^z_{r'}(\tau = 0) \rangle \tag{7}$$

which in contrast to the equal time correlation function also includes dynamical correlations between the spins. This correlation function fits an exponential with correlation length $\lambda \approx 0.6$ which is well below the lattice extension.

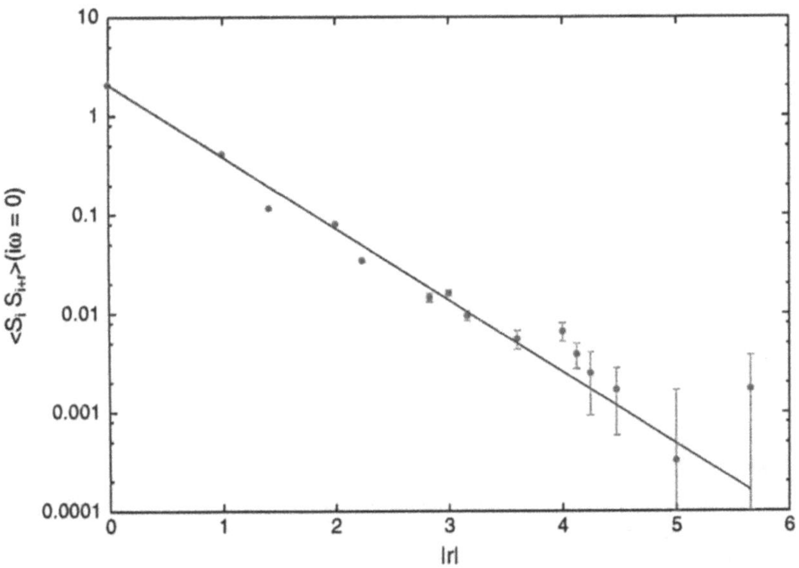

Fig. 16. Semilogarithmic plot of the staggered spin-spin correlation function for Matsubara frequency $i\omega = 0$ (defined in (7)) as obtained by QMC for an 8×8 lattice, $U = 8\,t$, $t' = 0.5\,t$. The fit to an exponential yields a correlation length of $\lambda \approx 0.6$ lattice units which is well below the lattice extension.

Secondly we note the absence of any reflected structures in the single particle spectral function in Fig. 15.

We proceed to the exact diagonalization study at $T = 0$ (Fig. 17): We note the metal-insulator transition happening at about the same critical interaction U_c between $U = 6\,t$ and $U = 7\,t$ as in the finite temperature case. Figure 17 shows that for $U = 6\,t$ at $k = (\pi/2, \pi/2)$ both inverse photoemission (IPES) and photoemission (PES) have structures very close to the fermi-energy whereas for $U = 7\,t$ IPES and PES are separated by a large gap in the whole Brillouin zone. However, in this case the insulator is of the "trivial" Mott-Heisenberg type, because reflected structures can be identified. Also an inspection of the spin-spin correlation function supports the Heisenberg insulator.

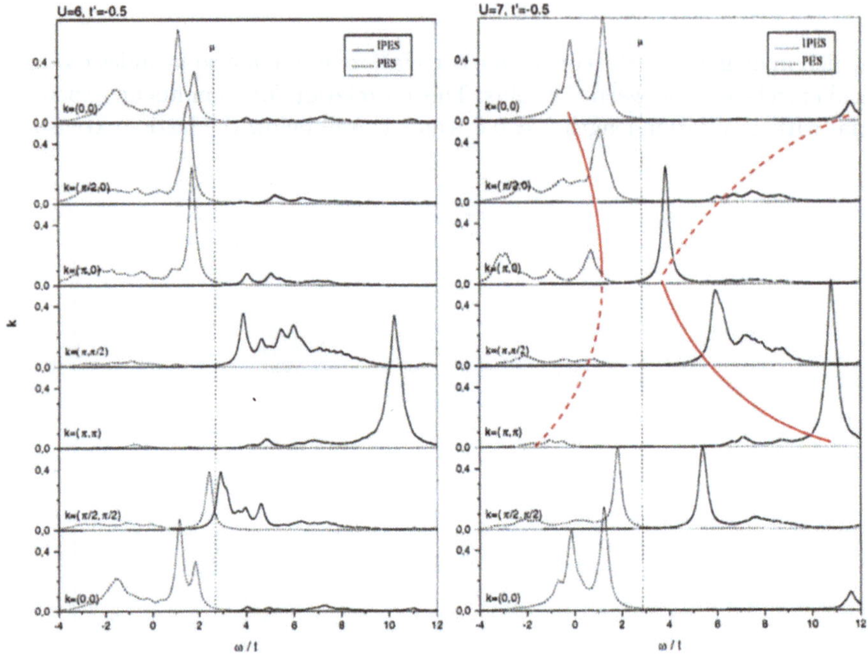

Fig. 17. $T = 0$ photoemission and inverse photoemission for a 4×4 cluster, $t' = 0.5\,t$, $U = 6\,t$ (left) and $U = 7\,t$ (right) as obtained by exact diagonalization. The system becomes insulating somewhere between $U = 6\,t$ and $U = 7\,t$. The insulator at $U = 7\,t$ is of the Mott-Heisenberg type because reflected structures (thick dashed lines in the right figure) being the result of the effective magnetic Brillouin zone can be identified (compare Fig. 4).

4 Summary

In summary we have found a metal-insulator transition of the Mott-Hubbard type in the two–dimensional Hubbard model with nearest neighbor hopping at half-filling for the relatively high temperature $T = 1/3t$ (neccessary for the suppression of long range antiferromagnetism). The critical interaction is $U_c \approx 4t$ which is about half the bandwidth of the noninteracting model. The transition towards the insulator is accompanied by pronounced spin- and charge-modes and the formation of four dispersive bands in the single particle spectral function.

Including a next nearest neighbor hopping term allowed us to study the problem also with exact diagonalization (ED) at zero temperature. Both the Quantum Monte Carlo results for the temperature $T = 1/3t$ and the ED results for $T = 0$ gave a metal-insulator transition with a critical interaction U_c between $U = 6t$ and $U = 7t$. However, whereas the finite temperature insulator is of the Mott-Hubbard type the zero temperature insulator is of the Heisenberg type. It is still unclear whether the same value of the critical interaction for both types of transitions is only a curiosity or systematic. Studies tracking this question are undergoing.

In comparison with the the infinite–dimensional Hubbard model our overall results support the second scenario of a continuous Mott-Hubbard metal-insulator transition as a crossover between the two phases. However, we have no means to *exclude* the possibility of the first scenario of a discontinous transition respectively a phase transition for finite temperatures because the temperature we have used in this study is rather high and therefore most likely *above* the critical temperature T_c below which the true phase transition occurs according to Georges *et al.*. Finally we note that the critical interaction U_c found for the two–dimensional case here are by a factor of two smaller than the typical critical interactions found in the infinite–dimensional case. This difference is likely to be due to the differences in the interaction-free density of states which serves as an input parameter for the DMFT treatment and is usually taken to be semielliptic.

5 Technical Details

The quantum Monte Carlo calculations were carried out on the T3E of the HLRS, Stuttgart. In order to achieve the for this project necessary accuracy, 128 processors on the Cray-T3E supercomputer were running parallel Monte-Carlo jobs. Each processor started with a different random field configuration. During the run the calculated mean values for physical observables on each processor were gathered to make a statistical analysis. In addition to that the raw mean values of each processor were written on disc to be able to perform a more complex analysis after the end of the calculation. With this scheme a total performance of 16 GFlops was achieved. This huge computing power enabled us – for the first time – to study the dynamics of a cluster containing 400 sites (Fig. 13).

The exact diagonalization studies were performed on the NEC SX4 of the HLRS. We used a matrix free Lanczos algorithm designed and optimized for vector computers like the SX4. The quantum mechanical base states are coded most efficiently as bit fields, where each bit describes the presence or absence of a spin-up or spin-down electron on a certain lattice site. Thus, the matrix free method has to provide an elaborated bookkeeping consisting of integer arithmetics, bit-shifts and bit-mask operations in order to find and calculate the non-vanishing matrix elements "on the run" whenever they are needed.

This is an atypical task for modern high performance computers which are optimized for rapid floating point operations. In spite of this inevitable performance bottleneck our codes reach a sustained Flop rate (counting only the floating point operations) of 0.5 GFlops on one processor of the NEC SX4, which corresponds to about 25 % of the peak performance. Our next aim is to parallelize the algorithm in order to make it suitable for T3E and comparable machines.

References

1. Ch. Kittel: "Introduction to Solid State Physics", John Wiley & Sons, New York (1986).
2. see, for example, S.R. White, D.J. Scalapino, R.L. Sugar, E.Y. Loh, J.E. Gubernatis, R.T. Scalettar, Phys. Rev. B **40**, 506 (1989).
3. F. Gebhard: "The Mott Metal-Insulator Transition", Springer, Berlin/Heidelberg/New York (1997).
4. M. Imada, A. Fujimori, Y. Tokura, Rev. Mod. Phys. **70**, 1039 (1998).
5. D. Vollhardt in "Correlated Electron Systems, Vol. 9, ed. by V. J. Emery, World Scientific, Singapore (1993).
6. W. Metzner, D. Vollhardt, Phys. Rev. Lett. **62**, 324 (1989).
7. E. Müller-Hartmann, Z. Phys. B **74**, 507 (1989).
8. W. Metzner, Z. Phys. B **77**, 253 (1989).
9. M. Jarell, Phys. Rev. Lett. **69**, 168 (1992).
10. A. Georges and G. Kotliar, Phys. Rev. B **45**, 6479 (1992).
11. A. Georges, G. Kotliar, W. Krauth, and M. J. Rozenberg, Rev. Mod. Phys. **68**, 13 (1996).
12. K. Yosida, K. Yamada, Prog. Theor. Phys. Suppl. **46**, 244 (1970).
13. K. Yosida, K. Yamada, Prog. Theor. Phys. **53**, 1286 (1975).
14. R. M. Noack, F. Gebhard, Phys. Rev. Lett. **82**,1915 (1999).
15. J. Schlipf, M. Jarrell, P. G. J. van Dongen, N. Blümer, S. Kehrein, Th. Pruschke, D. Vollhardt, cond-mat/9902267.
16. C. Gröber, M. G. Zacher, R. Eder, cond-mat/9810246.
17. C. Gröber, M. G. Zacher, R. Eder, cond-mat/9902015.
18. R. Preuss: Dissertation, Würzburg 1996.
19. R. Preuss, W. Hanke, C. Gröber, and H. G. Evertz, Phys. Rev. Lett. **79**, 1122 (1997).

Vibronic studies of adsorbate-covered semiconductor surfaces with the help of HPC

Ulrich Freking, Albert Mazur, and Johannes Pollmann

Westfälische Wilhelms-Universität Münster
Institut für Theoretische Physik II
48149 Münster

Abstract. The technologically relevant adsorbate systems S:Ge(001)-(1 × 1) and Se:Ge(001)-(1 × 1) are prototypes for the passivation of semiconductor surfaces. Over the last years, their electronic and structural properties have been studied by *ab-initio* methods in detail, while the description of vibronic properties was limited to semi-empirical methods. The use of high performance computing (HPC) allows for the first time the calculation of surface phonons for these particular systems from first principles within a reasonable time, for example with the help of density functional perturbation theory. This theory, its implementation on a parallel computer and results concerning the vibronic properties of the above-mentioned surfaces are presented in this article.

1 Introduction

Clean semiconductor surfaces do not exist in their ideal, i. e. bulk-like configuration. The existence of partially occupied states at the surface normally leads to structural rearrangements and, as a consequence, complicated reconstructions can occur. In some cases it is possible to restore the structure of the ideal surface by adsorption of appropriate adatoms. A prominent example is the adsorption of chalcogen atoms at semiconductor surfaces, for example sulfur or selenium at the Ge(001) surface. The clean surface shows a dimerized (2×1) reconstruction, while the adsorbate-covered surface exhibits an ideal (1×1) structure. In this way the reactivity of the surface can be reduced, the surface is passivated. Therefore, the investigation of the chemisorption of adatoms on semiconductor surfaces is of great interest both fundamentally and technologically. For many years primary research has focussed on the structural and electronic properties of such surfaces. In many cases the details of the atomic structure of clean and adsorbate-covered surfaces are now well understood. From the theoretical point of view, the density functional theory (DFT) in the local density approximation (LDA) has become a common and well-established descriptive method [Hoh64,Koh65]. This theoretical formalism allows the calculation of electronic and structural properties from first principles.

In contrast, information about the vibronic properties of surfaces is more difficult to obtain. This situation has changed during the past decade and the investigation of surface phonons has become more and more important in both experimental and theoretical physics. The refinement of several sample preparation techniques (e.g. molecular-beam epitaxy or metal-organic vapor phase epitaxy) and analyzing methods such as inelastic He-atom scattering or high-resolution electron energy-loss spectroscopy made it possible to measure surface vibrations for all wave vectors in the two-dimensional Brillouin zone and thus to deduce the complete phonon dispersions up to a high degree of precision. On the other hand there has been dramatic progress in the theoretical treatment of vibronic properties. Different methods were formulated to calculate surface phonons from first principles [Bar87,Gia91,Quo92,Sav96]. Baroni showed how to calculate force constants for bulk systems with the density functional perturbation theory (DFPT).

The main difficulty concerning the application of the DFPT to the description of surface phonons is the missing periodicity perpendicular to the surface plane. Since the theory is only applicable to periodic systems, a so called *repeated slab* geometry is used to model the surface. In this geometry thin crystal films (typical thickness about ten atomic layers) with infinite extensions in the two surface-parallel directions are regarded. These crystal films (slabs) are repeated in the third spatial direction and they have to be separated from each other by a sufficiently thick vacuum region (normally equivalent to about five layers). Since each slab has two surfaces, there will be two nearly degenerate surface modes of each kind. They become strictly degenerate in the case of an infinite number of atomic layers in the slab. In addition two surfaces of different slabs can interact through the vacuum region. In order to guarantee that the surfaces of the slabs are approximately decoupled, the thickness of both the crystal film and the vacuum region have to be chosen appropriately. The resulting system has a three dimensional periodicity and thus allows the application of the same calculational methods as for bulk crystals. Since the slab supercell is comparatively large in the direction perpendicular to the surface plane, many basis functions have to be taken into account for a convergent description of the system. For this reason the *ab-initio* calculation of surface phonons can only be carried out with high performance computing.

The purpose of this article is to present theoretical results for the vibronic properties of the adsorbate-covered surfaces S:Ge(001)-(1 × 1) and Se:Ge(001)-(1×1). Both systems are prototypes of the above mentioned passivation of semiconductor surfaces due to chemisorption of adatoms. To understand the necessity of using a parallel supercomputer like the IBM RS/6000 SP at the Scientific Supercomputing Center (SSC) in Karlsruhe and to show how to obtain a very high efficiency of a parallelized program, the article is organized as follows. In Section 2 the basic theory, especially the DFPT, is described briefly, and different aspects of the parallelizing are presented. In Section 3 our theoretical results are presented and discussed, and they are compared with other available theoretical data [Kru90,Kru91a,Kru91b,Pol91]. Section 4 gives the future prospects for possible further calculations of vibronic properties with the DFPT.

2 Theory

In order to calculate surface phonon dispersion curves in the repeated slab geometry several steps are necessary. First, the equilibrum adsorption geometry must be determined. Also the electronic properties like the one-particle eigenfunctions and the valence charge density must be calculated, before the vibronic properties can be studied in the BORN-OPPENHEIMER approximation.

2.1 Electronic and structural properties

In order to find the optimal geometry the total energy of the periodic slab system has to be minimized. The calculation is based on density functional theory (DFT) in the local density approximation (LDA) and it turns out that the total energy is given by

$$E_{tot}[n] = E_{bs} - E'_{Coul}[n] + \Delta E_{xc}[n(\mathbf{r})] + E_{Mad} + E_{rep} \tag{1}$$

Here, E_{bs} is the band structure energy, E'_{Coul} the non-divergent part of the COULOMB energy, ΔE_{xc} the exchange correlation correction, E_{Mad} the

MADELUNG energy and E_{rep} the repulsive part of the total engergy. By minimizing the total energy it's possible to derive effective one-particle equations for the electronic states, the so-called KOHN-SHAM equations [Hoh64,Koh65]

$$(-\nabla^2 + V_{eff}(\mathbf{r}, \mathbf{r}', [n]))\phi_{n,\mathbf{k}}(\mathbf{r}) = \epsilon_{v,\mathbf{k}}\phi_{n,\mathbf{k}}(\mathbf{r}). \tag{2}$$

In this effective one particle access the total valence charge density n is given by a sum over all occupied electronic valence bands v at all $\mathbf{k}$-points in the first Brillouin zone. This sum

$$n(\mathbf{r}) = 2\sum_{v,\mathbf{k}} |\phi_{v,\mathbf{k}}(\mathbf{r})|^2 \tag{3}$$

can be reduced to the consideration of a few *special points* as they were introduced by CHADI and COHEN [Cha73]. As the effective potential

$$V_{eff}(\mathbf{r}, [n]) = V_{ion}(\mathbf{r}) + 2\int d^3r' \frac{n(\mathbf{r}')}{|\mathbf{r} - \mathbf{r}'|} + \frac{\partial E_{xc}}{\partial n(\mathbf{r})} \tag{4}$$

depends on the valence charge density through the Coulomb (second term) and the exchange-correlation contribution (third term), the solution of (2) must be carried out iteratively up to self-consistency. In each iteration step the new input potential yields to a new valence charge density and this density enters the input potential of the next iteration loop. As a basis for the one particle functions $\phi_{v,\mathbf{k}}$ plane waves are chosen. Nonlocal pseudopotentials of KLEINMAN and BYLANDER [KB82] are used to desribe the crystal potential V_{ion}, and the exchange correlation energy E_{xc} is taken into account in the CEPERLEY-ALDER form [Cep80]. If self-consistency is achieved, the forces acting on the atoms - which are given as the first derivatives of the total energy by the ion positions - are calculated in accordance with the HELLMANN-FEYNMAN theorem [Fey37]. Then the atoms are moved in the corresponding directions. For each step of the force calculation the effective one particle equations have to be solved self-consistently. The positions of the ions are varied until all forces are smaller than $10^{-4}\frac{Ryd}{a.u.}$.

2.2 Vibronic properties

As we pointed out above, the knowledge of the electronic ground state density is sufficient for the calculation of the forces acting on the atoms. Additionally, the linear change of the electronic density has to be known for the calculation of the second order change of total energy, which is needed in the context of vibronic properties. Both statements are special cases of the so called $2n + 1$ *theorem*, whose validity for many particle systems was shown a few years ago [Gon91]. Thus, a combination of DFT and first-order perturbation theory allows the determination of harmonic force constants and dynamical matrixes.

Starting with the HELLMANN-FEYNMAN theorem [Fey37], the dynamical matrix for a given wave vector $\mathbf{q}$ is given as a sum of three different terms.

$$D_{\alpha,\nu}^{\alpha',\nu'}(\mathbf{q})^{total} = D_{\alpha,\nu}^{\alpha',\nu'}(\mathbf{q})^{ion} + D_{\alpha,\nu}^{\alpha',\nu'}(\mathbf{q})^{elec,(1)} + D_{\alpha,\nu}^{\alpha',\nu'}(\mathbf{q})^{elec,(2)} \tag{5}$$

The ionic part D^{ion} follows from the MADELUNG energy by EWALD summation. The first electronic part $D^{elec,(1)}$

$$D_{\alpha,\nu}^{\alpha',\nu'}(\mathbf{q})^{elec,(1)} = \frac{2}{N}\frac{1}{\sqrt{M_\nu M_{\nu'}}}\sum_{v,\mathbf{k}}\langle\phi_{v,\mathbf{k}}|\Delta_{\alpha,\nu,\mathbf{q}}\Delta_{\alpha',\nu',\mathbf{q}}V_{ion}|\phi_{v,\mathbf{k}}\rangle \tag{6}$$

turns out to be $\mathbf{q}$-independent and therefore needs only to be calculated once. The second electronic part

$$D_{\alpha,\nu}^{\alpha',\nu'}(\mathbf{q})^{elec,(2)} = \frac{4}{N}\frac{1}{\sqrt{M_\nu M_{\nu'}}}\sum_{v,\mathbf{k}}\langle\Delta_{\alpha,\nu,\mathbf{q}}\phi_{v,\mathbf{k}}|\Delta_{\alpha',\nu',\mathbf{q}}V_{ion}|\phi_{v,\mathbf{k}}\rangle \tag{7}$$

contains the screening of the electronic system as a response to the ionic displacement. The response function $|\Delta_{\alpha,\nu,\mathbf{q}}\phi_{v,\mathbf{k}}\rangle$ describes the answer of an electron in the state $|\phi_{\mathbf{k},v}\rangle$ to a change of the self-consistent potential $\Delta_{\alpha,\nu,\mathbf{q}}V_{SCF}$ by moving the ion ν in the direction α. The electron with the wave vector $\mathbf{k}$ is scattered by a phonon with wave vector $\mathbf{q}$ in a state indicated by $\mathbf{k}+\mathbf{q}$.

Perturbation theory of first order allows us to express $|\Delta_{\alpha,\nu,\mathbf{q}}\phi_{v,\mathbf{k}}\rangle$ as an expansion on the $\{|\phi_{v,\mathbf{k}}\rangle\}$-basis. As a consequence, sums over all electronic bands, both occupied and unoccupied, occur at all $\mathbf{k}$-points in the first Brillouin zone. Again, the number of $\mathbf{k}$-points considered can be reduced by the method of *special points*. The number of bands has the limiting influence on the possible size of the system. The projection onto the conduction band states via the operator $P_{c,\mathbf{k}+\mathbf{q}}$ was suggested by BARONI [Bar87,Gia91]. In this formulation, an explicit knowledge of these states is not necessary. With $H_{SCF}(\mathbf{r},[n]) = -\nabla^2 + V_{eff}(\mathbf{r},[n])$ the resulting STERNHEIMER equations

$$\left[\epsilon_{v,\mathbf{k}} - \hat{H}_{SCF}\right]|\Delta_{\alpha,\nu,\mathbf{q}}\breve{\phi}_{v,\mathbf{k}}\rangle = \hat{P}_{c,\mathbf{k}+\mathbf{q}}\Delta_{\alpha,\nu,\mathbf{q}}V_{SCF}|\phi_{v,\mathbf{k}}\rangle \tag{8}$$

$$\text{with}\quad |\Delta_{\alpha,\nu,\mathbf{q}}\phi_{v,\mathbf{k}}\rangle = \hat{P}_{c,\mathbf{k}+\mathbf{q}}|\Delta_{\alpha,\nu,\mathbf{q}}\breve{\phi}_{v,\mathbf{k}}\rangle \tag{9}$$

replace formally in the DFPT the one particle equations of KOHN and SHAM in the DFT. The density response in reciprocal space is given by

$$\Delta_{\alpha,\nu,\mathbf{q}}\tilde{n}(\mathbf{G}) = \frac{4}{\Omega}\sum_{v,\mathbf{k}}\langle\phi_{v,\mathbf{k}}|e^{-i(\mathbf{q}+\mathbf{G})\mathbf{r}}\hat{P}_{c,\mathbf{k}+\mathbf{q}}|\Delta_{\alpha,\nu,\mathbf{q}}\breve{\phi}_{v,\mathbf{k}}\rangle \tag{10}$$

Similar to the investigation of the ground state properties in DFT, the change of the effective potentials entering equation (8), depends on the change of the electronic density.

$$\Delta_{\alpha,\nu,\mathbf{q}} V_{SCF}(\mathbf{r},\mathbf{r}') = \Delta_{\alpha,\nu,\mathbf{q}} V_{ion}(\mathbf{r},\mathbf{r}') + e^2 \int d^3r' \frac{\Delta_{\alpha,\nu,\mathbf{q}} n(\mathbf{r}')}{|\mathbf{r}-\mathbf{r}'|} + \Delta_{\alpha,\nu,\mathbf{q}} n(\mathbf{r}) \left.\frac{\partial V_{xc}}{\partial n}\right|_{n=n_0(\mathbf{r})} \tag{11}$$

Therefore, the system of linear equations (8) needs to be solved self-consistently. If there are N atoms in the supercell, the density response must be determined for $3N$ different atomic movements. Again, plane waves are used as a basis set for both the electronic functions $|\phi_{v,\mathbf{k}}\rangle$ and the response functions $|\Delta_{\alpha,\nu,\mathbf{q}}\phi_{v,\mathbf{k}}\rangle$. Finally, these iteratively computed response functions allow to determinate the dynamical matrix and the phonon frequencies for a given wave vector $\mathbf{q}$.

2.3 Details of the Implementation

In our repeated slab calculations a minimum of at least eight atomic layers in a supercell, separated by four vacuum layers, must be considered in order to model the surface adequately. Because of the strength of the sulfur and the selenium potential it is necessary to take many plane waves into account for a sufficient convergence. In our calculations we considered about 1800 plane waves, which corresponds to a cut-off energy of 15 Rydberg for the kinetic energy. The sum over all electronic bands at all **k**-points could be reduced by the method of ***special points*** to a consideration of $N_s = 16$ points. With the self-consistent solutions of equation (8) it is possible to obtain the dynamical matrix for a given periodicity of the perturbation. In order to determine the vibronic properties for any wave vector $\mathbf{q}$ using the eigenfunctions for a few $\mathbf{q}_j$, force constants are determined by Fourier transformation of the corresponding N_j dynamical matrixes at $\mathbf{q}_j$. Here, the consideration of $N_j = 8$ proved to be sufficient. Finally, an inverse Fourier transformation allows an estimation of the dynamical matrix for any wave vector $\mathbf{q}$.

To discus the achievable efficiency of the parallelization we consider AMDAHL's law. Let t_{ser} be the time for a serial execution of the program, f with $0 < f < 1$ be the percentage of the computing time that can be parallelized with $0 < f < 1$ and N_{par} be the number of nodes that are used. Then the time for a parallel run of the program is given by

$$t_{par} = (1-f)t_{ser} + \frac{f}{N_{par}} t_{ser}. \tag{12}$$

The maximum possible speed up is given by

$$s = \frac{t_{ser}}{t_{par}} = \frac{1}{1-f+\frac{f}{N_{par}}} \tag{13}$$

It is not advisable to parallelize the loops over different atoms ν and different directions α when solving the inhomogeneous linear equation set (8). This is because the convergence for different distortions is not uniform. For the same reason, the simultaneous calculation of distortions with different wave vectors $\mathbf{q}$ is not advantageous from the computational point of view. This would correspond to a small parallelizing percentage and therefore yield only a minimal increase in speed. To realize a greater efficiency, inner loops have to be parallelized, namely the loops over different $\mathbf{k}$ points and/or different bands v in (7) and (8).

The parallelization is carried out with the message passing interface (MPI). Both following types of MPI routines have to be used.

- Environment Management Routines
- Collective Communication Routines

As the environment management routines are commonly used in every MPI program, it is not necessary to desribe them in detail. To distribute the solution of the equation set (8) on different nodes, the collective communication routines are used. The consideration of them is necessary, because in the change of the charge density a sum over different $\mathbf{k}$ points and different valence bands v occurs. That is, in every iteration loop a barrier synchronization of all nodes has to be created. In addition, the sum over the ***special points*** must be placed in all tasks. The remaining $\mathbf{k}$-independent operations, namely the estimation of the Coulomb and the exchange-correlation potential, contribute only negligibly to the time consumption. Therefore, they do not need to be distributed to different nodes. We should add, that the parallelization of the sums over $\mathbf{k}$ occuring in the electronic parts of the dynamical matrixes in accordance with (6) and (7) follow the same scheme.

For an exploitation of $N_{par} = N_s = 16$ nodes, that is to say for a number of processors corresponding to the number of ***special points***, we achieved a percentage of $f = 0.98$ and a speed up of $s = 12.3$. In view of the frequencies of the necessary local blocking communications of all nodes (several times per minute) this high efficiency is very considerable. According to AMDAHL's law, the relative increase in acceleration for a given parallelizing percentage decreases with an increasing number of nodes, so that we refrained from a further parallelization. On the IBM RS/6000 SP at the Supercomputing Center in Karlsruhe, the determination of the phonon modes with a given periodicity for a supercell containing eight atomic layers takes $t_{par} = 10$ hours. This time refers to the use of thin P2SC nodes with 120 MHz P2SC processors and each 512 MB main memory. All in all, to compute the complete dispersion curves for such a supercell $N_j N_{par} t_{par} = 1300$ hours are required.

As we will see in the next section, it is necessary to fill the slab with bulk layers for a better classification of the surface modes. We were able to develop a new slab-filling method that allows the consideration of interactions in the inner part of the slab over a great number of layers. The idea is to fill the dynamical matrix of the regarded supercell with bulk values that are gained by an unfolding of the germanium bulk crystal. Details of this method will be published elsewhere. In this method the computation of the dynamical matrix for the Germanium bulk has to be done on a three dimensional lattice in the Brillouin zone. In total, the determination of surface phonon dispersion curves for each S:Ge(001)-(1 × 1) or Se:Ge(001)-(1 × 1) takes less than 2000 hours CPU-time.

3 Results

Only the vibronic properties of the adsorbate-covered systems are studied by using the parallel computer IBM RS/6000 SP at the Scientific Supercomputing Center in Karlsruhe. Therefore, we concentrate on a description of the respective results. We refrain from a detailed presentation of the electronic and structural results, because they are in good agreement with earlier *ab-initio* calculations [Kru90,Kru91a,Kru91b,Pol91].

First, the equilibrum adsorption geometry of a supercell, consisting of eight atomic and four vacuum layers, is determined using the DFT in LDA. Concerning this geometry we just want to mention that the resulting optimized bond lengths mainly scale with the different covalent radii of germanium, sulfur and selenium. To illustrate this surface structure, it is shown in Fig. 1 together with the surface Brillouin zone (SBZ).

In a second step the dynamical matrices for this slab system are computed with the help of the DFPT. In order to resolve the surface modes clearly, the slab was enlarged by inserting 64 bulk layers, and the dynamical matrices for $N_j = 8$ **q**-points of the 72-layer slab were determined. The computation of force constants via fourier transformation allows the specification of the vibronic properties for any **q**. Then, the layer resolved spectral densities were computed for each **q**-point. For the surface phonon modes we are interested in these densities which show maxima on the layers close to the surface and hence allow their identification. They are shown in Fig. 2 as solid lines together with the projected bulk-phonon spectrum of germanium, which is represented by vertically dashed areas.

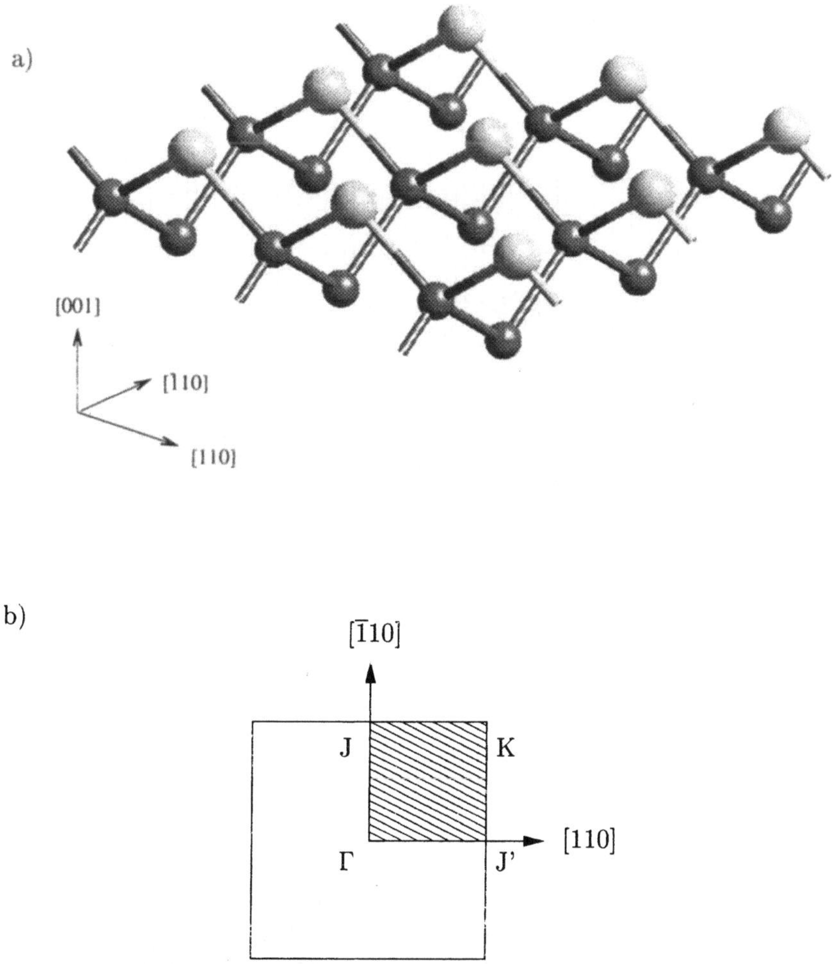

Fig. 1: a) Structur of the regarded adsorbate-covered surfaces S:Ge(001)-(1 × 1) and Se:Ge(001)-(1 × 1) and b) corresponding surface Brillouin zone, where we have shaded the irreducible part.

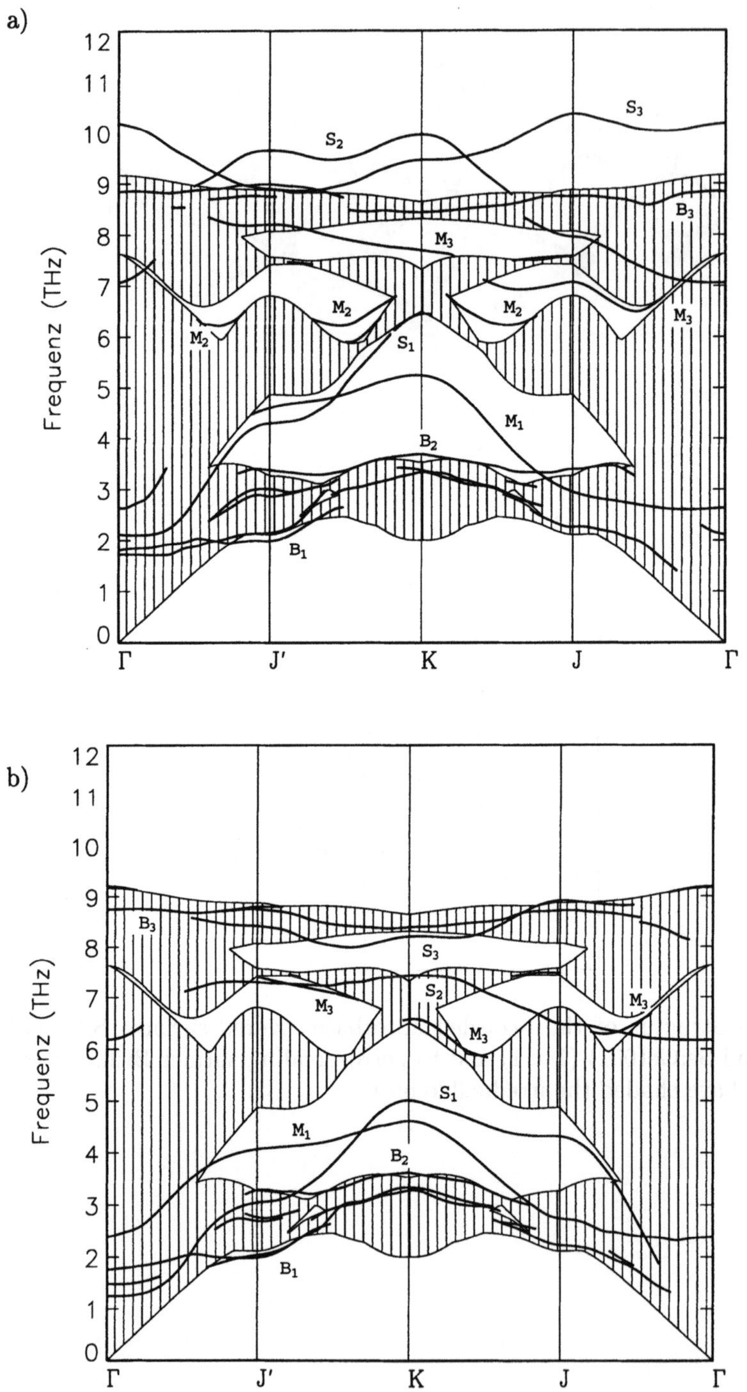

Fig. 2: Surface phonon dispersion curves of a) S:Ge(001)-(1 × 1) and b) Se:Ge(001)-(1 × 1). The projected bulk dispersion is represented by vertically dashed areas.

We see both localized and resonant surface modes. Three classes of modes can be distinguished. The modes indicated with S (surface) are mainly dominated by the adatoms sulfur and selenium. They correspond to movements of the atoms in the three different space directions. The displacement patterns of these modes at the Γ-point are shown in Fig. 3.

a)

b)

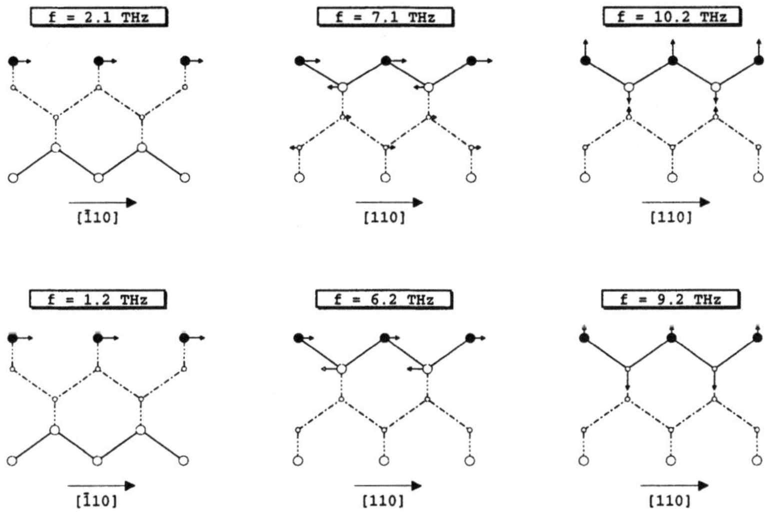

Fig. 3: Displacement patterns of the three surface modes S_1, S_2 and S_3 at the Γ-point in the structure optimized geometry for a) S:Ge(001)-(1×1) and b) Se:Ge(001)-(1×1).

The lengths of the arrows illustrate the contribution of the different layers to the vibration. The modes of type B (**b**ulk) describe movements, where only the substrate atoms are involved. Finally the modes that are indicated with M (**m**ixed) belong to movements, which both the adsorbate and the substrate atoms participate in.

The low-frequency mode S_1 describes at the Γ-point a wobbling of the surface chains perpendicular to their direction, that is in [$\bar{1}$10] direction. Along $\overline{\Gamma J'}$ the movement of S_1 only takes place in [$\bar{1}$10] direction and both for S:Ge(001)-(1×1) and for Se:Ge(001)-(1×1) the layer and symmetry resolved spectral density is localized on the adatom layer up to 99 %. Obviously, this mode does not yield a change of bond lengths, which explains its low energy. In the mode S_2 the atoms close to the surface move parallel to the surface chains in [110] direction. Along $\overline{\Gamma J}$ the atoms move exclusively in [110] direction, while the mode is sagittaly polarized along $\overline{\Gamma J'}$. Finally, S_3 describes the movement of the atoms perpendicular to the surface. This mode is also sagittaly polarized along $\overline{\Gamma J'}$. In contrast to the mode S_1 both S_2 and S_3 yield a change of the bond length between the adsorbate and the first layer of the Ge-substrate and therefore they are higher in energy. In the case of

S:Ge(001)-(1 × 1) the corresponding frequencies appear above the projected bulk dispersion, because in contrast to the case of selenium the mass of the sulfur adatom is much smaller than that of the substrate germanium. Already a simple model of a semi-infinite linear chain explains the existence of surface modes above the bulk continuum for a corresponding mass ratio.

For the Γ-point, ab-initio frequencies of the S-modes can be obtained by the determination of layer force constants with a BACHELET-HAMANN-SCHLÜTER pseudopotential in a basis set of gaussian functions [Pol91]. For the mode S_3 at the Γ-point our value of 10.2 THz is in good agreement with the result of 10.4 THz given in [Pol91], and for S_2 our frequency value of 7.1 THz agrees exactly with the value computed with the help of layer force constants. Only for the mode S_1 our value of 2.1 THz does not correspond to the value of 0.8 THz in [Pol91]. We are in the process of analyzing the origin of this discrepancy.

a)

b)

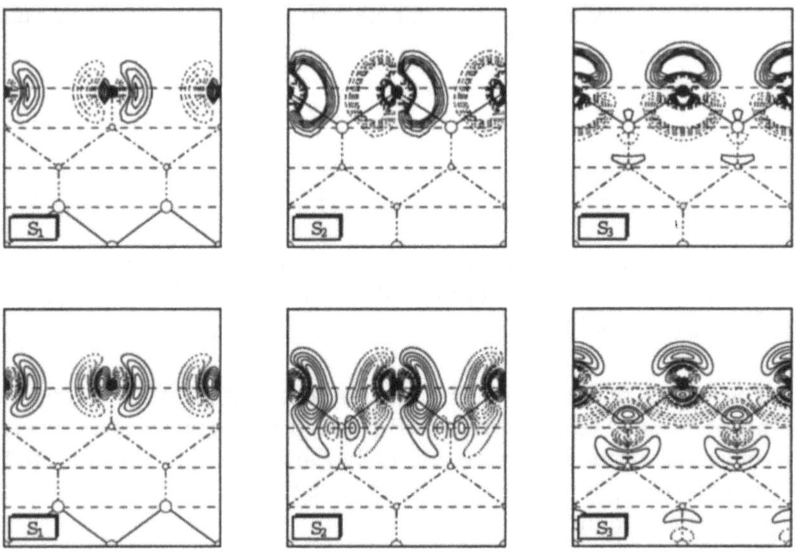

Fig. 4: Change of the density for the modes S_1, S_2 and S_3 at the Γ-point in the optimized geometry a) for S:Ge(001)-(1 × 1) and b) for Se:Ge(001)-(1 × 1)

The displacement patterns of the B- and M-modes cannot be described in detail, because the primary goal of this article is on computational aspects. Nevertheless, one interesting aspect concerning these modes should be mentioned. All surface modes nearly scale with the mass of the adatom. As a consequence, the modes indicated with B occur at nearly the same energy level on both surfaces, while the modes indicated with M or S lie energetically lower for the selenium covered surface. As a consequence the modes S_2 and S_3 are resonant with the bulk continuum in the latter case especially at the Γ-point and can hence couple to bulk modes. This results in a reduction of the localization of these modes on the surface layers, which is also reflected in the change of the valence charge density, which is shown in Fig. 4. From Fig. 4 it is clear that the localization of the modes S_2 and S_3 extends over a larger number of layers close to the surface if the adatom is selenium.

A further discussion of the vibronic properties and also of the electronic and structural properties will be published elsewhere.

4 Outlook

As we have shown, the use of high performance computing allows an *ab-initio* calculation of surface phonons in a reasonable time. The efficiency achieved with the parallelization is very high and independent of both the thickness of the supercell used in the repeated slab model and the number of plane waves that are taken into account in the basis set. The reason for this efficiency lies in the fact that almost the whole computation depends on *special points*, and the loops over these points can be parallelized comparably easily by using the message passing interface.

A comparable efficiency might be reached for other surfaces, so that the DFPT in a parallelized version facilitates an *ab-initio* determination of surface phonons for many interesting semiconductor surfaces, both clean and adsorbate-covered. Due to an extension of this method to include metals by GIRONCOLI [Gir95], corresponding surface phonons can be treated in the same manner. Apart from the description of vibronic properties of surfaces the DFPT is applicable to the determination of interface phonons as they occur in superlattices [Bar90]. Because of the techological importance of semiconductor nanostructures in fields such as microelectronics, the descpription of these vibronic properties is of great interest for future studies.

Acknowledgements
We thank Dr. W. Lu for helpful discussions about different aspects concerning high performance computing. In particular, it is our great pleasure to acknowledge the Supercomputing Center in Karlsruhe. Without the competent and friendly help of the co-workers of SCC, our calculations would not have been possible.

References

[Bar87] S. Baroni, P. Giannozzi and A. Testa,
Green's-Function Approach to Linear Response in Solids,
Phys. Rev. Lett. **58**, 1861 (1987).

[Bar90] S. Baroni, P. Giannozzi and E. Molinari,
Phonon Spectra of ultrathin GaAs/AlAs superlattices: An *ab-initio* calculation,
Phys. Rev. B **41**, 3870 (1990).

[Cep80] D. M. Ceperley and B. J. Alder,
Ground State of the Electron Gas by a Stochastical Method,
Phys. Rev. Lett. **45**, 566 (1980).
[Cha73] D. J. Chadi and M. L. Cohen,
Electronic Structure of $Hg_{1-x}Cd_xTe$ Alloys and Charge-Density Calculations Using Representative k Points,
Phys. Rev. B **7**, 692 (1973).
[Fey37] R. P. Feynman,
Forces in Molecules,
Phys. Rev. **56**, 340 (1939).
[Gia91] P. Giannozzi, S. de Gironcoli, P. Pavone and S. Baroni,
Ab-initio calculation of phonon dispersions in semiconductors,
Phys. Rev. B **43**, 7231 (1991).
[Gir95] S. de Gironcoli,
Lattice dynamics of metals from density-functional perturbation theory,
Phys. Rev. B **51**, 6773 (1995).
[Gon91] X. Gonze and J.-P. Vigneron,
Density functional approach to nonlinear-response coefficients of solids,
Phys. Rev. B **39**, 13120 (1989).
[Hoh64] P. Hohenberg and W. Kohn,
Inhomogeneous Electron Gas,
Phys. Rev. B **136**, 864 (1964).
[KB82] L. Kleinmann and D. M. Bylander,
Efficacious Form for Model Pseudopotentials,
Phys. Rev. Lett. **48**, 1425 (1982).
[Kru90] P. Krüger and J. Pollmann,
First-Principles Theory of Sulfur Adsorption on Semi-infinite Ge (001),
Phys. Rev. Lett. **64**, 1808 (1990).
[Kru91a] P. Krüger and J. Pollmann,
Self-consistent electronic structure of clean and adsorbate-covered Ge (001) surfaces,
Progr. Surf. Sci. **35**, 3 (1991).
[Kru91b] P. Krüger and J. Pollmann,
Self-consistent surface electronic structure of semi-infinite semiconductors from scattering theory,
Physica B **172**, 155 (1991).
[Koh65] W. Kohn and L. J. Sham,
Self-consistent equations including exchange and correlation effects,
Phys. Rev. A **140**, 1133 (1965).
[Pol91] J. Pollmann, P. Krüger, and A. Mazur,
Electronic, structural and vibronic proberties of chalcogen monolayers on (001) surfaces of elemental semiconductors,
Appl. Surf. Sci. **56-58**, 193 (1991).
[Quo92] A. A. Quong and B. M. Klein,
Self-consistent-screening calculation of interatomic force constants and phonon dispersion curves from first principles: Application to aluminium,
Phys. Rev. B **46**, 10734 (1992).
[Sav96] S. Y. Savrasov,
Linear-response theory and lattice dynamics: A muffin-tin-orbital approach,
Phys. Rev. B **54**, 16470 (1996).

Computational Methods in Chemistry and Molecular Biology

Bernd Artur Hess
Lehrstuhl für Theoretische Chemie,
Friedrich-Alexander-Universität Erlangen.
Egerlandstrasse 3, D–91058 Erlangen, Germany

In 1998, the Nobel prize in chemistry was awarded to two outstanding figures in the field of quantum molecular science, namely John A. Pople and Walter Kohn. Pople and Kohn had both made important contributions to the field over a few decades, and thus a substantial share of the work which has led to the state-of-the-art in Theoretical Chemistry and Computational Chemistry can be traced back to their influence.

Quantum molecular science aims at a microscopic description of molecular systems by means of a quantum-mechanical description of electrons in a molecule (or, more general, an assembly of molecules in a liquid, on a surface, or in the solid state). The theoretical means is the multiparticle Schrödinger equation for the electrons in the molecule or, if needed, for the nuclear degrees of freedom as well. Typically, a parameter-free ("exact") Hamiltonian is used, and the most sophisticated methods are called *ab initio*, this qualification characterizing methods which do not use empirical parameters and which make use of controlled approximations in the sense that a systematic improvement is possible. Less reliable methods (lacking the possibility of systematic improvement or making use of parameters adjusted to the experiment) still often provide important insight into chemistry, and often offer the only means for a theoretical treatment of large molecules.

Theoretical methods based on quantum molecular science seek to supplement experimental techniques of chemistry by means of the computational microscopic model of the molecules provided by the quantum-mechanical theory. In this way, it is possible not only to obtain a deeper understanding of chemical processes, but also to learn about unstable or otherwise inaccessible molecular species. As an almost elementary example, we mention the diatomic molecule AuF, which was considered non-existing and thus unstable until very recently. Quantum-mechanical calculations [1-2] showed, however, that it is perfectly stable with respect to dissociation into its constituents, and indeed it was finally prepared and experimentally characterized [3]. The reason that it was not synthesized before was the marked tendency of its constituents to react to another, even more stable species AuF_3.

Acknowledging the potentials of computational methods in chemistry, it is not surprising that an ever increasing number of publications in journals devoted to research in chemistry, and even biology, makes use of these methods.

Work along these lines comprises the development of new methodology, implementation into computer programs, and applications. In particular, development of methods to treat excited states of molecules and methods specially designed to treat large molecules are areas of current research. The implementation into computer programs is often focused on special modern techniques, as parallel computing, and we shall find a report on this type of work in the current issue. On the other hand, it was instrumental for the dissemination of theoretical methods, that special emphasis was given to an easily accessible user interface. This is important in particular for the non-specialist, who is mainly interested in applications. By and large, the most important program in this respect is the GAUSSIAN program, originally designed and written in John Pople's group. It can without doubt be termed the working horse of Computational Chemistry, and we shall encounter several interesting applications in the present volume.

[1] P. Schwerdtfeger, M. Dolg, W. H. E. Schwarz, G. A. Bowmaker, and P. D. W. Boyd, J. Chem. Phys. 91 (1989) 1762

[2] P. Schwerdtfeger, J. S. McFeaters, R. L. Stephens, M. J. Liddell, M. Dolg, and B. A. Hess, Chem. Phys. Lett 218 (1994) 362

[3] D. Schroeder, J. Hrusak, I. C. Tornieporth-Oetting, T. M. Klapoetke, and H. Schwarz, Angew. Chemie 106 (1994) 223

The multi-reference configuration interaction method on massively parallel architectures

P. Stampfuß and W. Wenzel

Institut für Physik, Universität Dortmund,
D-44221 Dortmund, Germany

Abstract. We report on a scalable implementation of the configuration-selecting multi-reference configuration interaction method for massively parallel architectures with distributed memory. Based on a residue driven evaluation of the matrix elements this approach allows the routine treatment of Hilbert spaces of well over 10^9 determinants as well as the selective treatment of triple and quadruple excitations with respect to the reference space. We demonstrate the scalability of the method for up to 128 nodes on the IBM-SP2. We elaborate on the specific adaptation of the transition residue-based matrix element evaluation scheme that ensures the scalability and load-balancing of the method and discuss the projects using this methodology that are presently under investigation: the elucidation of electronic dynamics near the conical intersection of NO_2, the calculation of potential energy surfaces and electronic spectra of transition metal compounds such as Cr_2 and the first row transition metal dihalides, the investigation of the electronic structure of aromatic compounds such as benzene, naphtalene and anthrazene and their derivatives and the *ab-initio* determination of microscopic spin- and spin-phonon coupling constants in quasi-onedimensional transition metal compounds, such as $CuGeO_3$ and NaV_2O_5 in the solid state.

1 Introduction

In the development of quantum chemical methods for complex molecules, a consensus has emerged that two important effects must be taken into account in a balanced and accurate fashion in order to arrive at quantitatively correct results. First, dynamical correlations, i.e. the mutual influence two electrons exercise on each other when they pass at close distance, must be accounted for. Secondly, one must be able to accommodate the multi-reference nature of the electronic states in many complex molecules. This effect is particularly important in many transition metal compounds where the close proximity of d-energy levels generates a large number of important multiplets. It is also relevant when one wants to describe an entire potential energy surface, where bond-breaking or bond-rearrangements can occur. An adequate treatment of multi-reference effects is mandatory for the quantitative treatment of electronically excited states. Meeting the needs of both these requirements simultaneously is responsible for the high cost of accurate quantum chemical calculations.

For many years the multi-reference configuration interaction method (MRCI) has been one of the benchmark tools for highly accurate calculations of the electronic structure of atoms and molecules [1–3]. Ever since the development of the direct CI algorithm [1], which obviates the explicit storage of the CI matrix, highly efficient implementations [4] have been used for a wide variety of molecules. The generic lack of extensivity of the MRCI method has at least been partially addressed with a number of *a posteriori* [5,6] corrections and through direct modification of the CI energy-functional [7–12].

Due to its high computational cost, however, applications of the MRCI method remain constrained to relatively small systems. For this reason the configuration-selective version of the MRCI-method (MRD-CI), introduced by Buenker and Peyerimhoff [13–15], has arguably become one of its most widely used versions. In this variant only the most important configurations of the interacting space of a given set of primary configurations are chosen for the variational wavefunction, while the energy contributions of the remaining configurations are estimated on the basis of second-order Rayleigh-Schrödinger perturbation theory [16,17]. A configuration is selected for the variational wavefunction if its perturbative energy contribution or coefficient is above a given threshold λ and the total energy (the sum of the variational and the perturbative contributions) is extrapolated to the limit $\lambda \to 0$. While this extrapolation is known to fail in isolated instances, it gives a remarkably good resolution of relative energies across the potential energy surface (PES) in the overwhelming majority of applications. Since the variationally treated subspace of the problem consists of only a fraction of the overall Hilbert space, the determination of eigenstates in the truncated space requires far less computational effort. Indeed, for typical applications the overwhelming majority of the computational effort is concentrated in the expansion loop, where the energy contribution of candidate configurations is computed.

Even within this approximation, the cost of MRCI calculations remains rather high. The development of efficient configuration-selecting CI codes [18,19,17,20–22] is inherently complicated by the sparseness and the lack of structure of the selected state-vector. In order to further extend the applicability of the method, it is thus desirable to employ the most powerful computational architectures available for such calculations. Here we report on the progress of the first massively parallel, residue-driven implementation of the MRD-CI method for distributed memory architectures. While efforts to parallelize standard MR-SDCI (all single and double excitations) on distributed memory architectures face significant difficulties rooted in the need to distribute the CI vectors over many nodes [23–26] — a parallel implementation of MRD-CI can capitalize on the compactness of its state representation. In our implementation the difficulty of the construction of the subset of nonzero matrix elements is overcome by the use of a residue-based representation of the matrix elements that was originally developed for the distributed memory implementation of MR-SDCI [26]. This approach allows us to efficiently evaluate the matrix elements both in the expansion loop as well as during the variational improvement of the coefficients of the selected vectors.

Going beyond somewhat beyond the standard MRD-CI, our implementation was specifically optimized to estimate the importance of triple- and quadruple excitations of the reference configuration. The energy arising from such configurations yields the overwhelming contribution to the energy difference between FCI and MR-SDCI and is thus of paramount importance for the development approximately extensive versions of the MRCI method [7–10]. Since the number of higher-than-doubly excited configuration rises so quickly with the system size, FCI as well as CI-SDTQ calculations are prohibitively expensive for all but the smallest systems. Configuration selecting CI thus provides a particularly effective, maybe the only viable, compromise between computational efficiency and accuracy. In this manuscript we focus on the details of the implementation of the method and provide timings for benchmark applications that demonstrate the scalability of the method for up to 128 nodes of an IBM-SP2 for Hilbert spaces of dimension up to 5×10^9 of which up to 5×10^6 elements were selected for the variational wavefunction[27]. The determinant based code we report here was developed in an object oriented implementation using C++ as the implementation language and is available with the Dortmund quantum chemistry package.

2 Methodology

In the following we will describe the key ingredients for the residue based parallel implementation of the configuration selecting MR-CI method. We begin with a description of the orbital partitioning scheme that allows a flexible treatment of the triple and quadruple excitations with regard to the active space. We then describe the principle of the residue-based matrix-element evaluation that is at the heart of our algorithm, followed by a more detailed treatment of the key difficulties in the two computationally expensive phases of the method: the expansion step and the iteration step. Next we present the results of benchmarks of the method for O_2, NO_2 and benzene as a function of the number of nodes used.

In order to facilitate the explicit treatment of triple and quadruple excitations from a given reference set, we partition the orbital space into five segments. An earlier investigation [28] into the hierarchy of TQ excitations demonstrated that in many cases the natural orbital occupations in the SD (single & double) segment of the configuration space is clearly separated in weight by the natural orbital occupation arising from the TQ segment. As a result the fraction of the overall natural orbital (NO) basis that contributes to important TQ configurations at some given threshold is much smaller than the fraction of the NO basis that contributes to SD configurations at the same threshold. This observation is illustrated in Figure (1) where we have separated the contributions to the diagonal elements of the single particle density matrix that arise from SD and

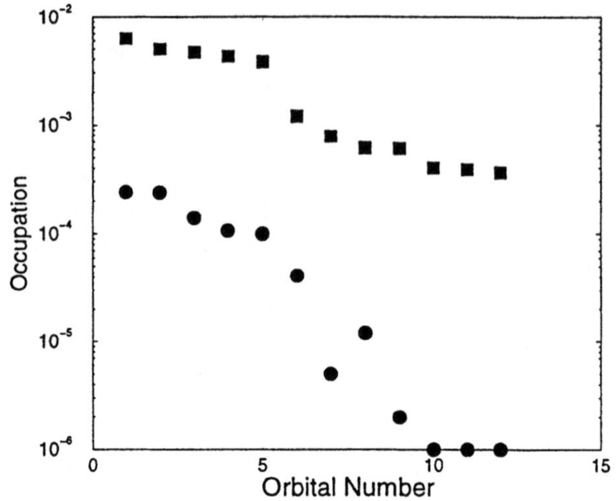

Fig. 1. Total contributions of individual orbitals (of a_g symmetry) to the diagonal matrix elements of the single particle density matrix that arise from single- and double excitations of the reference configurations (squares) or triple- and quadruple excitations (circles). The data were obtained in a CI-SDTQ calculation for O_2 in a cc-pVDZ basis at the experimental geometry. The details of the calculation are summarized in the benchmark section.

TQ excitations respectively. Note that for each orbital the total weight of all TQ excitations lies almost two orders of magnitude below the weight arising from SD contributions. This observation suggests that it is worthwhile to restrict the search of the overall TQ space to the most important orbitals of the SD space.

2.1 Residue Based Evaluation of Matrix Elements

In order to compute the matrix elements of the Hamilton operator we exploit an enumeration scheme in which each matrix element between two determinants (or configuration state functions) $|\phi_1\rangle$ and $|\phi_2\rangle$ is associated with the subset of orbitals that occur in both the target and the source determinant. This unique subset of orbitals is called the *transition residue* mediating the matrix element and serves as a sorting criterion to facilitate the matrix element evaluation on distributed memory architectures. For a given many-body state, we consider a tree of all possible transition residues as illustrated in Figure (2). For each such residue we build a list of *residue-entries*, composed of the orbital-pairs (or orbital for a single-particle residue) which combine with the residue to yield a selected configuration and a pointer to that configuration. While the number of transition residues is comparatively small, the overall number of residue-entries grows rapidly (as $N_{selected}\, n_e^2$) with the number of configurations $N_{selected}$ and the number of electrons n_e.

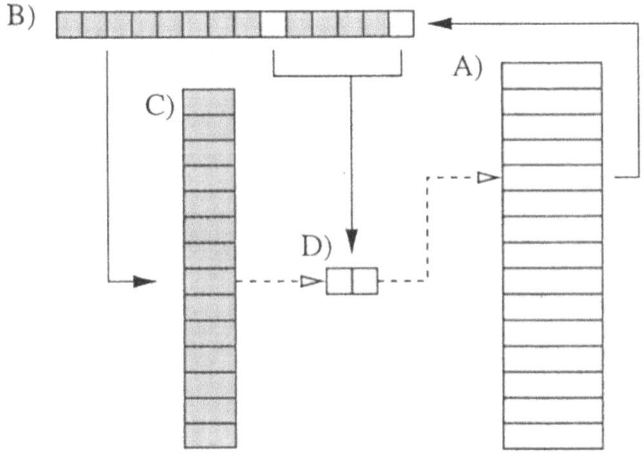

Fig. 2. Schematic representation of the two-particle residue-tree. For each element of the configuration-list (A) all possible two-particle residues are constructed. In the configuration illustrated in (B) each box represents one occupied orbital, the shaded region corresponds to the residue and the two white boxes to the orbital pair. The $(n_e - 2)$-electron residue configuration is looked up in the residue-tree (C), where an element (D) is added that encodes the orbitals that where removed, information regarding the permutation required and the index of the original configuration in the configuration list. Solid arrows in the figure indicate logical relations ships, dotted arrows indicate pointers incorporated in the data structure. The residue-list, along with all elements must be rebuilt once after each expansion loop, the effort to do so is proportional to product of n_e^2 with the number of configurations. The number of matrix elements encoded in a single element of the residue-tree is proportional to the *square of the number of entries* of type (D).

Once the residue tree is available the evaluation of the matrix elements is very efficient. In the *expansion step*, one must evaluate $q_i = \langle \phi_i | HP | \Psi \rangle$, where P projects on the part of the Hilbert space in which only inactive and active and low orbitals are occupied. This portion of the residue tree contains but a fraction of the overall residue tree and can be accommodated on all nodes. For each $|\phi_i\rangle$ we determine the required single- and two- particle residues, which are then searched for in the residue tree. In a SD calculation one can eliminate the search step by constructing the allowed excitations directly from the internal residues. If a match is found the information in the tree enables us to immediately compute *all matrix elements* associated with the given residue.

In the *iteration phase* the full residue tree for all selected configurations must be built, but a single copy of the tree can be distributed across all nodes. All matrix elements associated with a given transition residue can be locally evaluated if all associated orbital pairs are present on a unique node. We note that the residue tree itself (part B in Figure (3)) is not required at all, only the set of connected orbital pairs is needed. As a result no lookup operations

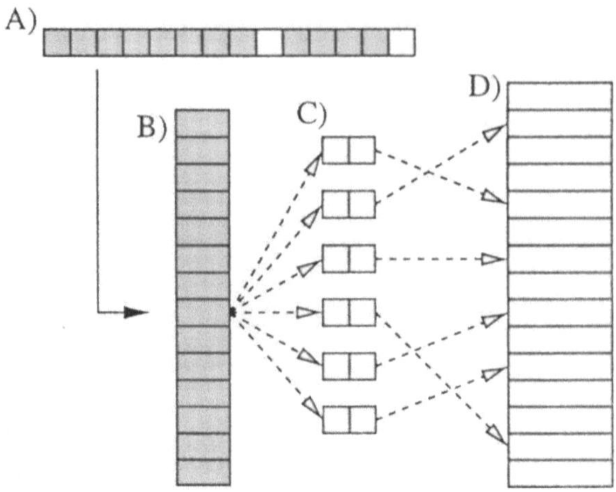

Fig. 3. Schematic representation of the computation of two-particle matrix-elements in the expansion step using the residue-tree. For a given configuration (A) we form all two-particle residues, which are looked up in the residue tree. In the configuration illustrated in (A) each box represents one occupied orbital, the shaded region corresponds to the transition residue and the two white boxes to the orbital pair. The $(n_e - 2)$-electron residue configuration is looked up in the residue-tree (B). Each orbital pair (C) associated with the residue encodes a matrix element with an element of the configuration list (D). The orbital indices of the required integral are encoded in the orbital pairs in (C), the coefficient of the source configuration is looked up directly in (D). Only one lookup operation is required to compute all matrix elements associated with the given transition residue and only the subset of matrix elements that lead to selected source-configurations are constructed.

are required in this step and one can simply loop over the locally available section of the orbital pair segments to evaluate all matrix elements that can be constructed for the present orbital sets. Since each matrix element is uniquely identified by its transition residue, the contributions to the many-body field can be simply collected at the end of this step on a single node to perform the Davidson iteration. This mechanism allows a rapid evaluation of all matrix elements while using the available core memory to its fullest extent.

2.2 Parallel Implementation

In a truly scalable implementation great care must be taken to divide all work equally across the participating nodes. A remaining non-scalable portion of 1% of the computational effort of a single processor application translates into a 100% overhead if the same task is distributed across 100 nodes. Our massively parallel algorithm for configurations-selecting MRCI is therefore based on a client server model that strictly separates the calculation from the communication steps. The latter were chosen to require only global communication directives of

	Phase		Type
I	Initialization	All nodes	Computation
	Expansion & Logic Steps		
D1	distribution of initial state	All nodes	Broadcast
R1	build section of restricted residue tree	All nodes	Computation
G1	gather and distribute residue tree	All nodes	Pairwise
E	expansion loop	All nodes	Computation
D2	distribute selected configurations	All nodes	All-to-All
R2	build one section of full residue tree	All nodes	Computation
G2	gather the residue tree	All nodes	All-to-All
	Iteration Steps		
D3	Distribute New Coefficients	Node 0	Broadcast
M	Evaluate matrix elements	All nodes	Computation
G3	Gather many body field	Node 0	Gather
X	Davidson iteration	Node 0	Iteration

Table 1. List of the distinct computational steps in the parallel implementation of the configuration-selecting MRCI procedure. There are three phases, associated with the initialization of the program and the expansion and iteration of the state-vector respectively. The details of the phases are discussed in the text. The fourth row of the table details which set of nodes is involved in each step, while the fifth row indicates the type of operation that dominates the step. Almost all the computational effort is concentrated in steps E and M, the next leading contribution arising from the logic steps D2,R2,G2. Only standard high-level communication routines were used to make the program as portable as possible.

the underlying MPI communication library which can be expected to execute efficiently on most modern parallel architectures.

According to the above outline the overall work can be broken into two distinct phases that require the same order of magnitude of computational effort. Table (1) summarizes the most important steps of the configurations-selecting CI procedure. In an *expansion step* we begin with the distribution of the current state-vector to all nodes (D1). Each node then builds the restricted residue tree for all the set of configurations it received in step (D1). The effort per node involved in this step (R1) is proportional to $n_e^2\ N_{\rm conf}/N_k$, it will therefore scale well with the number of nodes N_k. Next (step G1) the residue tree must be distributed to all nodes. Finally the information of all the nodes is distributed via an all-to-all communication across the entire machine. Now (step E) each node can run through a predetermined section of the search space to evaluate the energy contributions and to select the configurations for the variational subspace. Step E dominates the overall computational effort of the configuration-selection by a large margin.

The next three steps prepare the variational subspace for the iterations. Since the distribution of the selected configurations on the different nodes can be rather uneven, we first redistribute the configurations among the nodes(D2). Then each node constructs its portion of the full residue tree (R2). These contributions are gathered in analogy to step (G1) across all nodes, such that each node has all orbital pairs for its assigned transition residues. In contrast to step (G1), however, the entries are not distributed across the machine, but remain on the nodes for the local matrix element evaluation. The computational effort in the evaluation step of the matrix elements is proportional to the expectation value of the square of the number of orbital pairs over the transition residues. The hashing mechanism we use to assign transition residues to nodes, however, ensures only that there are approximately the same number of residues on each node. The computational effort can nevertheless vary quite significantly among the nodes, an effect that worsens with an increasing number of nodes. Such an imbalance in the work-distribution leads to the loss of scalability of the algorithm. It is therefore important to redistribute the workload among the nodes to achieve better performance. To this end we gather discretized histograms of the transition residue distribution on the server node, which uses this information to assign approximately even work-loads to all nodes. Based on this technique the theoretical deviations in the variation of the work-load can be reduced from over 50% to less than 12% of the overall average computational effort.

After the redistribution of the transition-residue table, the program can proceed with the iteration steps to converge the variational subspace. The four steps comprising an iteration are executed many times after each expansion loop, but require no further logic information. Almost all the work is concentrated in step M. We note in passing that in the expansion step only a fraction of all possible integrals are required on the nodes, a fact that will be exploited in future versions of the code.

3 Benchmark Calculations

In order to demonstrate the scalability of the implementation we have conducted benchmark calculations for typical applications of the program. The first example is concerned with the evaluation of the importance of the triple and quadruple excitations for the potential energy surfaces of the oxygen molecule and its anion[1]. The calculations were performed in a (sp)-augmented cc-pVTZ, cc-pVQZ and cc-pV5Z [30, 29] basis set in D_{2h} symmetry at the experimental geometries. In Hilbert spaces of dimension up to 5×10^9 containing triple and quadruple excitations we selected up to 5×10^6 determinants as a function of the threshold for the coefficients ranging from 10^{-3} to 10^{-6}. In the second example, more within the traditional applications of configuration-selecting CI, we have computed the ground state energy of benzene in a cc-pVDZ basis set using active spaces of 6 and 12 active orbitals. The latter calculation was motivated by the desire to test the program for very large Hilbert spaces, but the smaller active space is sufficient to adequately describe the chemistry of benzene. The calculation was performed in a C_{6v} geometry in D_{2h} symmetry resulting in Hilbert spaces of up to 3×10^9 determinants of which up to 2×10^6 were selected for the variational subspace.

[1] For a motivation of this calculation see section *Projects*.

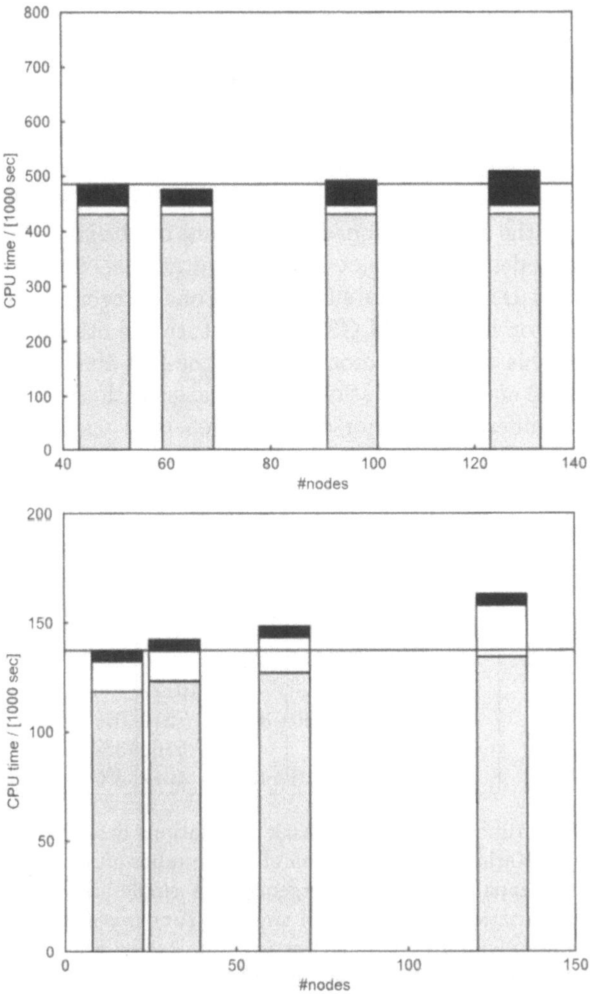

Fig. 4. Total CPU time in (sec) for the fully converged calculation of the ground state of the two benchmark calculations described in the text as a function of the number of nodes of the IBM-SP2. A straight line indicates perfect scaling of the computational effort with the number of nodes. The shaded areas in the bars, from top to bottom, indicate the contributions of the matrix element evaluation (steps D3,M,G3,X in Table (1)), logic (D1,R1,G1,D2,R2,G2) and the expansion loop (step E).

The most important consideration in the evaluation of the performance of a parallel program is its scalability with the number of processors used for a given calculation. For scaling purposes we selected a typical run with 10^9 determinants (1.8×10^6 selected) for O_2 and another with 1.3×10^9 determinants (1.6×10^6selected) for benzene respectively. For these cases we performed benchmark runs on the 256-node IBM-SP2 of the Karlsruhe supercomputer center.

Figure (4) shows the total computational effort (excluding the time to read the integral file) of the aforementioned scaling runs on the IBM-SP2 as a function of the number of nodes. In these plots, the computational effort for all logic-steps sections (D1,R1,G1,D2,R2,G2) are subsumed in one category, the expansion loop (E) and the iteration loop (D3,M,G3,X) constitute the other main components of the program. This division is motivated by the fact that the relative importance of these three main computational steps varies with the type of calculation performed and a different behavior of these steps will result in changes of the overall performance for different calculations.

	O_2		C_6H_6	
number of nodes	time (s/node)	loss	time (s/node)	loss
16	8308			
32	4374	5%		
48			10122	
64	2480	13%	7510	0%
96			5160	3%
128	1410	14%	4012	4%

Table 2. Total CPU times for the benchmark calculations described in the text on the IBM-SP2 of the RZ Karlsruhe as a function of the number of nodes. Given is the time in sec/node for the expansion and convergence of a single state in each calculation. The fractional computational loss between two test runs is defined as the ratio of the CPU-times per node divided by the perfect speedup factor given by the ratio of the nodes. The loss-data in the table always refer to successive entries.

For benzene we find almost perfect scaling from 48 to 128 nodes for the IBM-SP2. The total computational effort in the expansion loop, which dominates the overall computational effort, is constant to within 0.4% in going from the smallest to the largest number of nodes. In contrast, the effort associated with logic and communication grows somewhat with the number of nodes. This is to be expected, since the communication cost grows with the number of nodes and a total of 3.7 / 9.1 GB of data have to be transmitted across the machine for the small and large residue tables respectively. The overall speedup factor from 64 to 128 nodes is 1.86 (see Table (2)).

For the benchmark calculation of O_2 a more pronounced increase in the overall computational effort is observed in going from 32 to 128 nodes, in particular in with the last doubling from 64 to 128 processors. Again, the computational

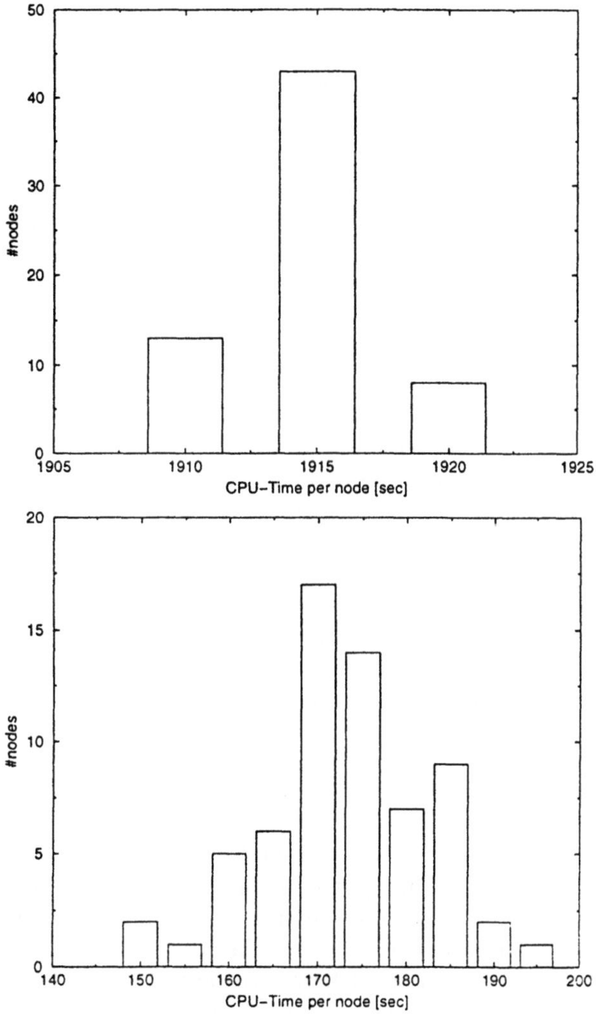

Fig. 5. Histogram of the CPU-time distribution of (a) an expansion step including TQ excitations and (b) an iteration step for the O_2 benchmark calculation described in the text on 64 nodes of the IBM-SP2. The computational effort in the expansion step is almost perfectly distributed that of the iteration step varies with a standard deviation of approximately 4% resulting in a loss of computational efficiency as "fast" nodes have to wait for the "slow" nodes to finish. Without the use of load-balancing the width of distribution for the iteration step increases significantly. Since the number of transition residues/node decreases with the number of nodes, fluctuations in the computational effort become more difficult to balance for a large number of processors.

effort in the expansion loop is relatively stable, increasing only 6.7% (6.3%) in going from 32 to 64 (64 to 128) nodes respectively. This is the result of the near perfect load-balancing in evidence in Figure (5), which shows a histogram of the total CPU time/node for the expansion step for a run on 64 nodes. The presence of triple and quadruple excitations significantly complicates the overhead associated with the generation of the residue trees. Given the relatively wide distribution of the computational effort per transition residue that results from the presence of the TQ excitations, it becomes more and more difficult to balance the computational load in the matrix element step. This results in a larger variation in the load (see Figure (5)) among the processors. This decreases the performance of the iteration step, since all processors must wait for the last node to finish. Because large amounts of data are required for the evaluation of the matrix elements it is difficult to go beyond the present implementation and to dynamically adjust the load while the iteration is in process.

The data for the test runs is summarized in detail in Table (2). For the IBM SP-2 we find near-perfect speedups for benzene. For O_2 the speedup factors associated with the doubling of the nodes are somewhat worse, but still warrant the use of a large number of nodes to perform the calculation in most circumstances. MRCI calculations require nontrivial communication steps on parallel machines, so that some loss of computational efficiency is unavoidable. We have verified that the overall performance of our implementations is comparable to that of other recent scalar implementations of MRD-CI[27].

4 Projects

Encouraged by the success of this initial implementation we are presently pursuing four extensions of the work reported here. Almost completed is the integral-driven version of the configuration selecting MRCI method. Using this code we are able to accomodate a much larger number of molecular orbitals and hence larger molecules than are reported here (see below). Secondly, we have begun with the implementation of a spin-adapted version of this code. Third, after the completion of these step we will continue with the implementation of the non-selecting version of the MRCI method on the basis of the residue driven matrix evaluation scheme. In MRCI, the main difficulty of the implementation shifts from the construction of the residue tree to the handling of the large state vectors present in the calculation. However, our experience with the present code indicates that MPI will be able to handle the required large data transfers associated with the exchange of vector elements among nodes with ease.

In addition, we will implement multi-reference second-order Brillouin-Wigner perturbation theory [16, 32–34], an approximation of MRD-CI that eliminates the selected variational subspace altogether. Such approximations rest on the assumption that the individual energy contributions of the selected configurations in MRD-CI are well estimated perturbatively. MRD-CI benefits nevertheless from their explicit treatment because the selected configurations generate a many-body field on the primary configurations that alters the relative importance of the latter. BW-MRPT incorporates this effect in the framework of perturbation theory and eliminates the selected subspace of MRD-CI. In comparison to MRD-CI, BW-MRPT thus discards the coupling among the selected configurations outside the primary space and the present program is an ideal framework to test the importance of these interactions for a wide variety of systems.

Since an efficient program is now available we have started with a number of investigations to help elucidate the electronic structure of challenging molecules. In the following we briefly summarize these projects:

- Previous work has established that the accurate calculation of the electron affinity of O_2 remains a formidable challenge even for present day quantum chemical techniques. At the level of a CAS-SCF description the adiabatic electron affinity of the oxygen molecule is predicted with the wrong sign even in the basis set limit. A careful study [29] concluded that strong differential dynamical correlation effects are most likely entirely responsible for the source of this discrepancy. In MRCI-SD calculations the correct sign for the electron affinity can barely be reached using aug-QZP quality basis sets. A semiquantitative agreement between experiment and theory was reached, when the multi-reference generalization of the Davidson correction [6] was applied to estimate the effect of higher excitations.
  O_2 is therefore one of the simplest molecules which challenges one of the central paradigms of modern quantum chemical correlation methods that rest on the assumption that the explicit treatment of single and double excitations of a chemically motivated reference set is sufficient to quantitatively account for dynamical correlation effects. This observation, as well as the desire to explicitely test approximations for extensivity corrections to MR-SDCI [7–10] motivated the development of the present code. Since the CAS+SDTQ Hilbert space of O_2 in a aug-QZP basis has dimension 32×10^9, this problem cannot be treated with any of the presently available MR-SDCI or MRD-CI implementations, but provides a suitable challenge for our parallel implementation.
  So far we have been able to demonstrate that the energies of both the neutral molecule and the anion converge smoothly with the threshold to their respective MRCI values. Using these results for basis sets up to doubly augmented quintuple zeta quality (using both cc-pV5Z basis sets and extending the sequence employed by Serrano et.al) we can demonstrate that the electron affinity of O_2 saturates at approximately -0.03 eV in the basis set limit for full valence [2s2p] CAS-MRD-CI calculations. The experimental electron affinity of 0.42 eV (corrected for vibrational and other non-electronic effects) thus cannot be reproduced at this level of calculation. In a previous study, agreement between experiment and theory could nevertheless be reached, when the bare MRCI results where corrected for the effect of higher excitations using a multi-reference version of the Davidson corrections *and* and by increasing the acrtive space to include unphysical [3s3p] orbitals. For smaller basis sets, where essentially the entire Hilbert space of triple and quadru-

ple excitations can be enumerated we have been able to demonstrate that the heuristic MR-Davidson correction *underestimates* the differential effect of the higher excitations by as much as 30 %. We are presently extending these calculations to the larger basis sets employing natural orbitals that are constructed from MRD-CI density matrices including triple- and quadruple excitations.

- A recent investigation[35] suggested that presently available potential energy surfaces of NO_2 are insufficient to quantitatively account for the dynamics near the conical intersection. In collaboration with this group we are therefore computing a three-dimensional potential energy surface of the relevant states of this molecule up to the dissociation limit in a cc-pVTZ basis, the largest basis set employed in such a calculation to date. Using the integral-driven version of the program the converged calculation of two states in C_s symmetry selecting upward of 6×10^5 determinants of a Hilbert space of dimension 2×10^6 takes only 84 sec on 64 nodes.
- For small molecules comprising only first- and second-row atoms, the application of benchmark methods such as MRCI has contributed significantly to the quantitative understanding of the chemistry of such materials. Because of their high computational cost, the application of such techniques to transition metal chemistry has been severely limited in the past. However, the large discrepancies between DFT and experiment motivate the use of high-precision methods also for these compounds. The family of the first row transition metal dihalides (MX_2, M = Sc,...,Zn, and X=F,Cl,Br) represent a simple model series that exhibits key problems associated with the theoretical description of transition metals in their chemical environment. Because of their interesting properties many experimental studies have been undertaken to characterize this family [36–46]. Despite equally active theoretical interest [47–55] many questions regarding their theoretical description remain unresolved. A number of theoretical investigations that at least partially account for orbital interaction and electron correlation effects[48, 53] have cast the assignments of the spectra based on ligand field theory (LFT) in doubt. A recent broad-based investigation of this series using density functional theory (DFT) [55] supported some (but not all) of these reassignments but showed significant qualitative and quantitative discrepancies to experiment and prior theoretical investigations for bond-lengths, dissociation and excitation energies.

The newly developed MRD-CI program will permit the the application of one such benchmark technique for a broad-based study of one interesting family of compounds that is at the threshold of present day capabilities. It will help elucidate not just the electronic structure of one particular molecule, but through the investigation of the entire series help to investigate the trends governing its behavior and help to understand its underlying principles. Since a significant multi-configurational character of the transition metal compounds can be expected, the MRCI family is a suitable choice for this study.

- We are presently benchmarking the new integral-driven code to calculate the electronic spectra of aromatic compounds. While traditional MRCI calculations are typically restricted to single-ringed molecules with simple substituents we a presently exploring the capabilities of our code to calculate the spectra of naphthalene (48 correlated electrons) and naphtalene (66 correlated electrons) which are difficult to treat on traditional architectures. In a similar vein, calculations to determine the ground- and exited states potential energy surfaces of the chromium dimer are presently under way.
- In a project funded by the DFG-Schwerpunkt "Quasi-one dimensional transition metal compounds" we will calculate the microscopic spin- and spin-phonone coupling constants of solids like $CuGeO_3$ and NaV_2O_5 which exhibit interesting magnetic properties. In $CuGeO_3$ quasi-onedimensional chains of copper-oxygen plaquettes give rise to effectively one-dimensional spin-1/2 Heisenberg chains with at least next-nearest neighbor interactions. The interactions of the copper spins are mediated via superexchange over the bridging oxygen molecules. Since the Cu-O-Cu angle along the plaquettes is close to 90°, where the superexhange coupling vanishes according to the Goodenough-Kanami-Anderson rules because of the orthognality of the oxygen p_x and p_y orbitals, the coupling constants in this compound are much smaller than those of the cuprate high-temperature superconductors. For this reason, the spin-phonon coupling constants can also be expected to be large. The goal of our project is to contribute to the understanding of the materials by offering an unambigious *ab-initio* determination of their microscopic coupling constants in a realistic chemical environment.

5 Discussion

Accurate benchmark methods for the treatment of dynamical correlation effects, such as MRCI, have made a significant impact in the development of quantum chemistry. Since their computational effort rises rapidly (as n_e^6) with the number of electrons, only the use of the most powerful computational architectures ensures their continued relevance to the field. Because massively parallel architectures with distributed memory will yield the highest computational throughput in the foreseeable future, it is worthwhile to pursue the use of these machines for quantum chemical benchmark calculations. The development of a scalable implementation of one of the most popular variants of the MRCI method family on such architectures is one important step in this direction. The present implementation allows the treatment of Hilbert spaces and systems that are larger than those that can be treated on traditional architectures, while significantly reducing the turnaround time for more moderate applications. With the ability to routinely treat Hilbert space exceeding 10 billion determinants or more many questions that require a delicate balance of dynamical and non-dynamical correlation effects, e.g in transition metal chemistry, become amenable to the MRCI method.

Acknowledgments: Part of this work was supported by DFG Grant KEI-164/11-2, the project to compute the microscopic coupling constants of quasi one-dimensional transition metal compounds is supported by DFG grant WE 1863/1. We greatfully acknowledge stimulating discussions with H. Lischka and I. Shavitt in the course of this work.

References

1. B. O. Roos. *Chem. Phys. Letters*, 15:153, 1972.
2. B. O. Roos and P. E. M. Siegbahn. The direct configuration interaction method. In H.F. Schaefer III, editor, *Methods of Electronic Structure Theory*, page 189. Plenum, New York, 1994.
3. I. Shavitt. In H. F. Schaefer III, editor, *Modern Theoretical Chemistry*. Plenum, New York, 1977.
4. R. Sheppard, I. Shavitt, R. M. Pitzer, D.C. Comeau, M. Pepper, H. Lischka, P. G. Szalay, R. Ahlrichs, F. B. Brown, and J. Zhao. *Int. J. Quantum Chem. Symp.*, 22:149, 1988.
5. S. R. Langhoff and E. R. Davidson. *Int. J. Quantum Chem.*, 8:61, 1974.
6. W. Butscher, S. Shih, R. J. Buenker, and S. D. Peyerimhoff. *Chem. Phys. Letters*, 52:457, 1977.
7. J. Cižek. *J. Chem. Phys.*, 45:4256, 1966.
8. R. J. Bartlett and I. Shavitt. *Chem. Phys. Letters*, 50:190, 1977.
9. R. Gdanitz and R. Ahlrichs. *Chem. Phys. Letters*, 143:413, 1988.
10. P. Szalay and R. J. Bartlett. *J. Chem. Phys.*, 103:3600, 1995.
11. J. P. Daudey, J.-L. Heully, and J. P. Malrieu. *J. Chem. Phys.*, 99:1240, 1993.
12. J. P. Daudey, J.-L. Heully, and J. P. Malrieu. *J. Chem. Phys.*, 99:1240, 1993.
13. R. J. Bunker and S. Peyerimhoff. *TCA*, 12:183, 1968.
14. R. J. Buenker and S. D. Peyerimhoff. *Theor. Chim. Acta*, 35:33, 1974.
15. R. J. Buenker and S. D. Peyerimhoff. *Theor. Chim. Acta*, 39:217, 1975.
16. Z. Gershgorn and I. Shavitt. *Int. J. Quantum Chem.*, 2:751, 1968.
17. B. Huron, J.P Malrieu, and P. Rancurel. *J. Chem. Phys.*, 58:5745, 1973.
18. R. J. Buenker and S. D. Peyerimhoff. *New Horizons in Quantum Chemistry*. Reidel, Dordrecht, 1983.
19. J. L. Whitten and M. Hackmeyer. *JCP*, 51:5548, 1969.
20. R. J. Harrison. *J. Chem. Phys.*, 94:5021, 1991.
21. S. Krebs and R. J. Buenker. *JCP*, 103:5613, 1995.
22. M. Hanrath and B. Engels. New algorithms for an individually selecting mr-ci program. *CP*, 225:197, 1997.
23. M.Schuler, T. Kovar, H. Lischka, R. Sheppard, and R. J. Harrison. *Theor. Chim. Acta*, 84:489, 1993.
24. H. Lischka, F. Dachsel, R. Shepard, and R.J. Harrison. Parallel computing in quantum chemistry-message passing and beyond for a general ab initio program system. In W. Gentzsch and U. Harms, editors, *High-Performance Computing and Networking. International Conference and Exhibition Proceedings. Vol.1: Applications*, page 203. Spinger, Berlin, 1994.
25. H. Dachsel, H. Lischka, R. Shepard, J. Nieplocha, and R.J. Harrison. *J. Comp. Chem.*, 18:430, 1997.
26. F. Stephan and W. Wenzel. *J. Chem. Phys.*, 108:1015, 1998.
27. P. Stampfuß, H. Keiter and W. Wenzel. *A Parallel Implementation of the Configuration-Selecting Multi-Reference Configuration Interaction Method*, preprint, 1999.
28. M. M. Steiner, W.Wenzel, J. W. Wilkins, and K. G. Wilson. *Chem. Phys. Letters*, 231:263, 1994.
29. R. Gonzáles-Luque, M. Merchán, M. P. Fülscher, and B. O. Roos. *Chem. Phys. Letters*, 204:323, 1993.
30. K. A. Peterson, R. A. Kendall, and T. H. Dunning. *J. Chem. Phys.*, 99:1930, 1993.

31. J.Olsen, P. Jorgensen, and J.Simons. *Chem. Phys. Letters*, 169:463, 1990.
32. W. Wenzel and K. G. Wilson. *Phys. Rev. Letters*, 68:800, 1992.
33. W. Wenzel and M. M. Steiner. *J. Chem. Phys.*, 108:4714, 1998.
34. W. Wenzel. *Int. J. Quantum Chem.*, page (accepted), 1998.
35. H. Kppel. L. Cedarbaum: *preprint*, 1998.
36. R. A. Kent, J. D. McDonald and J. L. Margrave, *J. Phys. Chem.*, 70:874, 1960.
37. C. E. Moore, *Atomic Energy Levels*, Vol. 2 ,NSRDS-NBS #35, US GPO, Washington, DC, 1971.
38. F. Ramondo, V. Rossi and L. Bencivenni, *Mol. Phys.*, 64:513, 1988.
39. F. Ramondo, L. Bencivenni and S. N. Cesaro, *J. Mol. Struc.*, 192:83, 1989.
40. V. P. Spiridonov, A. G. Gershikov and V. S. Lyutsarev, *J. Mol. Struc.*, 221:79, 1990.
41. O. V. Boltalina, A. Y. Borshchevskii and L. N. Sidorov, *Russ. J. Phys. Chem.*, 65:466, 1991.
42. O. V. Boltalina, A. Y. Borshchevskii and L. N. Sidorov, *Russ. J. Phys. Chem.*, 66:1223, 1992.
43. A. J. Ross, R. Bacis, A. J. Bouvier, S. Churassy, J. C. Coste, P. Crozet, and I. Russier, *J. Mol. Spectrosc.* 158:27, 1993.
44. M. P. Barnes, R. T. Carter, N. M. Lakin and J. M. Brown, *J. Chem. Soc. Faraday Trans.*, 89:3205, 1993.
45. D. L. Hildenbrand, *J. Chem. Phys.*, 103:2634, 1995.
46. D. L. Hildenbrand, *High Temp. Mater. Sci.*, 35:151, 1996.
47. S. Y. Shashkin and W. A. Goddard, *J. Phys. Chem.*, 90:225, 1986.
48. C. W. Bauschlicher and B. O. Roos, *J. Chem. Phys.*, 91:4785, 1989.
49. M. Kaupp, P. R. Schleyer, H. Stroll and H. Preuss, *J. Am Chem. Soc.*, 113:6012, 1991.
50. R. J. Deeth, *J. Chem. Soc. Dalton Trans.*, 1061,1993.
51. F. Rogemond, H. Chermette and D. R. Salahub, *Chem. Phys. Lett.*, 219:228, 1994.
52. D. M. Hassett and C. J. Marsden, *J. Mol. Struc.*, 346:249, 1995.
53. B. O. Roos, K. Andersson, M. P. Fulscher, P. A. Malmqvist, L. Serrano-Andres, K. Pierloot and M. Merchan, *Adv. Chem. Phys.*, 93:219, 1996.
54. C. Blanchet, H. A. Durante and D. R. Salahub, *J. Chem. Phys.* 106:8778, 1997.
55. S. G. Wang, W. H. E. Schwarz, *J. Chem. Phys.*, 109:7252, 1998.

Quantum Chemical Studies on Heterocyclic Rearrangements in Benzofuroxans: Reaction Paths, Vibrational Spectra, and Rate Constants

Guntram Rauhut and Frank Eckert

Universität Stuttgart, Institut für Theoretische Chemie, Pfaffenwaldring 55, 70569 Stuttgart, Germany.

Abstract. Quantum chemical calculations at the coupled-cluster level were used to investigate the reaction mechanism of the Boulton-Katritzky rearrangement and the ring chain tautomerism in benzofuroxans. The well known sensitivity of this molecular class to electron correlation effects could be overcome by the inclusion of triple excitations within the coupled-cluster approach. The computational results support experimental findings and reject a discussed intermediate in the Boulton-Katritzky rearrangement. The reactions of two prototype components were investigated, namely 4-nitrobenzofuroxan and 5-methyl-4-nitrobenzofuroxan.

1 Introduction

Benzofuroxans are of particular importance in pharmaceutical research [1], and due to their antibacterial and insecticidal properties, also in crop protection [2,3]. Although most of their reactions have been well known for many years (see e.g. Refs. [4–8]), their detailed reaction mechanisms are not yet understood completely [6]. In this context, the Boulton-Katritzky rearrangement (BKR) (cf. Fig. 1) is of particular importance since it applies to various classes of molecules [6,9].

Fig. 1. The bicyclic Boulton-Katritzky rearrangement of 5-R-4-nitrobenzofuroxan.

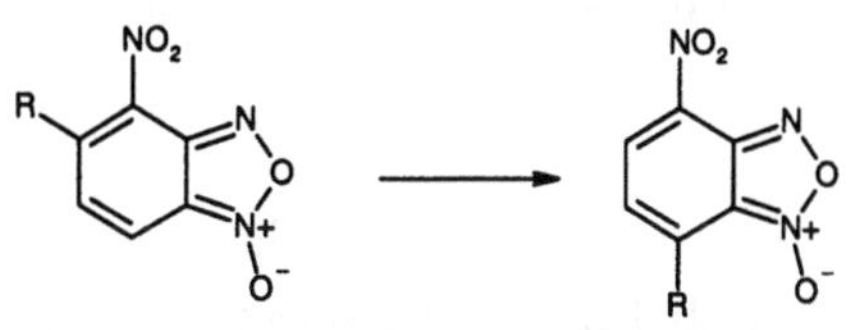

In order to gain a well-grounded understanding of this reaction three aspects need to be understood: (1) are intermediates involved in the mechanism, (2) what are the driving forces of the reaction, and (3) are there any competing reactions to the favored reaction mechanism? All these questions can be solved by quantum chemical calculations. However, previous studies showed that benzofuroxans are particularly sensitive to electron correlation effects [10]. Therefore, computationally very demanding coupled-cluster calculations are necessary to yield reliable energies. Due to the size of the structures, this task can only be solved by high performance computing.

2 Computational Details

Molecular structures as involved in the reaction mechanisms (i.e. intermediates, transition states, ...) were optimized at various levels of theory. Using density functional theory, the non-local B3-LYP exchange-correlation functional in combination with a valence double-ζ basis plus polarization functions on all atoms yielded geometries which are in very good agreement with experimental results. Unfortunately, this hybrid functional was not able to reliably describe all transition states under consideration. This conclusion was drawn from CCSD(T) single point calculations on top of B3-LYP and MP4(SDQ) geometries. While CCSD(T)//MP4(SDQ) energies (i.e. CCSD(T) single-point calculations performed on MP4(SDQ) optimized geometries) were able to reproduce the experimentally observed activation barrier of the ring-chain tautomerism in benzofuroxan, CCSD(T)//B3-LYP differed from experimental values by more than 5 kcal/mol. Therefore, very expensive MP4(SDQ) geometry optimizations were necessary, which indeed consistently corrected for the discrepancy between the calculations and the experimental results. MP4(SDQ) formally scales with the sixth power of the size of the system (usually measured in terms of basis functions). In order to perform geometry optimizations of systems of this size, analytical energy gradients are mandatory. Due to the availability of analytical MP4 gradients, the G94 program [11] was used for these calculations. Usually about 10 optimization cycles were necessary to achieve convergence, requiring about 15 GB scratch space and 3 CPU hours (C_1 symmetry, 1GB memory) per iteration.

It is well known that the coupled-cluster series of methods converges significantly faster with respect to the correlation energy than the Møller-Plesset series. For this very reason CCSD and CCSD(T) energy refinements were computed for most structures. Using the MOLPRO package of *ab initio* programs [12] on a single processor on the NEC SX-4, usually a speedup by a factor of 6 could be gained in comparison to an SGI PowerChallenge R10000/8 with 4 GB memory for the CCSD iterations, which included about 22×10^6 configuration state functions. Since the MOLPRO CCSD(T) code is completely written in terms of matrix operations, this program runs very efficiently on vector computers. The vectorization ratio of a CCSD(T) calculation is about 97% with an average vector length of about 150 for the systems studied here. The average performance on a single processor is about 1.2 GFlops. The (formal) seventh-order scaling of the triples-correction in the coupled-cluster approach resulted in extremely long computation times. Therefore, this correction could be computed for the smaller systems (benzofuroxan, 4-nitrobenzofuroxan) of this study only. For the larger systems, which are very similar to the small benchmark systems but show further functional groups connected to the 6-membered ring, the relative triples correction within the coupled-cluster approach was estimated from the smaller systems. We believe that most contributions of these substituents to electron correlation effects are already captured by single and double excitations as could be proven by the benchmark systems when the unsubstituted benzofuroxans were augmented by the nitro substituent. Therefore, computer time could be saved by this extrapolation scheme. In the same manner, effects on relative energies due to geometry changes between B3-LYP and MP4(SDQ) optimized structures were estimated. All structures of substituted 4-nitrobenzofuroxans were optimized at the B3-LYP level only. Subsequently, the activation barriers of all questionable transition states were corrected by the energy difference (B3-LYP to MP4(SDQ)) as observed in the benchmark systems. For details see references [13]. Savings in computer time due to both of these approximations are assumed to be more than 1000 CPU-hours.

3 Reaction Mechanisms

Although the BKR has extensively been used as a route for generating new compounds [6,8,14,15], the exact mechanism of this reaction was not clear. An intermediate involved in the mechanism was discussed in the literature [6] but its existence has been rejected recently [16]. This is in agreement with quantum chemical calculations performed in this study. Guided by these results, two possible reaction mechanisms were formulated. The first one is described by a single step rearrangement via a concerted transition state while the other

Fig. 2. Reaction mechanisms investigated as possible routes of the Boulton-Katritzky rearrangement. The second mechanism is symmetrical and thus one half of the complete reaction path is shown only.

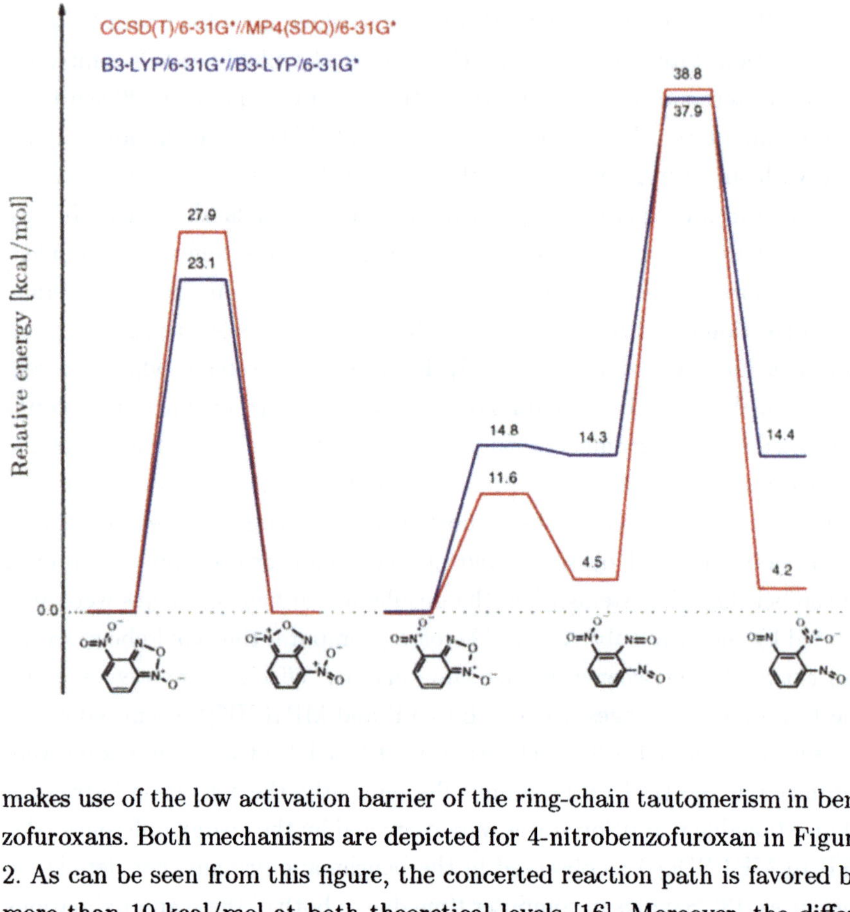

makes use of the low activation barrier of the ring-chain tautomerism in benzofuroxans. Both mechanisms are depicted for 4-nitrobenzofuroxan in Figure 2. As can be seen from this figure, the concerted reaction path is favored by more than 10 kcal/mol at both theoretical levels [16]. Moreover, the differences between the B3-LYP and the CCSD(T) calculations are comparably large. Consequently, in order to yield reliable data for the activation barriers the CCSD(T) level is mandatory. After correcting for zero-point vibrational energies the most reliable activation barrier found for the rearrangements of 4-nitrobenzofuroxan and 5-methyl-4-nitrobenzofuroxan were 27.9 and 25.4 kcal/mol, respectively. This is in agreement with experimental results which indicate a barrier higher than 20 kcal/mol for 4-nitrobenzofuroxan [17]. Since the rearrangement of the methylated compound is exothermic by about -3.6 kcal/mol, the observed fast rearrangement of this compound can readily be explained. The impact of substituents is significant and has thus been investigated in detail. Most substituents in the 5-position lead to negative reaction energies and thus to an acceleration of the reaction. As can be expected for these exothermic reactions the activation barrier is lowered as well. Unfortunately, there is no simple correlation between the reaction energy and the activation barrier, indicating that the occurence of an early or late transition state strongly depends on the nature of the substituent. The exothermicity is due to several reasons. In case of a methyl-substituent it is the unfavorable *ortho*-interaction between the methyl and the nitro group. This interaction leads to a pronounced bond alternation in the 6-membered ring and to a diminished π-conjugation between the nitro group and the π-system of the rings. For other substituents than methyl the interaction is even more complex and *para*-interactions between the substituent and the nitro group in the 7-R-isomer as well as interactions between the substituent and the furoxan ring become important. The influence of substituents on structural parameters is shown in Table I. In the educt it is the distance marked **b** that shows the largest variation in dependence on the substituent ($\Delta r = 0.268$Å) while it is the **c** "bond" in the transition state ($\Delta r = 0.058$Å). Bond lengths in the 7-isomer differ only slightly which indicates a much weaker interaction of the substituent with the furoxan ring and the nitro group. Interestingly, all transition states show C_s symmetry even though the 5-isomer can significantly be distorted. Consequently, the molecular orbitals which describe the breaking and forming bonds are also in plane without any π-contribution. An orbital plot of the corresponding MO as observed in the BKR of 5-methyl-4-nitrobenzofuroxan is shown in Figure 3. Since the relative energy of the transition state of the furoxan ring opening is significantly smaller than those of the concerted transition state of the BKR, a competing reaction leading to 6-R-7-nitrobenzofuroxan (shown in Figure 4) must be considered as well. This reaction is solely dominated by the energetically low transition state of the endocyclic bond cleavage and not by the rotation of the nitroso groups as indicated by B3-LYP calculations. Unfortunately, it is the structure of this stationary

Fig. 3. SCF Molecular orbital describing the bonding in the transition state of the Boulton-Katritzky rearrangement of 4-nitrobenzofuroxan.

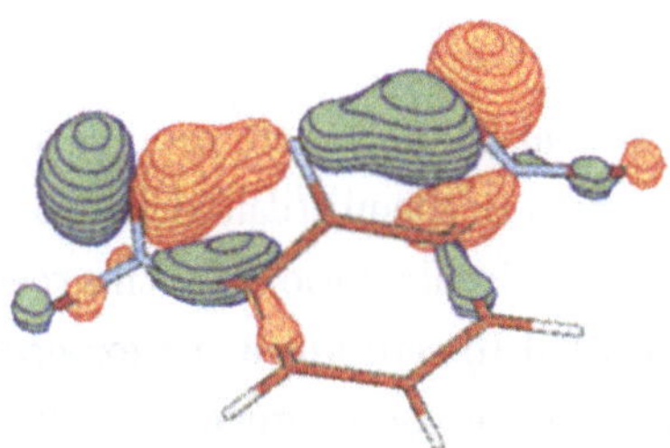

Table I. Selected bond lengths [Å] of the educt (5-R-4-nitrobenzofuroxan), the transition state, and the product (7-R-4-nitrobenzofuroxan) as computed at the B3-LYP/6-31G* level[a].

Structure	*R*	a	b	c	d
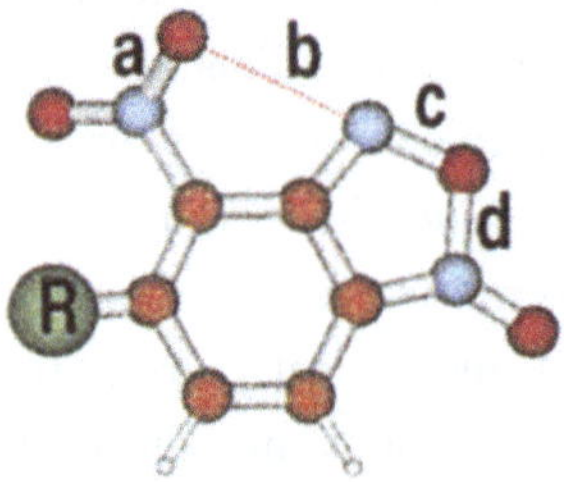	H	1.228	2.767	1.376	1.436
	F	1.229	2.782	1.377	1.434
	CH_3	1.228	2.741	1.376	1.436
	CHO	1.226	2.827	1.367	1.449
	CN	1.227	2.694	1.372	1.438
	OCH_3	1.231	2.809	1.381	1.431
	CF_3	1.225	2.962	1.367	1.449
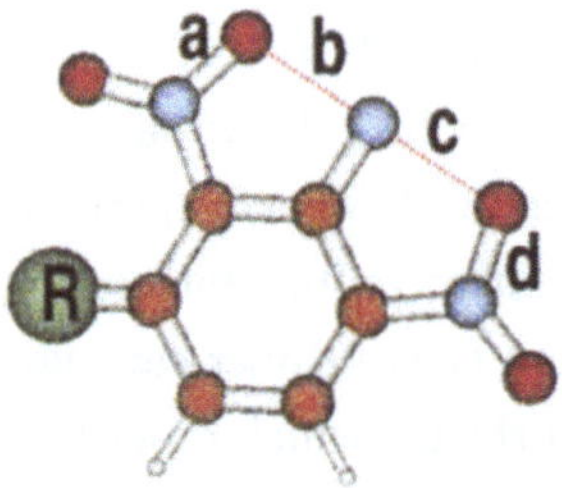	H	1.275	1.996	1.996	1.275
	F	1.278	1.985	1.978	1.278
	CH_3	1.274	1.989	1.963	1.279
	CHO	1.272	1.985	1.954	1.278
	CN	1.274	1.994	1.968	1.276
	OCH_3	1.278	1.997	1.951	1.282
	CF_3	1.272	1.995	1.938	1.279
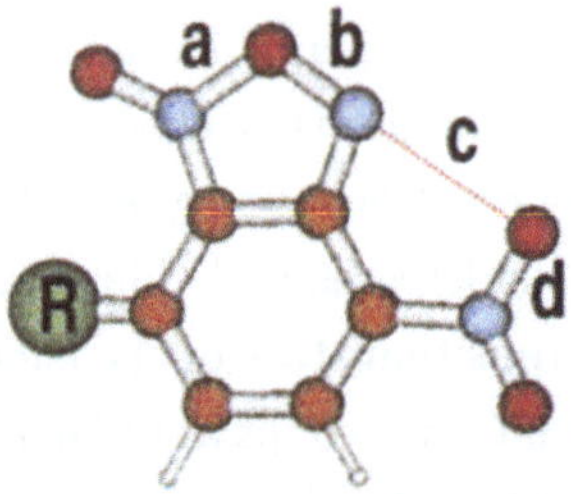	H	1.436	1.376	2.767	1.228
	F	1.442	1.373	2.762	1.228
	CH_3	1.432	1.375	2.762	1.229
	CHO	1.437	1.369	2.751	1.227
	CN	1.443	1.370	2.756	1.227
	OCH_3	1.442	1.374	2.767	1.230
	CF_3	1.441	1.370	2.750	1.227

[a] For R = OCH_3, CF_3 geometries were optimized at the B3-LYP/6-31G** level.

point that causes severe problems. The reaction barrier of the ring-chain tautomerism in unsubstituted benzofuroxan is known from NMR experiments. Benchmark calculations for this reaction showed that CCSD(T)//B3-LYP energies fail to reproduce the experimental data, while CCSD(T)/MP4(SDQ) values are in good agreement [13]. The triples correction has an extreme im-

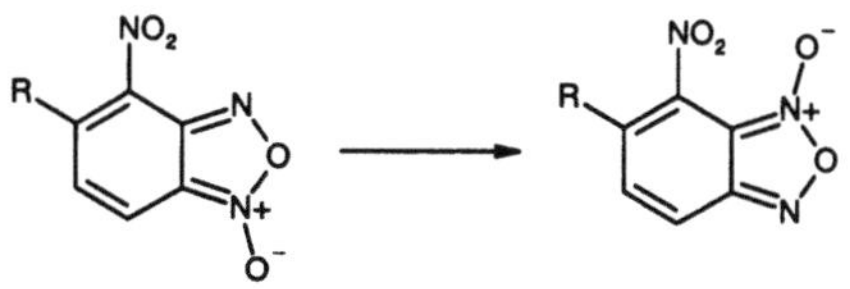

Fig. 4. The competing reaction toward 6-R-7-nitrobenzofuroxan.

pact on the relative energies. While CCSD(T)/MP4(SDQ) calculations could be afforded for 4-nitrobenzofuroxans they were prohibitive for the more interesting methyl compounds. However, as described above, due to the similarity of the structures the triples correction and the geometry effects could be estimated for these systems. The activation barrier for the reaction of 5-methyl-4-nitrobenzofuroxan toward 6-methyl-7-nitrobenzofuroxan has thus been estimated to be 12.1 kcal/mol. The reaction is endothermic by about 1.7 kcal/mol. From this it follows that this reaction is extremely fast and consequently will compete with the BKR in principle, but the equilibrium will be on the side of the educt. However, in order to detect this reaction experimentally (for instance by reactive trapping), a temperature must be chosen that guarantees that the BKR will not take place, otherwise the reaction will definitely lead to the energetically favourite 7-methyl-4-nitrobenzofuroxan.

4 Vibrational Spectra

Vibrational spectra of benzofuroxan, 4-nitrobenzofuroxan, 5-methyl-4-nitrobenzofuroxan, and 7-methyl-4-nitrobenzofuroxan were computed at the B3-LYP/6-31G* level and subsequently corrected for basis set effects, anharmonicity, and electron correlation effects by applying Pulay's scaled quantum mechanics method (SQM) [18]. Within this approach the cartesian force constants are transformed into natural internal coordinates. These coordinates are then sorted into groups of related types (i.e. heavy atom stretchings, etc.) for which individual scaling factors are established. Consequently, the force constant matrix is usually scaled by about 10 scaling factors. This scaling, which is significantly superior to uniform frequency scaling by a single factor, of course affects not only the position of the fundamentals but also their intensities, since the normal modes depend directly on the force constants. Transferable scaling factors, which can be applied to any common organic molecule, were used throughout this work [19]. For details see [13].

A comparison showed that the computed IR spectra are in excellent agreement with experimental values [13]. The computed spectrum of 4-nitrobenzofuroxan is shown in Figure 5. Based on these spectra, assignments were established for all fundamentals. Benzofuroxans are characterized by two very strong peaks between 1590 and 1640 cm^{-1} (see Figure 5). These frequencies can be assigned to the NO-stretchings of the furoxan moiety and to in-plane deformations of the 6-membered ring.

Fig. 5. Calculated (SQM/B3-LYP/6-31G*) IR vibrational spectrum of 4-nitrobenzofuroxan.

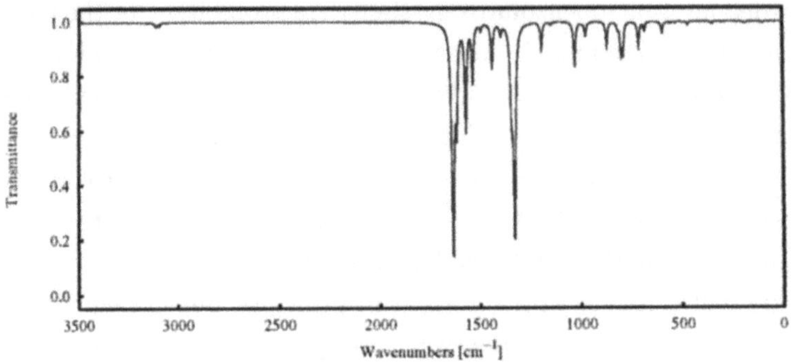

5 Reaction Rates

Reaction rates were calculated with the canonical variant of variational transition state theory (VTST) [20]. Quantum effects (tunneling) were included *a posteriori*, using the semiclassical small curvature tunneling (SCT) correction [20]. The reaction rate calculations were done with the POLYRATE program [21]. The minimum energy path (MEP) of the reaction entering the VTST calculations was computed pointwise at B3-LYP/6-31G* level, using a quadratic steepest descent (QSD) algorithm [22] as implemented in MOLPRO [12]. Starting from the tightly optimized saddle point of the BKR, using a step size of $s = 0.2$ a_0, the complete MEP was traced out by 28 points. The Hessian was calculated at 12 geometries along the path, selected automatically by the dynamic regeneration scheme of the QSD algorithm [22]. Additionally, the energy was computed on CCSD(T)/6-31G* level at the stationary points and at one additional point on the MEP, located at a distance of $s = 0.1$ a_0 from the saddle point in the direction of the reactant. Using the

B3-LYP/6-31G* reaction path data plus the CCSD(T)/6-31G* energies, it was possible to obtain an approximation of the reaction rates at the CCSD(T) level using the interpolated VTST method (IVTST) [23].

The reaction rates thus computed show similar trends to those reported for 5-methyl-4-nitrobenzofuroxan [13]: For a given level of *ab initio* or density functional theory, the different levels of VTST lead to quite similar results. The reaction rates obtained from simple transition state theory, form VTST, and from IVTST differ less than by 8 %. The increase of the rates due to tunneling effects is comparably small. Less than 20 % of the total rate constant is due to tunneling, even for temperatures as low as 200 K. In addition, the rate constants show pure Arrhenius behaviour, as is visible from the Arrhenius-plot of the rate constants in Fig. 6. There is no deviation from the linearity of the plot. The reaction rates have a much greater dependency on the level

Fig. 6. Arrhenius-plot of the reaction rates of the Boulton-Katritzky rearrangement of 4-nitrobenzofuroxan.

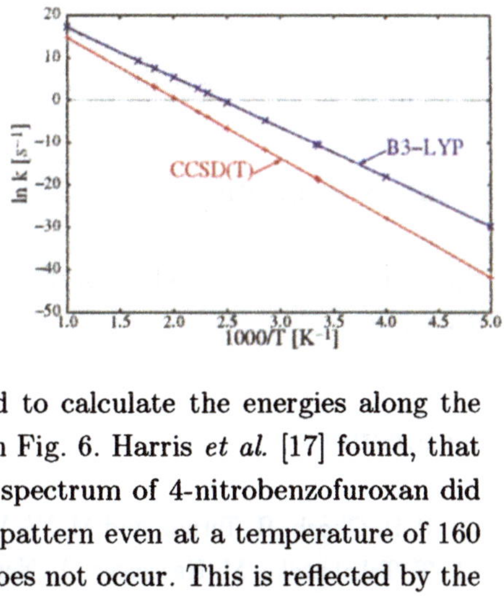

of the electronic structure theory used to calculate the energies along the reaction path, as can also be seen from Fig. 6. Harris *et al.* [17] found, that the ABC pattern of the proton NMR spectrum of 4-nitrobenzofuroxan did not alter toward a time-averaged AB_2 pattern even at a temperature of 160 ^oC, indicating that fast tautomerism does not occur. This is reflected by the CCSD(T) rates allowing a rapid reaction at temperatures over 500 K. The B3-LYP values, which would allow the reaction at around 400 K, must be considered too large. Thus, it can be concluded that the reaction rate of the BKR of 4-nitrobenzofuroxan is dominated by the reaction barrier. Recrossing and tunneling effects are small. However, this also results in the necessity of a very high level of *ab initio* theory (i.e. CCSD including triples corrections) on which the reaction barrier needs to be computed, making the study of the reaction rates of the BKR a genuine high performance computing task.

6 Summary and Conclusion

The reaction mechanism of the Boulton-Katritzky rearrangement was investigated by using high level *ab initio* methods. All results indicate that the reaction proceeds in one step via a planar transition state (at least for benzofuroxans). Substituents in the 5-position have significant impact on the reaction by slightly lowering the activation barrier and by resulting in an exothermic reaction. Alternative reactions, as for instance the ring opening of the furoxan ring, show lower activation barriers but result in endothermic reactions and are thus unlikely to be observed. All computed results are in agreement with experimental data and indicate that electron correlation effects are extremely important for this molecular class. Especially, it is the transition state of the ring opening of the furoxan ring that requires at least triples corrections at the coupled-cluster level for calculating energies and methods beyond currently available density functionals and MP2 theory for computing the geometry. Vibrational spectra and rate constants were computed for the Boulton-Katritzky rearrangement of 4-nitrobenzofuroxan and 5-methyl-4-nitrobenzofuroxan.

Acknowledgements

Computer time on the NEC SX-4 of the HLRS Stuttgart is kindly acknowledged.

References

1. P. B. Ghosh, B. Ternai, and M. W. Whitehouse, *Med. Res. Rev.* **1**, 159 (1981).
2. T. Takabatake, M. Hasegawa, T. Nagano, and M. Hirobe, *Chem. Pharm. Bull.* **38**, 128 (1990).
3. The patent literature, especially of the last years, contains many references to benzofuroxans as insecticides, fungicides, algicides and antibacterial agents.
4. A. J. Boulton and P. B. Ghosh, *Adv. Heterocycl. Chem.* **10**, 1 (1969).
5. A. Gasco and A. J. Boulton, *Adv. Heterocycl. Chem.* **29**, 251 (1981).
6. A. R. Katritzky and M. F. Gordeev, *Heterocycles* **35**, 483 (1993).
7. W. Friedrichsen, *Arenofurazan-1-oxide*, volume E8c, p. 769, in Houben-Weyl, Methoden der Organischen Chemie, edited by E. Schaumann, (Thieme, Stuttgart, 1994).

8. R. M. Paton, *1,2,5-Oxadiazoles*, volume 4, p. 229–265, in Comprehensive Heterocyclic Chemistry II, edited by A. R. Katritzky, C. W. Rees, and E. F. V. Scriven, (Pergamon Press, New York, 1995).
9. A. J. Boulton, P. B. Ghosh, and A. R. Katritzky, *J. Chem. Soc. (B)* 1004 (1966).
10. G. Rauhut, *J. Comput. Chem.* **17**, 1848 (1996).
11. Gaussian 94, Revision D.1, M. J. Frisch, G. W. Trucks, H. B. Schlegel, P. M. W. Gill, B. G. Johnson, M. A. Robb, J. R. Cheeseman, T. Keith, G. A. Petersson, J. A. Montgomery, K. Raghavachari, M. A. Al-Laham, V. G. Zakrzewski, J. V. Ortiz, J. B. Foresman, J. Cioslowski, B. B. Stefanov, A. Nanayakkara, M. Challacombe, C. Y. Peng, P. Y. Ayala, W. Chen, M. W. Wong, J. L. Andres, E. S. Replogle, R. Gomperts, R. L. Martin, D. J. Fox, J. S. Binkley, D. J. Defrees, J. Baker, J. P. Stewart, M. Head-Gordon, C. Gonzalez, and J. A. Pople, Gaussian, Inc.,Pittsburgh PA, 1995.
12. MOLPRO is a package of *ab initio* programs written by H.-J. Werner and P. J. Knowles, with contributions from R.D. Amos, A. Berning, D.L. Cooper, M.J.O. Deegan, A.J. Dobbyn, F. Eckert, C. Hampel, T. Leininger, R. Lindh, A. W. Lloyd, W. Meyer, M.E. Mura, A. Nicklass, P. Palmieri, K. Peterson, R. Pitzer, P. Pulay, G. Rauhut, M. Schütz, H. Stoll, A.J. Stone, and T. Thorsteinsson. Version 98.2, University of Birmingham, UK, 1998 (see `http://www.tc.bham.ac.uk/molpro/`).
13. F. Eckert, G. Rauhut, A. Katritzky, and P. J. Steel, *J. Am. Chem. Soc.* **121**, (1999) in press.
14. I. M. Takakis and P. M. Hadjimihalakis, *J. Heterocycl. Chem.* **28**, 1373 (1991).
15. I. M. Takakis and P. M. Hadjimihalakis, *J. Heterocycl. Chem.* **29**, 121 (1992).
16. F. Eckert and G. Rauhut, *J. Am. Chem. Soc.* **120**, 13478 (1998).
17. R. K. Harris, A. R. Katritzky, S. Øksne, A. S. Bailey, and W. G. Paterson, *J. Chem. Soc.* 197 (1963).
18. P. Pulay, G. Fogarasi, G. Pongor, J. E. Boggs, and A. Vargha, *J. Am. Chem. Soc.* **105**, 7037 (1983).
19. G. Rauhut and P. Pulay, *J. Phys. Chem.* **99**, 3093 (1995).
20. D. G. Truhlar, B. C. Garrett, and S. J. Klippenstein, *J. Phys. Chem.* **100**, 12771 (1996), and references therein.
21. R. Steckler, Y. -Y. Chuang, P. L. Fast, E. L. Coitiño, J. C. Corchado, W. -P. Hu, Y. -P. Liu, G. C. Lynch, K. A. Nguyen, C. F. Jackels, M. Z. Gu, I. Rossi, S. Clayton, V. S. Melissas, B. C. Garret, A. D. Isaacson and D. G. Truhlar, POLYRATE Version 7.3, University of Minnesota, Minneapolis 1997.
22. F. Eckert and H. J. Werner, *Theor. Chem. Acc.* **100**, 21 (1998).
23. A. Gonzales-Lafont, T. N. Truong, and D. G. Truhlar, *J. Phys. Chem.* **95**, 8875 (1991).

High Level Quantum-Chemical Computations on the Cyclizations of Enyne Allenes

Matthias Prall and Peter R. Schreiner

Institute of Organic Chemistry, Georg-August-University, Tammannstr. 2, 37077 Göttingen, Germany

1. Introduction

The ring-closure reactions of the enediyne and enyne-allene moieties of the natural antitumor antibiotics neocarzinostatin (**1**), calicheamicin (**4**), and dynemicin (**5**) are held responsible for the DNA-cleavage abilities of these potent pharmacophors. Neocarzinostatin, first isolated from *Streptomyces carzinostaticus* in 1961 [1], consists of the key chromophore **1** (Fig. 1.1) bound to a 113-amino acid apoprotein. [2,3] The enediyne moiety of **1** is activated by a thiol nucleophile, rearranges to enyne-cumulene **2**, and cyclizes to the highly reactive biradical **3** (Fig. 1.1), a reaction akin to the Myers-Saito cyclization. [4] Since **1** binds to the minor groove of DNA, [5-7] **3** is able to abstract hydrogens from adenine or thymine moieties [8] leading to cell damage. While calicheamicin **4** and dynemicin **5** (Fig. 1.1) undergo Bergman cyclizations to give rise to 1,4-didehydrobenzene biradicals, the neocarzinostatin chromophore **1** reacts differently.

Myers and independently Saito *et al.* demonstrated that enyne-allenes **7** show reactivity patterns similar to the enyne-cumulene core **6** (Fig. 1.2). [9-12] The C^2-C^7 cyclization, now known as the "Myers-Saito cyclization," of **7** yields α,3-didehydrotoluene (**8**) and toluene derivatives (**9**) after abstraction of hydrogens from suitable H-donors such as DNA. In contrast to the endothermic (8.5$\pm$1.0 kcal mol^{-1} [13]) Bergman cyclization, the Myers-Saito reaction is quite exothermic (-15$\pm$3 kcal mol^{-1} or less, *vide infra* [10]). In addition, the cyclization products **8** are partially conjugated σ,π-biradicals [10] which are likely to be less reactive than the σ,σ-biradicals obtained from Bergman-type cyclizations. [12,14]

Although the conjugated π-system of enyne-allenes **7** may offer a wide variety of cyclization modes, only the Myers-Saito products **8** were obtained until Schmittel *et al.* demonstrated that **7** also undergoes a novel C^2-C^6-ring closure leading to methyl fulvene derivatives **10** (Fig 1.3). [15-19] For convenience, we will henceforth refer to the C^2-C^6 cyclization as the "Schmittel" cyclization. Although the Myers-Saito reaction is generally energetically

[1] Adapted with permission from *J. Am. Chem. Soc.* **1999**, *121*, 8615-8627. Copyright 1999 American Chemical Society.

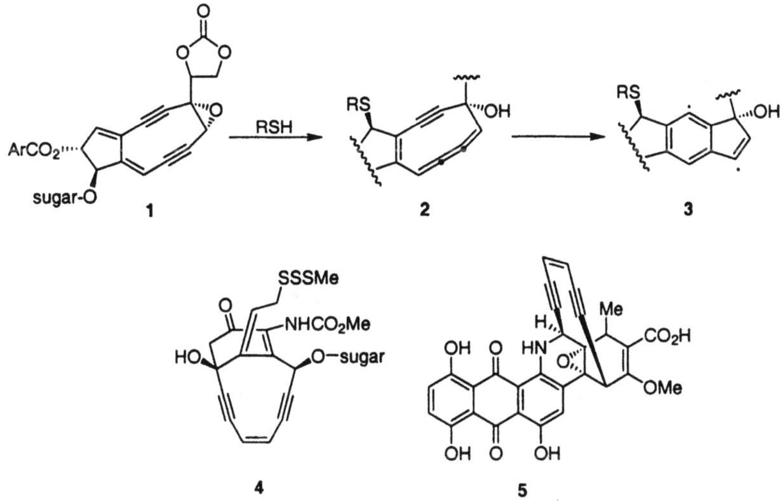

Fig. 1.1. Cyclization reaction of the neocarzinostatin chromophore **1** and the enediyne antibiotics calicheamicin **4** as well as dynemicin A **5**.

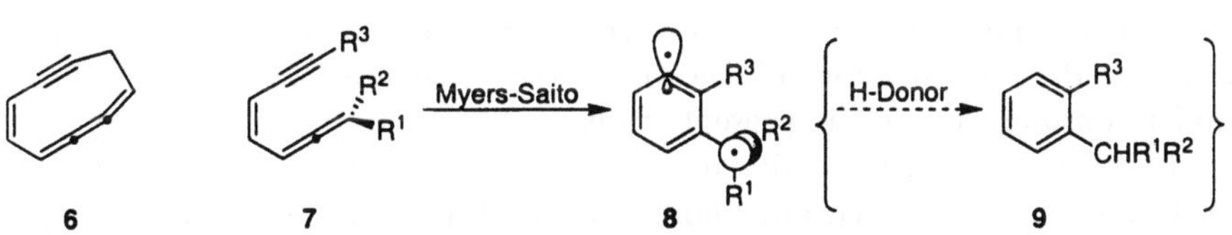

Fig. 1.2. Enyne-cumulene core **6** and the Myers-Saito cyclization reaction of enyne-allenes **7**.

much more favorable, Schmittel *et al.* showed that the methyl fulvenes **10** can be obtained as the major products when the alkyne terminal hydrogen is replaced by phenyl [19] or bulky groups like *tert*-butyl or trimethylsilyl. [20,21] Experimental analyses suggest that the Schmittel products **10** also are σ,π-biradicals which are able to cleave DNA. [16,17]

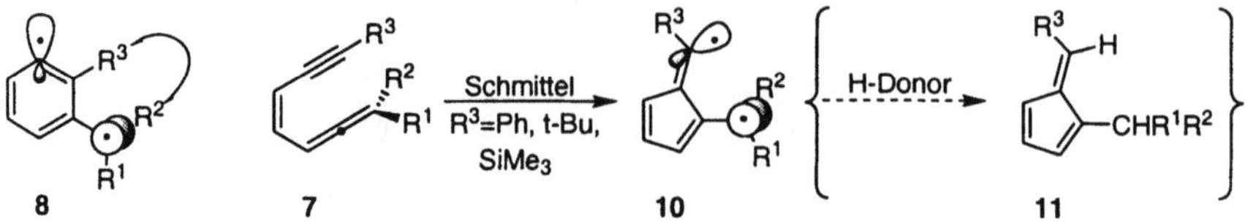

8 **7** **10** **11**

Fig. 1.3. C^2-C^6-cyclization ("Schmittel"-cyclization) of substituted enyne-allenes **7**.

In the present study a suitable approach for computational studies on Myers-Saito as well as the Schmittel cyclizations will be identified. Since limited experimental data are available for the Myers-Saito reaction, we use these together with high-level single point energies at the CCSD(T) and BCCD(T) levels with cc-pVDZ basis sets for the parent reactions to gauge the quality of other high-level methods and of the much less time-consuming BLYP/cc-pVTZ//BLYP/6-31G(d) computations.

Furthermore, highly accurate barriers and reaction enthalpies for the title reactions of the parent systems were computed. High-level predictions for the heat of formation of the Myers-Saito (**8**) and Schmittel products (**10**) follow. While the first ΔH_f^0 is experimentally known, the error bars are relatively large (± 3 kcal mol^{-1}); the ΔH_f^0 of **10** is, at least to our knowledge, unknown. Finally, the unpaired spin distributions of the key species were examined to aid in the prediction of their hydrogen abstraction ability; this also includes computations of the singlet-triplet separations of all biradicals under consideration in the present study.

In order to model and rationalize ring strain effects in enyne-allene analogs of the neocarzinostatin chromophore **1**, we used the carbocycles **12** to analyze the regioselectivity of the ring closures (Myers-Saito *vs.* Schmittel) and the energetic consequences for the formation of the biradical intermediates (Fig 1.4).

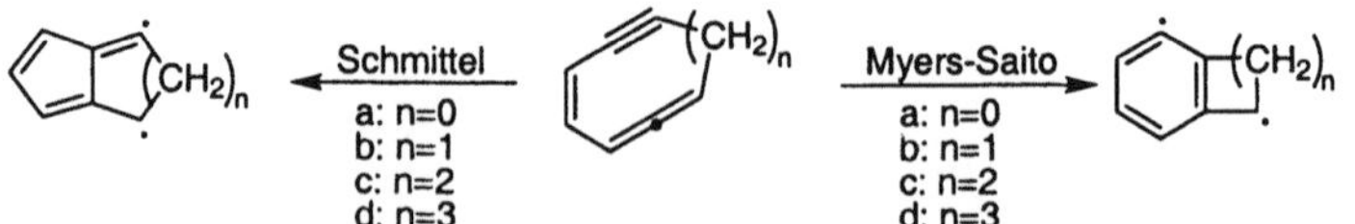

14 a-d **12 a-d** **13 a-d**

Fig. 1.4. Ring-closures of **12 a-d** and their cyclization products.

2. Computational Methods

Geometries of all stationary points were optimized using analytical energy gradients within density functional theory (DFT), [22] utilizing the BLYP (Becke-Lee-Yang-Parr) combination of gradient-corrected functionals, [23,24] as implemented in the Gaussian 94 program package. [25] For comparison, some computations were also carried out with Becke's three-parameter hybrid functional (B3) [26] and with Perdew's functional (BPW91). [27] Three basis sets were employed: 6-31G(d) for geometry optimizations as well as Dunning's correlation-consistent valence double-ζ (cc-pVDZ) and triple-ζ (cc-pVTZ) basis sets [28,29] for single point energy evaluations using the BLYP/6-31G(d) geometries. Closed-shell singlet states were computed using restricted wavefunctions. For open-shell singlets we utilized unrestricted broken-spin-symmetry wavefunctions (UBS). Triplet states were evaluated with an unrestricted approach (U). All stationary structures were identified as minima (number of imaginary frequencies, NImag = 0) or transition structures (NImag = 1) by computation of analytic vibrational frequencies, which were also used to compute zero-point vibrational energies (ZPVE) and thermal enthalpy contributions ($\Delta H_{298} - \Delta H_0$).

For the most critical structures, we evaluated single point energies utilizing the coupled-cluster with single and double excitations plus perturbatively included triplet excitations, CCSD(T) [30,40] as well as the Brueckner-doubles (BD) [39] coupled cluster approach, BCCD(T) [31-33] with Hartree-Fock reference wavefunctions. All coupled-cluster calculations were carried out with the cc-pVDZ basis set.

3. Results and Discussion

3.1 Choice of theoretical approach

The choice of level of theory to study the title reactions is critical. Pure DFT approaches appear to be superior to hybrid Hartree-Fock functionals as they offer a reasonable compromise to study biradical systems such as the Bergman cyclizations. [34-37] Since both **8** and **10** are σ,π-biradicals with small ΔE_{ST}'s, and are almost planar, an unrestricted broken-spin symmetry (UBS) wavefunction where

$$\mathrm{HOMO}_\alpha = \frac{\mathrm{HOMO} + \mathrm{LUMO}}{\sqrt{2}} \text{ and } \mathrm{HOMO}_\beta = \frac{\mathrm{HOMO} - \mathrm{LUMO}}{\sqrt{2}}$$

should be a reasonable approximation of the true wavefunction. [37] A well-written detailed discussion of this approach may be found in Cramer's work on *para*-benzyne chemistry. [37] The expectation values for the spin operator $<S^2>$ for Slater determinants for such wavefunctions is not very meaningful due to apparent singlet-triplet mixing; we note that restricted wavefunctions for **8** and **10** are unstable with respect to symmetry breaking.

Most of the above problems are not encountered when pure DFT wavefunctions are used (we have tested BLYP and BPW91, *vide infra*). This conclusion, reached independently by one of us, [34,35] Chen *et al.*, [36] and Cramer *et al.* [37,38] seems generally be true for biradicals as well as carbenes. As Cramer pointed out, it is not the case that the HF-DFT functionals are intrinsically flawed but is likely that the added HF character decreases the quality of the Kohn-Sham solutions for these electronically difficult systems. [37]

The high degree of multireference character (in the classical HF formalism) also implies that a high-level coupled-cluster wavefunction may suffer from symmetry breaking due to the fact that the reference orbitals may not be optimal for the state under consideration. As suggested by Cramer, [37] the use of Brueckner orbitals [39] may salvage this approach for low-symmetry systems. The BD-orbitals are determined in the presence of the correlation perturbation so that the amplitudes of the single excitations vanish. [40]

3.2 Energies and structures of the parent system (Table 3.1)

Similar to the transition state of the Bergman cyclization, [35,37] the transition structures for the two cyclizations virtually display no biradical character. This is evident from the fact that the energies of the transition structures computed at restricted and unrestricted levels of theory are essentially identical, and that the spin operator expectation value for unrestricted wavefunctions is virtually zero. Hence, the experimental barrier (*via* $\mathbf{TS_{MS}}$) for the Myers-Saito cyclization should be well reproduced at the levels of theory employed here. [44,53] Myers remarked [10] that the heat of formation of the parent enyne-allene (**7**, 118 kcal mol^{-1}) may have be corrected downward by 1-2 kcal mol^{-1}, [41] so that the activation barrier for the Myers-Saito cyclization should be about 23 kcal mol^{-1}, a value which is very well reproduced at BLYP, CCSD(T), and BCCD(T). The agreement of the different levels for the Schmittel cyclization barrier (*via* $\mathbf{TS_S}$; experimentally not known, computed: 29-35 kcal mol^{-1}) is equally good.

The energies of the open-shell products are much more difficult to compute; the experimental heat of formation of **8** also has a relatively large error bar ($\pm$3 kcal mol^{-1}) [42] which adds to the 1-2 kcal mol^{-1} error for **7**. Hence, the experimental reaction enthalpy for the Myers-Saito cyclization is in the range of -13$\pm$4 kcal mol^{-1}. While the BCCD(T) result agrees within 1 kcal mol^{-1} with the experimental value, CCSD(T) suffers from symmetry breaking. The agreement with BLYP employing the 6-31G(d) basis set is reasonably good and shows a systematic error of about 6 kcal mol^{-1}. The reaction enthalpies for the Schmittel reaction computed at the various levels center around 14$\pm$5 kcal mol^{-1}, whereby BCCD(T) lies at the lower and BLYP at the high end.

Table 3.1. Comparison of computed Gibbs free activation and reaction enthalpies for the Myers-Saito and Schmittel (C^2-C^6-cyclization of enyene-allenes) at 298 K at various levels of theory.

level of theory	Myers-Saito cycl. $\Delta^{\ddagger}G_{298}$ (MS)	$\Delta_R G_{298}$ (MS)	Schmittel cycl. $\Delta^{\ddagger}G_{298}$ (S)	$\Delta_R G_{298}$ (S)
CASSCF(10,10)/6-31G(d) [44]	29	4	37	18
MR-CI+Q [44][a,b]	25	-21	35	12
BPW91/6-31G(d)	16.6	-13.8	28.6	9.2
BPW91/cc-pVTZ//BPW91/6-31G(d)	18.2	-9.4	30.2	13.0
BLYP/6-31G(d)	21.7	-7.8	31.5	14.2
BLYP/cc-pVTZ//BLYP/6-31G(d)[c]	22.7	-3.7	34.4	19.3
CCSD(T)/cc-pVDZ//BLYP/6-31G(d)[c]	22.2	-26.0	34.7	17.3
BCCD(T)/cc-pVDZ//BLYP/6-31G(d)[c]	22.2	-13.6	34.8	10.0
Experiment	21.8±0.8[d]	-15±3[d]	n.a.	n.a.

[a] Using an Huzinaga's (9s5p) AO DZ basis set in Dunning's [4s2p] contraction with d-functions added ($\alpha d = 0.7$) on carbon and a (4s) AO —> [2s] contracted basis for hydrogen. [43] [b] These values are not ZPVE- or thermally corrected (*i.e.*, H);
[c] Thermal corrections from BLYP/6-31G(d) vibrational frequency calculations.
[d] ref. [10]; Myers *et al.* note that the heat of formation for the parent enyne-allene may have to be corrected downward by 1-2 kcal mol^{-1} so that $\Delta^{\ddagger}G_{MS}$ and ΔG_{MS} are more likely around 23.3±1 and -13.5±4 kcal mol^{-1}, respectively.

The heat of formation for **10** has, to the best of our knowledge, not been determined experimentally or computationally. To convince ourselves that an isodesmic approach works well for biradical species, we first derived the experimentally known heat of formation of **8** by equation 1 (Table 3.2). [42] Although the agreement between theory and experiment is just satisfactory, the derived value of 107 kcal mol^{-1} for the ΔH_f^0 of **8** sheds new light on the finding of Wenthold *et al.* that bond energy additivity calculations, which give ΔH_f^0 (**8**) = 107.6±1.7 kcal mol^{-1}, do not agree with the experimental heat of formation. As Wenthold's MCSCF calculations also predict a somewhat larger value (105-106 kcal mol^{-1}), it is likely that the ±3 kcal mol^{-1} error bar has a positive sign and that the ΔH_f^0 of **8** should indeed be slightly higher.

Equation 3 (Table 3.4) was then used to derive the heat of formation of **10** in a similar fashion using 45.4 kcal mol^{-1} as the ΔH_f^0 of 2-methyl fulvene (isodesmic equation 2; Table 3.3). Hence, our best estimate for ΔH_f^0 (**10**) is 129 ±4 kcal mol^{-1}. With the two heats of formation for **8** and **10** at hand, we compute the difference of the two cyclization reactions as about 26±4 kcal mol^{-1}, which agrees well with the computed values at BLYP (23 kcal mol^{-1}) and BCCD(T) (24 kcal mol^{-1}).

An important unanswered question is where the spin densities in **8** and **10** are located, *i.e.*, which of the spin-resonance structures (**8a - d** and **10a**

Table 3.2. Comparison of the computed and experimentally derived heat of formation of **8** *via* eq. 1. Absolute energies in -au, ΔH values in kcal mol^{-1}. The experimental heats of formation in kcal mol^{-1} are: 19.7 (benzene [46]), 79.1±2 (phenyl radical [42]), and 48±2 (tolyl radical [45]).

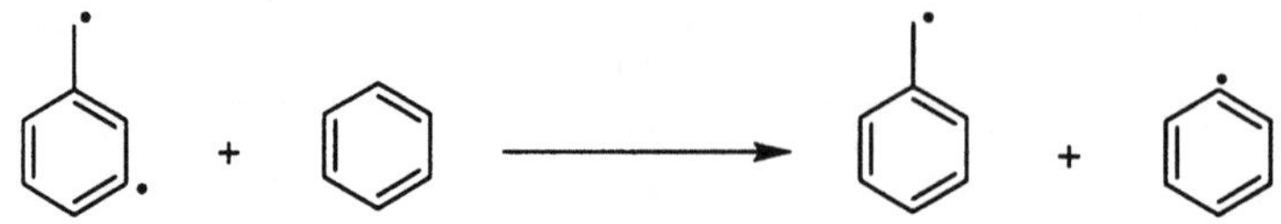
(1)

8

ΔH_0 at	$\Delta_R H_0$	ΔH_f^0 (**8**)
BLYP/cc-pVTZ[a]	+0.6	107.0
CCSD(T)/cc-pVDZ[a]	+17.1	90.2
BCCD(T)/cc-pVDZ[a]	+0.6	107.0
exptl.	-4.4	103±3 [42]

[a] Using the geometries and ZPVE-corrections at BLYP/6-31G(d).

Table 3.3. Determination of the heat of formation of 2-methylfulvene *via* eq. 2 to derive the heat of formation of **10**. Absolute energies in -au, ΔH values in kcal mol^{-1}. The experimental heats of formation in kcal mol^{-1} are: 53.6 (fulvene), -25.3 (methylcyclopentane), and -18.4 (cyclopentane). [46]

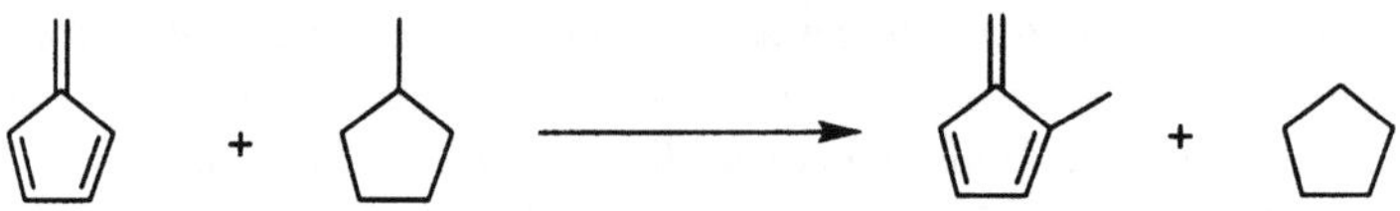
(2)

ΔH_0 at	$\Delta_R H_0$	ΔH_f^0 methylfulvene
BLYP/cc-pVTZ[a]	-0.2	44.7
CCSD(T)/cc-pVDZ[a]	-1.2	45.5
BCCD(T)/cc-pVDZ[a]	-1.3	45.4

[a] Using the geometries and ZPVE-corrections at BLYP/6-31G(d).

Table 3.4. Determination of the heat of formation of **10** *via* eq. 3. Absolute energies in -au, ΔH values in kcal mol^{-1}. The experimental heats of formation in kcal mol^{-1} are: 103±3 (**8**), 12.0 (toluene), and 45.4 (methylfulvene).

(3)

8 **10**

ΔH_0 at	$\Delta_R H_0$	ΔH_f^0 (**10**)
BLYP/cc-pVTZ[a]	-9.7	126.7
CCSD(T)/cc-pVDZ[a]	12.3	148.7
BCCD(T)/cc-pVDZ[a]	-7.4	129.0

[a] Using the geometries and ZPVE-corrections at BLYP/6-31G(d).

- d, Fig. 3.1) are important. This, together with the ΔE_{ST}, determines the hydrogen abstracting ability of these biradicals in DNA-cleavage reactions. Since the singlet-triplet separations are relatively small, the coefficients of the distributed spin densities of the open-shell singlets and triplets are expected to be qualitatively very similar (Fig. 3.2 depicts the triplet spin densities for **8** and **10**). [47]

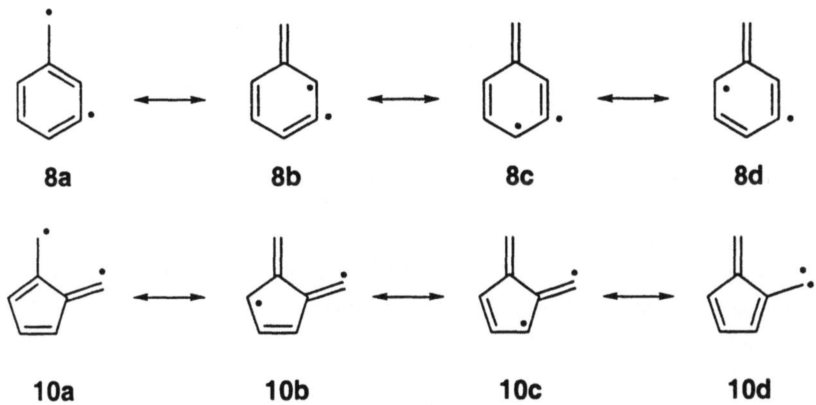

Fig. 3.1. Spin resonance structures for Myers-Saito and Schmittel cyclization biradical products.

While the σ-radical site located in an sp^2 orbital is highly localized and cannot interact with the bonds or other orbitals, the π-radical site is delocalized as found in, *e.g.*, benzylic or allylic systems. The consequences for **8** are not dramatic because the aromatic benzene electron sextet does not effectively interact with the benzylic radical; spin resonance depiction **8a** followed by **8c** are the most appropriate representations of the spin density for the Myers-Saito cyclization product. The Schmittel product, on the other hand, shows a rather different pattern where **10a** is the least important as it does not account for the true distribution of the π-radical density which is best presented by **10c**. This explains why the difference in the heats of formation between **8** and **10** (26 kcal mol^{-1}) is less than expected if **10** were not stabilized. The theoretical difference should be about 13 + 21 = 33 kcal mol^{-1} for benzylic- and aromatic stabilization of **8**, [48,49] but is smaller because **10** also benefits from allylic resonance. Hence, while **8** is best described as a benzyl/σ-diradical, **10** has an allyl/σ-structure.

3.3 Effect of ring size on cyclization (Table 3.5, Fig. 3.3, 3.4)

Given the excellent agreement with experiment of BLYP and BCCD(T) for the cyclization barriers and the sizable deviations in the reaction enthalpies for the parent reactions, we will argue quantitatively for the barriers, but will refrain to relative energy comparisons for the reaction enthalpies (BLYP/6-31G(d) Gibbs free enthalpies are used) in the following sections.

As found for the Bergman-cyclization, [34-36] ring strain has a large effect on the cyclization barriers and enthalpies for the Myers-Saito and Schmittel reactions. Hence, this is a way of adjusting the barriers for synthetic anti-tumor pharmacophors based on cyclic enyne-allenes, and the ratio of Schmittel over Myers-Saito cyclizations. Since the Bergman and Schmittel cyclizations are endothermic, the strain of the products should dominate the barriers for the ring closures of cyclic reactants; the exothermic Myers-Saito reaction should be sensitive to ring strain effects of the reactants. [50]

First of all, cyclic enyne-allenes **12a-d** have smaller ring-closure barriers than the parent systems due to the inherent strain of the starting materials. The Myers-Saito cyclizations of these are endothermic for the small rings **12a** as well as **12b** and about equally exothermic for **12c** and **12d**. The enthalpies for the Schmittel cyclizations of **12a-d** parallel these findings: the cyclizations are less or as endothermic as the parent reaction, and the reverse barriers are considerably smaller (9 - 14 kcal mol^{-1} vs. about 24 kcal mol^{-1} for the parent system). Temperature and entropic effects generally lower the cyclization barrier by about 1 kcal mol^{-1} but have virtually no effect on the reaction enthalpies (Table 3.5).

As expected, the cyclization reactions of the highly strained seven-membered ring enyne-allene **12a** are highly unfavorable. Since the reverse barrier is negligible, the Myers-Saito cyclization of **12a** *via* $\mathbf{TS_{MS}}$ will not occur. The corresponding Schmittel reaction has a lower barrier (24 *vs.* 34 kcal mol^{-1}) but about the same reaction enthalpy (about 14 kcal mol^{-1}) as the parent system. Hence, if cyclization occurs and if suitable hydrogen donors are present, the seven-membered enyne-allene **12a** (if it can be prepared) should give Schmittel products only.

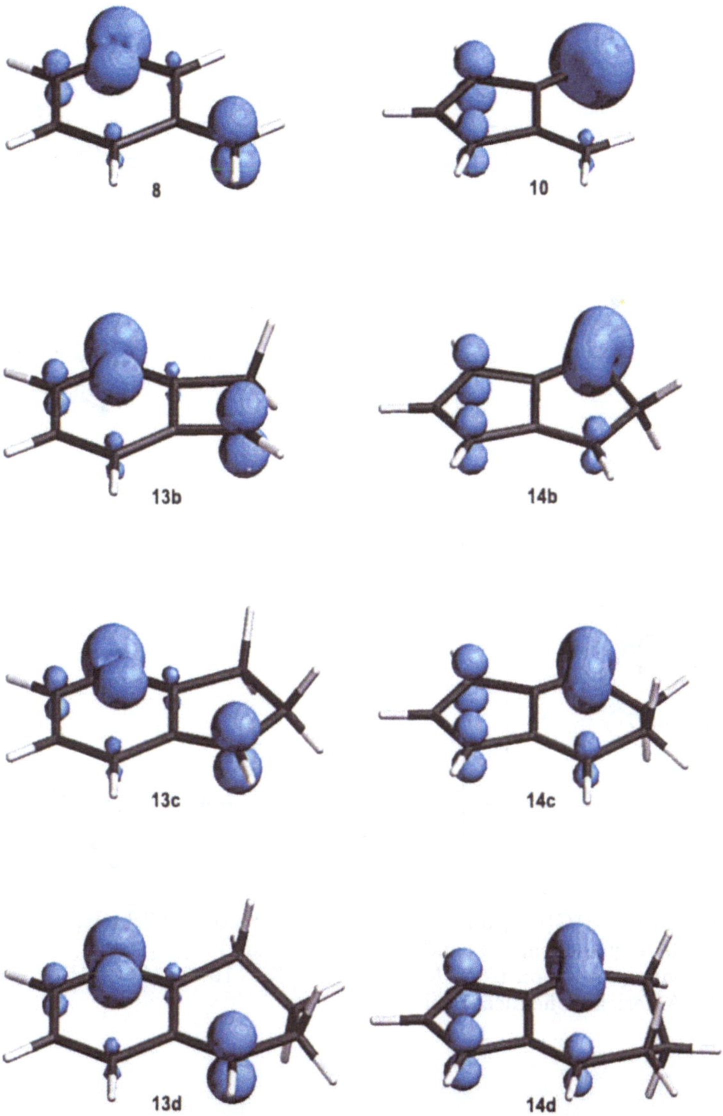

Fig. 3.2. Open-shell spin densities of the Myers-Saito and Schmittel cyclization products. The depicted spin density coefficients were computed for the triplet states. As the singlet-triplet separations are very small (see text), these spin densities are also a very good qualitative approximation of the open-shell singlet spin distributions.
Reprinted with permission from *J. Am. Chem. Soc.* **1999**, *121*, 8615-8627. Copyright 1999 American Chemical Society.

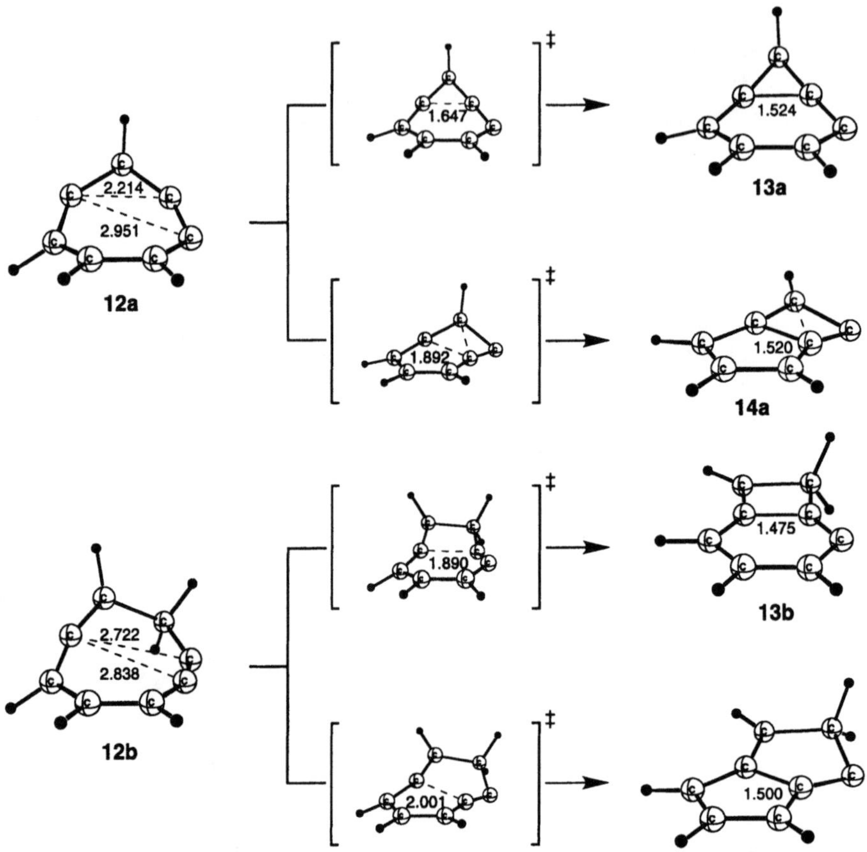

Fig. 3.3. BLYP/6-31G(d) geometries of the Myers-Saito and Schmittel cyclizations of monocyclic enyne-allenes (7- and 8-membered Ring).

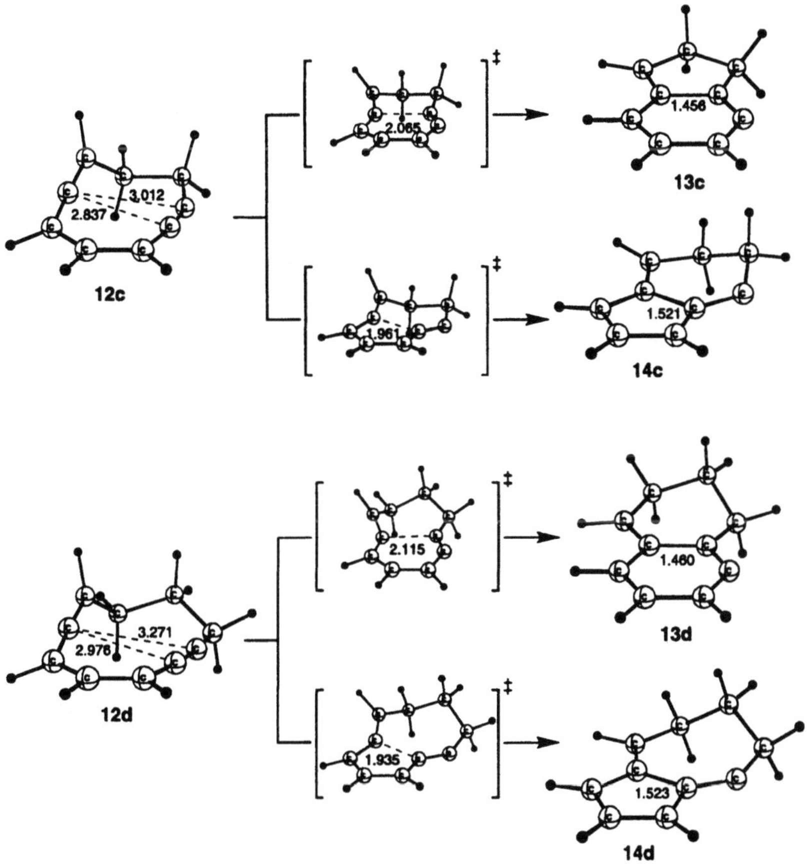

Fig. 3.4. BLYP/6-31G(d) geometries of the Myers-Saito and Schmittel cyclizations of monocyclic enyne-allenes (9- and 10-membered Ring).
Reprinted with permission from *J. Am. Chem. Soc.* **1999**, *121*, 8615-8627. Copyright 1999 American Chemical Society.

Table 3.5. Gibbs free activation and reaction enthalpies for the cyclization of cyclic enyne-allenes **12a-d** in kcal mol^{-1} at BLYP/6-31G(d); PG = point group; NImag = number of imaginary frequencies.

Ring Size	Structure	ΔH	ΔG_{298K}	PG/NImag
7	**12a**	0.0	0.0	$C_1/0$
	TS$_{MS}$12a	10.0	9.0	$C_1/1$
	13a	9.7	9.0	$C_1/0$
	TS$_S$12a	25.4	24.0	$C_1/1$
	14a	14.3	14.0	$C_1/0$
8	**12b**	0.0	0.0	$C_1/0$
	TS$_{MS}$12b	18.4	17.8	$C_1/1$
	13b	3.8	3.8	$C_S/0$
	TS$_S$12b	15.7	15.2	$C_1/1$
	14b	2.2	2.6	$C_1/0$
9	**12c**	0.0	0.0	$C_1/0$
	TS$_{MS}$12c	14.7	14.2	$C_1/1$
	13c	-11.8	-11.8	$C_S/0$
	TS$_S$12c	12.8	12.3	$C_1/1$
	14c	4.1	4.2	$C_1/0$
10	**12d**	0.0	0.0	$C_1/0$
	TS$_{MS}$12d	16.6	16.6	$C_1/1$
	13d	-12.6	-11.5	$C_1/0$
	TS$_S$12d	23.0	22.2	$C_1/1$
	14d	7.2	7.3	$C_1/0$

The eight-membered enyne-allene **12b** has a significantly lower barrier (15 *vs.* 34 kcal mol^{-1}) and is much less endothermic (3 *vs.* 14 kcal mol^{-1}) for the Schmittel reaction, making it the thermodynamically most favorable one in this series. Since the Myers-Saito cyclization for this ring size shows about the same endothermicity (4 kcal mol^{-1}) but a higher barrier (18 kcal mol^{-1}), the Schmittel reaction should be the favored cyclization mode for the eight-membered monocyclic enyne-allene **12b**. These findings can be rationalized by the difference in olefin strain energies for the cyclobutene (31 kcal mol^{-1}) and cyclopentene (7 kcal mol^{-1}) [51] moieties in the biradical products **13b** and **14b**, respectively.

Remarkably, the nine-membered ring, which is also a part of neocarzinostatin **1** and the hydrocarbon **6**, is associated with the lowest barriers to cyclization for both reactions (14 and 12 kcal mol^{-1}; Table 3.5 and Fig. 3.5). Moreover, the Myers-Saito reaction enthalpy for this ring size is the lowest of all enyne-allenes considered in the present study. Part of the driving force for the cyclizations of **12c** originates from the small olefin ring strain in the cyclopentene and cyclohexene moieties formed in the two reactions. Another reason is the short distance between the reacting centers in the ring which is considerably longer in the parent system (3.195 and 3.755 Å in **7** *vs.* 2.837 and 3.012 Å in **12c**, for the Schmittel and Myers-Saito pathways, respectively), resulting in higher cyclization activation energies for **7**. The Schmittel cyclization barrier increases much more than the Myers-Saito barrier on going to the ten-membered ring, favoring the latter transformation. Hence, ring strain effects and distances of the reacting carbon centers not only are highly important in these cyclization reactions but may also be used to design enyne-allene containing antitumor antibiotics which cyclize under physiological or other specific conditions.

As found for the parent system, the ΔE_{ST}'s are very small for the cyclic biradicals (Table 3.6); the Myers-Saito products have a slight preference for a singlet ground state, while the Schmittel products apparently favor the triplet state. However, taking into consideration that the singlet states are generally computed to high in energy (due to the neglect of the second configuration), this conclusion may not hold true for higher level computations. Nevertheless, the hydrogen-abstracting ability and hence anti-tumor reactivity of both biradical types should therefore be rather high. [52] The triplet (and hence the qualitatively very similar open-shell singlet) spin densities are also depicted in Fig. 3.2.

4. Conclusions

The Myers-Saito and C^2-C^6-cyclizations (termed "Schmittel-cyclization") of parent and monocyclic enyne-allenes employing various high-level *ab initio* and density functional theory approaches were studied. As was found for the related Bergman-cyclization of enediynes, pure (non-hybrid) DFT methods

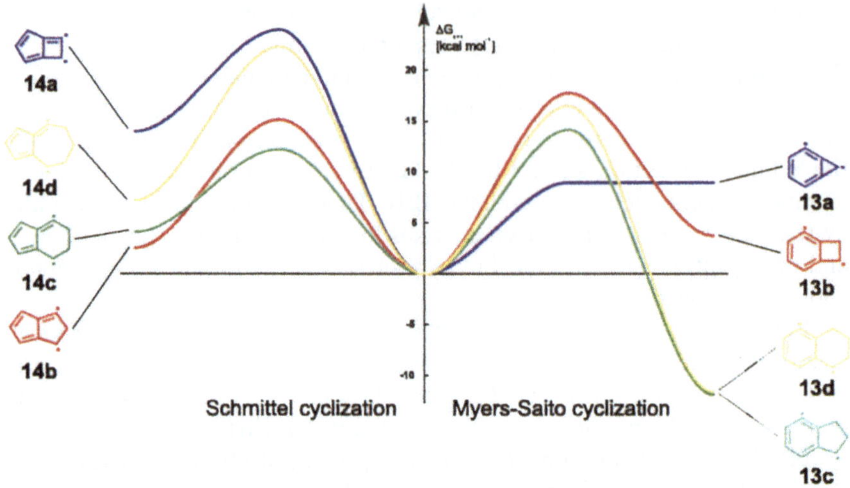

Fig. 3.5. Comparison of the barriers and enthalpies of the Myers-Saito and Schmittel cyclizations of monocyclic enyne-allenes **12a - d**.
Reprinted with permission from *J. Am. Chem. Soc.* **1999**, *121*, 8615-8627. Copyright 1999 American Chemical Society.

Table 3.6. Singlet-triplet separations (ΔE_{ST}, *i.e.*, a positive value denotes a triplet ground state) of Myers-Saito and Schmittel cyclization products of monocyclic enyne-allenes at two levels of theory.[a]

level of theory	BLYP/6-31G(d)	BLYP/cc-pVTZ//BLYP/6-31G(d)[b]
13a	-4.3[c]	-4.8[c]
13b	-0.8	-0.7
13c	-0.8	-0.7
13d	-0.8	-0.8
14a	-8.3[c]	-8.9[c]
14b	2.3	1.4
14c	-0.4	0.5
14d	1.8	1.5

[a] An unrestricted broken-spin symmetry wavefunction was used for the open-shell singlet structures; the triplets were computed with an unrestricted approach.
[b] ZPVEs taken from BLYP/6-31G(d) frequency computations.
[c] No open-shell singlet was located, see text
Reprinted with permission from *J. Am. Chem. Soc.* **1999**, *121*, 8615-8627. Copyright 1999 American Chemical Society.

(*e.g.*, BLYP and BPW91) are qualitatively suitable to study these types of systems.

The cyclization barriers are exceptionally well reproduced relative to experimental data owing to the single-reference character of the corresponding transition structures. Best results for the parent system are obtained with high-level coupled-cluster methods in conjunction with Brueckner-Orbitals {BCCD(T)} and correlation-consistent basis sets (cc-pVDZ). The CCSD(T) method performs similarly but suffers from symmetry breaking problems for the Myers-Saito product giving unacceptably low energies.

The computed enthalpy of formation (ΔH_f^0) of the Myers-Saito product (107±4 kcal mol^{-1}) compares relatively well to the experimental value (103±3 kcal mol^{-1}) where an upward correction within the error bars seems indicated. Using the same approach the ΔH_f^0 of the Schmittel-product is predicted to be 129±4 kcal mol^{-1}. A spin-density analysis shows that the Myers-Saito product is a benzyl/σ-diradical, whereas the Schmittel-biradical has an allyl/σ-structure.

Ring strain effects on the two reaction modes were examined at BLYP/6-31G(d) for the cyclizations of monocyclic enyne-allenes (ring sizes = 7-10 carbons). All considered cyclization reactions have smaller barriers and reaction enthalpies relative to the parent system due to inherent ring strain in the starting materials. The formation of the Schmittel product is kinetically slightly favored for the eight-membered enyne-allene. The Myers-Saito cyclization of the nine-membered ring is associated with the smallest reaction barrier and the highest exothermicity.

As the singlet-triplet separations of all cyclization products are very small (1 kcal mol^{-1}), the hydrogen-abstracting ability of such cyclic systems should be rather high.

Acknowledgement. This work was supported by the Fonds der Chemischen Industrie (Liebig-Fellowship for PRS), the Deutsche Forschungsgemeinschaft (Schr 597/1-1 and 597/3-1). We are grateful to Prof. A. de Meijere for his support and the Höchstleistungsrechenzentrum Stuttgart (HLRS) for generous allotments of computer time. MP thanks A. Wittkopp and C. Rinderspacher for their help.
Parts of this report are taken from a paper accepted for publication in the *J. Am. Chem. Soc.* **1999**, *121*.

References and Notes

1. Ishida, N.; Miyazaki, K.; Kumagai, K. M.; Rikimaru, M. *J. Antibiot.* **1965**, *18*, 68-76.
2. Nicolaou, K. C.; Dai, W.-M. *Angew. Chem.* **1991**, *103*, 1453-1481.
3. Wisniewski-Grissom, J.; Gunawardena, G. U.; Klingberg, D.; Huang, D. H. *Tetrahedron* **1996**, *52*, 6453-6518.
4. Brückner, R.; Suffert, J.; Abraham, E.; Raeppel, S. *Liebigs Ann.* **1996**, 447-456.

5. Lee, S. H.; Goldberg, I. H. *Biochemistry* **1989**, *28*, 1019-1026.
6. Goldberg, I. H. *Free Radicals Biol. Med.* **1987**, *3*, 41.
7. Povirk, L. F.; Dattagupta, N.; Warf, B. C.; Goldberg, I. H. *Biochemistry* **1981**, *20*, 4007-4014.
8. Kappen, L. S.; Goldberg, I. H. *Biochemistry* **1983**, *22*, 4872-4878.
9. Myers, A. G.; Kuo, E. Y.; Finney, N. S. *J. Am. Chem. Soc.* **1989**, *111*, 8057-8059.
10. Myers, A. G.; Dragovich, P. S.; Kuo, E. Y. *J. Am. Chem. Soc.* **1992**, *114*, 9369-9386.
11. Saito, K.; Watanabe, T.; Takahashi, K. *Chem. Lett.* **1989**, 2099-2102.
12. Saito, I.; Nagata, R.; Yamanaka, H.; Murahashi, E. *Tetrahedron Lett.* **1990**, *31(20)*, 2907-2910.
13. Roth, W. R.; Hopf, H.; Horn, C. *Chem. Ber.* **1994**, *127*, 1765-1769.
14. Nicolaou, K. C.; Maligres, P.; Shin, J.; de Leon, E.; Rideout, D. *J. Am. Chem. Soc.* **1990**, *112*, 7825-7826.
15. Schmittel, M.; Strittmatter, M.; Kiau, S. *Angew. Chem.* **1996**, *108(16)*, 1952-1954.
16. Schmittel, M.; Kiau, S.; Siebert, T.; Strittmatter, M. *Tetrahedron Lett.* **1996**, *37(43)*, 7691-7694.
17. Schmittel, M.; Steffen, J.-P.; Auer, D.; Maywald, M. *Tetrahedron Lett.* **1997**, *38(35)*, 6177-6180.
18. Schmittel, M.; Keller, M.; Kiau, S.; Strittmatter, M. *Chem. Eur. J.* **1997**, *3(5)*, 807-816.
19. Schmittel, M.; Strittmatter, M.; Kiau, S. *Tetrahedron Lett.* **1995**, *36(28)*, 4975-4978.
20. Gillmann, T.; Hülsen, T.; Massa, W.; Wocadlo, S. *Synlett* **1995**, 1257-1259.
21. Rodriguez, A.; Garcia, J. G.; Ramos, B.; Pratt, L. M. *Tetrahedron Lett.* **1995**, *36(41)*, 7391-7394.
22. Parr, R. G.; Yang, W. *Density Functional Theory of Atoms and Molecules*; Oxford University Press: New York, 1989.
23. Lee, C.; Yang, W; Parr, R. G. *Phys. Rev. B* **1988**, *37*, 785-789.
24. Becke, A. D. *J. Chem. Phys.* **1993**, *98*, 5648-5652.
25. Frisch, M. J.; Trucks, G. W.; Schlegel, H. B.; Gill, P. M. W.; Johnson, B. G.; Robb, M. A.; Cheeseman, J. R.; Keith, T.; Petersson, G. A.; Montgomery, J. A.; Raghavachari, K.; Al-Laham, M. A.; Zakrzewski, V. G.; Ortiz, J. V.; Foresman, J. B.; Cioslowski, J.; Stefanov, B. B.; Nanayakkara, A.; Challacombe, M.; Peng, C. Y.; Ayala, P. Y.; Chen, W.; Wong, M. W.; Andres, J. L.; Replogle, E. S.; Gomperts, R.; Martin, R. L.; Fox, D. J.; Binkley, J. S.; Defrees, D. J.; Baker, J.; Stewart, J. P.; Head-Gordon, M.; Gonzalez, C.; Pople, J. A., 1995, Revision C.3, Gaussian 94, Pittsburgh PA, Gaussian, Inc.
26. Becke, A. D. *J. Chem. Phys.* **1993**, *98*, 5648-5652.
27. Perdew, J. P.; Burke, K.; Wang, Y. *Phys. Rev. B* **1996**, *54*, 16533-16539.
28. Dunning Jr., T. H.; *Basis Sets: Correlation-Consistent;* in *The Encyclopedia of Computational Chemistry*, Schleyer, P. v. R., Allinger, N. L., Clark, T., Gasteiger, J., Kollman, P. A., Schaefer III, H. F. and Schreiner, P. R. (Ed.); John Wiley & Sons: Chichester, 1998, 88-115.
29. Dunning Jr., T. H.; *J. Chem. Phys.* **1989**, *90*, 1007-1023.
30. Scuseria, G. E. *Chem. Phys. Lett.* **1991**, *176*, 27.
31. Scuseria, G. E. *Chem. Phys. Lett.* **1994**, *226*, 94.
32. Scuseria, G. E.; Schaefer III, H. F. *Chem. Phys. Lett.* **1987**, *142*, 354-358.
33. Handy, N. C.; Pople, J. A.; Head-Gordon, M.; Raghavachari, K.; Trucks, G. W. *Chem. Phys. Lett.* **1989**, *164*, 185-192.
34. Schreiner, P. R. *J. Chem. Soc. Chem. Commun.* **1998**, 483-484.

35. Schreiner, P. R. *J. Am. Chem. Soc.* **1998**, *120*, 4184-4190.
36. Chen, W.-H.; Chang, N.-Y.; Yu, C.-H. *J. Phys. Chem.* **1998**, *102*, 2584-2593.
37. Cramer, C. J. *J. Am. Chem. Soc.* **1998**, *120*, 6261-6269.
38. Cramer, C. J.; Truhlar, D. G.; Falvey, D. E. *J. Am. Chem. Soc.* **1997**, *119*, 12338-12342.
39. Brueckner, K. A. *Phys. Rev.* **1954**, *96*, 508.
40. Gauss, J. *Coupled-Cluster Theory*; in *The Encyclopedia of Computational Chemistry*, Schleyer, P. v. R., Allinger, N. L., Clark, T., Gasteiger, J., Kollman, P. A., Schaefer III, H. F. and Schreiner, P. R. (Ed.); John Wiley & Sons: Chichester, 1998, 615-636.
41. Benson, S. W.; Garland, L. J. *J. Phys. Chem.* **1991**, *95*, 4915.
42. Wenthold, P. G.; Wierschke, S. G.; Nash, J. J.; Squires, R. R. *J. Am. Chem. Soc.* **1993**, *115*, 12611-12612.
43. Dunning, T. H. *J. Chem. Phys.* **1970**, *53*, 2823-2833.
44. Engels, B.; Hanrath, M. *J. Am. Chem. Soc.* **1998**, *120*, 6356-6361.
45. McMillen, D. F.; Golden, D. M. *Annu. Rev. Phys. Chem.* **1982**, *33*, 493-532.
46. Lias, S. G.; Bartmess, J. E.; Liebman, J. F.; Holmes, J. L.; Levin, R. D.; Mallam, W. G. *J. Phys. Chem. Ref. Data* **1988**, *17 (S1)*, 1-872.
47. Wavefunction, I., 1998, 5.1.1, Spartan, Irvine, Wavefunction, Inc.
48. Roth, W. R.; Ruf, G.; Ford, P. W. *Chem. Ber.* **1974**, *107*, 48-52.
49. Schleyer, P. v. R.; Jiao, H. *Pure Appl. Chem.* **1996**, *68*, 209-218.
50. Hammond, G. S. *J. Am. Chem. Soc.* **1955**, *77*, 334-338.
51. Schleyer, P. v. R.; Williams, J. E.; Blanchard, K. R. *J. Am. Chem. Soc.* **1970**, *92*, 2377-2386.
52. Chen, P. *Angew. Chem.* **1996**, *108*, 1584-1585.
53. Engels, B.; Lennartz, C.; Hanrath, M.; Schmittel, M.; Strittmatter, M. *Angew. Chem.* **1998**, *110*, 2067-2070

MD Simulation of a Phospholipid Bilayer

Ralf Schmelter, Alfons Geiger

Physikalische Chemie, Universität Dortmund

Abstract. We report first results of a molecular dynamics simulation of a fully hydrated dipalmitoyl-*sn*-glycero-phosphatidylcholin (DPPC) bilayer using the NγT-ensemble. Because of the large size of the simulated system, a parallel version of the simulation package MOSCITO [1] has been developed, which employs a dynamic loadbalancing algorithm to ensure uniform workload among the processors.

1 Introduction

Phospholid bilayers can be regarded as models of eukaryontic cell membranes and therefore have been heavily investigated by experimental and theoretical methods [2] [3] [4]. In recent years it became feasible to perform molecular dynamics (MD) and Monte-Carlo (MC) simulations of such systems, by which one hopes to gain unterstanding of the structural and dynamic properties of the membranes at the atomic level. While the first simulations suffered from short simulation times, too small system sizes or improper treatment of the long range electrostatic interactions, the more recent simulations had overcome these limitations, mainly because of the rapid growth in available computer power [5] [6].

Nevertheless there are some open questions regarding the simulation of a fully hydrated phospholipid bilayer and one of the most important and most discussed topics is the choice of the ensemble to be used [7] [12]. Many of the first simulations were carried out in an NVT ensemble. In this case one has to choose the box dimensions carefully, which means to select a proper area per lipid and the correct multilamellar repeat distance. Unfortunately there are many different experimental values for this area [8] (they range between 0.56 nm^2 and 0.72 nm^2) and the multilamellar repeat distance. Another problem is the validation of the forcefield, because there are only few experimental results which can be easily and unambiguously compared with the results of the simulation. One of the most important parameters to be compared to the experiment is the deuterium order parameter S_{CD} of the alkyl chains, which almost exclusively depends (at least in the simulation) on the area per lipid [9], so that after a proper choice of the area even a totally inadequate forcefield can give sensible results for the order parameters, while the other properties may be far from reality.

Consequently simulations using the NPT ensemble were carried out, in which the box dimensions are adjusted during the simulation to produce a given pressure. To take the anisotropy of the membrane system into account, the box was allowed to change its dimensions independently in each direction. But the question arose, whether an isotropic pressure is an adequate choice or whether different pressure values parallel und perpendicular to the bilayer normal have to be used. This led to the important question of the surface tension of a phospholipid bilayer [10]. While some argued, that a bilayer should have a vanishing surface tension [11], Feller and Pastor showed [13] that because of the periodic boundary conditions and the relatively small size of a simulation box these arguments are not valid for a MD–simulation. They gave an approximation of the surface tension to be applied, based on the statistical theory of bilayers. Chiu [14] on the other hand applied a surface tension which was taken from a lipid-monolayer–air-interface and which was therefore much higher than the surface tension proposed by Pastor and Feller. So in the moment there is no consensus on the value of the surface tension to be applied. Nevertheless the NγT ensemble is mostly used now for bilayer simulations. In this work γ is obtained from the approximation by Pastor and Feller.

Another crucial point in the simulation setup is the treatment of the electrostatic interactions. In the early days of bilayer simulation more or less elaborate cutoff-schemes were used, because the computer power was too limited to use more appropriate schemes like the Ewald summation. Some authors even argued, that Ewald summation would enhance the artifacts of the periodic boundaries [7]. But in the last years the computers have become faster and new methods for the efficient treatment of the long range electrostatic interactions, like the smooth particle mesh Ewald summation technique (SPME) [15] used in our simulation have emerged. Thus the computational advantages of a simple cutoff-scheme is not significant anymore (unfortunately, this is not really true for parallel simulations, because of the increased ammount of communication required by the more advanced schemes, which reduce the scalabilty of such algorithms).

2 Simulation Setup

The bilayer consists of 200 molecules of dipalmitoyl–*sn*–glycero–phosphatidylcholin (DPPC, see fig. 1) and 6125 water molecules, which corresponds to about 30 water molecules per lipid and a water weight fraction of c = 0.43, which is above the experimentally determined value of a fully hydrated multilamellar DPPC system in the L_α phase (where values of 0.36 and 0.40 are reported [7]). The AMBER forcefield [16] was used for the DPPC, except for the alkyl chains, where the all-atom OPLS [17] forcefield was employed. The missing parameters for the improper dihedral of the ester link were taken from CHARMM [18], which should cause no problems, because of the inflexibility of this region. The charges were calculated by the RESP procedure [19], fitting them to the electrostatic potential of dipropyl–*sn*–glycero–phosphatidylcholin. For the alkyl chains the charges from the OPLS forcefield

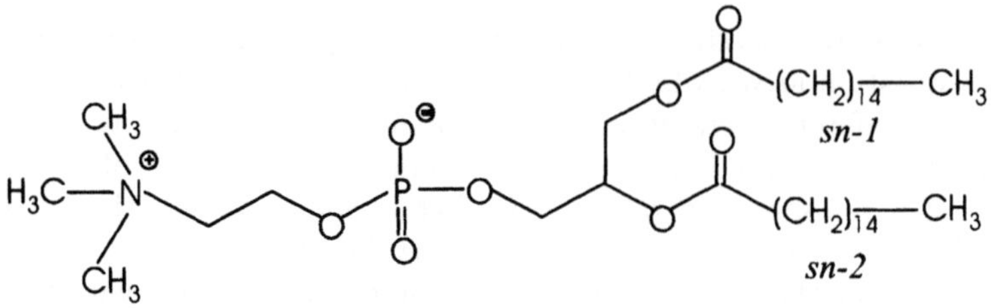

Fig. 1. The structure of the DPPC molecule

were used. The quantum chemical calculations of the electrostatic potential where performed on the HF-6-31G*-level with Gaussian 94 [20]. The water model was flexible SPC/E [21], which differes from the normal SPC/E model [22] in that is has an angle bending term. A simulation timestep of 1 fs was chosen. The pressure and temperature was controlled by a Berendsen thermostat [23] with a coupling constant of 2.0 ps for the pressure and 0.2 ps for the temperature during the equilibration procedure and 100 ps and 1 ps for the rest of the simulation. The three box dimensions were allowed to change independently. The simulation was carried out at a temperature of 50 ^{0}C and a pressure of 1 atm in the direction of the bilayer normal (the y-axis) and -11 atm in the direction of the x-, and z-axis, which corresponds to a surface tension of $9.12 \cdot 10^{-3}\mathrm{Nm}^{-1}$. The electrostatic interactions were calculated by the SMPE method with an interpolation order of 4 and a grid spacing of 0.1 nm. A cutoff distance of 0.9 nm was chosen for the real space part of the interactions.

The system itself was prepared as follows: A single DPPC molecule was energy minimized, while having a slight repulsive force applied between the N-atom and the terminal C-atoms of the alkyl chains to bring the molecule in a linear form. Then the bilayer was build by inserting the minimized DPPC molecules (randomly rotated around the y-axis) on a $10 \times 10 \times 2$ grid in such a manner, that there were no unfavourable interactions between the molecules. This system was then simulated for 10 ps at a temperature of 10 K and an isotropic pressure of 1.0 atm. After this a water cap constisting of 6125 molecules (where the x- and z-dimensions were the same as the dimensions of the bilayer box) was added. This system was then equilibrated for 1.2 ns at the temperatur and pressure given above. After this equilibration a production run consisting of 600 ps was performed.

3 Computational Details

Because of the exceeding system size of 44375 interaction sites, the MOSCITO [1] simulation program was parallelized on the basis of the replicated data scheme. This can easily be incorporated into an existing simulation code.

Moreover the SPME method more or less requires a replicated data approach. Unfortunately this approach does not scale very well to a large number of processors, because the amount of communication per processor does not decrease with increasing number of processors (in fact it's slightly increasing because of the additional communication for the temperature and pressure scaling).

The replicated data approach works as follows: At the beginning of a timestep each node has the coordinates of all atoms , so each node could calculate all forces. This allows to partition the force calculation in any way that seems appropriate and this flexibility is the the main advantage of the replicated data scheme. After the forces are calculated the trajectory integration step and the SHAKE [24] algorithm need to be done. Accordingly we assign to each node a subset of molecules for which it has to know the forces and velocities. After the pair force calculations, each node sends a proper subset to any other node, so that at the end of this step each node knows all the forces on all atoms for which it has accomplish the integration and SHAKE step. After the integration is done, each node sends its subset of new coordinates to any other node, so that finally each node has the new coordinates of all atoms and the next timestep can be performed. As one can see, the amount of data that has to be sent or received by a node does not depend on the number of nodes but scales linearly with the number of atoms (to be more precise we have to say that it tends toward a constant as the number of nodes is increased). This means, that this approach is not feasible for a calculation on many nodes (see [27] for another approaches, which scales better, but is more complicated to implement). The maximum number of nodes one can use efficiently, strongly depends on the speed of the underlying network and the speed of the processors (the higher the network speed or the slower the processors, the higher the number of nodes that can be used).

The simple picture given above is only true, if one uses a cutoff-based treatment of the electrostatic interactions. The use of more appropriate methods, like the SPME technique or the Fast Multipole Method (FMM) [26] complicates this picture significantly, because additional communication during the force caculation is required. As we use the SPME method in our program, we will shortly discuss this method from a more communication centred point of view (for a more detailed discussion see [25]). The SPME method, like the normal Ewald summation, splits the electrostatic interactions in a real space part (which can be calculated with a normal cutoff-scheme) and a sum in the reciprocal space. For the calculation of the reciprocal space part the charges are 'interpolated' on a grid by cardinal B-splines of a given order. Then a 3-D FFT is performed, the grid points are modified and a back 3-D FFT transformation is added. After these steps the forces on the atoms due to the reciprocal part can be interpolated from the values of the grid. The first part of this scheme, the filling of the grid, is relativly easy to parallelize (in fact it requires no communication at all if one accepts a few redundant calculations, what we have done in our case). We assigned to each node incorporated in the calculation of the reciprocal part of the SPME (nodes N_{rec}) a region of the grid with a z-coordinate in a given range (see fig. 2(a)). Because of the

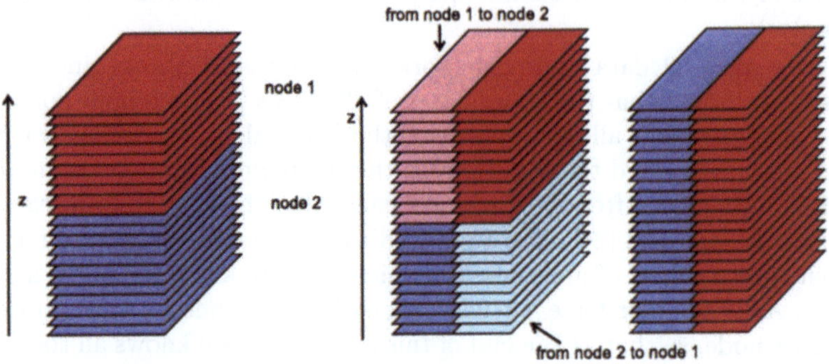

(a) Initial partitioning (b) Communication

Fig. 2. The parallelization of the recprocal part of the SMPE

use of cardinal B-splines for the grid assignment, one has to calculate only the B-spline coefficients of the atoms which are in the partition of the node or in its nearest neighbourhood. After the end of this step, every node N_{rec} has the values of all grid points in its slice. The next step is the 3-D Fast Fourier Transformation, which is split into two parts. The first part consists of a 2-D FFT in the the x-y-plane. This can be done locally on each node, because only grid points calculated by the node in the first step are involved. But for the second part, a FFT along the z-axis, parts of the grid must be exchanged between the nodes (see fig. 2(b)). This step requires additional und huge communication between the nodes N_{rec} and this is the limiting factor in our simulation. The next step involves a modification of the grid values and can be done without communication at all. After this a back 3-D FFT is performed analogous to the first 3-D FFT (and with an equal ammount of communication) and in the last step the forces are calculated from the grid coefficients, which requires only little communication between some nodes to exchange the grid coefficients in the border region. As one can see, it is desirable to do the reciprocal part of the SPME on as few nodes as possible, because of the additional communication required. In fact it can be advantageous to do it only on one node and therefore have no additional communication, even when this calculation takes longer than the other nodes need for the calculation of the other forces (see below).

A problem that arises when performing parallel calculations is that of load-balancing. Because the time required for a timestep is limited by the slowest node (which can be the slowest, because it has a weaker CPU or because it has to do more work than the other nodes). Therefore it is necessary to partition the work in a manner, that ensures that all nodes need the same time for their calculations. In the case of the MD-Simulation it is slightly more complicated, because there are two times to consider. The first time is the time needed to calculate the forces. This time has to be equal among all nodes, because each node has to wait for all other nodes in order to get the forces needed to do the integration and SHAKE step. The second time is the time needed to accomplish these steps, which are normally much shorter than the time needed for the force calculation, so that an unbalance during these steps is not that critical und is not considered here further. On the other hand it is not known a priori, what runtime characteristics a given partition will have. In our case this is mainly due to the fact, that the number density in the simulation box varies to some degree and therefore the number of nearest neighbours varies and with it the number of forces to calculate. Because of this we employed a dynamic loadbalancing scheme, which works as follows: First, all nodes are assigned to calculate the force acting on an equal number of atoms (because we use Newton's third law, some forces for a particular atom may in fact be computed on a different node than that node, which was assigned to the atom). One node additionally has to calculate the reciprocal SPME term. Now the time t_{real} required for the calculation of the real space part of the forces is taken on each node, as is the time t_{rec} to calculate the reciprocal part of the SPME. Next the partitioning is adjusted accoring to the deviation of $t_{real} + t_{rec}$ of each node from the mean value (if the node needed more time than the average it will get fewer forces to calculate and vice versa). Note that only the partition for the calculation of the real space part (which affects only t_{real}) is modified during this step. The partitioning of the reciprocal SPME part is treated seperately, because of the two additional communication steps needed. This requires, that all the nodes N_{rec} have to reach these two communication steps at the same time, because otherwise the other nodes are idle, until the last node is ready. But this goal is hardly ever reached when the particle distribution along the z-axis is not uniform. Fortunately a rather uniform particle distribution along the z-axis is given in our simulation, so this load balancing step works analogous to the balancing of the real space force calculation.

The assignment of the number of nodes to the calculation of the reciprocal force requires a special treatment. This is due to the fact, that an increase in this number can give an increase in the time each node needs for the overall calculation. This can be illustrated by the following example: Let's assume that we are working on three nodes. Node 1 is doing only the reciprocal force calculations, while the others are doing the rest. Node 1 needs 5.0 seconds for its calculations and node 2 and 3 need 4.5 seconds. Now it seems a good idea to split the work of the reciprocal force calculation among node 1 and node 2. But this means, that node 1 and 2 have to do two additional communication steps. If we assume, that the additional communication takes 1 second per node (this value is reasonable on computers like an SP2), after load balancing we get the following result: Node 1 and 2 need 5.33 seconds (2.5 seconds for the SPME, 1 second for the additional communication and 1.833 seconds for the calculations of the other forces) and node 3 needs 5.33 seconds for the calculation of the other forces. As one can see, the overall time increased. If the third node is also assigned to do a part of the SPME calculation the communication time per node in the reciprocal part will stay approximately constant. This assumption may seem questionable, because the ammount of data each node must communicate is slightly decreasing by about 10 percent, but on the other hand the number of communications per node doubles, and therefore the time lost due to the latency time needed to start a communication. Another factor is that the chance of load unbalace increases, because now three nodes instead of two have to reach the communication steps at the same time. With a reasonable value for the communication time of 1 seconds every node will need 5.67 seconds (1.67 seconds for the SPME, 1.0 seconds for the communication + 3 seconds for the calculation of the other forces). As one can see, the overall time increased instead of decrased as the SPME calculation was spread across more nodes and this would even be the case if the communication time per node would be decreased by the above mentioned 10 percent, which would result in an overall time of 5.57 seconds. This example shows that one has to use specific heuristics for the assignment of nodes to the reciprocal force calculation in addition to the normal load balancing mechanism, because these effects can not be captured by a simple scheme, especially when the needed communication times can not be predicted a priori.

Because of the not optimal scaling of the replicated data algorithm itself and the additional problems of the parallelization of the SPME method, we normally used only 10 nodes in our calculations. Table 1 and fig. 3 shows the speedups depending on the number of nodes used. As one can see it makes not much sense to use more than 10 nodes, especially if one has only a limited amount of computer time available and every derivation from the optimal speedup means a loss of valuable computer time.

4 Results

This simulation run was finished recently. Since then we are evaluating the trajectory (see fig. 4 for a snapshot of the simulation box). First results are given below:

The system equilibrated to an area per lipid of 0.56 nm^2 which is at the lower end of the experimentally determined values. But as Nagle [8] argued, the most likely experimental value of the area per lipid is around 62 nm^2, so the simulation value is significantly too low (about 10 percent). Another indication that this area is too small arises from the deuterium order parameters -S_{DC}, which strongly depend on the area per lipid [9] and proved to be too high in our simulation (see fig. 5). This implies further, that the mul-

nodes	speedup	time per timestep
1	1	43.2
2	1.95	22.1
3	2.9	14.9
4	3.8	11.4
6	5.6	7.7
8	7.3	5.9
10	8.9	4.8
12	9.8	4.4
14	10.6	4.1
16	11.2	3.9

Table 1. Dependence of the speedup on the number of nodes

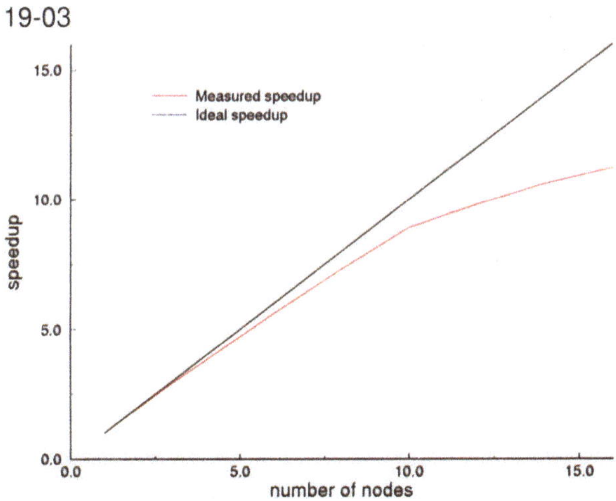

Fig. 3. Dependence of the speedup on the number of nodes

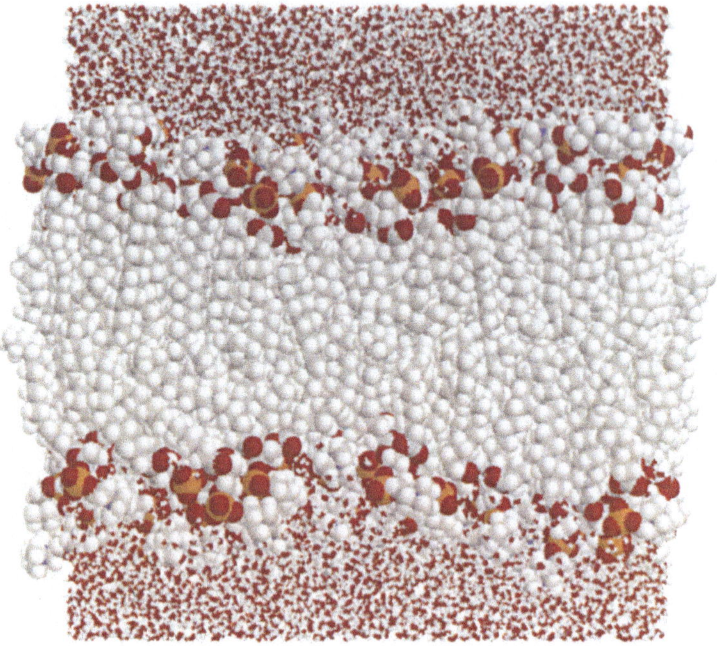

Fig. 4. Snapshot of the simulation box

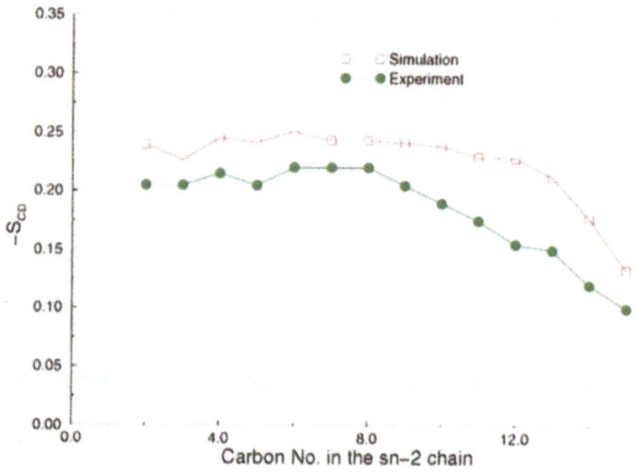

Fig. 5. Experimental and simulated S_{CD} parameters for the sn-2 chain

tilamellar repeat distance is too large (7.6 nm) and clearly shows, that the applied forcefield has some deficits. But the basic properties of a phospholipid membrane should nevertheless be captured by this simulation.

To ensure that the reduce area per lipid is not the result of a decoupling of the temperature of the DPPC molecules from the temperature of the water molecules, the temperatures of both molecule types were determined (see fig. 6). As one can see, the mean temperature of the two molecule types

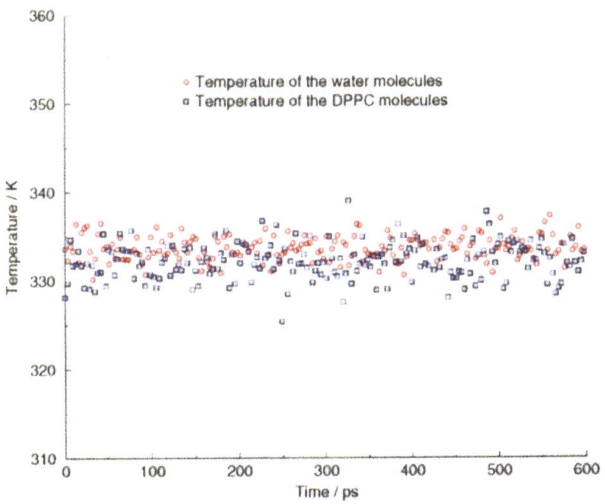

Fig. 6. Temperature of the water and the DPPC molecules throughout the simlation

are slightly different (the average temperature of the DPPC is about 1.5 K smaller than the temperature of the water molecules). But this difference is not large enough to explain the deminished value of the area per lipid as an effect of the lower temperature of the DPPC molecules (this would require a temperature difference of 10 K or more).

Fig. 7 shows the mean square displacement (MSD) of the center of mass (COM) of the DPPC molecules in y direction (this is parallel to the bilayer normal) and in the xz plane, which is perpendicular to the bilayer normal. Apart from the first 50 ps the curves are rather linear (see the dashed lines which are calculated by a linear regression in the range between 100 ps and 350 ps). The calculated diffusion constants are $D_y = 4.45 \times 10^{-7} cm^2 s^{-1}$ in the y direction and $D_{xz} = 2.67 \times 10^{-7} cm^2 s^{-1}$ in the xz plane. It may seem surprising that the lateral diffusion coefficient D_{xz} is smaller than the transversal diffusion coefficient D_y, because experiments suggest the opposite behaviour. But one has to remember that the simulation time is too short to

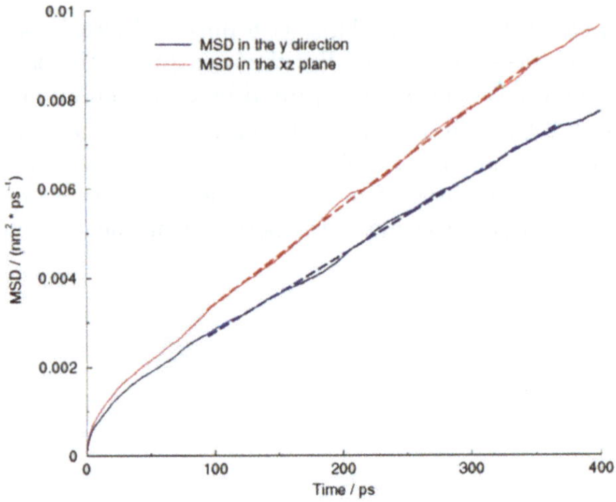

Fig. 7. Mean square displacement of the center of mass of the DPPC molecules in y direction and the xz plane

observe a real transversal diffusion and the value of D_y is only determind by undulation like motions, so that D_y will be significantly lower given a much longer simulation time. The value of D_{xz} is in excellent aggrement with a value recently obtained by Essman [5] from a 10 ns simulation.

Fig. 8 shows the mean square displacement of the COM of the water molecules in x, y and z direction and in the xz plane. As expected the mean square displacements in x and z direction are practically identical. The diffusion in y direction on the other hand is significantly lower especially for times above 30 ps, which results in a small curvature in the mean square displacement curve. This can easily explained by the fact, that the diffusion in y direction is hindered by the bilayer (see also the dependence of the overall diffusion coefficient on the position discussed below). The diffusion in the xz plane shows no peculiarities, as the MSD in x and z direction already suggest. To examine the diffusional behaviour further, the diffusion coefficient of the water molecules were calculated as a function of the position along the y-coordinate. The results are given in fig. 9. In this figure we also show the particle density of the water molecules as a function of the y coodinat. As one can see the diffusion constants drop from a value of about 3.35 cm^2s^{-1} in the bulk water between the bilayer to very low values for water molecules in the interior of the bilayer. It has to be mentioned that the values of the diffusion coefficients in regions with low water content could not be determined with high accuracy, because of the small ammount of water molecules found in this region. Another point to mention is the fact that no water molecules are found in the region of the alkyl chains. The low diffusion coefficients in the

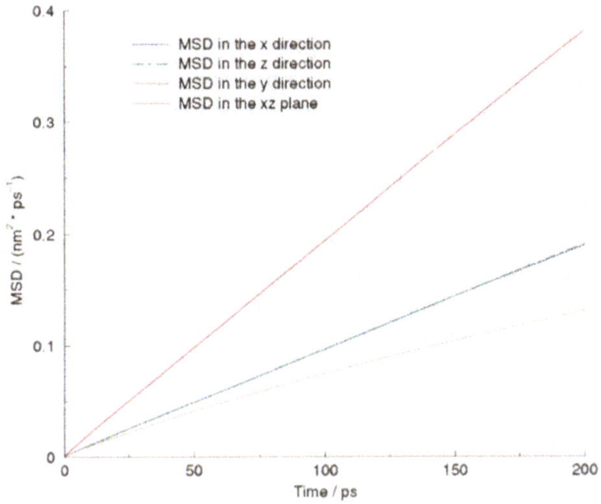

Fig. 8. Mean square displacement of the center of mass of the water molecules in x, y and z direction and the xz plane

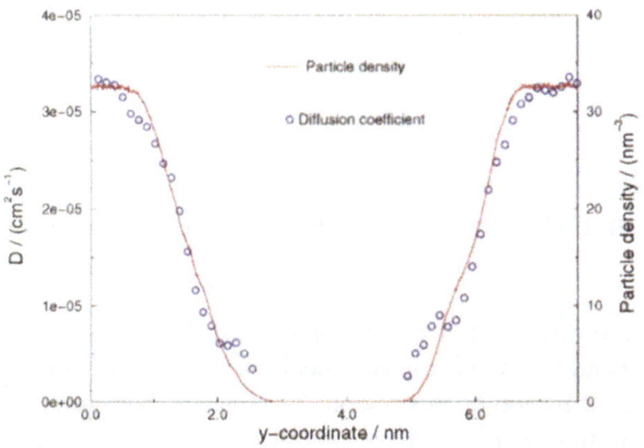

Fig. 9. Diffusion coefficient and particle density of water as a function of its y-coordinate

bilayer can be explained by the strong electrostatic interactions of the water molecules with the choline and phosphate groups. The plateau region visible at y coordinates of about 2 nm and 5.5 nm respectivetly can be explained by a 'trapping' of a single water molecule between two ester groups, but this has to be examined further to give a definite explanation. Another phenomenon that has to be examined further is the strong correlation of the water density and the diffusion coefficient.

As a last result some probability distributions of intramolecular distances between specific atoms are presented. In fig. 10 the probability density of the distance between the nitrogen atom in the choline group and the phosphorous atom in the phosphate group is shown. It can be seen, that there are

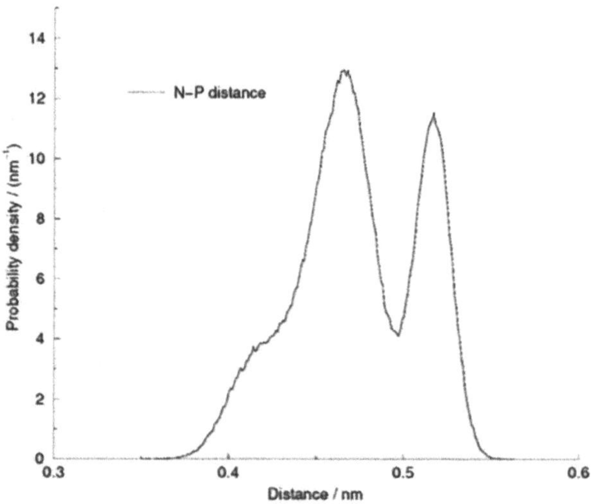

Fig. 10. Probability density of the intramolecular N–P distance

two peaks in the distribution which correspond to 2 different configurations of the DPPC headgroup. The second peak at about 0.52 nm corresponds to a stretched conformation of this region. Because of the positive net charge of the choline group and the negative net charge of the phosphate group this conformation may seem energeticly unfavorable in contrast to the conformation which corresponds to the first peak at 0.47 nm, where the charged centers are closer to one another. But one has to take into account, that the conformation of the second peak allows a greater contact with the surrounding water molecules, so that both configurations can be observed in comparable ammounts. It will be interesting to determnine the lifetime of the two configurations and the mechanism of the transformation between them. Such investigations are underway.

In fig. 11 distance distributions with a different qualitative behaviour are shown. Here the distances of phosphorous atom to the three carbon atoms

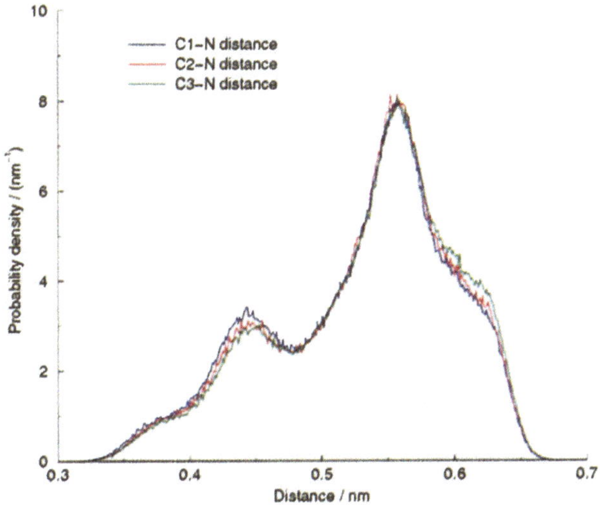

Fig. 11. Probability density of the intramolecular distance of the N-atom to the three carbon atoms of tetramethylammonium group

in the choline group are considered. As can be seen, the curves for the three different carbon atoms are virtually the same, which is a hint that the rotation of this group is not hindered very much. Once again there are two main peaks in this distribution, but this time the area under the first peak is less than half the area under the second peak. This is an indication for a conformation where one of the three carbon atoms is directed to the phosphorous atom, while the other two are showing in the opposite direction.

The last distance probability distributions, shown in fig. 12, give an impression of the overall shape of the DPPC moleules. The distributions of the distance between the nitrogen atom in the headgroup and the terminal carbon atoms of the two alkyl chains show that the DPPC molecule normally exist in a stretched conformation. But a considerable probability for a much shorter N-C distance can be observed. These shorter distanances correspond to configurations where the alkyl chains are no longer side by side, but where one (or even both) chains are orientated more perpendicular to the bilayer normal. This can be verified by the distribution of the distance between the two terminal carbon atoms, which is also shown in fig. 12. While the peak at about 0.5 nm suggests that the chains are close to each other most of the time, there are also conformations where this is no longer the case. This shows that there is a considerable ammount of disorder in the alkyl chains,

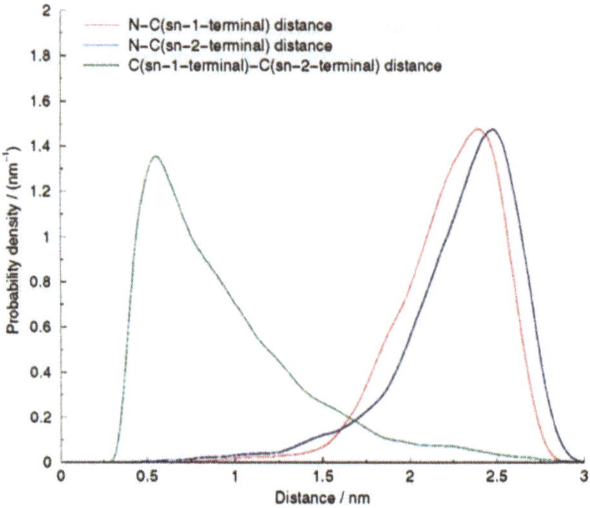

Fig. 12. Probability density of the intramolecular distance of the N-atom to the terminal carbon atoms of the sn-1 and sn-2 alkyl chains

which would for example not be the case, if the bilayer was in the gel phase, which is the stable form of the membrane at temperatures lower than about 40 ^{0} C.

References

1. Paschek, D.: Performing Moleculat Dynamics Simulations with MOSCITO. University of Dortmund, 1998
2. Pastor, R.: Molecular dynamics and Monte Carlo simulations of lipid bilayers. Curr. Opin. Struct. Biol. **4** (1996) 486–492
3. Tieleman, D. P., Marrink, S. J., Berendsen, H. J. C.: A computer perspective of membranes: molecular dynamics studies of lipid bilayer systems. Biochim. Biophys. Acta. **133** (1997) 235–270
4. Nagle, J. F. R., Zhang, R. , Tristam-Nagle, S., Sun, W., Petrahce, H., Suter, R. M.: X-ray structure determination of fully hydrated L_α phase DPPC bilayers. Biophys J. **70** (1996) 1419–1431
5. Essmann, U., Berkowitz, M. L.: Dynamic Proprties of Phospholipid Bilayers from Computer Simulation. Biophys. J. **76** (1999) 2081–2089
6. Husslein, T., Newns, D. M., Pattnaik, P. C., Zhong, Q. Moore, P. B., Klein, M. L.: Constant pressure and temperature molecular-dynamics simulation of the hydrated diphytanolpshophatidylcholine lipid bilayer. J. Chem. Phys. **109** (1998) 2826–2832
7. Tieleman, D. P., Berendsen, H. J. C.: Molecular dynamics simulation of a fully hydrated dipalmitoylphosphatidylcholine bilayer with different macroscopic boundary conditions and parameters. J. Chem. Phys. **105** (1996) 4871–4880)

8. Nagle, J. F.: Area/lipid of bilayers from NMR Biophys. J. **64** (1993) 1476–1481
9. Feller, S. E., Venable, R. M., Pastor, R. W.: Computer Simulations of a DPPC Phospholipid Bilayer: Structural Changes as a Function of Molecular Surface Area. Langmuir **13** (1997) 6555–6561
10. Roux, B.: Commentary: Surface Tension of Biomembranes. Biophys. J. **73** (1996) 1346–1347
11. Jähning, F.: What is the Surface Tension of a Lipid Bilayer Membrane? Biophys. J. **73** (1996) 1348–1349
12. Jakobsson, E., Subramaniam, S., Scott, H. L.: Strategic Issues in Molecular Dynamics Simulations of Membranes. Merz, K., Roux, B. (editors): Biological Membranes, Birkhäuser Boston 1996
13. Feller, S. E., Pastor, R. W.: On simulating lipid bilayers with an applied surface tension: periodic boundary conditions and undulations. Biophys. J. **71** (1996) 1350–1355
14. Chiu, S. -W., Clark, M., Balaji, V., Subramaniam, S., Scott, H. L., Jakobsson, E. : Incorporation of Surface Tension into Molecular Dynamics Simulation of an Interface: A Fluid Phase Lipid Bilayer Membrane. Biophys. J. **69** (1995) 1230–1245
15. Essmann, U., Perera, L., Berkowitz, M. L., Darden, T., Lee, H., Pdersen, L. G.: A smooth particle mesh Ewald method. J. Chem. Phys. **103** (1995) 8577–8593
16. Cornell, W. D., Cieplaj, P., Bayly, C. I., Gould, I. R., Merz, K. M., Ferguson, D. M., Spellmeyer, D. C., Fox, T., Caldwell, J. W., Kollman, P. A.: A Second Generation Force Field for the Simulation of Proreins, Nucleic Acids, and Organic Molecules. J. Am. Chem. Soc **117** (1995) 5179–5197
17. Kaminski, G., Duffy, E. M., Matsui, T., Jorgensen, W. L.: Free Energies of Hydration and Pure Liquid Properties of Hydrocarbons from the OPLS All-Atom Model. J. Phys. Chem. **98** (1994) 13077–13082
18. Schlenkrich, M., Brickmann, J., MacKerrel, A. D., Karplus, M.: An Empirical Potential Energy Function for Phospholipids: Criteria for Parameter Optimization and Applications. K. Merz, Roux, B. (editors): Biological Membranes Birkhäuser Boston 1996
19. Bayly. C. I., Cieplyk, P., Cornell, W. D., Kollman, P. A.: A well-behaved electrostatic potential based method using charge restraints for deriving atomic charges: The RESP model. J. Phys. Chem **97** (1993) 10269–10280
20. Frisch, M. J., Trucks, G. W., Schlegela, H. B., Gill, P. M. W., Johnson, B. G., Robb, M. A., Cheeseman, J. R., Keith, T., Petersson, G. A., Montgomery, J. A., Raghavachari, K., Al-Laham, M. A., Zakrzewski, V. G., Ortiz, J. V., Foresman, J. B., Cioslowski, J., Stefanov, B. B., Nanayakkara, A., Challacombe, Peng, M. C. Y., Ayala, P. Y., Chen, W., Wong, M. W., Andres, J. L., Replogle, E. S., Gomperts, R., Martin, R. L., Fox, D. J., Binkley, J. S., Defrees, D. J., Baker, J., Stewart, J. P., Head-Gordon, M., Gonzalez, C., Pople, J. A.: Gaussian 94, Revision C.3 Gaussian, Inc., Pittsburgh PA, 1995.
21. Alejandre. J., Tildesley, D. J., Chapela, G. A.: Molecular dynamics simulation of the orthobaric densities and surface tension of water. J. Chem. Phys. **102** (1995) 4574–4583
22. Berendsen, H. J. C., Grigera, J. R., Straatsma, T. P.: The missing term in effective pair potentials. J. Phys. Chem. **91** (1987) 6269–6271
23. Berendsen, H. J. C., Postma, J. P. M., van Gunsteren, W. F., DiNiola, A., Haak, J. R.: Molecular dynamics with coupling to an external bath. J. Chem. Phys. **81** (1984) 3684–3690

24. Ryckeart, J. -P., Cicotti, G., Berendsen, H. J. C.: Numerical Integration of the Cartesian Equations of Motion of a System with Constraints: Molecular Dynamics of n-Alkanes. J. Comp. Phys. **23** (1976) 327–341
25. Crowley, M.F., Darden, T.A.; Cheatham, T.E., III, Deerfield, D.W., II Adventures in improving the scaling and accuracy of a parallel molecular dynamics program. J. Supercomp. **11** (1997) 255–278
26. Greengard, L., Rokhlin, V.: A FastAlgorithm for Particle Simulations. J. Comp. Phys **73** (1987) 325–348
27. Lim, K. -T., Brunett, S., Iotov, M., McClurg, R. B., Vaidehi, N., Dasgupta, S., Taylor, S, Goddard III, W. A. Molecular Dynamics for Very Large Systems on Massively Parralel Computers: The MPSim Program. J. Com. Chem. **18** (1997) 501-521

Three-Dimensional Organization of Chromosome Territories and the Human Cell Nucleus

Tobias A. Knoch, Christian Münkel, and Jörg Langowski

Division Biophysics of Macromolecules, German Cancer Research Center, Im Neuenheimer Feld 280, D-69120 Heidelberg, Germany

Abstract. Despite the successful linear sequencing of the human genome its three-dimensional structure is widely unknown. However, the regulation of genes — their transcription and replication — has been shown to be closely connected to the three-dimensional organization of the genome and the cell nucleus. On the bases of polymer physics we have recently developed detailed and quantitative structural models for the folding of the 30 nm chromatin fiber within the human interphase cell nucleus. A quantitative test of several plausible theories resulted in a best agreement between computer simulations of chromosomes, cell nuclei and experiments for the so called Multi-Loop-Subcompartment (MLS) model. Results concern the following properties: overlap of chromosome territories, -arms, -bands, 3D spatial distances between genomic markers as function of their genomic separation in base pairs, fractal analyis of simulations, mass distribution of chromatin in cell nuclei and the fragmentation distribution of cellular DNA after irradiation with carbon ions. Thus in an anology to the Bauhaus principle that "form follows function", analyzing in which form DNA is organized might help us to understand genomic function.

1 Introduction

Research on the three dimensional structures of genomes has been obstructed by the incapabilities of imaging systems: Methods with high spatial resolution (e. g. electron microscopy) partialy destroy the structure while in light microscopy where the structure is preserved only low spatial resolution is reached. At the turn of the century, Carl Rabl (1885) and Theodori Boveri (1909) had already postulated a separation of chromosomes into distinct territories in the interphase, i. e. between cell divisions, from studies envolving light microscopy. But later, especially since the advent of electron microscopy, the cell nucleus has been viewed as a 'spaghetti soup' of DNA bound to various proteins without much internal structure (Comings et al., 1968) except during metaphase, i. e. cell division, when chromosomes are condensed into separate entities. Only recently has it become apparent that chromosomes occupy distinct 'territories' also in the interphase (Zirbel et al., 1993; see also Fig. 2). This can be granted to the development of fluorescent in situ hybridization (FISH) and confocal laser scanning microscopy (CLSM). In FISH cell nuclei are fixed and after melting the DNA double helix a complementary probe of single stranded and fluorescently labeled DNA is added.

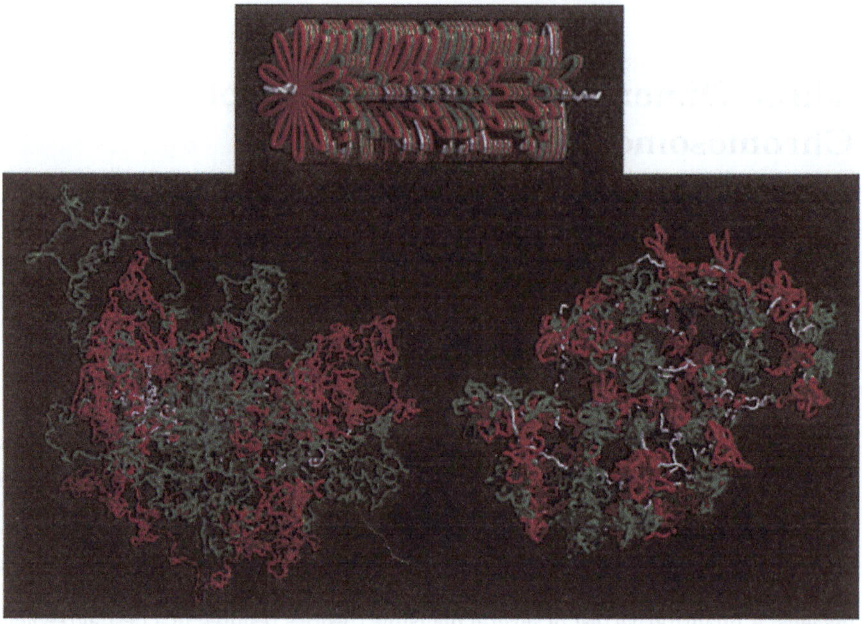

Fig. 1. Visualization of chromosome 15 simulations with the Random-Walk/Giant-Loop-(RWGL)Model and the Multi-Loop-Subcompartment-(MLS)Model. Consecutive loops or rosettes are painted in red and green; the linker is painted in blue. Above: Starting configuration resembling a metaphase chromosome. Left: The RW/GL - model, 5 Mbp loop size, after ~80,000 Monte Carlo and 1000 relaxing Browninan Dynamics steps. Large loops intermingle freely thus forming no distinct features like in the MLS model. Right: The MLS model, loop size 126 kbp and linker size 126 kbp, after ~50,000 Monte Carlo and 1000 relaxing Brownian Dynamics steps. Here rosettes form subcompartments as separated organizational and dynamic entities. The big holes visible are filled by other chromosomes in a whole nucleus. Nevertheless, in usual cell nuclei 80% of the volume is filled with water.

While diffusing through the nucleus the probe finds its complementary target sequence and binds with high affinity. The territories thought to be rather compact regions of the nucleus, the hypothesis has been forwarded that active genes are transcribed mainly at the periphery, and macromolecules (i. e. transcripts and proteins) are transported in a hypothetical space between adjacent and nonoverlapping territories. However, the postulated intrachromosomal domain (ICD, Cremer et al., 1993) channels have never been directly visualized.

The distribution of the chromatin fiber within a chromosome territory is also far from random. Several theoretic descriptions of chromosome structure based on polymer-chain models of the chromatin fiber have been postulated. These can be compared to experimental data obtained by marking small genomic targets with FISH and measuring the spatial distance between these

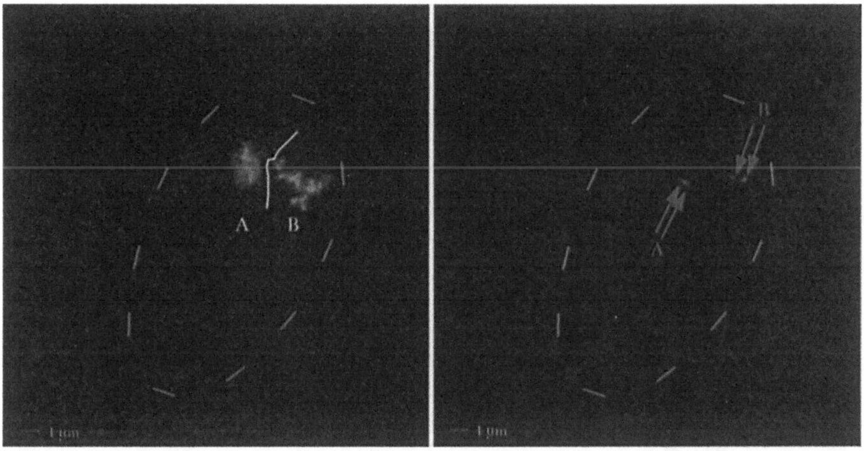

Fig. 2. Fluorescent in situ hybridization (FISH) of chromosome 15 and FISH marked target sequences in the Prader-Labhart-Willi/Angelmann (PLW/A) region in the same nucleus. Left: Projection of confocal image series with 'painted' chromosome 15 territories (A and B). Territories are placed behind each other and detected with a green fluorescent colour. Right: Projection of confocal image series with two target sequences with a genomic separation of 1 Mbp and marked with a red fluorescent colour in each of the territories. Territory A, two signals slightly behind each other, spatial distance 550nm. Territory B, spatial distance 840nm.

markers as a function of their genomic (i. e. linear sequence) distance. Based on such experiments, Sachs proposed the so-called Random-Walk/Giant-Loop (RW/GL) model (Sachs et al., 1995). In the RW/GL-model big chromatin loops with a base pair content of around 3 to 5 million are bound to a nuclear matrix. However, whether a nuclear matrix exists or not is still an unresolved question. We developed the so-called Multi-Loop-Subcompartment (MLS) model for solving some contradictions of the RW/GL-model (Münkel et al., 1998). Here the chromatin fiber folds into ~120 kbp sized loops which again are forming rosettes of in total 1-2 Mbp. These rosettes are interconnected by a piece of chromatin of similar basepair content so that no protein matrix is needed for structural support. Rosettes in the MLS-model correspond to the size of chromosomal interphase (ideogram) band domains. This model leads also to an easy mechanism for decondensation (or condensation) of a chromosome from metaphase: A loop is opened (or formed) at its base (e. g. protein mediated connection). This model also agrees with the metaphase organization as proposed by Pienta and Coffey (1984).

2 Simulation Methods

The energetics and dynamics of the above mentioned models can be explored numerically using Monte Carlo and Brownian Dynamics methods. Various

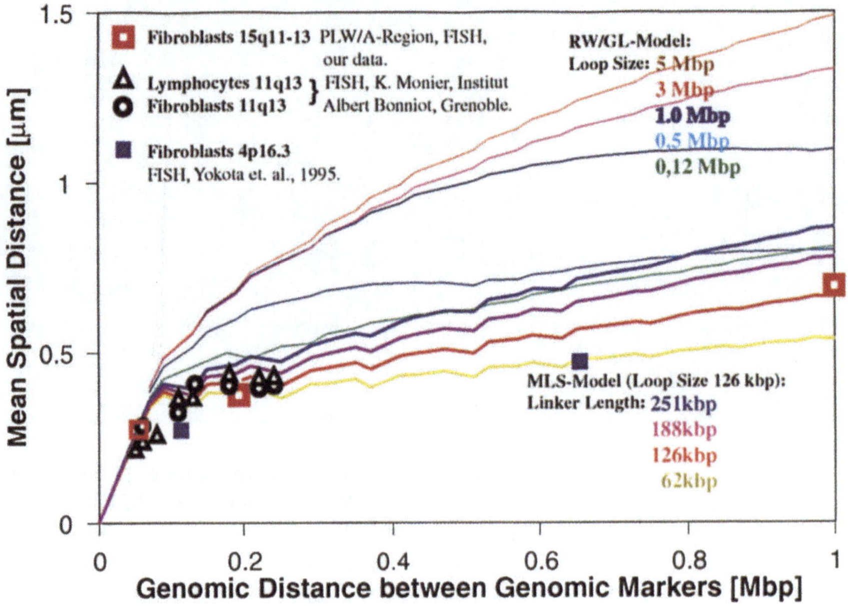

Fig. 3. Comparison between experimental and simulated spatial distances versus genomic distances between genomic markers (Knoch, 1998). MLS-model with a fixed loop size of 126 kbp, linker length of 63 kbp, 126 kbp, 190 kbp, and 250 kbp and in the RW/GL-model loop sizes of 5 Mbp, 4 Mbp, 3 Mbp, 2 Mbp, 1 Mbp, 500 were simulated. The linker length was adjusted in such a way that the global behaviour leads to comparable results in both models. Red rectangles: Our experiments (FISH) between genomic markers in the Prader-Labhart-Willi/Angelmann-(PLW/A) region, chromosome 15q11-13 (Knoch, 1998). Triangels/Circles: Fluorescence in situ hybridization (FISH) data for both lymphocyte and fibroblast cells, chromosome 11q13, f from K. Monier, Institut Albert Bonniot, Grenoble, France. Blue rectangles: Fluorescence in situ hybridization (FISH data for lymphocyte cells, chromosome 4p16.3, from Yokota.

models of human interphase chromosome 15 assuming a flexible polymer chain were simulated. To each polymer segment we assigned a harmonic stretching potential and between two segments a harmonic bending potential. To avoid self-crossing of the polymer chain a short ranged excluded volume potential was introduced. Its potential barrier can be changed to facilitate chain disentanglement. In vivo this is mediated by the protein Topoisomerase II. The starting configuration of a chromosome has the approximate form and size of a metaphase chromosome decondensation into interphase resembles the natural process (Fig. 1, above). To decrease computation time we started with polymer segments of 300 nm (31,000 bp) corresponding to 3300 segments for chromosome 15. Typically 400,000 Monte Carlo steps are needed to generate

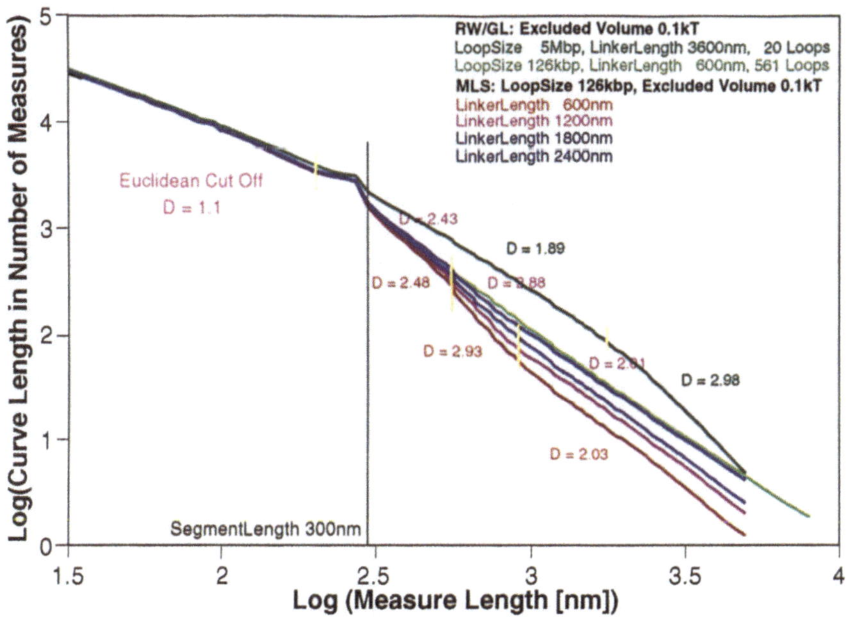

Fig. 4. Fractal analysis of the Random-Walk/Giant-Loop model and the Multi-Loop-Subcompartment-model. In agreement with porous network research fractal analysis shows multifractal behaviour in simulations of chromosome 15. Different fractal dimensions mean different process-dynamics in these scale region. Thus chromosome territories show a higher implicit degree of determinism than previously thougt. Standard errors of fractal dimensions are below 0.3%.

enough statistically independent configurations at equilibrium. For comparison with experimental data at small genomic distances the Monte Carlo configurations are taken as starting points for relaxation at higher spatial resolution by Brownian Dynamics methods. The segment length is decreased to 50 nm (5000 bp), corresponding to 20,000 segments for chromosome 15. 2000 Brownian Dynamics steps are performed until equilibration is reached again (Fig. 1, left and right).

For the simulation of a whole nucleus, 46 of the above mentioned starting configurations are put in a spherical potential representing the nuclear membrane. Depending on the segment length a nucleus consists of 200,000 to 1,200,000 segments (Fig. 5). The simulations programe "VirtNucSim" was written in C++ and because of the large numbers of segments the algorithms had to be parallelized using Message Passing Interface (MPI). One Brownian Dynamics step for nuclei with 1,200,000 segments takes arround 23 min and a comparable Monte Carlo step arround 45 min. The simulations presented here had a degree of parallization of 77% [CPU-parallel/CPU-total] using 16 processors R6000-70MHz and took in total about 180.000 h of CPU time

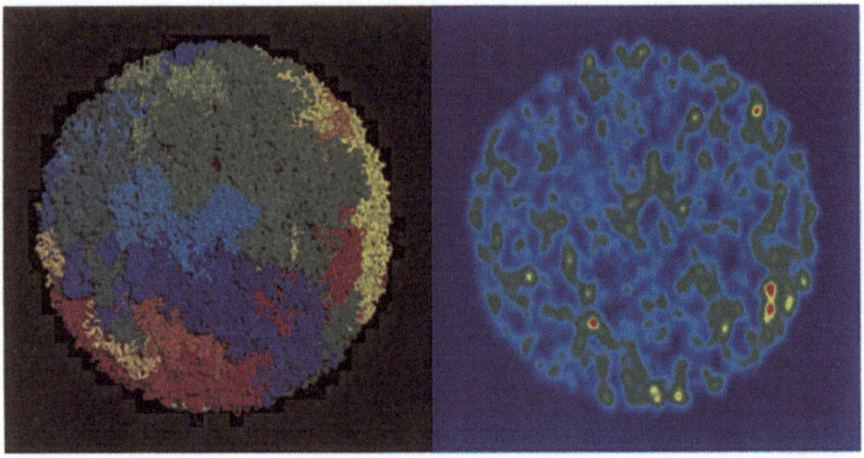

Fig. 5. Virtual Human Cell Nucleus and simulated confocal section. The Nucleus is simulated assuming a flexible polymer chain, modelling the 46 chromatin fibers with in total 1,248,794 50 nm = 5.2 kbp segments. Visualizations are shown after 0,5 ms Brownian Dynamics simulation, one step taking 10 s. As starting configuration a metaphase nucleus was chosen, i. e. each chromosome was chosen es metaphase chromosome (Fig. 1, above) and homologous chromosome were placed randomly but next to each other into the nucleus. Left: Simulation of an nucleus with the MLS model. The different chromosomes are painted with different colours. The forming of territories of chromosomes is clearly visible. Right: Simulated confocal section of the left nucleus in agreement with experiment. False colour representation.

on a single processor (normalized to the IBM SP2 in Karlsruhe). The platforms used were the IBM-SP2-DKFZ, the HP Convex-(S-class)-DKFZ, the IBM-SP2-SCC-Karlsruhe, and the Cray T3E-HWW-Stuttgart.

3 Experimental Methods

For measuring spatial distances between genomic markers as a function of their genomic separation we chose target sequences of chromosome 15 within the region responsible for the Prader-Labhart-Willi/Angelmann (PLW/A) syndrom (c15q11.2-13.3), which is believed to be caused by a structural change in the genome, in contrast to the common base pair mutation. We used probes of about 11 kbp with separations of 18 kbp to 1 Mbp (in cooperation with K. Bütig and Prof. B. Horsthemke, Essen, Germany). Human fibroblast cells were grown on coverslips to layers, and assumed to be in the same phase of the cell cycle. The cells were fixed with paraformaldehyde in an isotonic environment, labeled with biotin and digoxigenin and were detected with fluorescent dyes bound to anti-digoxigenin and anti-biotin antibodies. Typically 25 sections per image series were then taken with a confocal laser scanning microscope (CLSM) Leica TCS NT (in cooperation with J.Rauch, H. Bornfleth,

C. Cremer, Heidelberg, Germany). With the CLSM distance measurements between the center of mass of genomic markers with an accuracy of 35 nm are possible if each marker is detected with a different fluorescent colour and a correction is made for chromatic shift (Fig. 2).

4 Results

For the determination of the spatial distance as a function of genomic distance from the simulations, the spatial distance was calculated for genomic distances randomly positioned within the chain of segments and for statistically independent configurations. Therefore the mean is taken for marker positions with no definate relative position to loops, rosettes or linker between rosettes. Segments in the linker between the rosettes are not as much spatially constrained as segments in a loop. The mean taken over the spatial distances resulted in a standard error of the mean below 1%. In the experiments the mean was taken over 120 nuclei for each distance, and including controls 1600 nuclei were imaged. Best agreement between simulation and experiment (Fig. 3) is reached for a Multi-Loop-Subcompartment model with a loop size of 126 kbp and a linker length of again 126 kbp testing against our and K. Moniers data (Monier et al., 1999). Thus the distance measurements of the genomic regions used here do not support a Random-Walk/Giant-Loop model. Due to the lack of experimental data it is, however, not yet possible to make clearer statements about the positioning of the experimental markers relative to loops, rosettes or linker between rosettes.

For calculating more general properties of chromosomes the fractal dimension of the chromatin fiber was determined from the simulations. An exact-divider analysis was used, were the divider-measure (i. e. the length of the divider) covers four orders of magnitude. The starting point of the divider was randomly positioned within the chain of segments and then the number of dividers necessary to reach both ends of the chain of segments was computed. To account for biased starting points for bigger divider measures a mean over several starting points was taken as well as for independent configurations. The fractal dimension was calculated as the negative slope from a linear regression in plotting the logarithm of the number of dividers used to measure the chromosome as function of the logarithm of the length of the divider measure. The standard error for the fractal dimension was smaller than 0.3%. The fractal analysis resulted in different and distinct multifractal behaviour for the MLS- and the RW/GL-model (Fig. 4) in good agreement with predictions drawn from porous network research (Avnir, 1989; Mandelbrot, private communications). The scale region on which the various fractal dimensions are valid defines the borders of the underlying structure (loops, rosettes and the whole chromosome for the MLS and loops and the whole chromosome for the RW/GL-model, yellow border lines in Fig. 4; Knoch, 1998).

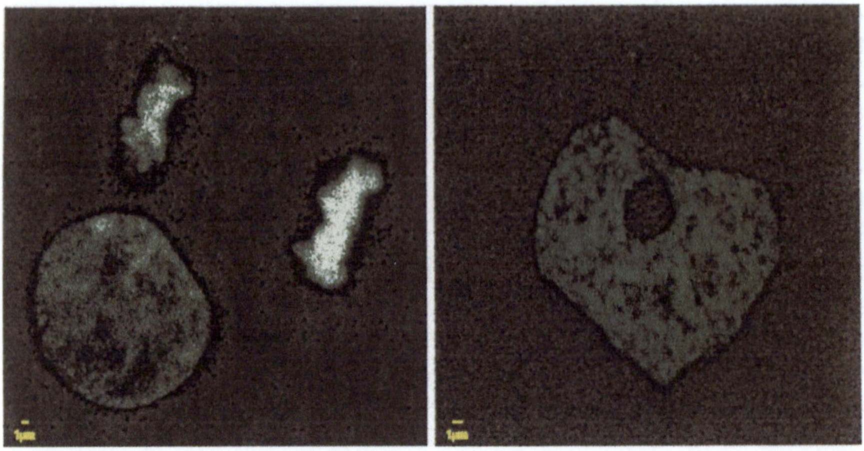

Fig. 6. Mapping of Histone H2B-GFP and H10-GFP distribution in vivo. The Histone-GFP (Green-Fluorescent-Protein) Fusion Protein reflects the distribution of chromation in interphase, because the labeled histone is built into the nucleosme. The structure visible in the images is similar to those found in simulations (see Fig. 5). Left: HeLa cells stabaly transfected with H2B-GFP (image provided by K. Sullivan, Scripps Institute (Kanda et al., 1998)). Confocal in vivo section of a cell nucleus and a mitosis. Right: Cos7 cells stabaly transfected with H10-GFP (cells provided by A. Alonso, DKFZ). Confocal in vivo section.

The simulation of a whole human cell nucleus in connection with the simulation of single chromosomes resulted in the formation of distinct chromosome territories as predicted(Fig. 5). To compute the overlap of chromosomes, chromosome arms and chromosome subcompartments (i.e. rosettes) from simulations the simulated chromatin massdistribution of the respective structures was folded with an experimental point spread function (PSF) of a confocal microscope and filtered with a 3*3*3-median filter. The overlap was callculated using an autocorralation approach (Münkel et al., 1998). In contrast to the RW/GL-model the MLS-model leads to low overlap between chromosome territories as well as chromosome arms, in agreement with overlap analysis of confocal image series in which the overlap was callculated with the same autocorrelation approach (Münkel et. al., 1999). The available experimental data show similar behaviour throughout the whole cell nucleus. Overlaps are the result of the small scale folding of the chromatin fiber within chromosomes and the whole cell nucleus. In connection with the measurement of spatial distances between genomic markers as a function of their genomic separation one could conclude that throughout the cell nucleus an MLS-model might best aggree with experimental data.

The density distribution of the chromatin fiber from simulations can also be compared to the distribution of Histon-GFP fusion proteins which are built into the chromatin fiber in living cells (Fig.6). A morphologic compar-

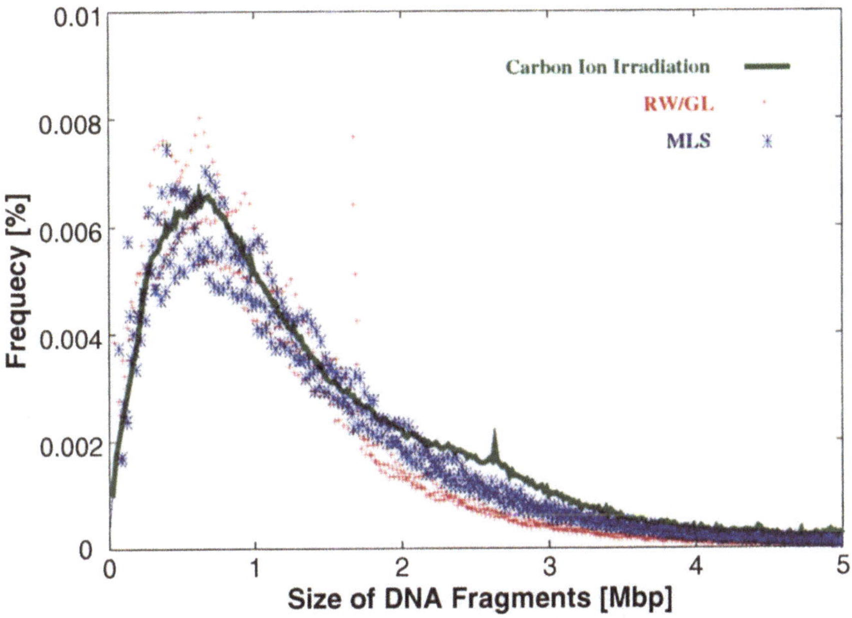

Fig. 7. DNA fragmentation by irradiation with carbon ions - comparison between experimental and simulated distributions. Irradiation with carbon ions results in DNA double strand breackage. The length of the fragments follow distributions depending on the spatial arrangement of the 30 nm chromatin fiber in the nucleus. The carbon ion irradion was simulated on the basis of our chromosome simulations by P. Quicken (data provided by P. Quicken, GSF, Munich).

ison between simulations (Fig. 5, right) and experiments lead to a similar distribution of the 30 nm chromatin fiber. The "clumps" visible have sizes between 300 and 600 nm which might correspond to the subcompartments in agreement with the MLS-model.

Another test of the model is the length distribution of DNA fragments after irradiating cell nuclei with carbon ions causing double strand breaks in the DNA double helix. The length distribution is characteristic for the folding of the chromatin fiber. P. Quicken from the GSF in Munich used our computed configurations to simulate the length distribution of DNA fragments to compare it to his experimental results. Preliminary results favour again the Multi-Loop-Subcompartment model (MLS) (Fig. 7).

Acknowledgements

Special thanks to M. Wachsmuth for helping in the preparation of this manuscript and many inspiring discussions. This Research is part of the German Human

Genome Program (DHGP) funded by the BMBF under grant No. 01 KW 9620.

References

Avnir, D., ed.: The Fractal Approach to Heterogeneous Chemistry John Wiley & Sons, 1989.

Boveri, T.: Die Blastomerenkerne von Ascaris meglocephala und die Theorie der Chromosomenindiviualitt Archiv fr Zellforschung 3, 181-268, 1909.

Cremer, T., A. Kurz, R. Zirbel, S. Dietzel, B. Rinke, E. Schreck, M. R. Speicher, U. Mathie, A. Jauch, P. Emmerich, H. Scherhan, T. Ried, C. Cremer and P. Lichter: Role of Chromosome Territories in the Functional Compartmentalization of the Cell Nucleus. Cold Spring Harbor Symp. Quant. Biol. 58:777-792, 1993.

Comings, D. E.: The rationale for an ordered arrangement of chromatin in the interphase nucleus. American Jum. of Genetics 20, 440, 1968.

Knoch, T. A.: Three-Dimensional Organization of Chromosome Territories in Simulation and Experiments (German). Diploma thesis, German Cancer Research Center Heidelberg, Faculty for Physics und Astronomy, University of Heidelberg, 1998.

Monier, K., T. A. Knoch et al.: Distinct distributions and morpholgy of chromosome territories in normal lymphocyte and fibroblast nuclei during the cell cycle. (in preparation 1999).

Münkel, C. and J. Langowski: Chromosome structure predicted by a polymer model. Phys. Rev. E 57#5:5888-5896, 1998.

Münkel, C., Eils, R., Dietzel, S., Zink, D., Mehring, C., Wedemann, G., Cremer, T. and Langowski, J.: Compartimentalization of Interphase Chromosomes Observed in Simulation and Experiment. J. Mol. Biol. 285, 1053-1065, 1999.

Pienta, Kenneth J. and Coffey, Donald S.: A structural analysis of the role of the nuclear matrix and DNA loops in the organization of the nucleus and chromosome. In: Higher Order Structure in the Nucleus, edited by P. R. Cook and R. A. Laskey, Journal of Cell Science, Supplement I:123-135, 1984.

Rabl, C.: Über Zellteilung. Morphologisches Jahrbuch 10, 214-330, 1885.

Sachs, R. K., G. van den Engh, B. Trask, H. Yokota and J. E. Hearst: A random-walk/giant-loop model for interphase chromosomes. Proceedings of the National Academy of Sciences 92:2710-2714, 1995.

Kanda, Teru, Kevin F. Sullivan and Geoffrey M. Wahl: Histone-GFP fusion protein enables sensitive analysis of chromosome dynamics in living mammalian cells. Current Biology 8#7, 377-385, 1998.

Zirbel, R. M., U. Mathieu, A. Kurz, T. Cremer and P. Lichter: Evidence for a nuclear compartment of transcription and splicing located at chromosome domain boundaries. Chromosome Research 1:92-106, 1993.

Computational Fluid Dynamics (CFD)

S.Wagner

Institut für Aero- und Gasdynamik, Universität Stuttgart
70550 Stuttgart

Many products in industry, e.g. space vehicles, airplanes, helicopters, trains, cars, engines, single crystals, machines in process engineering and nuclear engineering, are extremely complex systems that combine and integrate various highly advanced technologies. Their definitions and developments are carried out in complex and time consuming processes requiring a lot of high skilled people .

Fluid dynamics and aerodynamics are two disciplines playing a major role among others in developing these advanced products. The interdisciplinary connections of these two disciplines with other major disciplines, e.g. structural dynamics and acoustics are more and more important and become evident in nomenclatures like aeromechanics, aeroelasticity or aeroacoustics.

In order to reduce cost of development, level of risks, time of development and to increase the quality and performance of products high performance computing is more and more used as a powerful tool.

Universities and other research institutions are involved in the process of developing new tools to support technology. They perform basic research to gain more insight into the mechanics of complicated processes. In CFD there are many problems like the laminar-turbulent transition of boundary layers, separated flows, unsteady and vortical flows, two phase flows, crystal growth, mixing problems, transonic flows around helicopter rotors and airplanes, fluid-structure interference, aeroacoustics, etc.

The development of new methods to numerically solve the governing equations that are in many cases systems of nonlinear partial differential equations, the development of algorithms and the extreme progress in computer power have created a completely new basis to get insight into flow phenomena that were impossible ten years ago or only partially possible by very expensive and long lasting experiments.

In the following chapter 9 papers are presented being selected for presentation among the many research projects in the area of CFD running presently on the high performance computers of HLRS. A still unsolved problem in fluid dynamics is the complete understanding of turbulent flows. There is no theory that allows to exactly calculate this type of flow that is highly unsteady. The governing equations, i.e. the Navier-Stokes equations, are usually time averaged and to mathematically close the problems turbulence models have to be used. These so-called Reynolds averaged Navier-Stokes equations (RANS) are widely used in industry despite many shortcomings (e.g.

computing separated flows) because there is no replacement for them at the moment.

Even the nowadays available high performance computers are too small to solve practical problems by Direct Numerical Simulations (DNS) which means the numerical solution of the unsteady three-dimensional Navier-Stokes equations without any modeling of turbulence. One approach to get more insight into the mechanism of turbulence is to use DNS and to study the process of laminar-turbulent transition of boundary layers along simple configurations and at low Reynolds numbers because the number of grid points required for numerical solution is $O(Re^{\frac{9}{4}})$. Contributions to this problem are presented in the papers by Manhart and by Maucher et al. The code of the latter paper was run on the NEC SX/4-32 using 16 processors. The achieved performance was 14 GFOPS which is 50 per cent of the theoretical peak performance per node. This gives a hint about the computer power required to study problems of this kind.

The next approach to capture the unsteady character of turbulent flow and at the same time to reduce computer running time is Large Eddy Simulation (LES) or Very Large Eddy Simulation (VLES). In this case the large eddies are simulated by procedures comparable to DNS and the small scales are simulated by a model. The number of grid points in LES is then approximately $O(Re^2)$. This approach is used in the papers by Fröhlich et al., by Schmid et al. and by von Lavante et al., whereas Hüttl et al. are using both DNS and LES.

Rieber et al. studied the interface dynamics in incompressible two-phase flows using the Navier-Stokes equations. The parallel code was run both on the NEC SX/4 and the Cray T3E-900/512. Rieber et al. show that the code should be run on the Cray T3E only for problems that are too large for the NEC SX/4.

Enger et al. investigated the fluid flow and heat transfer in industrial Crochralski melt to produce large single crystals. They show that turbulence models are not appropriate to study this kind of problem. A more rigorous approach would be necessary such as LES or DNS.

Finally, Buchtala et al. simulated the aeroelastic behaviour of a helicopter rotor in forward flight. A sophisticated three-dimensional visualization system is used for the interpretation of the complex rotor wake structures, available in the CUBE of HLRS.

Parallel Computation of Interface Dynamics in Incompressible Two-Phase Flows

Martin Rieber and Arnold Frohn

ITLR · Institut für Thermodynamik der Luft- und Raumfahrt,
Universität Stuttgart, Pfaffenwaldring 31, 70550 Stuttgart, Germany

Abstract. A numerical method for the direct numerical simulation of incompressible flows with interfaces between two fluids with totally different density and viscosity is developed. The method is applied to binary droplet collisions and to the droplet impact on liquid films. Corresponding simulation running times are reported. The single-processor performance and the parallel performance of the code on NEC SX-4 and Cray T3E/512-900 at HLRS are analyzed. It is concluded, that simulations with this code should be run on the Cray T3E/512-900 only for problems, which are too large for the NEC SX-4 .

1 Introduction

The numerical prediction of two-phase flows is of increasing importance in many technical disciplines such as process engineering, nuclear engineering, or aerospace and automotive engineering. Applications in these fields are for example spray cooling in steel industries, safety systems in various types of nuclear reactors and heterogeneous combustion. For a better modeling of such systems more knowledge about elementary phenomena like droplet-droplet and droplet-wall interactions is needed. While most of the results on these phenomena are still obtained experimentally, direct numerical simulation on supercomputers will provide additional insight in the complex dynamics of these systems.

2 Numerical Method

The flows considered here are governed by the local instant single-field equations of incompressible viscous two-phase flow without phase change [1,2]

$$\nabla \cdot \mathbf{u} = 0 , \tag{1}$$

$$\frac{\partial \rho}{\partial t} + \nabla \cdot (\rho \mathbf{u}) = 0 , \tag{2}$$

$$\frac{\partial (\rho \mathbf{u})}{\partial t} + \nabla \cdot [(\rho \mathbf{u}) \otimes \mathbf{u}] = \nabla p + \nabla \cdot \mu [\nabla \mathbf{u} + (\nabla \mathbf{u})^T] + \nabla \cdot \mathbf{T} . \tag{3}$$

The last term in the momentum equation accounts for surface tension effects. The capillary stress tensor $\mathbf{T}$ is defined only at the phase interface. The density and the viscosity are constant inside the gas and the liquid, but vary discontinuously at the sharp interface separating liquid and gas. Thus inside each phase, these equations reduce to the incompressible Navier-Stokes equations for a fluid with constant kinematic viscosity, which may be interpreted in the strong sense. However, in order to be valid at the same time in both phases and across the interface, (1-3) have to be interpreted in the weak, integral sense.

In the well-known volume-of-fluid (VOF) method the density and the viscosity are related to the volume fraction f of the liquid by

$$\rho = \rho_g + (\rho_l - \rho_g) f \tag{4}$$

$$\mu = \mu_g + (\mu_l - \mu_g) f \,, \tag{5}$$

where the subscripts l and g refer to liquid and gas, respectively. The advection of the volume fraction and thus of the discontinuity is governed by the transport equation

$$\frac{\partial f}{\partial t} + \nabla \cdot (f \, \mathbf{u}) = 0 \,, \tag{6}$$

which follows from (2) using (4).

In the framework of the volume-of-fluid method all flow discontinuities can be captured in a straightforward way using conservative finite volume discretizations for (1), (3) and (6). For reasons of robustness of the method all discretizations are based on a Cartesian MAC grid. Specifically, we use a second-order conservative Godunov projection method. A similar method on a collocated grid has been described in Ref. [3]. In contrast to the method mentioned before the formulation of momentum advection is fully conservative.

The computation of volume fluxes is based on a piecewise linear reconstruction of the interface (PLIC) [4]. In this way the interface remains sharp even for long time simulations. Like momentum and volume fraction transport, surface tension and momentum diffusion are treated explicitely, while the pressure is determined by an implicit Poisson equation in the so-called MAC projection step. Discontinuous coefficients in the Poisson equation due to the discontinuous density field cause standard multigrid solvers to fail. We use an efficient cell-centred multigrid method, which is rather insensitive to the density ratio [5]. Because separate coarsening in the three coordinate directions is possible, this multigrid method works even for computational domains with high aspect ratios. This property is especially important for the parallelization of the code on the parallel vector computer NEC SX-4 , where subdomains with high aspect ratios lead to a better vector register utilization. The program FS3D (Free Surface-3D) has been parallelized using domain decomposition and the communication library MPI, which results in highest portability of the parallel program [6].

3 Results

The parallel program FS3D has been used to simulate binary droplet collisions [7,8] and splashing droplets [2,9]. The main dimensionless parameters for these phenomena are the Weber number $We = \rho_l D V^2/\sigma$, the Reynolds number $Re = \rho_l D V/\mu_l$ and the non-dimensional time $T = tV/D$, where D, V and σ are droplet diameter, relative velocity and surface tension, respectively. In case of the splash additionally the non-dimensional film thickness $H = h/D$ is taken into account.

The simulation of a head-on droplet collision in Fig. 1 shows a good quantitative agreement with experimental results. The simulation needed 22801 time steps on a $128 \times 64 \times 64$ grid using three symmetries. Explicit treatment of viscous stresses is the reason for the large number of time steps. To resolve the thin film formed in the collision plane around $T = 3.31$, the grid was refined by a factor of two in the direction perpendicular to the collision plane. With 8 Processors of the NEC SX-4 the simulation had an elapsed time of 3 h 55 min.

The impact of a single drop on a liquid film (splash) is another challenging CFD-problem because the effects of transport, diffusion and surface tension interact in a strong way, while liquid structures of very different length scales develop. Thus very fine grids are needed to resolve small liquid structures. In Figure 2 two splash simulations with different resolution are compared. For both cases two symmetries where used in the simulation, thus only one quarter of the visualized splash was really simulated. The simulation on the left-hand side needed 685 time steps on a 128^3-grid. This splash will serve as test case in the following section. A careful inspection of the results shows some unphysical effects due to low resolution of the computational grid. The simulation on the right-hand side shows more reliable results using a 320^3-grid and 2553 time steps. This high resolution simulation had an elapsed time of 4 h 30 min using 250 processors of the Cray T3E/512-900 . Due to high resolution of the problem, the formation of the crown with fingers ejecting small droplets (splashing) is in good agreement with experimental observations [2,10]. For higher impact energies, more and smaller droplets are formed at the free rim on top of the crown, as shown by comparison of the splash simulation for $We = 250$ and a simulation for $We = 598$ (Fig. 3).

Most recently, it has been discovered in experiments performed at ITLR that splashing occurs also in binary droplet collisions with high impact energies [8]. The numerical results in Fig. 4 show typical node and finger formation at the free rim of the flat, circular disk, which is formed immediately after head-on impact of the droplets. The simulation needed 40405 time steps on a $256 \times 256 \times 64$ grid using three symmetries. With 128 Processors of the Cray T3E/512-900 the simulation had an elapsed time of 10 h 50 min.

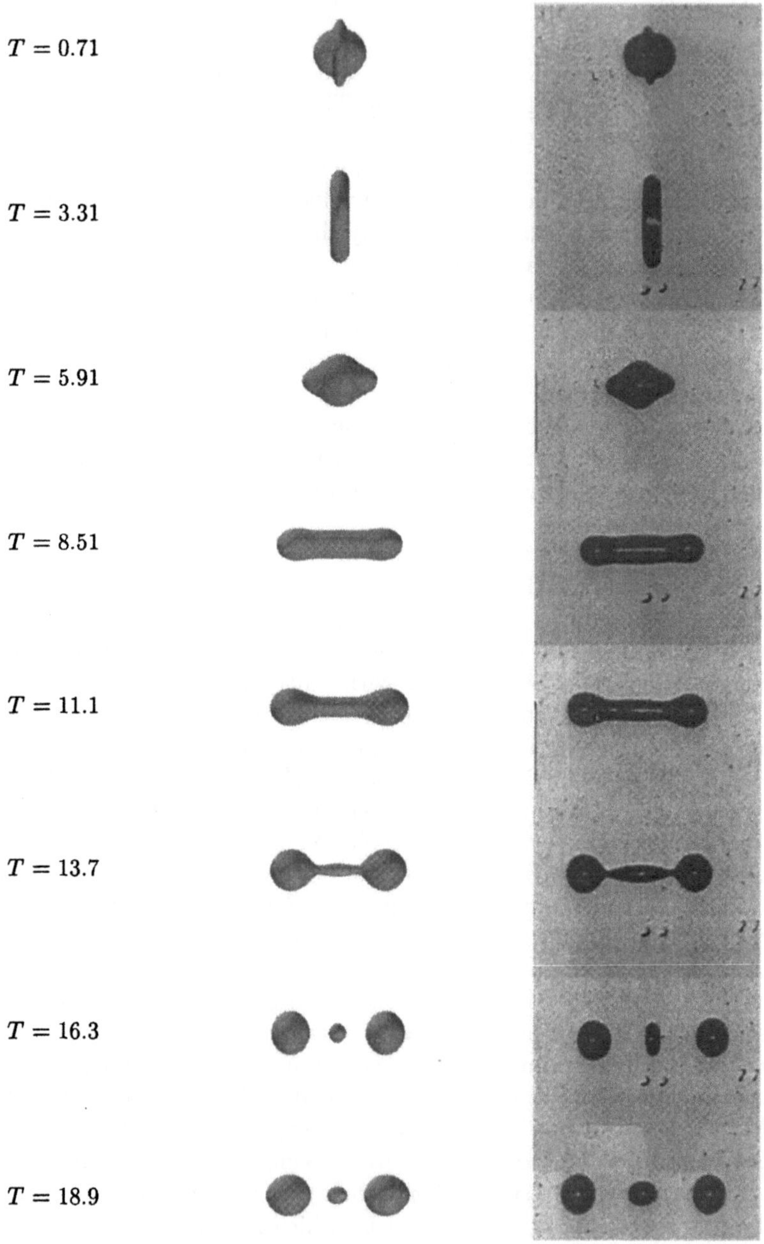

Fig. 1. Head-on droplet collision with $We = 70$ and $Re = 193$ leading to the formation of a satellite droplet. Comparison with experiments performed at ITLR.

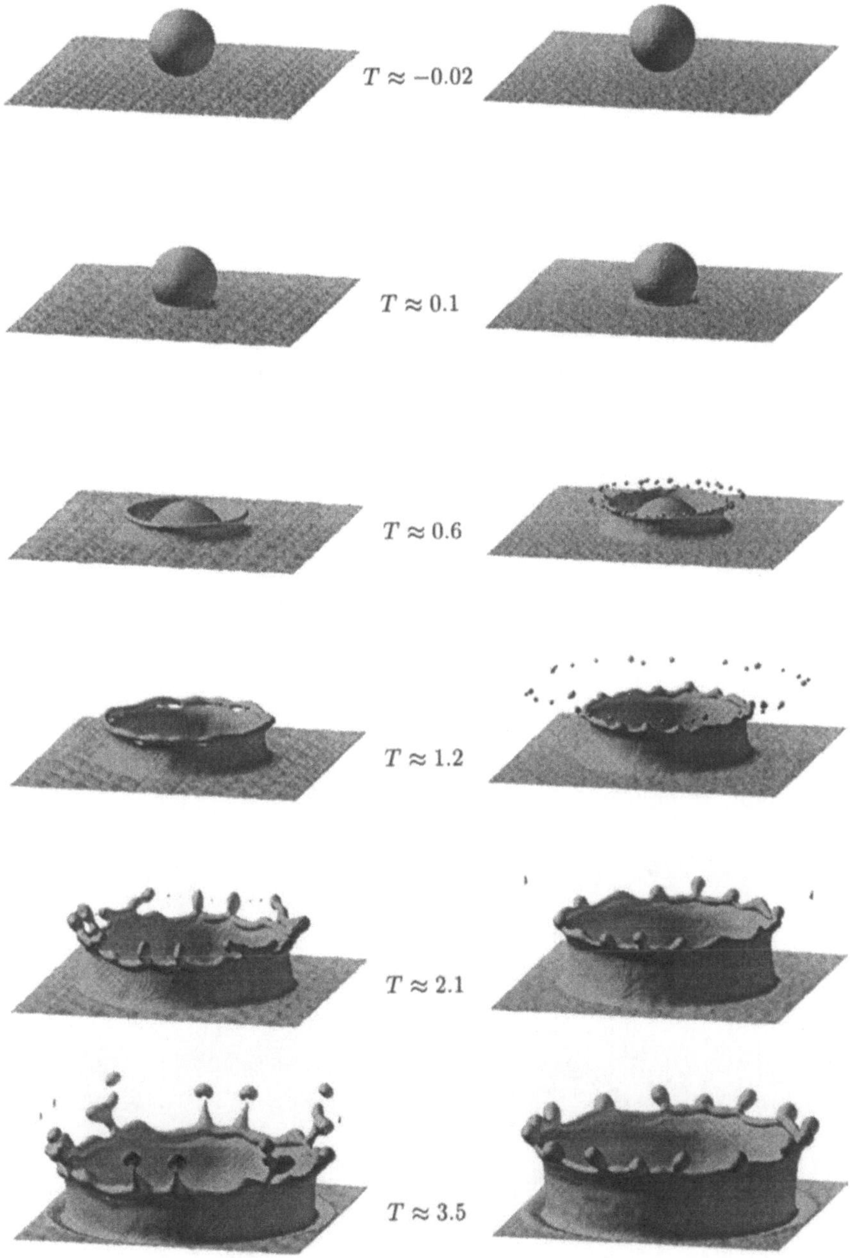

Fig. 2. Disintegration of the splashing lamella for $We = 250$, $H = 0.12$ and $Re \approx 10000$. Comparison of simulations on a 128^3-grid (left-hand side) and on a 320^3-grid (right-hand side).

Fig. 3. Shape of the disintegrating splashing lamella at $T = 2.43$ after impact of a droplet on a liquid film. Same simulation as on the right-hand side of Fig. 2 (top) and simulation with $We = 598$, $Re \approx 18000$, $H = 0.12$ (bottom). The images have been generated using the ray tracing software POV-Ray 3.1 (`http://www.povray.org`).

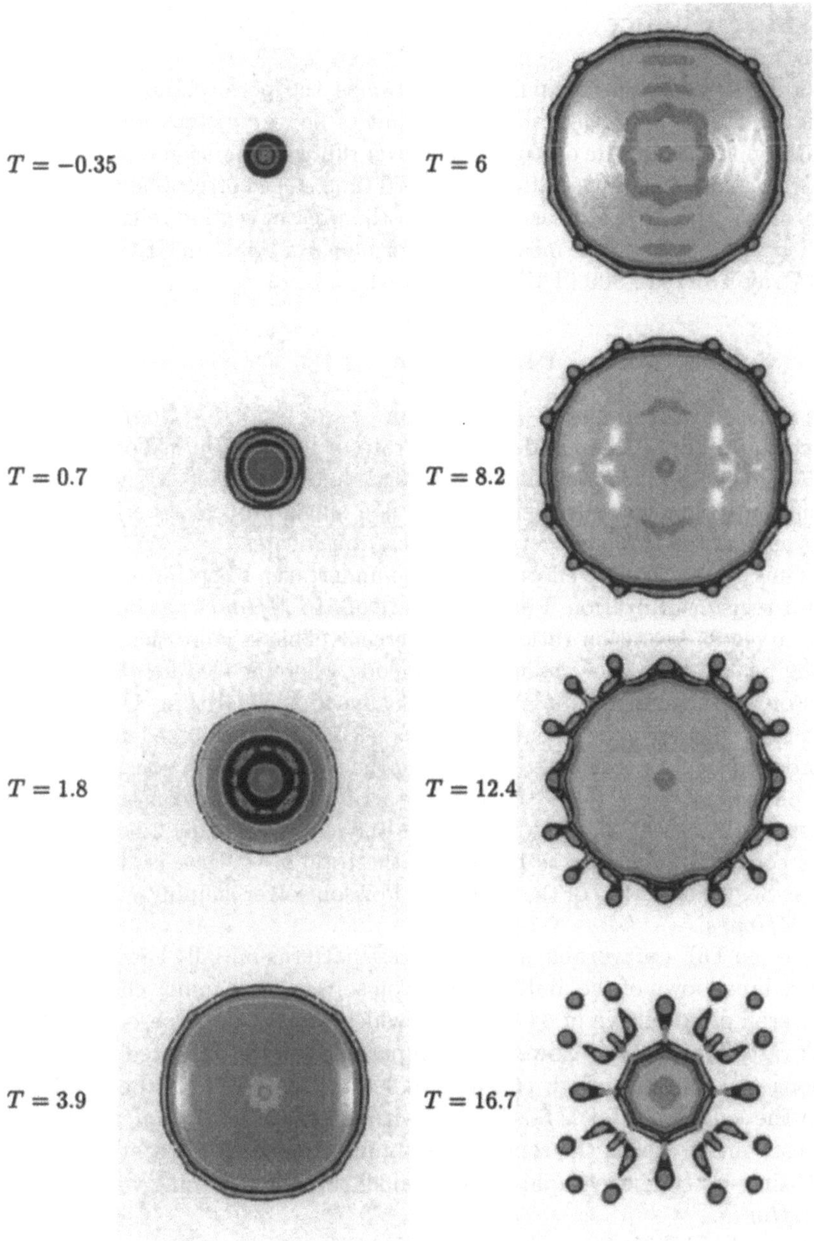

Fig. 4. High energy head-on droplet collision with $We = 849$ and $Re = 1015$. As in Fig. 3, the images are generated using POV-Ray 3.1 and a tool to compute a smooth interface triangulation from the VOF data.

4 Performance

In order to measure, compare and optimize the performance of FS3D on different HLRS platforms, the first 70 out of 685 time steps needed for the splash on left-hand side of Fig. 2 have been run with additional performance-analyzing code instrumentation. These 70 time steps correspond to the physical time $T = 0.22$. All measurements in the present section refer to this test case. In the following, the measured performance of FS3D on NEC SX-4 (SX4) and Cray T3E/512-900 (T3E) is discussed.

4.1 Single-Processor Performance

The peak performance of a single SX4-processor is 2000 $Mflop/s$, while one processor of the T3E provides a peak rate of 900 $Mflop/s$. For computers with high communication- and/or memory bandwidth like T3E and SX4, the main key to high parallel performance is a single-processor performance as near as possible to these (very theoretical) peak rates.

Thus on the SX4 big efforts have been undertaken to optimize the code for vector register utilization. Finally the rate of 515 $Mflop/s$ has been achieved with a vector operation ratio of 98.6 percent. This is more than 25 percent of the peak rate. Major changes in the code where needed for the multigrid Poisson solver and for the PLIC VOF advection algorithm. Out of 25000 lines of the source code about 1000 lines needed to be changed substantially in order to achieve vectorization. Among the routines which consume at least one percent of the total CPU time the PLIC interface reconstruction runs fastest with a rate 1460 $Mflop/s$, while the slowest routine among the routines consuming at least 10 percent of the total CPU time is the red-black Gauss-Seidel smoother of the multigrid Poisson solver running with a rate of 210 $Mflop/s$.

On the T3E unfavorable memory access patterns initially caused a performance breakdown of the multigrid Poisson solver. Some minor changes led to an overall performance of 44 $Mflop/s$, which is less than 5 percent of T3E's peak rate. This low rate however is comparable to rates reported in Ref. [11]. Among the routines which consume at least one percent of the total CPU time the calculation of ∇f is fastest with a rate of 93 $Mflop/s$, while the slowest routine among the routines consuming at least 10 percent of the total CPU time is the lexicographic Gauss-Seidel smoother running with a rate of 44 $Mflop/s$.

The relative time spent in performing different computational tasks like momentum advection and diffusion, surface tension calculation, MAC projection (Poisson solver) and VOF advection is of the same order on the two machines compared, as depicted in Fig. 5. It is concluded, that no severe performance breakdown exists on either of the machines.

In Figure 6 the absolute single-processor CPU times on SX4 and T3E are compared for the whole program and for the different computational tasks

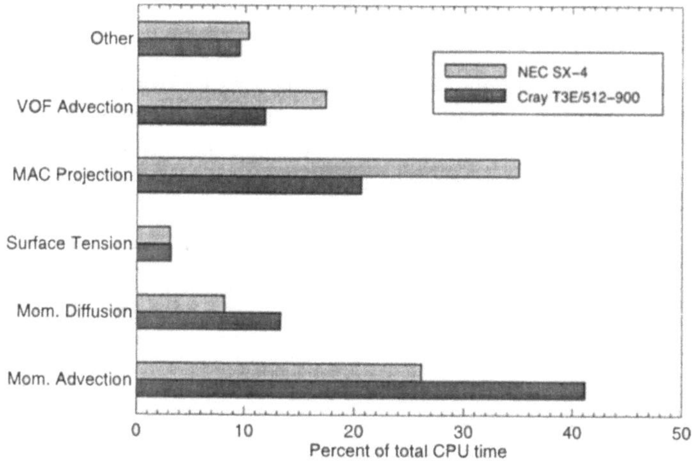

Fig. 5. Comparison of relative single-processor CPU times for different computational tasks on NEC SX-4 and Cray T3E/512-900 using the splash test case.

mentioned above. The test case needs 1525 s of CPU time on the SX4 and 13733 s on the T3E, thus the local work is done by a factor of 9 faster on the SX4. It should be noted that the test case needed at least 16 T3E nodes. Thus in addition to the local work, which is taken into account in this subsection, a communication overhead of 4022 s leads to an overall factor of 12 between the two machines. Because actually the SX4 CPU-time costs are by a factor of 8.3 higher the test case is by a factor of 1.44 more expensive on the T3E.

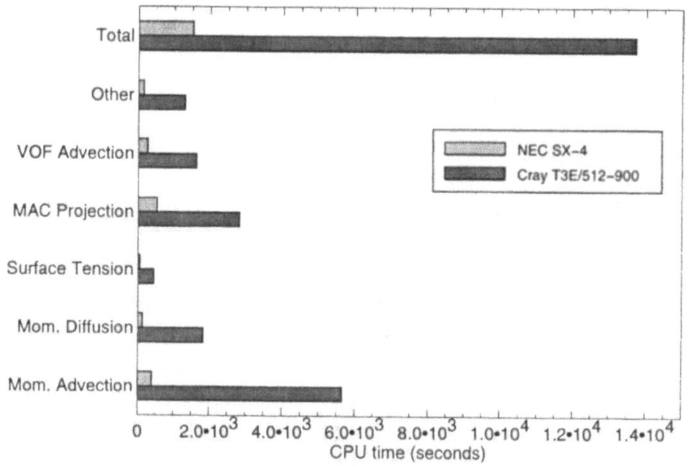

Fig. 6. Comparison of the single-processor CPU time on NEC SX-4 and Cray T3E/512-900 for different computational tasks of the splash test case.

4.2 Parallel Performance

First of all, we are again mainly interested in the time FS3D spends in running the test case. This time is depicted in Fig. 7 for different numbers of processors P. Notice, that for the relatively small test problem the T3E cannot be faster than the SX4. In order to compare the parallel performance of FS3D in more

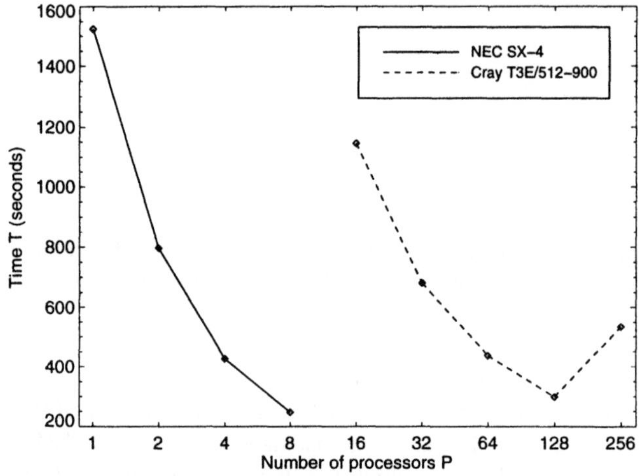

Fig. 7. Comparison of total time versus number of processors on NEC SX-4 and Cray T3E/512-900 for the splash test case.

detail, we need two definitions of speedup S and efficiency E at a time. On the SX4 we use

$$S = \frac{T(1)}{T(P)} \quad \text{and} \quad E = \frac{T(1)}{PT(P)} , \tag{7}$$

while on the T3E we use

$$S = P\frac{t(P) - t_{overhead}(P)}{t(P)} \quad \text{and} \quad E = \frac{t(P) - t_{overhead}(P)}{t(P)} , \tag{8}$$

because it is not possible to run the test problem on a single T3E processor. In these equations for a program running on P processors, $T(P)$ is the active time of the program, $t(P)$ is the accumulated CPU time and $t_{overhead}(P)$ is the accumulated parallel overhead time, which consists of the CPU times needed for communication and for double calculations of any kind at the subdomain boundaries. The latter definitions of S and E require the explicit evaluation of the parallel overhead time by the help of some performance analysis tool. Speedup and efficiency on the two platforms considered are depicted in Figs. 8 and 9. In either of the diagrams, the behavior of the program on the T3E continues the behavior indicated on up to 8 processors of the SX4. With respect to the times T depicted in Fig. 7 it is better to use the SX4 as long as the problem is not too large for this machine. For problems larger than that tested, the T3E will be the only possible platform at HLRS. It should be noted, that for larger subdomains the efficiency on the T3E will be higher.

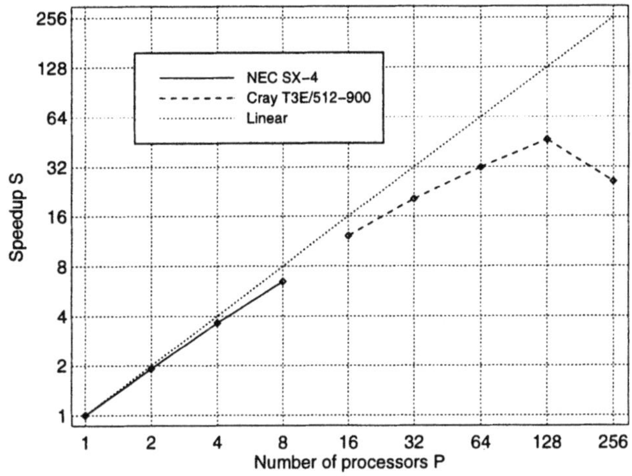

Fig. 8. Comparison of speedup versus number of processors on NEC SX-4 and Cray T3E/512-900 for the splash test case.

5 Conclusion

An efficient numerical model for the prediction of interface dynamics in two-phase flows has been described. Promising results of parallel simulations demonstrate the capability of the numerical model to resolve fine liquid structures accurately. In case of the splash, the T3E enabled us to perform high resolution simulations, which predicted details of the splash phenomenon never captured in a computer simulation before. However, simulations should be run on the T3E only for problems, which are too large for the NEC.

References

1. Kataoka, I.: Local Instant Formulation of Two-Phase Flow. Int. J. Multiphase Flow, Vol. 12, pp. 745-758, 1986.
2. Rieber, M. and Frohn, A.: A Numerical Study on the Mechanism of Splashing. Int. J. Heat and Fluid Flow, Vol. 20, October 1999, pp. 455-461, 1999.
3. Puckett, E. G., Almgren, A. S., Bell, J. B., Marcus, D. L. and Rider, W. J.: A high-order projection method for tracking fluid interfaces in variable density incompressible flows. J. Comp. Phys., Vol. 130, pp. 269-282, 1997.

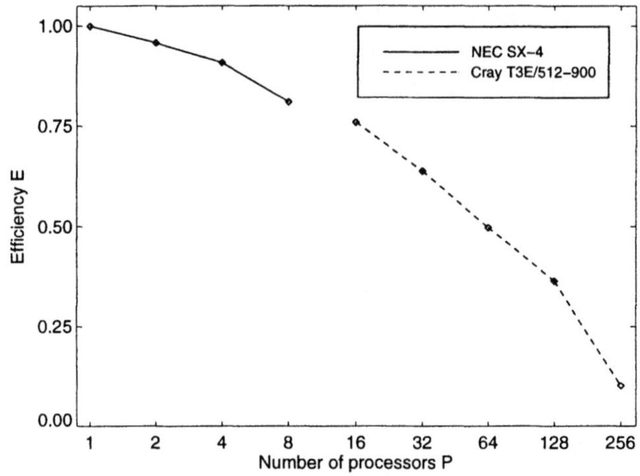

Fig. 9. Comparison of efficiency versus number of processors on NEC SX-4 and Cray T3E/512-900 for the splash test case.

4. Rider, W. J. and Kothe, D. B.: Reconstructing Volume Tracking. J. Comp. Phys., Vol. 141, pp. 112-152, 1998.
5. Wesseling, P.: Cell-Centred Multigrid for Interface Problems. J. Comp. Phys., Vol. 79, pp. 85-91, 1988.
6. Meyer, Daniel.: Parallelisierung eines Zweiphasencodes durch Gebietszerlegung. Studienarbeit am Institut für Thermodynamik der Luft- und Raumfahrt, Universität Stuttgart, 1997.
7. Rieber, M. and Frohn, A.: Navier-Stokes Simulation of Droplet Collision Dynamics. Proc. 7th Int. Symp. on CFD, Beijing, China, pp. 520-525, 1997.
8. Roth, N., Rieber, M. and Frohn, A.: High Energy Head-on Collision of Droplets. Proc. 15th Int. Conference on Liquid Atomization and Spray Systems, 5th-7th July 1999, Toulouse, France, 1999.
9. Schelkle, M., Rieber, M. and Frohn, A.: Numerische Simulation von Tropfenkollisionen. Spektrum der Wissenschaft, Januar 1999.
10. Cossali, G. E., Coghe, A. and Marengo, M.: The Impact of a Single Drop on a Wetted Solid Surface. Experiments in Fluids, Vol. 22, pp. 463-472, 1997.
11. Anderson, E., Brooks, J. and Hewitt, T.: The Benchmarkers Guide to Single-Processor Optimization for T3E Systems. Cray Research Inc., 1997.

Acknowledgments

This work was supported by the *Deutsche Forschungsgemeinschaft DFG* through grant Fr235/43-1. We want to thank Dipl.-Ing. Daniel Meyer (now IAG, University of Stuttgart), who helped us to parallelize FS3D.

Numerical Simulation of Fluid Flow and Heat Transfer in an Industrial Czochralski Melt Using a Parallel-Vector Supercomputer

Sven Enger, Michael Breuer, and Biswajit Basu*

Institute of Fluid Mechanics, Friedrich-Alexander-University of Erlangen-Nuremberg, Cauerstr. 4, D-91058 Erlangen, Germany

Abstract. In this paper, the fluid flow and heat transfer in an industrial Czochralski melt was analyzed by solving the time-dependent three-dimensional Navier-Stokes equations on curvilinear boundary-fitted grids in a rotating frame of reference. In order to represent the ellipsoidal crucible, a grid with 720,896 control volumes was generated in six blocks. Using the natural advantage of block-structuring, computations were carried out on a parallel-vector machine (NEC SX-4) with an optimal load-balancing efficiency of 100% using four processors. Simulations of the flow field were performed with and without the $k-\varepsilon$ turbulence model. It was seen that the turbulence model suppresses the fluid mechanical instabilities leading to an axisymmetric flow and thermal field, while the simulations without the turbulence model were found to predict the three-dimensional time-dependent features of the melt flow well. A total performance of 2.95 and 2.81 GFlops on four processors was reached for the simulation with and without turbulence model, respectively.

1 Introduction

Czochralski (CZ) crystal growth is one of the most popular methods of producing large single crystals for electronic and photonic devices. Most industries producing silicon crystals, III–V crystals and crystal of oxide materials follow the CZ process. Due to the importance for the single crystal production and its technological impact on semi-conductor industry, there have been extensive research studies in the past couple of decades for thorough fundamental understanding of mechanisms governing the Czochralski process and, in turn, to improve as well as optimize the process further for larger crystals of high quality.

It has been pointed out by several researchers, that fluid flow in the Czochralski melt plays a major role on the crystal quality; fluid flow in the Czochralski melt is the single most important phenomenon controlling transport of species like oxygen and, in turn, purity of the crystal. As mentioned by Langlois [1], the flow in the CZ melt is governed by several interacting forces such as buoyancy, Coriolis and centrifugal forces, as well as surface tension. The liquid in the melt is maintained above the freezing point through controlled heating of the crucible which results in thermal buoyancy in the melt.

* On leave from Tata Research Development & Design Centre, Pune, India.

In order to smear out the thermal asymmetries resulting from heating, the crucible is rotated. The uniformity at the crystal/melt interface is also improved by rotating the crystal while pulling to generate von Kármán-Cochran centrifugal flow with its constant boundary layer thickness.

The most important and interesting aspect of the flow in a CZ melt is the breakdown of the axisymmetric nature of the field leading to a host of interesting alternatives involving non-periodic time-dependent flows.

There have been a number of experimental studies on model fluids for the Czochralski melt with different geometry approaches as well as measuring the thermal field in the real silicon melt to understand the nature of the flow field. These investigations are difficult and time-consuming, thus, three-dimensional numerical simulation of the flow in the melt has become an attractive alternative to fully characterize the melt flow. A number of attempts have been reported in the recent past on this subject. Due to several interacting forces, different length scales exist for viscous and convective transport. Furthermore, due to the turbulent nature of the flow a characterization of flow regimes in the Czochralski melt through time-dependent three-dimensional simulations is a difficult task and therefore a challenging subject of research in fluid mechanics.

Over more than two decades, there has been a large number of numerical studies of the Czochralski crystal growth process. Different types of numerical simulations can be broadly classified into two categories namely global and local simulations. In the global simulation studies, the entire system including heater, crucible, puller rod, etc. is modeled first based on conductive-convective-radiative transport of energy to estimate the thermal environment of the melt in the crucible (i.e., in terms of crucible wall temperature and temperature of the free surface of the melt). Subsequently, the transport in the melt is simulated based on fluid flow and heat transfer. In the local simulations, detailed studies are carried out on the three-dimensional flow field of the melt only based on measured wall and environment temperatures and/or boundary conditions obtained from global simulations.

An overview of the most significant three-dimensional simulations of fluid flow and heat transfer in the Czochralski melt is given in Tab. 1 in terms of problem definition, dimensionless numbers, applied turbulence model, and numerical details. From this table, it is evident that most of the three-dimensional simulations are restricted to simple cylindrical geometries and Rayleigh numbers below 10^6, except Hirata and Hoshikawa [6] and Koai et al. [8], who considered a real geometry of the crucible with smaller dimensions leading to a low Rayleigh number ($\simeq 10^5$) and Tanaka et al. [11], who considered a simple cylindrical crucible but a high Rayleigh and Grashof number (about 10^7 and 10^9, respectively). Although the flow fields are distinctly turbulent for many of the reported studies, the Navier-Stokes equations without any turbulence model are applied in all simulations. While finite volume methods with structured grids are followed in most of the cases [2–4,6,7,12,11,13], finite element methods are adopted only by a few authors [9].

Table 1. Overview of physical data used in previous simulations of the flow in a crucible.

No	Reference	Pr	Ra 10^{-5}	Geom.	Turb. Model	Nodes
1	Mihelčić and Wingerath [2]	0.01	0.644–5.21	Cyl.	No	144,000
2	Bottarao and Zebib [3]	0.01	→0.1	Cyl.	No	34,560
3	Mihelčić et al. [4]	0.01	1.54, 2.57	Cyl.	No	307,800
4	Mihelčić et al. [5]	0.01	0.32, 2.57	Cyl.	No	63,000
5	Hirata and Hoshikawa [6]	0.01	24.5	Real	No	40,000
6	Leister and Perić [7]	10.45	4×10^3	Cyl.	No	81,920
7	Koai et al. [8]	0.068	3.33, 11.1	Real	No	22,400
8	Xiao and Derby [9]	8.0	21.0	Cyl.	No	266,987
9	Wagner and Friedrich [10]	0.018	17.5	Cyl.	DNS	1,250,000
10	Tanaka et al. [11]	0.01	187	Cyl.	No	326,106
11	Seidl et al. [12]	0.01	0.9	Real	No	12,800

A detailed review of two- and three-dimensional, local or global simulations of the Czochralski process is given by Enger et al. [14,15]. It is important to note, that none of the previous studies consider large industrial crucible as can be seen from Tab. 1.

Three-dimensional simulations require long time integration, small time steps and a fine grid to achieve convergence and to resolve the fine structures of the flow. This kind of simulations needs supercomputers, and even then, the simulation time will be in order of days rather than hours.

This paper presents simulations of the three-dimensional fluid flow and heat transfer in an industrial Czochralski melt. The simulations are carried out applying a block-structured finite volume method using a vector-parallel supercomputer.

2 Mathematical Formulation

The flow and heat transfer in a crucible of a Czochralski system is governed by the three-dimensional Navier-Stokes equations for an incompressible fluid expressing the conservation of mass, momentum and energy. Buoyancy is taken into account by the Boussinesq approximation. A general conservation equation for all transport variables in a Cartesian coordinate system is given by

$$C_\Phi \left[\frac{\partial(\rho\Phi)}{\partial t} + \frac{\partial}{\partial x_i}(\rho u_i \Phi) \right] - \frac{\partial}{\partial x_i}\left(\Gamma_\Phi \frac{\partial \Phi}{\partial x_i} \right) = S_\Phi , \qquad (1)$$

where Φ denotes a general transport variable, x_i the i-th Cartesian coordinate, u_i the i-th Cartesian velocity component in the rotating reference frame, T the temperature of the fluid, ρ the density, Γ_Φ the diffusivity of

Table 2. Transport variable Φ, diffusivity Γ_Φ, parameter C_Φ and source term S_Φ for the mass, momentum and energy equation, including the $k-\varepsilon$ turbulence model. Additional relations and constants not mentioned in the text: $\mu_t = c_\mu \rho \frac{k^2}{\varepsilon}$, $c_\mu = 0.09$, $\sigma_k = 1.0$, $\sigma_\varepsilon = 1.3$, $c_{\varepsilon 1} = 1.44$, $c_{\varepsilon 2} = 1.92$, $c_{\varepsilon 3} = 1.0$.

Conservation variable	Φ	C_Φ	Γ_Φ	S_Φ
Mass	1	1	0	0
Momentum (j-th comp.)	u_j	1	μ	$-\frac{\partial p}{\partial x_j} + \frac{\partial}{\partial x_i}\left(\mu \frac{\partial u_i}{\partial x_j}\right) - \rho_{\text{ref}} g_j \beta (T - T_{\text{ref}})$ $+2\epsilon_{mnj} u_m \Omega_n - \epsilon_{mnj} \Omega_m (\epsilon_{pqn} \Omega_p x_q)$
Turbulent kinetic energy	k	1	$\mu + \frac{\mu_t}{\sigma_k}$	$\underbrace{\mu_t \left(\frac{\partial u_i}{\partial x_k} + \frac{\partial u_k}{\partial x_i}\right) \frac{\partial u_i}{\partial x_k}}_{P_k} - \rho\varepsilon + \underbrace{\rho \beta g_j \frac{\mu_t}{\sigma_t} \frac{\partial T}{\partial x_j}}_{G_k}$
Dissipation rate of k	ε	1	$\mu + \frac{\mu_t}{\sigma_\varepsilon}$	$\frac{\varepsilon}{k} c_{\varepsilon 1} \left[\mu_t P_k + c_{\varepsilon 3} G_k\right] - c_{\varepsilon 2} \rho \frac{\varepsilon^2}{k}$
Energy	T	c_p	λ	$\mu \left(\frac{\partial u_i}{\partial x_j} + \frac{\partial u_j}{\partial x_i}\right) \frac{\partial u_i}{\partial x_j}$

Φ (dynamic viscosity μ or thermal conductivity λ), and S_Φ the source term of Φ. Tab. 2 gives the values of Φ, C_Φ, Γ_Φ and S_Φ for all transport equations applied in the model. Here, c_p denotes the specific heat at constant pressure, β the coefficient of volumetric thermal expansion, T_{ref} the reference temperature for the Boussinesq approximation, and g_j the gravitational acceleration in j-direction. The flow is clearly in the turbulent regime [16], and a buoyancy extended $k-\varepsilon$ turbulence model is applied. The simulations are carried out using a boundary-fitted grid in a rotating frame of reference, i.e., the coordinate system rotates with the crucible. The extensions needed for the rotating coordinate system are limited to the last two terms in the momentum equations in Tab. 2, where Ω is the rotation vector and ϵ_{mnj} the Levi-Civita tensor.

For fluid flow simulations with the $k-\varepsilon$ turbulence model, the standard wall functions are applied for the velocity at the walls and at the crystal/melt interface. The rotation of the crystal is realized by setting the appropriate wall velocity. In the simulation without turbulence model, no-slip boundary conditions are applied. A measured temperature distribution for the crucible wall and a fixed value for the crystal/melt interface are used as boundary conditions for the energy equation (see [14] for details).

The free surface of the melt is assumed to be plane and free of shear stresses. Accordingly, the normal velocity at the surface is set to zero. A symmetry condition is applied for the other velocity components. Marangoni convection due to surface tension gradient is not taken into account. Heat loss from the surface of the melt in the industrial crucible takes place by radiation which is modeled according to the Stefan-Boltzmann law [14].

3 Numerical Method

The solution domain is divided into a finite number of control volumes (CV). Fig. 1 shows the grid used in the simulations. The above equations are approximated for each CV using a flux blending scheme consisting of the second-order accurate central difference scheme (CDS) and the first-order upwind difference scheme (UDS) for the convective terms. For the time integration a second-order fully implicit method was applied. The discretization results in a large system of equations consisting of seven main diagonals which, due to the non-linearity and the coupling of the variables, must be found iteratively using the strongly implicit procedure (SIP) of Stone [17]. More details of the solver can be found in [15,18,19].

4 High-Performance Computing Approach

Each year, the 500 most powerful computer systems installed worldwide are ranked by Dongarra et al. [20]. The computers listed can be divided into two different categories depending on the architecture: massively parallel processing (MPP) systems and moderate vector-parallel systems (e.g., NEC SX-4, actually rank 29). MPP systems are built up by an extremely large number of comparatively "cheap and slow" microprocessors whereas vector-parallel processing (VPP) systems apply the fastest individual processors available. Although VPP systems actually rank below MPP computers, we prefer to use parallel-vector computers for our CFD simulations because they show much better performance characteristics for engineering applications.

In order to achieve high performances on vector processors the finite volume code was adapted to this machine architecture independent on the specific vendor with the exception of a few vector directives. With respect to Amdahl's law, which describes the speedup of a code by vectorization, it is extremely important to achieve a high vectorization rate. In the present CFD code three different types of do-loops have to be distinguished for the vectorization purpose. The first are all loops which calculate the source terms and coefficients of the conservation equations for the internal flow region. Because in general no data dependencies or recursions exist for this kind of loops, they were easily vectorized. The vector length of these loops is determined by the number of internal control volumes in one block which is typically larger than the optimal vector length. The second group of loops which calculate the boundary conditions for the computational domain leads to much shorter vector lengths because of the restriction to 2-D surface data. Also in this case, typically no recursions prevent vectorization. Larger efforts had to be made to vectorize the implicit part of the code, namely the SIP solver mentioned above for solving linear equation systems which typically requires at least half of the computing time of the entire CFD code. Due to a recursive data structure, the SIP solver is not vectorizable in a straightforward manner. However, Leister and Perić [21] showed that vectorization of the SIP solver can be achieved by avoiding data dependencies through indirect addressing and sweeping through the computational domain along diagonal planes, so called ***hyper-planes***. Thus one sweep through the entire domain consists of hyper-planes having different vector length. Due to this variable vector length and the indirect addressing used, the performance of the vectorized SIP solver is slightly lower than the other parts of the code.

The parallelization strategy used in FASTEST-3D is based on grid partitioning with explicit message passing and consequently works on all kind of machines independently of the memory architecture (distributed or shared memory). The massage-passing-interface (MPI) software is used as communication library. The grid partitioning technique is directly related to the block-structuring. The principle idea is to transform the block-structure which results from the requirements to model the geometry (geometric block-structure) by some suitable mapping process to a new block-structure (parallel block-structure). The resulting blocks are assigned to the available processors. Fig. 2 shows an example of the partitioning process. The blocks of the parallel block-structure, in addition to the geometric ones, also have to meet the requirements for efficient implementation on a parallel computer. Although there are several requirements, the most important one is to obtain similar numbers of control volumes on each processor in order to ensure good load-balancing on parallel machines. A measure for this is the load-balancing efficiency, which is the ratio of the total number of control volumes to the number of processors times the largest number of control volumes of one processor. If the control volumes are equally distributed to the processors, the load-balancing efficiency is 100%. More details on this parallelization technique are given by Durst et al. [18] and Durst and Schäfer [19].

5 Problem Description and Numerical Details

The flow and thermal field in an industrial crucible with height $H_c = 0.107$ m, radius $R_c = 0.17$ m and crystal radius $R_s = 0.05$ m was simulated. The geometry can be in Fig. 1 (left). The crucible and the crystal rotate opposite to each other with a rotation rate of $\Omega_c = 10$ rpm and $\Omega_s = -15$ rpm, respectively. Silicon was used as working fluid, and the thermo-physical properties can be found in Enger et al. [14]. The dimensionless numbers for the simulations, based on a vertical temperature difference ΔT_{ref} of 35 K, the crucible height H_c and the rotation rate of the crucible Ω_c are given in Tab. 3.

A time step of $\Delta t = 0.05$ s was used in the simulations to ensure a maximum Courant number less than one in each control volume and time

Table 3. Dimensionless numbers for the simulation of the flow and heat transfer in an industrial crucible.

Dimensionless number	Symbol	Definition	Value
Rayleigh number	Ra	$\mathrm{Pr}\, g\, \beta\, H_c^3\, \Delta T_{\mathrm{ver}}/\nu^2$	7.2×10^6
Taylor number	Ta	$4\Omega_c^2\, R_c^4/\nu^2$	6.31×10^9
Rossby number	Ro	$g\, \beta\, H_c\, \Delta T_{\mathrm{ver}}/(\Omega_c^2\, R_c^2)$	25.8

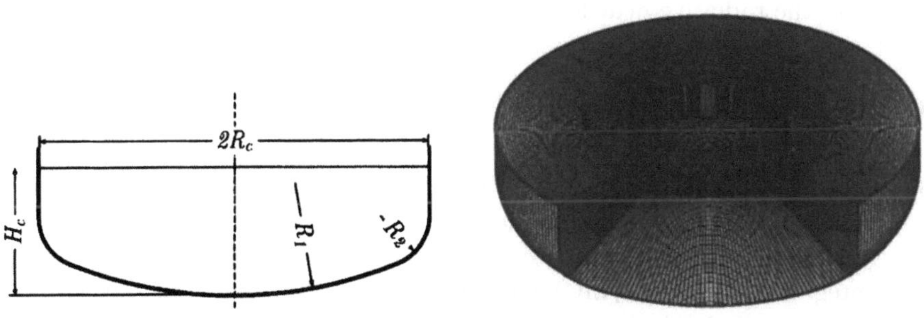

Fig. 1. Geometry and the corresponding grid of an industrial crucible.

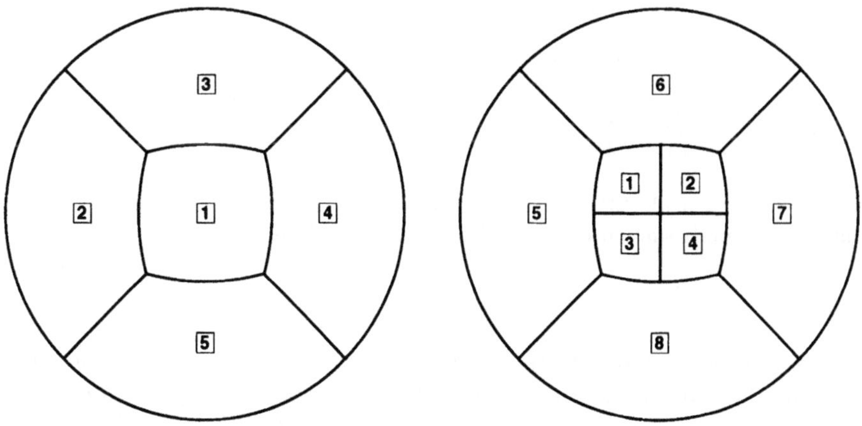

Fig. 2. Geometric (left) and parallel (right) block-structure for optimal load-balancing on four processors used for the simulation of the Czochralski crucible. The sixth, bottom block is not visible in this view.

step. Since a typical pulling process lasts for several hours, only a part of this can be simulated within reasonable computational time. In order to compare the numerical results with experimental data, the simulations were continued up to five minutes real time. In each time step, the iteration process was continued until the residual values of all variables had reached convergence, i.e., dropped six orders of magnitude.

A uniform solid body rotation was assigned as initial condition for the velocity field in the melt. The temperature was set to a constant value ($T_{\text{ini}} = 1700\ K$) and the turbulent kinetic energy and the dissipation rate were zero for the whole melt. An environment temperature of 1300 K was used to calculate the radiative heat loss at the free melt surface.

6 Results and Discussion

Figs. 3 and 4 present the velocity field and the isotherms, respectively, at a vertical plane through the middle of the crucible after 90 seconds of time integration, starting from initial condition. The flow field is visualized by the line integral convolution (LIC) technique. As can be seen from the figures, the isotherms and velocity results show an axisymmetric thermal and flow field.

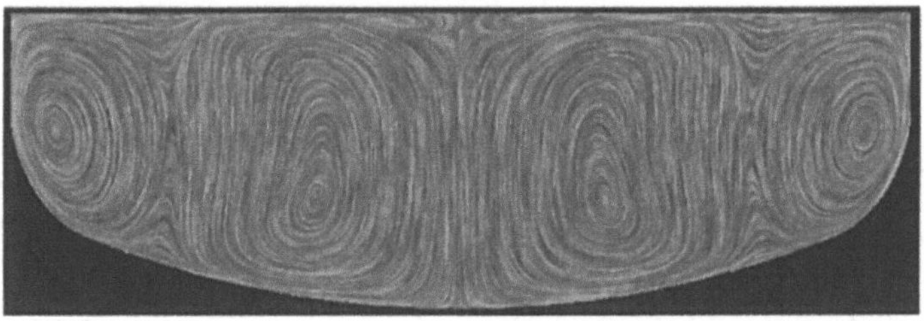

Fig. 3. Velocity field (LIC) in a cut through the middle of the crucible after 1 minute 30 seconds simulation time with the $k - \varepsilon$ turbulence model.

The flow field shows presence of three vortices in the cutting plane, i.e., one near the central axis due to the rotation of the crystal, one caused by buoyancy close to the wall, and a small one just below the crystal near the top. The last vortex is restricted to a small region of the order of the boundary layer thickness [22]. Due to the central vortex, the isotherms can be seen to be stretched downwards near the center and upwards in the region between the centre and the outer wall.

Using the $k - \varepsilon$ turbulence model, the effect of different parameters (e.q., rotation rates) on the overall features of the flow and thermal field was studied

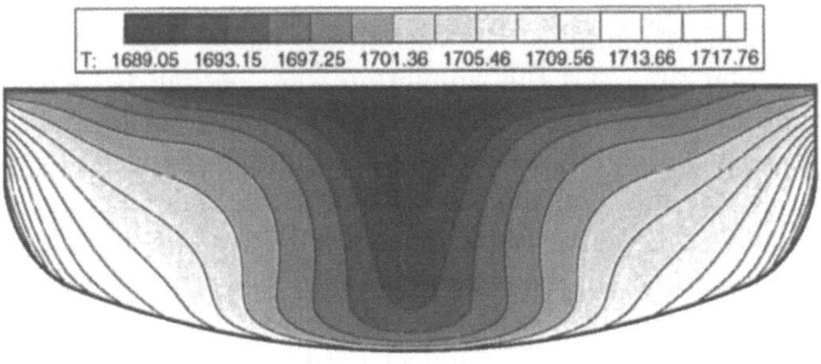

Fig. 4. Isotherms in a cut through the middle of the crucible after 1 minute 30 seconds simulation time with the $k-\varepsilon$ turbulence model.

and the relative dominance and interaction of different convection generating forces on the bulk transport phenomena in the crucible was established. From Fig. 5, which shows the isotherms at the top of the melt after 1 minute 30 seconds real time, it can be seen that the flow is perfectly axisymmetric, a situation which is *not* known from experiments. Several tests were performed with the $k-\varepsilon$ turbulence model in order to confirm its weakness in three-dimensional simulations of the buoyancy driven flow in the silicon melt. Finally, it has to be concluded that the $k-\varepsilon$ turbulence model *fails* to predict the three-dimensional time-dependent nature of the flow and thermal field.

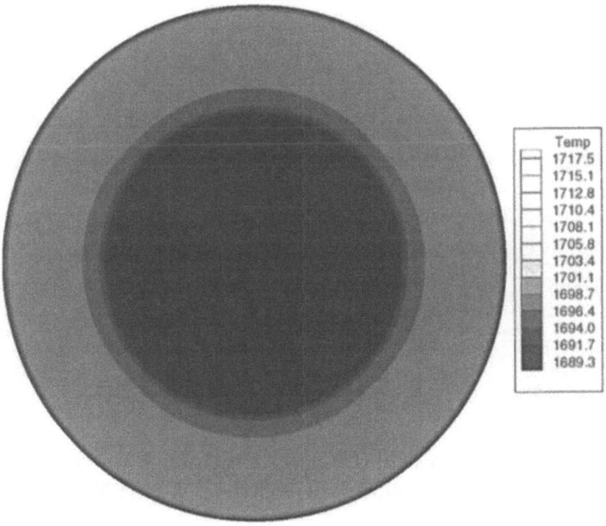

Fig. 5. Isotherms at the top of the crucible after 1 minutes 30 seconds simulation time with the $k-\varepsilon$ turbulence model.

In order to study the quantitative effect of the $k-\varepsilon$ turbulence model on prediction of instabilities, the turbulence model was switched off after 300 seconds real time and the simulation was continued. Using a deferred correction approach, several simulations were carried out to find the optimal flux blending. A too high percentage of upwind differences tends to damp all instabilities in the melt; on the other hand, a too high CDS percentage can lead to non physical results. Based on numerical tests, a flux blending factor of 80% CDS was used for the simulation without turbulence model in order to achieve meaningful prediction by suppressing high frequency turbulent fluctuations. A strongly time-dependent nature of the flow and thermal field can be noticed in Figs. 6 and 7, which show the velocity field and isotherms in a cut through the middle of the crucible after about 8 minutes.

The three-dimensional nature of the flow and thermal field can be clearly seen from these predictions where there are multiple waves and vortices as well as buoyant plumes, which can be observed from the thermal field in Fig. 7. Also from the isotherms close to the surface in Fig. 8, the buoyant plumes can be recognized as hot spots near the crucible wall. It is interesting to note that Tanaka et al. [11] reported buoyant plumes at the free surface of a model crucible based on their experimental work with a crucible rotation rate of $\Omega_c = 10$ rpm.

Starting from an axisymmetric and stable flow and thermal field obtained from the simulation with the $k-\varepsilon$ turbulence model, simulations without turbulence model show a time-dependent three-dimensional flow and thermal fields (as observed from experiments with the Czochralski melt [23]). This clearly demonstrates the drawback of the $k-\varepsilon$ turbulence model in terms of smearing out the fluid mechanical instabilities.

Fig. 6. Instantaneous velocity field (LIC) in a cut through the crucible at 7 minutes 50 seconds. The $k-\varepsilon$ turbulence model was switched off after 300 seconds simulation time.

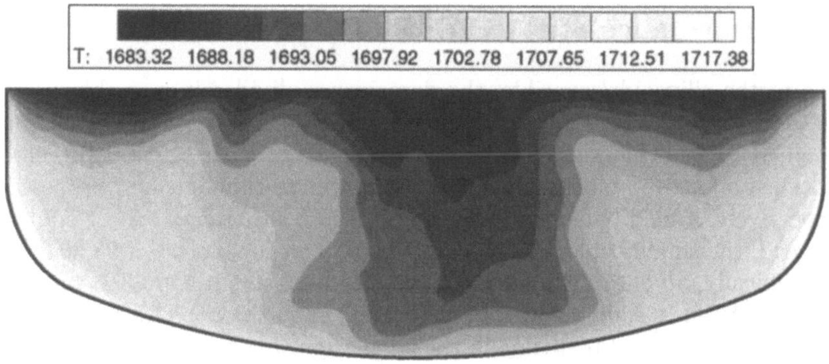

Fig. 7. Instantaneous isotherms in a cut through the crucible at 7 minutes 50 seconds. The $k-\varepsilon$ turbulence model was switched off after 300 seconds simulation time.

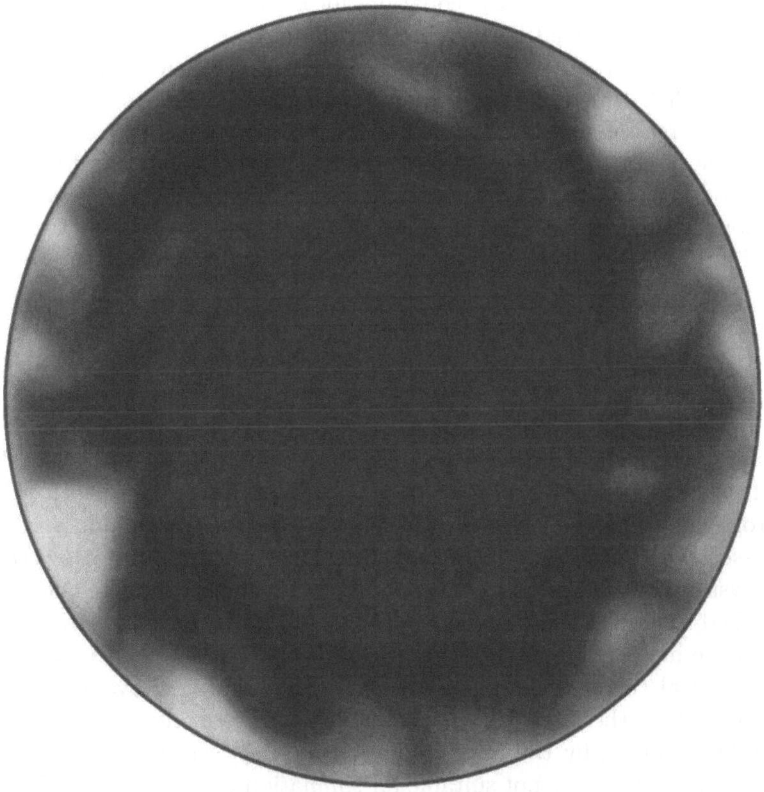

Fig. 8. Isotherms close to the surface at 7 minutes 50 seconds. The $k-\varepsilon$ turbulence model was switched off after 300 seconds simulation time.

7 Performance of FASTEST-3D

Due to the ellipsoidal crucible, the domain was divided into six blocks (see Fig. 1) with a total of 720,896 CVs. Each of the identical (outer) blocks were assigned to four different processors. By splitting each of the two central blocks into four sub-blocks which are distributed to the same four processors, respectively, a load-balancing efficiency of 100% is obtained. After splitting the grid, the smallest block consists of 32,768 and the largest of 98,304 CVs.

Typically, an entire simulation takes a total of about 640 CPU hours on the NEC SX-4 machine, i.e., 160 CPU hours on each of the four processors, to reach the required real time in the simulation. In the simulation with the turbulence model, a total performance of about 2.95 GFlops is reached on four processors, i.e., about 738 MFlops on each processor. The vector operation ratio is 99.58% and the memory requirement is about 150 MB for each of the four processors, resulting in a total of 602 MB.

When the turbulence model is switched off, the total memory requirement decreases to 550 MB, since the equations for k and ε are not solved anymore. For this case a somewhat lower vector performance is achieved, i.e., a total of 2.81 GFlops. The reason for this can be explained as follows: A large percentage of the CPU-time is typically spent for the internal SIP iterations for the pressure correction approach as well as for the temperature. As explained in Sec. 4, the performance of the vectorized SIP solver is less than for the rest of the code. Therefore, solving additional scalar equations with less internal iterations such as k and ε leads to a decreasing percentage of CPU-time required for the SIP solver and thus to an increasing performance of the entire code.

8 Conclusions

The three-dimensional and time-dependent turbulent flow field and heat transfer of the melt in an industrial Czochralski melt was predicted applying an efficient block-structured finite volume method. Using a grid with 720,896 control volumes in six blocks, simulations were carried out on four NEC SX-4 processors with 100% load-balancing efficiency for a counter-rotating crucible and crystal with an buoyancy extended $k - \varepsilon$ turbulence model. The simulation failed to predict the time-dependent three-dimensional flow and thermal field in the melt. Furthermore, simulations without a turbulence model predicted typical cellular features of the melt representing buoyant plumes. It was found, that the $k - \varepsilon$ turbulence model suppresses the fluid mechanical instabilities caused by the interaction of buoyancy, centrifugal and Coriolis forces completely and is not suitable for simulation of the turbulent flow field in a Czochralski crucible. In order to capture different scales of instabilities along with turbulence, it is necessary to adopt more rigorous approaches such as Large-Eddy Simulation (LES) or Direct Numerical Simulation (DNS). The present work is a first step in this direction.

Acknowledgements

Financial support by the Bayerische Staatsministerium für Wirtschaft, Verkehr und Technologie (Förderprogramm 'Neue Werkstoffe') and the Bayerische Forschungsstiftung in the Bavarian Consortium of High-Performance Scientific Computing (FORTWIHR II) is gratefully acknowledged. The simulations were carried out on the NEC SX-4 at the High-Performance Computing Center, Stuttgart. This support is also gratefully acknowledged.

References

1. W.E. Langlois. Buoyancy-Driven Flows in Crystal-Growth Melts. *Ann. Rev. of Fluid Mechanics*, 17:191–215, 1985.
2. M. Mihelčić and K. Wingerath. Instability of the Buoyancy Driven Convection in Si Melts During Czochralski Crystal Growth. *Journal of Crystal Growth*, 97:42–49, 1989.
3. A. Bottarao and A. Zebib. Three-Dimensional Thermal Convection in Czochralski Melt. *Journal of Crystal Growth*, 97:50–58, 1989.
4. M. Mihelčić, H. Wenzi, and K. Wingerath. Flow in Czochralski Crystal Growth Melts. Technical Report, Institut für Festkörperforschung, Forschungszentrum, Jülich, Germany, December 1992.
5. M. Mihelčić, K. Wingerath, and C. Pirron. Three-Dimensional Simulation of the Czochralski Bulk Flow. *Journal of Crystal Growth*, 69:473–488, 1984.
6. H. Hirata and K. Hoshikawa. Three-Dimensional Numerical Analyses of the Effect of a Cusp Magnetic Field on the Flows, Oxygen Transport and Heat Transfer in a Czochralski Silicon Melt. *Journal of Crystal Growth*, 125:181–207, 1992.
7. H.J. Leister and M. Perić. Numerical Simulation of a 3D Czochralski-Melt Flow by a Finite Volume Multigrid Algorithm. *Journal of Crystal Growth*, 123:567–574, 1992.
8. K. Koai, A. Seidl, H.J. Leister, G. Müller, and A. Köhler. Modelling of Thermal Fluid Flow in the Liquid Encapsulated Czochralski Process and Comparison with Experiments. *Journal of Crystal Growth*, 137:41–47, 1994.
9. Q. Xiao and J.J. Derby. Three-Dimensional Melt Flows in Czochralski Oxide Growth: High-Resolution, Massively Parallel, Finite Element Computations. *Journal of Crystal Growth*, 152:169–182, 1995.
10. C. Wagner and R. Friedrich. Direct Numerical Simulation of Turbulent Flow in an Idealized Czochralski Crystal Growth Configuration. In *Proc. of French-German Workshop on "Computation of Three-Dimensional Vortical and Turbulent Flow"*, December 1996.
11. M. Tanaka, H. Hasebe, and N. Saito. Pattern Transition of Temperature Distribution at Czochralski Silicon Melt Surface. *Journal of Crystal Growth*, 180:487–496, 1997.
12. A. Seidl, G. McCord, G. Müller, and H.J. Leister. Experimental Observations and Numerical Simulation of Wave Pattern in a Czochralski Silicon Melt. *Journal of Crystal Growth*, 137:326–334, 1994.

13. Y.C. Yon, K. Kakimoto, and H. Ozoe. Transient Three-Dimensional Numerical Computation for Silicon Melt under a Cusp-Shaped Magnetic Field. In *Proc. of the 5th ASME/JSME Joint Thermal Engineering Conference*. ASME, March 1999.
14. S. Enger, B. Basu, and M. Breuer. Numerical Simulation of the Three-Dimensional Turbulent Fluid Flow in a Crucible during Czochralski Growth. Technical Report LSTM 587/I 16.04.1999, Institute of Fluid Mechanics, University of Erlangen-Nuremberg, Germany, 1999.
15. B. Basu, S. Enger, and M. Breuer. Three-Dimensional Simulation of Flow and Thermal Field in a Czochralski Melt Using a Block-Structured Finite-Volume Method. *Submitted to Journal of Crystal Growth*, 1999.
16. J.R. Ristorcelli and J.L. Lumley. Instabilities, Transition and Turbulence in the Czochralski Crystal Melt. *Journal of Crystal Growth*, 116:447–460, 1992.
17. H.L. Stone. Iterative Solution of Implicit Approximations of Multidimensional Partial Differential Equations. *SIAM J. Num. Anal.*, 5:530–557, 1968.
18. F. Durst, M. Schäfer, and K. Wechsler. Efficient Simulation of Incompressible Viscous Flows on Parallel Computers. In *Flow Simulation with High Performance Computers II, Notes on Numerical Fluid Mechanics*, 52, pages 87–101. Vieweg, Braunschweig, 1996.
19. F. Durst and M. Schäfer. A Parallel Block-Structured Multigrid Method for the Prediction of Incompressible Flows. *Int. J. Num. Methods Fluids*, 22:549–565, 1996.
20. J.J. Dongarra, H.W. Meuer, and E. Strohmaier. TOP500 Supercomputer Sites, 13th Edition. http://www.top500.org/index.html, June 1999.
21. H.J. Leister and M. Perić. Vectorized Strongly Implicit Solving Procedure for a Seven-Diagonal Coefficient Matrix. *Int. J. Heat Fluid Flow*, 4:159–172, 1993.
22. A.D.W. Jones. Scaling Analysis of the Flow of a Low Prandtl Number Czochralski Melt. *Journal of Crystal Growth*, 88:465–476, 1988.
23. O. Gräbner and A. Mühe. Personal communication, 1998.

Numerical flow simulation in cylindrical geometries

T.J. Hüttl[1], M. Smieszek[2], M. Fröhlich[3], M. Manhart[1], R.J.D. Schmidt[4] and R. Friedrich[1]

[1]) Lehrstuhl für Fluidmechanik, Technische Universität München
Boltzmannstr. 15, 85748 Garching, Germany
[2]) Brandenburgische Technische Universität Cottbus
Karl-Marx-Str. 17, 03044 Cottbus, Germany
[3]) Institut für Informatik, Technische Universität München
Arcisstr. 21, 80333 München, Germany
[4]) Lehrstuhl für Thermodynamik A, Technische Universität München
Boltzmannstr. 15, 85748 Garching, Germany

Abstract. The CFD-code MGLET, which is optimized for direct and large-eddy simulation on high performance computers, has been extended by a cylindrical version allowing to compute flow problems in cylindrical geometries. Laminar pipe flow and cylindrical Couette flow have been simulated to test the implementation. Preliminary results of Taylor-Couette flow and turbulent boundary layer flow along cylinders are shown to demonstrate the applicability of the new code-version.

1 Introduction

It is the objective of current and future investigations to compute flow in pipes, Taylor-Couette flow and boundary layer flow along cylinders, by means of Large-Eddy simulation (LES) and direct numerical simulation (DNS). These simulation techniques have been used successfully since many years and several CFD-Codes have been developed by different groups, mostly for cartesian coordinates. For efficiency reasons a blockstructured, parallel code is required, which is optimized for turbulence simulation on high performance computers. The code MGLET, originally designed for cartesian coordinates has been selected for extension to cylindrical coordinates because of its various favorable features, such as: efficient Poisson-solver, parallel version, high vectorization level. Before applying the new version in production runs, the implementations and modifications have to be verified. The accuracy has to be tested and the code has to be installed successfully on a parallel computer. The results of these test simulations and of preliminary turbulence simulations will be shown and discussed below.

2 Governing equations

With s representing the axial, r the radial and θ the circumferential direction, the non-dimensional incompressible Navier-Stokes equations in a cylindrical coordinate system read:

continuity equation:

$$r\frac{\partial u_s}{\partial s} + \frac{\partial}{\partial r}(r u_r) + \frac{\partial u_\theta}{\partial \theta} = 0 \tag{1}$$

s-momentum:

$$\begin{aligned}
&\frac{\partial u_s}{\partial t} + \frac{\partial}{\partial s}(u_s u_s) + \frac{1}{r}\frac{\partial}{\partial r}(r u_s u_r) + \frac{1}{r}\frac{\partial}{\partial \theta}(u_s u_\theta)\\
&= -\frac{\partial p}{\partial s} + \frac{1}{\mathrm{Re}}\left[2\frac{\partial}{\partial s}\left(\frac{\partial u_s}{\partial s}\right)\right.\\
&\left.+\frac{1}{r}\frac{\partial}{\partial r}\left(r\frac{\partial u_s}{\partial r} + r\frac{\partial u_r}{\partial s}\right) + \frac{1}{r}\frac{\partial}{\partial \theta}\left(\frac{1}{r}\frac{\partial u_s}{\partial \theta} + \frac{\partial u_\theta}{\partial s}\right)\right]
\end{aligned} \tag{2}$$

r-momentum:

$$\begin{aligned}
&\frac{\partial u_r}{\partial t} + \frac{\partial}{\partial s}(u_s u_r) + \frac{1}{r}\frac{\partial}{\partial r}(r u_r u_r) + \frac{1}{r}\frac{\partial}{\partial \theta}(u_r u_\theta) - \frac{u_\theta u_\theta}{r}\\
&= -\frac{\partial p}{\partial r} + \frac{1}{\mathrm{Re}}\left[\frac{\partial}{\partial s}\left(\frac{\partial u_s}{\partial r} + \frac{\partial u_r}{\partial s}\right) + \frac{2}{r}\frac{\partial}{\partial r}\left(r\frac{\partial u_r}{\partial r}\right)\right.\\
&\left.+\frac{\partial}{\partial \theta}\left(\frac{1}{rr}\frac{\partial u_r}{\partial \theta} + \frac{\partial}{\partial r}\left(\frac{u_\theta}{r}\right)\right) - \frac{2}{rr}\left(\frac{\partial u_\theta}{\partial \theta} + u_r\right)\right]
\end{aligned} \tag{3}$$

θ-momentum:

$$\begin{aligned}
&\frac{\partial u_\theta}{\partial t} + \frac{\partial}{\partial s}(u_s u_\theta) + \frac{1}{r}\frac{\partial}{\partial r}(r u_r u_\theta) + \frac{1}{r}\frac{\partial}{\partial \theta}(u_\theta u_\theta) + \frac{1}{r}u_r u_\theta\\
&= -\frac{1}{r}\frac{\partial p}{\partial \theta} + \frac{1}{\mathrm{Re}}\left[\frac{\partial}{\partial s}\left(\frac{1}{r}\frac{\partial u_s}{\partial \theta} + \frac{\partial u_\theta}{\partial s}\right) + \frac{2}{rr}\frac{\partial}{\partial \theta}\left(\frac{\partial u_\theta}{\partial \theta} + u_r\right)\right.\\
&\left.+\frac{1}{r}\frac{\partial}{\partial r}\left(\frac{\partial u_r}{\partial \theta} + rr\frac{\partial}{\partial r}\left(\frac{u_\theta}{r}\right)\right) + \frac{1}{rr}\frac{\partial u_r}{\partial \theta} + \frac{\partial}{\partial r}\left(\frac{u_\theta}{r}\right)\right]
\end{aligned} \tag{4}$$

These equations contain the hydrodynamic pressure p and the velocity vector $(u_s, u_r, u_\theta)^T$, with components in the cylindrical coordinate directions. Curvature terms appear, rendering the set of equations partly non-conservative. The pressure is fully determined by the velocity field.

3 Numerical method

A finite volume method on staggered grids is used to discretize the spatial derivatives and source terms in the governing equations. It leads to central formulations of second order accuracy for the mass and momentum fluxes across the cell faces. The structured grid is equidistant in axial and circumferential direction, but it can be chosen non-equidistant in radial direction.

Except for turbulent pipe flow an explicit time-integration scheme is used. Here, the convection terms are integrated in time with a second order accurate leapfrog-step. An averaging step all the 50 time steps avoids possible $2\Delta t$-oscillations. Diffusive terms are treated with a first-order Euler backward step. The size of the time step is selected according to a linear stability argument. Only for turbulent pipe flow, where the grid size in circumferential direction becomes very small near the axis, a semi-implicit time-integration scheme is taken. It treats all those diffusion terms implicitly which contain derivatives in θ-direction. Then, the time step can be chosen about 10 times higher than for the explicit time integration scheme.

The use of a projection step leads to a 3D Poisson problem for the pressure correction, which is solved by an iterative point-wise velocity-pressure iteration like that described in Hirt et al. [2]. The maximum divergence is chosen in order to keep the maximal velocity error below $\Delta u_{max} \leq 10^{-5} u_{ref}$ (according to the relation $div_{max} = \Delta u_{max}/\Delta x_{min} \cdot l_{ref}/u_{ref}$). Here u_{ref} and l_{ref} are the characteristic global velocity and length scales, respectively.

4 Boundary conditions

Depending on the flow case, different boundary conditions have to be used. In homogeneous directions (axial direction for pipe- and cylindrical Couette-flow, circumferential direction for pipe-, cylindrical Couette-, Taylor-Couette- and boundary-layer-flow) periodic boundary conditions are used. At walls impermeability and no-slip boundary conditions are realized. For moving walls (cylindrical Couette-flow), the wall-velocity can be specified. For pipe flow, velocity components are needed on the pipe axis. They are obtained by interpolation across the axis.

For boundary layer flow, we are imposing a time-averaged velocity profile of a zero-pressure-gradient boundary layer at the inlet plane that matches the Reynolds number desired, [7]. The fluctuations are extracted from a position downstream and superposed onto the inlet profile. This procedure gives similar results than the method proposed by Richter et al. [5] for LES of boundary layers and the method of Lund et al. [4] applied to a zero-pressure-gradient boundary layer. At the outflow, a Neumann condition for the velocity and a Dirichlet condition for the pressure is used. The negative effect of this simple formulation is visible up to about two boundary layer thicknesses upstream.

In a pipe, the flow is driven in the axial direction by means of a pressure gradient $\Delta P/\Delta s$, which must balance the viscous friction along the pipe wall.

5 Laminar Test cases

5.1 Laminar pipe flow

Simple pipe flow has been computed, because it can easily be compared with an analytical solution for fully developed pipe flow. By using the following reference quantities

$$u_{ref} = u_b \qquad \text{and} \qquad l_{ref} = R \tag{5}$$

for normalization, the velocity profile becomes:

$$u_s = 2\left(1 - r^2\right) \qquad u_r = u_\theta = 0 \quad . \tag{6}$$

The normalized pressure gradient is:

$$\frac{dp}{ds} = -\frac{8}{\mathrm{Re}}. \tag{7}$$

Several cases with a different number of grid points in radial direction $n_r = 8, 16, 24, 32, 40$ have been computed. For all these cases, the same Reynolds number $Re = u_b R/\nu = 100$ and the same number of grid points in axial ($n_s = 4$) and circumferential direction ($n_\theta = 4$) have been used. For the steady solutions the radial and circumferential velocity are of the order of machine accuracy.

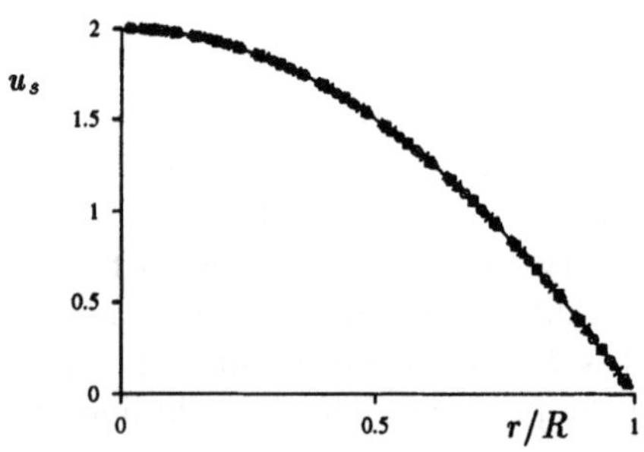

Fig. 1. Axial velocity component u_s over the radius r/R for laminar pipe flow ($Re = 100$): ——— analytical solution (64 grid points), + simulation with $n_r = 8$, $\times$ $n_r = 16$, $\square$ $n_r = 24$, $\bigcirc$ $n_r = 32$, $\triangle$ $n_r = 40$.

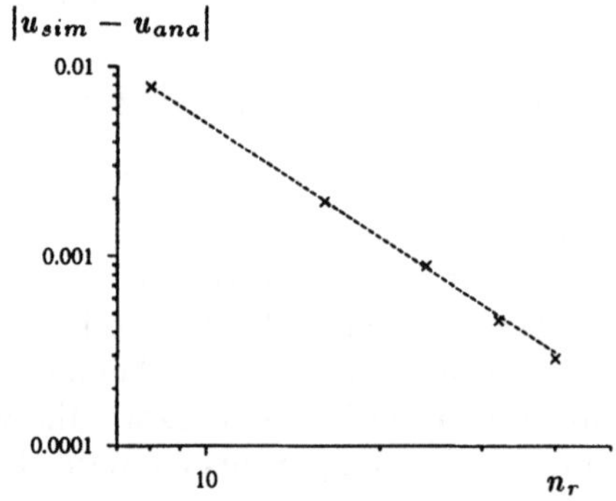

Fig. 2. Difference between simulation and analytical solution drawn over n_r for fully developed laminar pipe flow ($Re = 100$): – – – – curve of second order ($y = \alpha/n_r^2$), $\times$ simulations.

Figure 1 shows the profle of the axial velocity over a radial cut. The perfect matching between analytical and numerical solution can be seen. The absolute error between simulation and analytical solution $|u_{sim} - u_{ana}|$ is almost constant along the radius for each case, [3]. The difference between simulation and analytical solution decreases with increasing number of grid points, as shown in Figure 2. The diagram shows that the results match with a curve $y = \alpha/n_r^2$ ($\alpha = 0.4992$). This proofs that a second order accurate discretisation scheme is used.

5.2 Cylindrical Couette-flow

The flow between two rotating concentric cylinders (cylindrical Couette-flow) has been studied as a second test case. The geometry can be seen in Figure

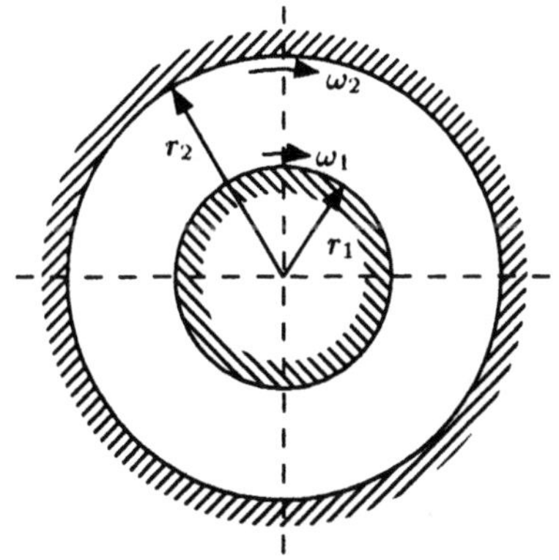

Fig. 3. Geometry of the cylindrical Couette-flow.

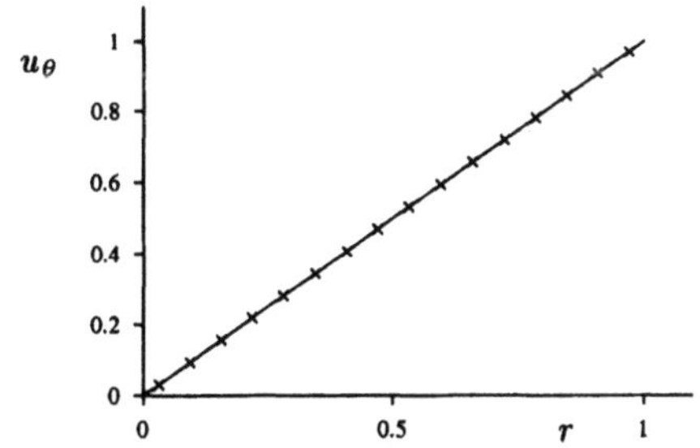

Fig. 4. Circumferential velocity u_θ between two rotating cylinders with $r_1 = 0.0$, $u_1 = 0.0$, $r_2 = 1.0$, $u_2 = 1.0$: —— — analytical solution (16 grid points), $\times$ simulation with $n_r = 16$.

3. The inner cylinder with radius r_1 rotates with a constant angular velocity of ω_1. The outer cylinder with radius r_2 rotates with a constant angular velocity of ω_2. The velocites $u_{\theta,1}$ and $u_{\theta,2}$ at the cylinder surfaces can be calculated from $u_{\theta,1} = r_1\omega_1$ and $u_{\theta,2} = r_2\omega_2$. For cylindrical fully-developed Couette-flow, an analytical solution of the flow equations can be derived, where the velocity components in axial and radial directions vanish and the remaining circumferential velocity component only depends on the radial direction, [6,8,10]:

$$u_\theta = Ar + \frac{B}{r} \tag{8}$$

with:

$$A = \frac{\omega_2 r_2^2 - \omega_1 r_1^2}{r_2^2 - r_1^2} \quad \text{and} \quad B = \frac{(\omega_1 - \omega_2)\, r_1^2 r_2^2}{r_2^2 - r_1^2} \tag{9}$$

The velocity profile has the form $u_\theta = Ar + \frac{B}{r}$ and can be seen as a combination of a solid body rotation Ar with a potential vortex $\frac{B}{r}$. Several different test cases can be defined:

Testcase C_A Provided the inner cylinder radius reduces to zero ($r_1 = 0$), the velocity profile becomes the profile of a solid body rotation.

$$u_\theta = \omega_2 r \tag{10}$$

A simulation with the following parameters shows excellent agreement with the analytical solution, see Figure 4.

- grid points: $n_s = n_\theta = 4$, $n_r = 16$

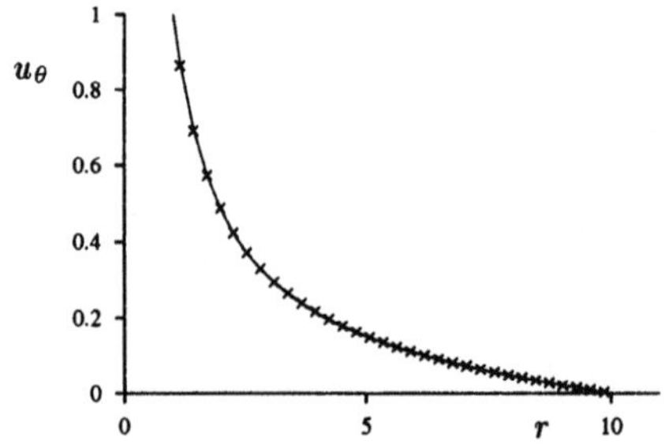

Fig. 5. Circumferential velocity u_θ between two rotating cylinders with $r_1 = 1.0$, $u_1 = 1.0$, $r_2 = 10.0$, $u_2 = 0.0$: —— analytical solution (32 grid points), × simulation with $n_r = 32$.

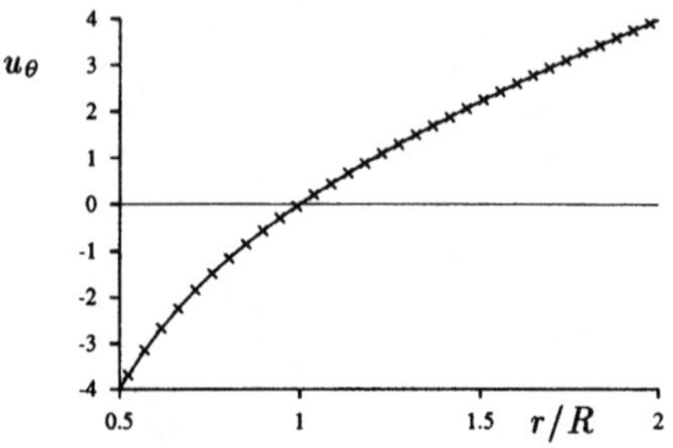

Fig. 6. Circumferential velocity u_θ between two rotating cylinders with $r_1 = 0.5$, $u_1 = -4.0$, $r_2 = 2.0$, $u_2 = 4.0$: —— analytical solution (32 grid points), × simulation with $n_r = 32$.

- geometry: $r_1 = 0$, $r_2 = 1.0$
- angular velocities: $\omega_1 = 0.0$, $\omega_2 = 1.0$
- velocities: $u_{\theta,1} = 0.0$, $u_{\theta,2} = 1.0$
- kinematic viscosity: $\nu = 0.01$
- time steps: $n_t = 6667$, $\Delta t = 0.03$, $t_p = n_t \Delta t = 200$
- Reynolds number: $Re = \frac{u_2 r_2}{\nu} = 100$

Testcase C_B If the outer cylinder does not rotate and its radius r_2 goes to infinity, the velocity profile matches the curve of a potential vortex

$$u_\theta = \frac{u_1^2 \omega_1}{r} \tag{11}$$

A simulation with the following parameters shows good results again, see Figure 5.

- grid points: $n_s = n_\theta = 4$, $n_r = 32$
- geometry: $r_1 = 1.0$, $r_2 = 10.0$
- angular velocities: $\omega_1 = 1.0$, $\omega_2 = 0.0$
- velocities: $u_{\theta,1} = 1.0$, $u_{\theta,2} = 0.0$
- kinematic viscosity: $\nu = 0.01$
- time steps: $n_t = 100000$, $\Delta t = 0.1$, $t_p = n_t \Delta t = 10000$
- Reynolds number: $Re = \frac{u_1 r_1}{\nu} = 100$

Testcase C_C Besides the cases C_A and C_B, where one of the two cylinders does not rotate, a third test case has been defined, where both cylinders are counter-rotating with a constant angular velocity. Here a shear flow is built up with a change in sign for the circumferential velocity, see Figure 6.

For this case a higher viscosity has been taken to get a steady solution. The parameters are:

- grid points: $n_s = n_\theta = 4$, $n_r = 32$
- geometry: $r_1 = 0.5$, $r_2 = 2.0$
- angular velocities: $\omega_1 = -8.0$, $\omega_2 = 2.0$
- velocities: $u_{\theta,1} = -4.0$, $u_{\theta,2} = 4.0$
- kinematic viscosity: $\nu = 0.1$
- time steps: $n_t = 10000$, $\Delta t = 0.003$, $t_p = n_t \Delta t = 30$
- $Re_1 = \frac{|u_1| r_1}{\nu} = 20$, $Re_2 = \frac{u_2 r_2}{\nu} = 80$, $Re_\delta = \frac{|u_2 - u_1|(r_2 - r_1)}{\nu} = 120$

6 Numerical investigations

6.1 Taylor-Couette-flow

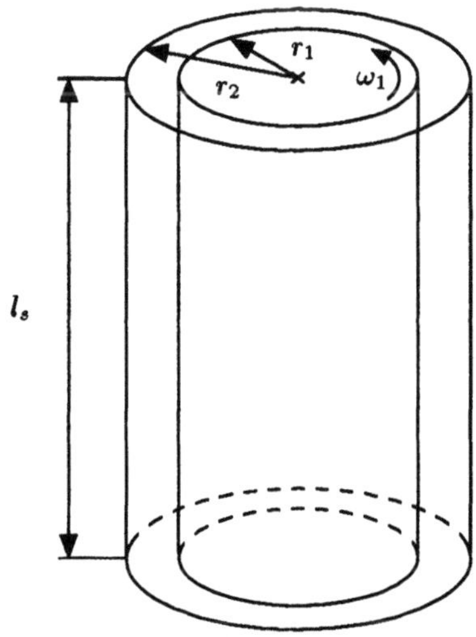

Fig. 7. Geometry of the Taylor-Couette flow experiment.

The geometry of the experimental configuration for the investigation of Taylor-Couette flow can be seen in Figure 7. It consists of two concentric cylinders of height l_s. The inner cylinder of radius r_1 rotates with a constant angular velocity ω_1, while the outer cylinder of radius r_2 does not move. At the top and the bottom the gap between the two cylinders are closed by solid, nonmoving walls. The experimental parameters are:

- geometry: $l_s = 975$ mm, $r_1 = 100$ mm, $r_2 = 144$ mm
- Fluid: Water
- viscosity: $\nu = 1.0076$ mm^2/s
- angular velocities: $\omega_1 = 0.03...0.09$ 1/s, $\omega_2 = 0$
- velocities: $u_{\theta,1} = 3...9$ mm/s, $u_{\theta,2} = 0$

A 2D simulation is currently computed on the Cray T3E with the same parameters ($l_{ref} = r_2 - r_1$, $u_{ref} = u_1$). Further simulation parameters are:

- grid points: $n_s = 1152, = n_\theta = 2, n_r = 64$
- time step width: $\Delta t = 0.002$
- Reynolds number: $Re = (r_2 - r_1)u_1/\nu = 131$
- number of processors: $n_{proc} = 32$
- CPU-time requirement (40 pressure iterations): $t_{CPU} = 0.81$ s per Δt

Preliminary results of the experiment and the simulation are shown in Figures 8 to 10. A secondary motion with pairs of Taylor vortices is induced. The size and form of the induced Taylor-vortices are qualitatively the same.

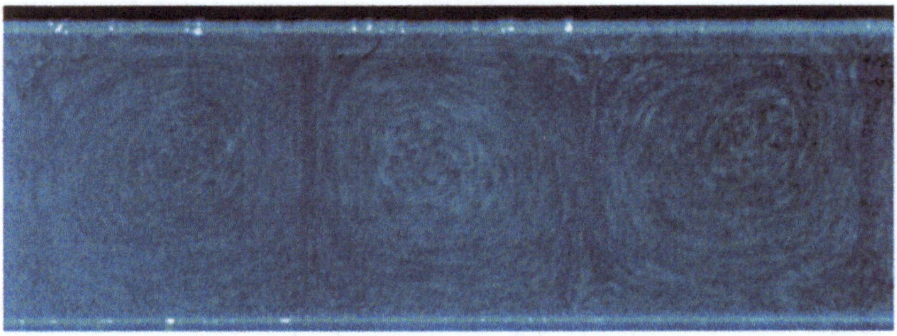

Fig. 8. Experimental investigation of Taylor-Couette flow.

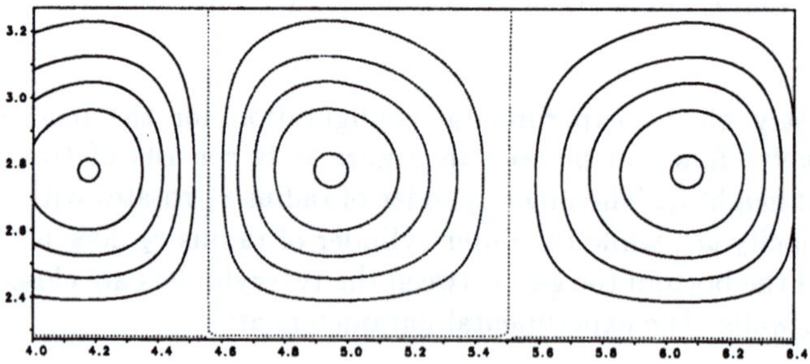

Fig. 9. Streamlines of the secondary motion of Taylor-Couette flow.

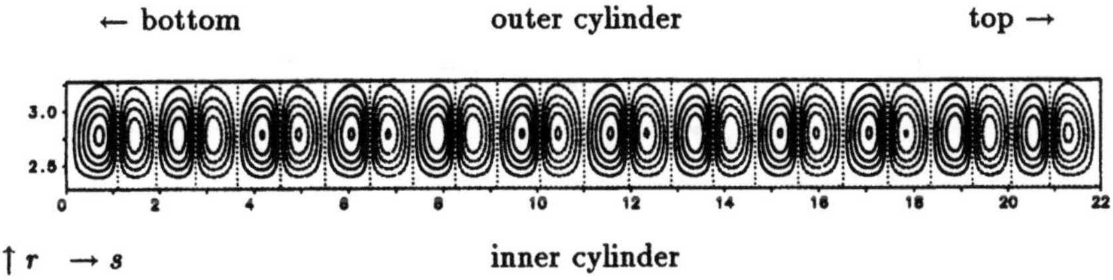

Fig. 10. Streamlines of the secondary motion in the whole computational domain of Taylor-Couette flow.

6.2 Turbulent boundary layer flow along cylinders

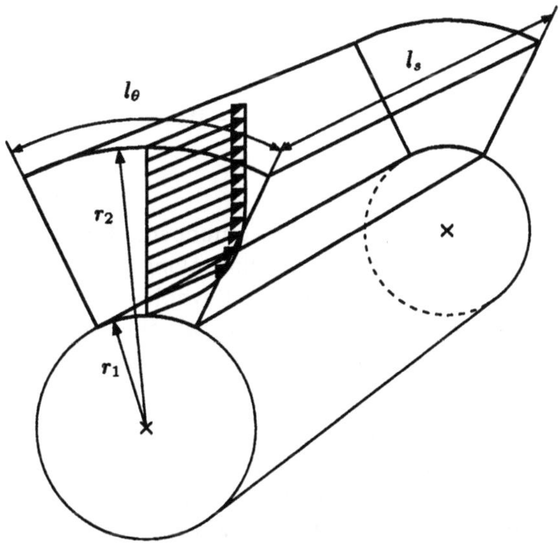

Fig. 11. Description of the computational domain of the boundary layer flow simulation along cylinders.

Turbulent boundary layer flow is very important in many branches of engineering. The effect of transverse curvature of the wall is estimated to be small for a large ratio of cylinder radius to boundary layer thickness. Then the boundary layer flow along cylinders (Figure 11) is similar to that along a plain wall. But if the cylinder radius is of the order of the boundary layer thickness, curvature effects cannot be neglected any more. Actual simulations with different values for the cylinder radius r_1 will show the effects of transverse curvature. Here, preliminary results for the case 2, $r_1 = 10$ are shown. While the mean axial velocity profile (Fig. 12) and the rms values of

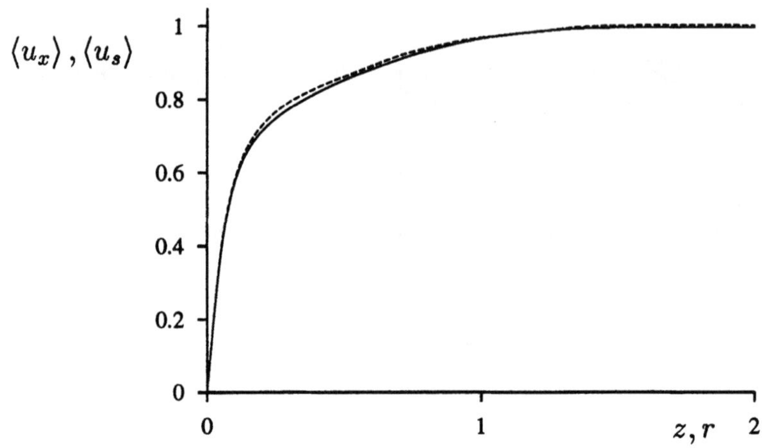

Fig. 12. Profile of the mean streamwise (axial) velocity component over the wall-normal (radial) direction at the position $x = 10\delta_0$, $(s = 10\delta_0)$: ——— Boundary layer simulation along a plane wall, - - - - Boundary layer simulation along a cylinder with $r_1 = 10\delta_0$.

the circumferential and radial velocity component (Fig. 13) match the corresponding values of the boundary layer flow along a plane wall very well, the rms values of the axial velocity and the Reynolds shear stress (Fig. 14) are quite different. Better statistics will show, whether this tendency is due to statistical effects or a real effect of tranverse curvature.

- grid points: $n_s = 256$, $n_\theta = 144$, $n_r = 96$
- kinematic viscosity: $\nu = 0.00033333$
- time steps: $\Delta t = 0.002$, $t_p = O(100)$
- Reynolds number: $Re_\delta = 3000$, $Re_\theta = 300$
- mesh size: $\Delta l_s = 0.05682$, $r_1 \Delta\theta = 0.02574$, $\Delta r = 0.00196$
- geometry case 1: $l_s = 14.54592$, $r_{1,1} = 100$
- geometry case 2: $l_s = 14.54592$, $r_{1,2} = 10$

The boundary layer flow was used to test the efficiency of the parallelisation of MGLET with MPI on a Cray T3E. Six short jobs with 100 time steps were run and a minimum number of $n_{proc} = 8$ processors were needed. Parallelization only in s-direction shows that the CPU time per job t_{CPU}/n_{proc} decreases with increasing number of blocks in axial direction $n_{b,s}$. A single job runs faster, but it costs more CPU time $t_{CPU}/\Delta t$. A good compromise is $n_{b,s} = 32$, because the job runs four times faster but it only needs 11% more CPU time.

Two dimensional parallelization shows that an odd number of blocks in circumferential direction $n_{b,\theta}$ should be used. With an even number of blocks, more CPU-time per time step is required, due to the specific implementation of the message passing routines in the programmes.

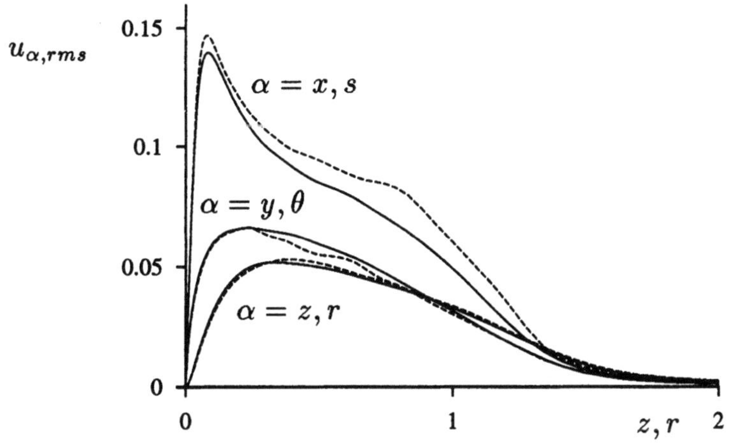

Fig. 13. Profile of the rms-values of the velocity components over the wall-normal (radial) direction at the position $x = 10\delta_0$, ($s = 10\delta_0$): ——— Boundary layer simulation along a plane wall, - - - - Boundary layer simulation along a cylinder with $r_1 = 10\delta_0$.

Table 1. Test of parallelisation efficiency (boundary layer flow):

$n_{b,s}$	$n_{b,r}$	$n_{b,\theta}$	n_{proc}	t_{CPU}/n_{proc}	$t_{CPU}/\Delta t$	$(t_{CPU} - t_{CPU,8})/t_{CPU,8}$
8	1	1	8	2810.7	224.86	0 %
16	1	1	16	1458.2	233.31	4 %
32	1	1	32	783.41	250.69	11 %
64	1	1	64	457.92	293.07	30 %
32	1	2	64	544.53	348.50	55 %
32	1	3	96	312.58	300.08	33 %

7 Conclusions

The implementation of cylindrical coordinates into the CFD-code MGLET has been done in order to investigate flow in cylindrical geometries. Good agreement with analytical solutions was obtained for simulations of laminar test cases, like pipe flow or cylindrical Couette flow. The installation of the CFD-code on a parallel high performance computer (Cray T3E) has successfully been done. Now, the investigation of more complex flows like Taylor-Couette flow, turbulent pipe- or boundary layer flow is in progress and the systematic evaluation of the results will be done soon.

References

1. BRUN, C. and FRIEDRICH, R. - A-priori tests of SGS stress models in

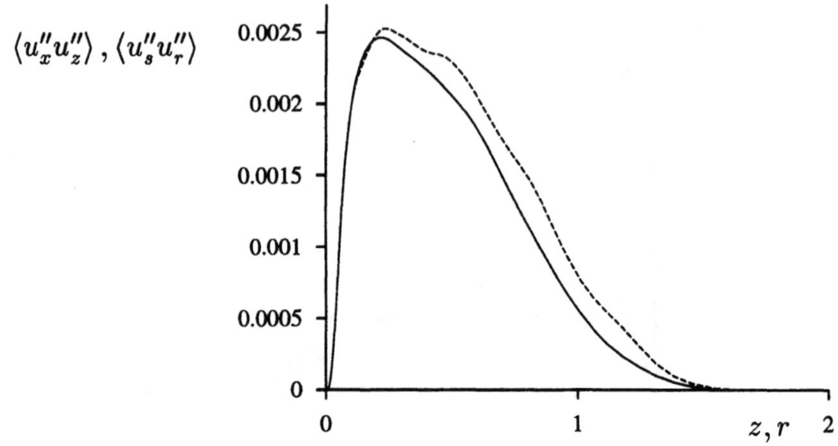

Fig. 14. Profile of the Reynolds shear stress over the wall-normal (radial) direction at the position $x = 10\delta_0$, $(s = 10\delta_0)$: ——— Boundary layer simulation along a plane wall, - - - - Boundary layer simulation along a cylinder with $r_1 = 10\delta_0$.

fully developed pipe flow and a new local formulation, Isaac Newton Institute/ERCOFTAC workshop on direct and large-eddy simulation, Cambridge, UK, to appear (1999).

2. Hirt, C., Nichols, B. and Romero, N. - SOLA – A numerical solution algorithm for transient fluid flows. In: Los Alamos Sci. Lab. Los Alamos, 1975.
3. HÜTTL, T.J., FRÖHLICH, M., SMIESZEK, M., MANHART, M. - Erweiterung des CFD-Codes MGLET zur Verwendung von Zylinder- und Toruskoordinaten, internal report, Lehrstuhl für Fluidmechanik, TU München, (1999).
4. LUND, T., WU, X. and SQUIRES, K. - Generation of turbulent inflow data for spatially-developing boundary layer simulations, J. Comp. Phys., 140, 233-258, 1998.
5. RICHTER, K., FRIEDRICH, R., SCHMITT, L. - Large-eddy simulation of turbulent wall boundary layers with pressure gradient, In: 6th Symposium on Turbulent Shear Flows, Toulouse, pp. 22/3/1-22/3/7, 1987.
6. SCHLICHTING, H. - Boundary–Layer Theory, McGraw-Hill, 6th edition, (1968).
7. SPALART, P. - Direct simulation of a turbulent boundary layer up to $R_\theta =$ 1410, J. Fluid. Mech. 187, 61-98, (1988).
8. TRITTON, D.J. - Physical Fluid Dynamics, Oxford Science Publications, 2nd edition, (1988).
9. UNGER, F. - Numerische Simulation turbulenter Rohrströmungen, phd-thesis, Technische Univerität München (1994).
10. WHITE, F.M. - Viscous Fluid Flow, McGraw-Hill, 1974.

DNS of Laminar-Turbulent Transition in Separation Bubbles

U. Maucher, U. Rist, M. Kloker, and S. Wagner

Institut für Aerodynamik und Gasdynamik, Universität Stuttgart,
Pfaffenwaldring 21, 70550 Stuttgart, Germany

Abstract. A Laminar Separation Bubble (LSB) is created under the influence of an adverse (positive) pressure gradient along a wall by laminar separation, laminar-turbulent transition and turbulent re-attachment of the flow to the wall. Direct Numerical Simulations (DNS) of the flow in an airfoil boundary layer which are based on solving the full Navier–Stokes equations exhibit good quantitative agreement of the mean-flow data with wind-tunnel experiments, but only examinations of the unsteady results are able to reveal the underlying physics. Thus, a hitherto unknown temporal amplification of three-dimensional small-amplitude disturbances is observed and explained by the entrainment of three-dimensional fluctuations by the roll-up of the detached boundary layer. Once the three-dimensional disturbances become saturated this mechanism leads to a rapid breakdown of the laminar flow into regions of small-scale turbulence which are organized in a quasi two-dimensional coherent manner.

1 Introduction

When an object is placed in a fluid flow, a boundary layer develops close to its surface. Drag is caused by the friction due to fluid viscosity in the boundary layer and lift is influenced by the thickness of the boundary layer. For small disturbance environments, such as flight in the atmosphere, a laminar boundary layer develops starting from the stagnation region. But further downstream it becomes turbulent by what is called laminar-turbulent transition. The present work is part of ongoing research on laminar-turbulent transition using the Direct Numerical Simulation (DNS) technique (solution of the complete Navier–Stokes equations) in the HLRS-project LAMTUR. Results for transition behind a periodically forced point source on a wing have already been presented in [8]. In the present paper the focus will be on transition in a so-called Laminar Separation Bubble (LSB) on an airfoil section of a glider plane.

Figure 1 shows a typical wing section together with its surface pressure distribution for a certain angle of attack and Reynolds number where LSBs on the upper and lower surface of the airfoil lead to characteristic pressure plateaus (pointed out by circles). In these regions the boundary layer separates and undergoes transition. Because of the enhanced transversal momentum exchange caused by turbulence the flow is able to re-attach as a turbulent boundary layer at the position where the pressure increases (note the direction of the pressure coefficient c_p in Fig. 1). A snapshot of the instantaneous flow in the vicinity of a LSB is shown in Fig. 2. The dark line emanating from the lower left corner is the separating shear layer which starts to oscillate and to roll up into discrete vortices in the right part of the figure. Apparently, transition is connected to the instability of the shear layer.

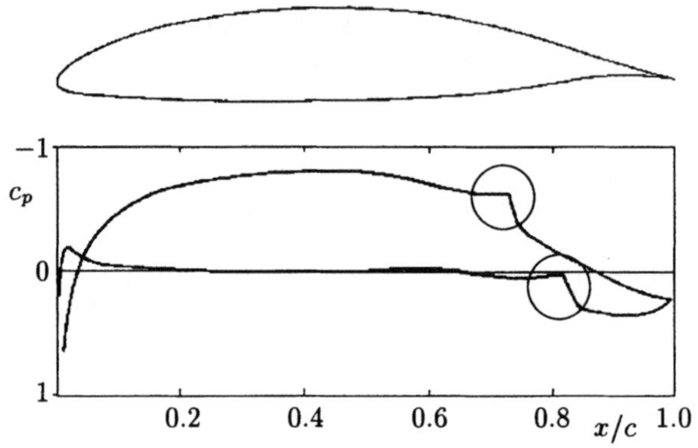

Fig. 1. Airfoil at Re = 700,000 and $\alpha = -0.5^o$ angle of attack showing the influence of laminar separation bubbles on the surface pressure (circles).

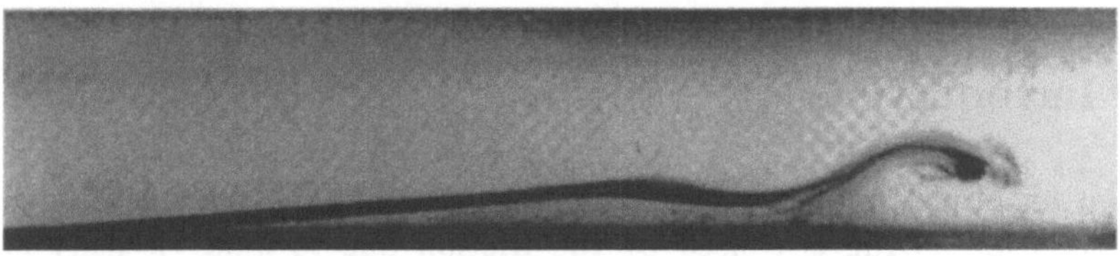

Fig. 2. Experimental flow visualization of the detached boundary layer in a laminar separation bubble.

In practice, LSBs are difficult to predict from wind tunnel data as well as by means of computations. We attribute this to a lack of knowledge and understanding of the underlying physics which are closely related to laminar-turbulent transition of the separated boundary layer. Thus, prior to the beginning of the present research in 1993 no three-dimensional (3D) simulations have been performed for flows with LSBs simply because of the lack of large and fast enough computers. It was known that the separated boundary layer develops into a kind of free shear layer which is highly unstable with respect to two-dimensional (2D) disturbance waves (cf. Fig. 2) and it was speculated that a sequence of secondary and higher-order instabilities with respect to 3D disturbances will lead to the inherently three-dimensional turbulence. Whether this is true and whether the reverse-flow region underneath the detached shear layer has a non-negligible influence which might lead to a temporal disturbance growth, a so-called absolute instability, was completely unknown. Wind-tunnel experiments intended to answer such questions are extremely difficult to perform because of the high sensitivity of the flow with respect to disturbances caused by the tunnel and the measuring devices. To date, there are almost no experimental data available which describe the instantaneous 3D flow field. Theoretical investigations need to solve the complete Navier–Stokes equations in order to obtain a reliable base flow for investigations based on linear stability theory (LST).

The present project which relies on DNS and careful stability calculations using LST has already made several contributions to a better understanding of instability and transition in LSBs. In [6] and [7] it was shown that LST can be reliably applied for two- and three-dimensional disturbances if the base flow profiles computed in DNS are analyzed. The supposed secondary instability plays a less important role than originally assumed because of two reasons: The secondary disturbance amplification rates are barely higher than the very large primary amplification rates (according to LST) and an unexpected reduction of the 3D growth rate occurs after saturation of the 2D disturbance. Yet, the nonlinear interaction of weakly-oblique traveling waves was found to be the fastest route to turbulence in these earlier investigations. At the same time this mechanism produces longitudinal vortices in the reattachment zone which have also been detected in some earlier experiments.

When the Reynolds number based on the displacement thickness at laminar separation was increased from approximately $Re_{\delta_{1,s}} = 1250$ to $Re_{\delta_{1,s}} = 2500$, a new mechanism of secondary instability was observed: In the presence of a large-amplitude 2D-wave, 3D-modes are temporally amplified. This mechanism has been firstly reported by Maucher *et al.* [4]. The investigations presented here focus on this mechanism and its impact on the transition in a LSB.

2 Numerical Method

The numerical method was developed in the research-group "Transition and Turbulence" of the ***Institut für Aerodynamik und Gasdynamik, IAG.*** It is based on the complete Navier–Stokes equations in vorticity-velocity formulation for an incompressible fluid. All variables are non-dimensionalized by a reference length $\hat{L}$, by the velocity $\hat{U}_\infty$, and the variables in wall-normal direction y and v are stretched with the square-root of the Reynolds number $Re = \hat{U}_\infty \hat{L}/\hat{\nu}$, where $\hat{}$ denotes dimensional variables and $\hat{\nu}$ is the kinematic viscosity

$$x = \hat{x}/\hat{L};\ y = \sqrt{Re}\,\hat{y}/\hat{L};\ z = \hat{z}/\hat{L};\ u = \hat{u}/\hat{U}_\infty;\ v = \sqrt{Re}\,\hat{v}/\hat{U}_\infty;\ w = \hat{w}/\hat{U}_\infty. \tag{1}$$

This leads to the definition of the non-dimensionalized vorticity components

$$\omega_x = \frac{1}{Re}\frac{\partial v}{\partial z} - \frac{\partial w}{\partial y}, \quad \omega_y = -\frac{\partial u}{\partial z} + \frac{\partial w}{\partial x}, \quad \omega_z = \frac{\partial u}{\partial y} - \frac{1}{Re}\frac{\partial v}{\partial x}. \tag{2}$$

A flat-plate boundary-layer is subjected to an adverse pressure gradient by prescribing the edge velocity distribution u_e at the upper boundary of a rectangular integration domain (Fig. 3). Basically, u_e may be an arbitrary function of the streamwise coordinate x but here it is taken from experimental measurements. A refined, very accurate boundary-layer interaction model (Maucher *et al.* [5]) captures the displacement effect (index ν) of the boundary layer (in particular of the LSB) and models its influence on the initially prescribed potential velocity distribution u_p yielding an instantaneous edge-velocity distribution:

$$u_e(x,t) = u_p(x) + u_\nu(x,t). \tag{3}$$

From the experiments, two boundary-layer edge-velocity distributions are available (Fig. 3b). The first one (squares) refers to a flow with separation bubble and for the second, turbulent one (triangles), separation has been suppressed by fixing a turbulator upstream of the separation bubble. For the computations, the turbulent distribution has been prescribed as initial condition u_p at the free-stream boundary. Maucher *et al.* [5] show that the displacement effects are captured correctly by their new viscous-inviscid interaction model, and finally the edge velocity distribution u_e comes very close to the experimental measurements with separation bubble.

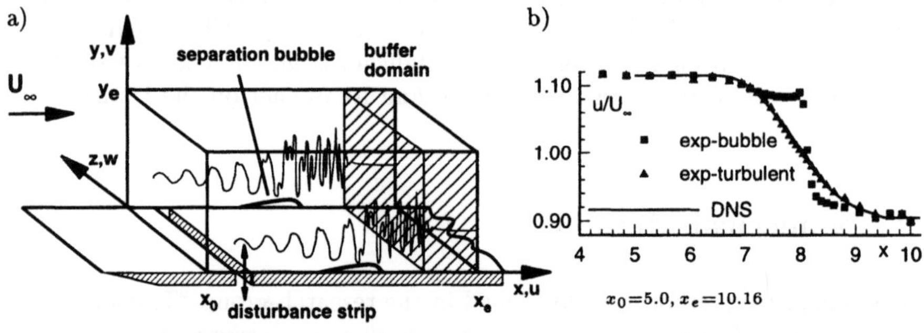

Fig. 3. Integration Domain (a) and comparison of edge-velocity distribution between experiment and DNS (b).

Steady flow is assumed at the inflow boundary and in a disturbance strip at the wall, artificial 2-D and 3-D disturbances can be excited by periodic or pulse-wise, wall-normal suction and blowing. Upstream of the outflow boundary a buffer domain is applied, which damps the unsteady component of the flow smoothly to zero [3].

2.1 Numerical and Computational Aspects

In order to reduce the computational effort, the flow field is assumed to be periodic in spanwise direction z and symmetric initial and boundary conditions are exploited by calculating a symmetric flow field with respect to z=0. A Fourier spectral ansatz with a number of $(2K+1)$ modes is used to transform the variables and equations into Fourier space and vice versa:

$$f(x,y,z,t) = \sum_{k=-K}^{K} F_k(x,y,t)e^{ik\gamma_0 z}\,, \tag{4}$$

where f, and F_k designate the variables in physical, and in spectral space, respectively. The basic spanwise wave number γ_0 for the fundamental Fourier mode (k=1) is related to the spanwise width λ_z of the integration domain through $\gamma_0 = 2\pi/\lambda_z$. The remaining spatial dimensions are discretized by fourth-order accurate finite differences which are employed in alternating upwind/downwind manner for the streamwise convective terms in order to increase numerical stability for long-time integration where time integration is based on a fourth-order Runge-Kutta scheme (cf., Kloker [2]). Thus, the spanwise Fourier ansatz principally reduces the 3-D problem in physical space to a set of $(K+1)$ 2-D problems in Fourier space enabling a largely parallel computation in Fourier space. However, the modes are coupled by the nonlinear convective terms of the vorticity transport equations and are transformed to physical space for the calculation of the nonlinear vorticity terms ("pseudo-spectral method" with de-aliasing procedure), which in turn are parallelized in streamwise direction.

The code has been optimized and run on the NEC SX-4 supercomputer cluster of the hww GmbH, Stuttgart (NEC SX-4/32 & NEC SX-4A/4, 8 GB RAM each). For the fine-resolution simulation described further down in Sect. 3.3 a uniform equidistant grid containing 2,754 × 193 × 89 (K = 44) points in (x,y,z)-directions (NX, NY, $2K+1$) has been employed needing 4 GB main memory, and 1 GB XMU memory. The computation of 3,600 time steps on the NEC SX-4/32 took 321,950 s, i.e., 3.7 μs per grid point, Fourier mode, and time step in the average. According to the built-in job statistics of the SX-4 the vector operation ratio was 98.7% with an average vector length of 183 points. Using 15 concurrent CPUs, the parallelized and vectorized code ran at 14,287 concurrent MFLOPS, and based on the total number of floating-point operations and the total execution time the job statistics indicated 1,063 MFLOPS per CPU. Compared to 2 GFLOPS theoretical peak performance per CPU this last number appears rather satisfactory. The performance, however, depends heavily on the load of the computer and the data just described were obtained on a day with a partly empty computer.

The y-discretization is the crucial point in resolution requirements for the present simulations. This is due to large-amplitude fluctuations impinging on the wall in the re-attachment zone. Therefore, in the high-resolution case, the grid spacing was $\Delta x = 0.001875$, $\Delta y/\sqrt{Re} = 0.000982$ and $\Delta z = 0.004412$ ($\gamma_0 = 16$, $K = 44$).

3 Numerical Results

The present simulations are closely related to experiments in the so-called Laminar Wind Tunnel of the institute, where a LSB on a wing section with a chord-length of $\hat{c} = 0.615\,m$ was investigated (Würz *et al.* [9]). For this case the chord Reynolds number is $Re_c = 1.2 * 10^6$ and the free-stream velocity $\hat{U}_\infty = 29.3\frac{m}{s}$. For the computations the reference length $\hat{L}$ is chosen to be $6.15\,cm$, i.e., $1/10\,\hat{c}$. Hence, the Reynolds number for the simulation is 120,000 and $c = \hat{c}/\hat{L} = 10$. The computational grid, memory requirements and computation times have been already discussed in the previous section. At the disturbance strip a 2D disturbance, a so-called Tollmien–Schlichting (TS) wave, with non-dimensional frequency $\beta = 10$ and wall-normal amplitude $v/\sqrt{Re} = 10^{-6}$ is periodically forced. One wave length of the TS wave is discretized with approximately 160 grid points in x-direction and the time-wise resolution is 600 time steps per wave cycle.

3.1 Mean Flow

Averaging the resulting flow field over one or more disturbance cycles yields the time-averaged or mean flow. Streamlines and iso-vorticity contours of the averaged flow field are shown in Fig. 4. Note that the y-axis is enlarged by a factor of 10 with respect to x so that the actual separation bubble is shallower than it appears here. According to the streamlines the flow separates at $x \approx 7.1$ and re-attaches at $x \approx 8.15$. From separation to $x \approx 7.8$ there is a large dead-air region with only minor recirculating velocity which ends in a vortex centered around $x \approx 8.05$. The vorticity nicely portrays the separation of the boundary layer from the wall forming a quasi-free shear layer which somehow disintegrates at $x \approx 8.0$. Apart from this mean-flow effect the Tollmien–Schlichting wave and any other fluctuations are absent from the mean-flow data, of course. After re-attachment the wall vorticity and hence skin friction becomes very large.

Direct quantitative comparisons of these results with the experiments are presented in Figs. 5 and 6. The mean-flow profiles show a very good agreement with the experimental measurements, except for the last two stations close to the wall. Initially, the flow is attached from station A to B. The shear layer then lifts from the wall (C – E) and the profiles exhibit an inflection point away from the wall. If, as in the present experiments, a single hot wire is applied, it cannot be detected whether reverse flow is present. Moreover, introducing a probe into the bubble disturbs the flow, and the accuracy of the measurements deteriorates when large-amplitude fluctuations around a small mean amplitude occur. All this is true for the experimental data at stations E, F, and G.

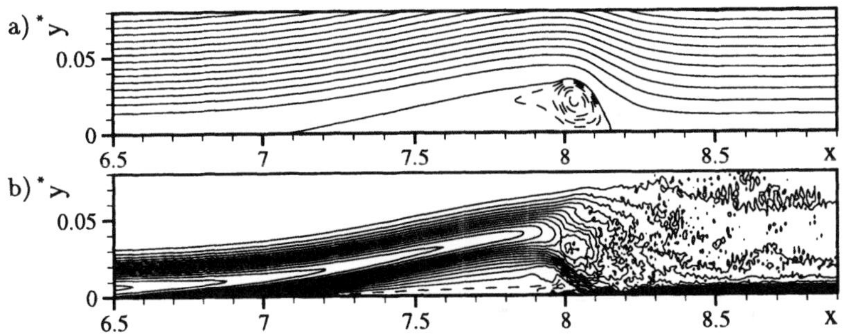

Fig. 4. Streamlines (a) and vorticity (b) of the mean flow; $y^* = y/\sqrt{Re}$.

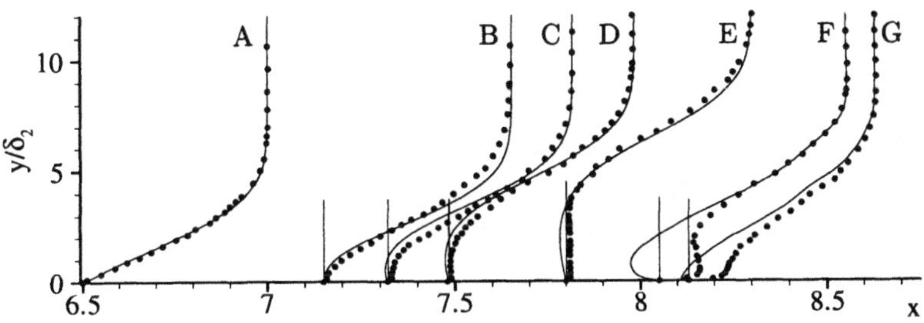

Fig. 5. Comparison of mean-flow profiles at different streamwise stations with the experiment. Symbols = experiment and lines = DNS. A – G = stations $x = 6.5$, 7.15, 7.32, 7.48, 7.8, 8.05, and 8.15, respectively.

Comparing the boundary-layer edge velocity u_e in Fig. 6 with the experiment shows the benefits of using the viscous-inviscid interaction model described in Maucher *et al.* [5]. The experimental data (circles) have been prescribed as potential edge velocity u_p in (3) and the simulation results yield an edge velocity u_e which is in very close agreement with the experiment (crosses).

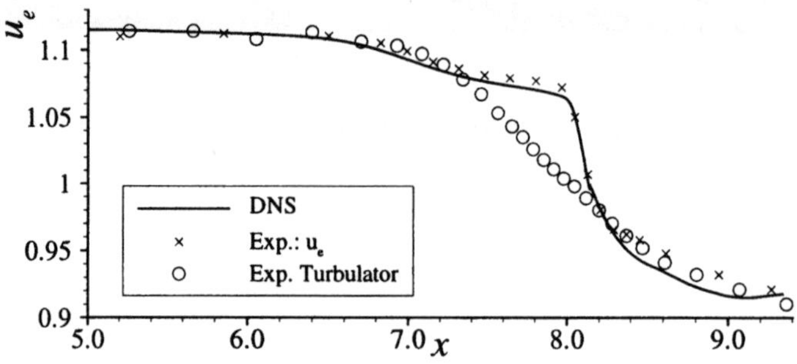

Fig. 6. Comparison of the boundary-layer edge velocity with the experimental results.

Actually, a rapid disintegration of the shear layer occurs starting between station E and F in Fig. 5, as already mentioned above. This is demonstrated by two visualizations of the instantaneous vorticity in the re-attachment region in Fig. 7. The momentary flow is strongly different from the averaged flow in Fig. 4 (note the absence of the strong recirculation vortex in the instantaneous data). Thus, the time-averaged data yield a completely wrong picture of the re-attachment zone. They have been used here only for comparison of the numerical results with the experiment because in most experimental realizations of the problem so far, only mean flow and statistical *rms*-data have been obtained. Since these data do not describe the actual flow physics in an adequate manner it is impossible to understand the underlying mechanisms and it appears plausible that an accurate modeling of the flow based on 2D time-averaged data or equations is at least critical.

3.2 Secondary Temporal Amplification

At sufficiently high Reynolds numbers 3D-disturbances are temporally amplified in the re-attachment region of the LSB once they are present due to interactions with a forced 2D wave. Similar to the mechanisms of secondary instability theory according to Herbert [1], Maucher *et al.* [4] found 3D amplification with subharmonic and fundamental frequency with respect to the forced 2D-wave for different spanwise wave numbers. In recent investigations the secondary temporal growth of 3D-modes in a large range of spanwise wave

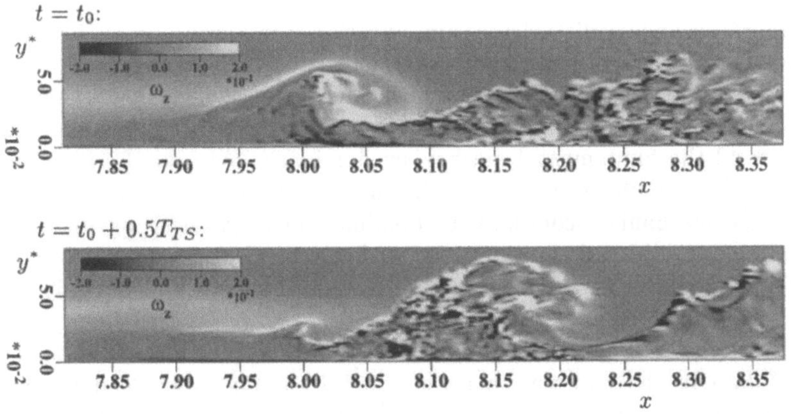

Fig. 7. Spanwise vorticity ω_z along the centerline ($z = 0$) in the vicinity of reattachment at two time-instances separated by one half forcing period T_{TS}; $y^* = y/\sqrt{Re}$.

numbers was investigated. Short pulse-like 3D excitations with very low amplitude and fixed spanwise wave number γ have been used to probe the 3D instability in DNS with a periodically forced 2D TS wave with an amplitude $A_{wall} = 10^{-5}$, i.e., 10 times larger than in the reference case. However, this did not enforce secondary instability by itself. Rather in the contrary: large 2D disturbances produce smaller and hence more stable LSBs.

For low γ there is no temporal secondary instability but up to very high values of γ 3D modes grow exponentially with the temporal growth rate $\beta_i = \frac{\partial}{\partial t}[\ln A(t_0 + t) - \ln A(t_0)]$. Figure 8 shows the temporal amplification rates β_i obtained from these DNS. The open and the filled symbols denote amplification of disturbances with subharmonic respectively fundamental frequency relative to the forcing.

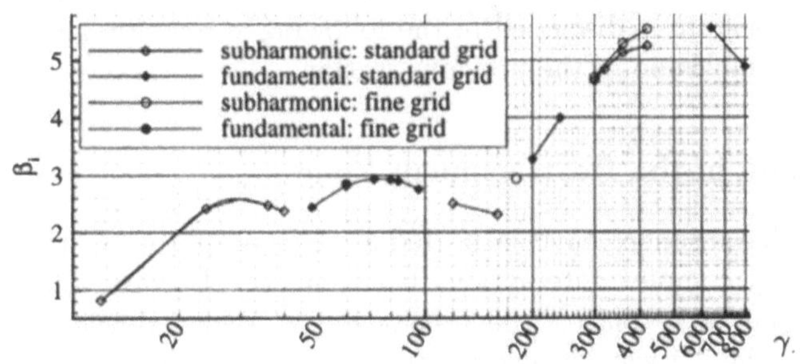

Fig. 8. Secondary temporal amplification rate β_i versus spanwise wave number $\gamma = 2\pi/\lambda_z$ for different discretizations.

Temporal secondary disturbance amplification starts around $\gamma \approx 10$ which corresponds to a ratio of wave numbers $\gamma/\alpha_{TS} \approx 0.5$, where $\alpha_{TS} = 2\pi/\lambda_{TS}$ ($\lambda_{TS} \approx 0.3$) is the streamwise wave number of the TS wave. For low spanwise wave number γ the subharmonic mechanism is observed. If γ is increased above $\gamma \approx 42$ the fundamental mechanism dominates. At $\gamma \approx 160$ a strong increase of the amplification rate sets in and either the fundamental or the subharmonic mechanism continues to dominate in an alternating manner. For validation, additional computations with a refined grid were performed (240 grid points per TS wave length, circles). Grid influence appears only for large $\gamma > 300$.

As already mentioned, the examination and interpretation of the time-averaged data in the rear-ward part of the LSB is highly misleading and unable to describe the flow physics since a phase with strong reverse flow alternates with a phase of positive u-velocity during each TS cycle. It turned out, that the phase with reverse flow is decisive for the secondary temporal amplification. The instantaneous vorticity field shows the roll-up of the 2D free shear layer in the re-attachment region (emphasized by contours of high vorticity in Fig. 9a). Apparently the shear layer is unstable with respect to 3D perturbations in the regions indicated by boxes, because 3D u' maxima occur at these positions, as is shown in Fig. 9b). The 3D disturbance maxima remain almost at a constant place during approximately one half of each TS period T_{TS}, i.e., as long as reverse 2D flow is present in the re-attachment region and their amplitude grows with respect to time during this phase. In the second phase the 2D velocity becomes positive and the uppermost part of the 3D perturbations is convected downstream together with the shear layer they are riding on (right column in Fig. 9). Thus, the onset of three-dimensionality happens inside the separation bubble where 3D perturbed fluid remains present from the previous TS period and where instantaneous high-shear layers are strongly unstable with respect to temporally growing 3D modes. In addition, it is highly conceivable that this re-entrainment of disturbances from the previous cycle contributes to the continuous temporal amplification of 3D modes. Finally, the amplitudes of secondary 3D modes saturate and an equilibrium state between 2D and 3D modes ends the transient phase. This state is discussed in the following subsection again for the fine-resolution simulation of the previous section.

3.3 Transition in the Reattachment Zone

When the 3D disturbances have saturated at large amplitudes the temporal growth ends and complex interactions between the 2D wave and the secondary 3D modes take place. The re-attachment region develops towards turbulence and the 2D wave is also affected compared to the transient case. Finally, the 2D wave attains its amplitude maximum in the re-attachment zone around $x \approx 8.0$ (Fig. 10 - thick solid line and triangles). From then on it decays towards the level of the 3D disturbances. Note the very good

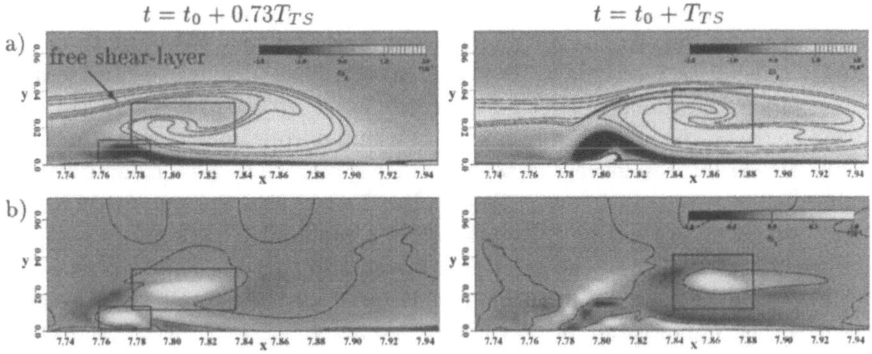

Fig. 9. Comparison of instantaneous 2D vorticity (a) with instantaneous 3D u'-distribution for $\gamma = 160$ (b) in the re-attachment zone; T_{TS} = fundamental 2D forcing period.

agreement of the amplification rate of the fundamental disturbance wave with linear stability theory (LST). However, this is not the case for the 2D higher harmonic (2, 0) because of nonlinear effects. The occurrence of many modes in the frequency/spanwise-wave-number spectrum means a disintegration of the flow structures (cf. Fig. 7). Apart from the dominance of the 2D disturbance such a picture is typical for laminar-turbulent transition.

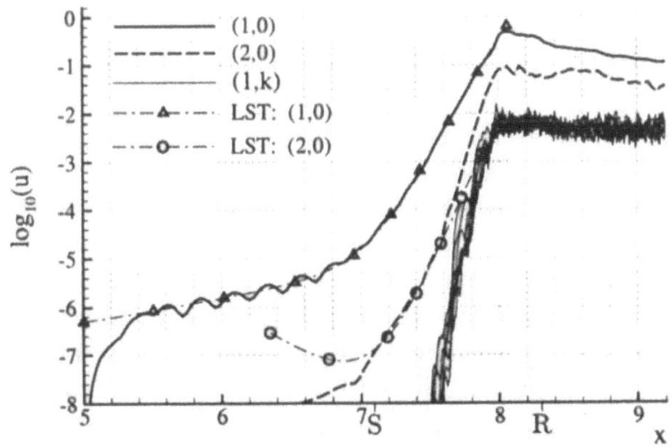

Fig. 10. Amplitude of the forced 2D wave (1,0), its first higher harmonic (2,0), and self-sustaining 3D modes (1,k), $\gamma = k\gamma_0$. Symbols indicate results of LST.

The breakdown of the free shear layer is visualized by iso-surfaces of the spanwise vorticity at four time instances during one TS cycle in Fig. 11. The x-domain shown begins well inside the separation zone and the iso-surface

$\omega_z = 0.1$ encloses the free shear layer as well as the vorticity maxima of the ensuing structures. At $t = T_0$ (a) the shear layer is almost 2D until $x = 8.05$. Afterwards, it is pierced by 3D perturbations from inside the reverse-flow zone and longitudinal vorticity structures build up at $x \approx 8.0$ ($t = T_0 + 0.25T_{TS}$, b) in the shear layer. Spanwise oriented 'rolls' which consist of fine-scale 3D structures leave the separation bubble (c+d) and comparatively calm phases appear in-between these rolls. Both phases are connected to the observations in subsection 3.2, each roll is a remainder of the roll-up of the 2D free shear layer and the calm phase occurs during the build-up phase of a new roll with new 3D disturbances. Because of their quasi 2D character despite their composition of very fine-scaled 3D flow structures the signature of these rolls in the frequency/spanwise-wave-number spectrum in Fig. 10 is a large amplitude 2D mode for $x > 8$. Finally it takes a remarkably long way downstream, until fully developed turbulence occurs and the dominance of the 2D wave (respective of the turbulent rolls) vanishes and an equilibrium turbulent boundary layer is reached.

The breakdown of the shear layer is further illustrated in cross-stream (y, z) cuts of the vorticity field ω_z in Fig. 12. Here the cutting plane is moving downstream with the evolving shear layer according to Fig. 11. The 2D free shear layer is centered around $y/\sqrt{Re} \approx 0.05$ and initially well above the 3D structures (a). But as the 3D disturbances get pushed up (b) the shear layer disintegrates and widens in wall-normal direction (c – d). The resulting structures are highly complicated and perhaps not yet grid independent for the present resolution. This must be investigated in further simulations.

4 Conclusions

The Navier–Stokes equations have been solved using a fourth-order accurate combined finite-difference spectral method to investigate laminar-turbulent transition in a laminar boundary layer that separates from the wall and re-attaches under the influence of turbulence, thus forming a so-called Laminar Separation Bubble (LSB). Without a well-adapted numerical scheme that exhibits good supercomputer performance through vectorization and parallelization, it would have been impractical to perform such simulations because of the large memory and computer time requirements caused by the small spatial and temporal step sizes needed for resolving the fine-scale unsteady fluid motions in the re-attachment zone of the bubble. Specifically, the code was run at 14 GFLOPS with 50% of the theoretical peak performance per node of the NEC SX/4-32 for regular computations.

The temporal mean flow of the present results agrees extremely well with experimental measurements performed in the laminar-flow wind tunnel of the institute but it is unable to describe the underlying flow physics. For the first time, a temporally growing secondary instability has been observed that amplifies 3D disturbances under the action of a 2D large-amplitude TS wave.

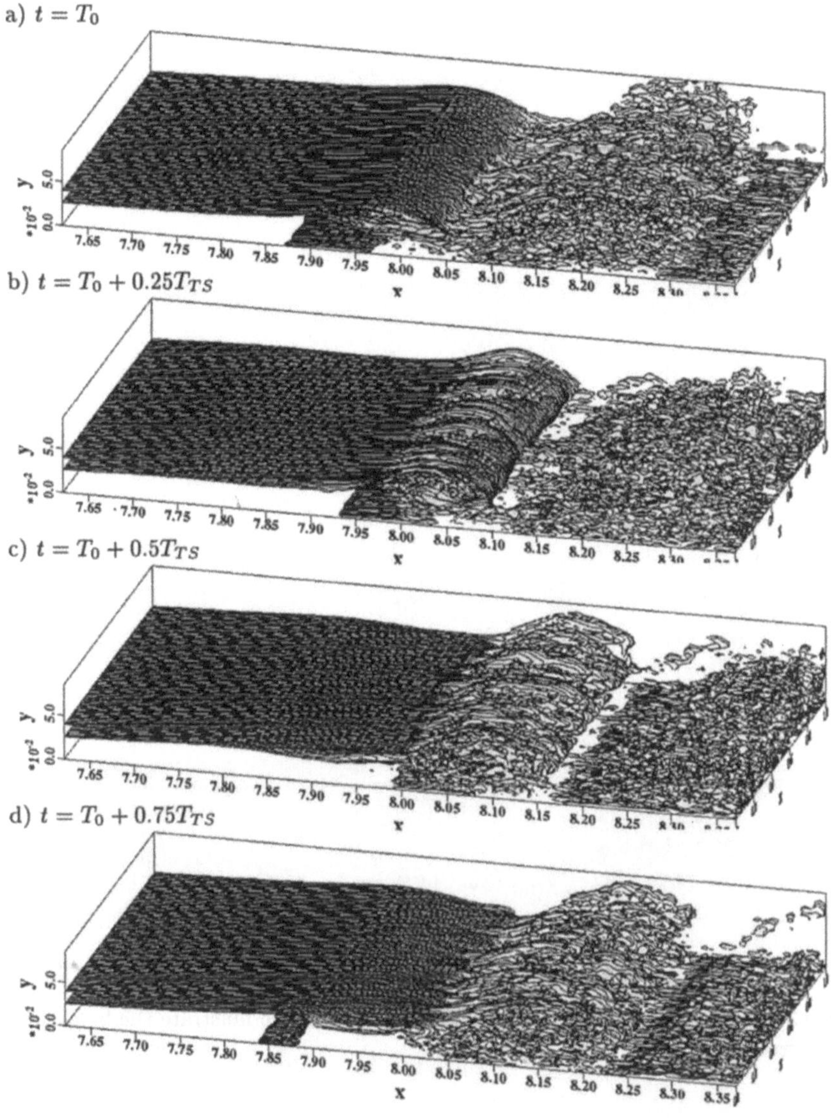

Fig. 11. Breakdown of the free shear layer into rolls of fine-scale turbulence. Isosurfaces of the spanwise vorticity at four instances of the TS-cycle T_{TS}.

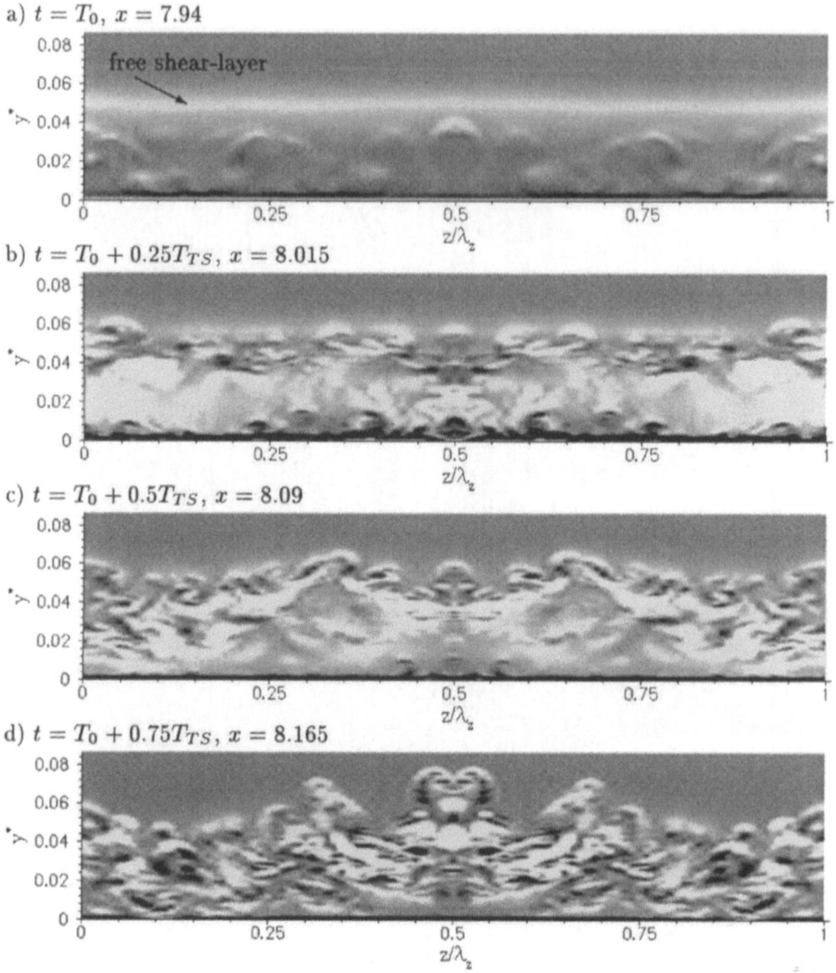

Fig. 12. Cross-stream cuts of the spanwise vorticity ω_z to illustrate the breakdown of the free shear layer.

The physical mechanism that leads to this instability has been identified as well: the entrainment of 3D disturbances by the roll-up of the 2D shear layer. Finally, high-resolution simulations have revealed the breakdown of the separated shear layer in the re-attachment zone. Again, 3D disturbances are entrained by the shear layer. They are carried away from the wall and pierce the undisturbed layer from underneath. This leads to an extremely rapid breakdown of the laminar flow into small-scale turbulence. However, the 2D disturbance remains visible for a certain distance downstream of the LSB, either in the frequency spectrum or as coherent structures in the ensuing boundary layer flow.

Acknowledgments

The financial support of this research by the Deutsche Forschungsgemeinschaft DFG under grant Ri 680/1 and by the Universität Stuttgart is gratefully acknowledged, as well as the provision of computer time by HLRS within the project LAMTUR.

References

1. T. Herbert. Secondary instability of boundary layers. In *Ann. Rev. of Fluid Mech.* **20** (1988) 487–526.
2. M. Kloker. A robust high-resolution split-type compact FD scheme for spatial direct numerical simulation of boundary-layer transition. *Appl. Scientif. Res.* **59** (1998) 353–377.
3. M. Kloker, U. Konzelmann, and H. Fasel. Outflow boundary conditions for spatial Navier-Stokes simulations of transitional boundary layers. *AIAA J.* **31** (1993) 620–628.
4. U. Maucher, U. Rist, and S. Wagner. Secondary instabilities in a laminar separation bubble. In H. Körner and R. Hilbig (Eds.) *New Results in Numerical and Experimental Fluid Mechanics*, Vieweg *NNFM* **60** (1997) 229–236.
5. U. Maucher, U. Rist, and S. Wagner. A refined method for DNS of transition in interacting boundary layers. AIAA 98-2435 (1998).
6. U. Rist. Nonlinear effects of 2D and 3D disturbances on laminar separation bubbles. In S.P. Lin (Ed.) *Proc. IUTAM-Symposium on Nonlinear Instability of Nonparallel flows*, Springer, New York (1994) 324–333.
7. U. Rist, U. Maucher, and S. Wagner. Direct numerical simulation of some fundamental problems related to transition in laminar separation bubbles. In Désidéri *et. al.* (Eds.), *Computational Methods in Applied Sciences '96*, John Wiley & Sons Ltd (1996) 319–325.
8. C. Stemmer, M. Kloker, U. Rist, and S. Wagner. DNS of point-source induced transition in an airfoil boundary-layer flow. In E. Krause and W. Jäger (Eds.), *High Performance Computing in Science and Engineering '98*, Springer, Berlin, Heidelberg (1998) 213–222.
9. W. Würz and S. Wagner. Experimental investigations of transition development in attached boundary layers and laminar separation bubbles. In H. Körner and

R. Hilbig (Eds.) *New Results in Numerical and Experimental Fluid Mechanics*, Vieweg *NNFM* **60** (1997) 413–420.

Numerical Simulation of Supersonic Hydrogen-Air Combustion

Ernst von Lavante, Michael Kallenberg, and Dirk Zeitz

University of Essen, Lehrstuhl für Strömungsmaschinen, Schützenbahn 70, D-45127 Essen, Germany

Abstract. In this investigation, unsteady, three-dimensional, supersonic flow with nonequilibrium chemistry in a square channel with transverse hydrogen injection was numericaly analyzed. To this end, the concepts of large eddy simulation (LES) were applied to a model supersonic combustion chamber using the three-dimensional solver of the compressible Navier-Stokes equations with chemical reactions called "ACHIEVE". The time accurate computation was accelerated by an implicit method.

1 Introduction

After a seemingly slow pace of research and development of high speed flight vehicles, supersonic or hypersonic, the speed of scientific development in this area of interest is becoming faster, again. The recent increase of research activities can be attributed to the reemerging interest in the concept of the Scramjet. Several nations are planning uninhabited hypersonic research or military vehicles, for which the Scramjet is the logical choice as a means of propulsion. A typical representative of these research vehicles is the highly promissing Hyper-X. It is a NASA propulsion demonstration program intended to validate design tools which could be used in future development of hypersonic propulsion technology, [7], [8]. The concequence is stronger involvement in the scientific area of supersonic combustion. Moreover, the concepts and solution methods, developed for supersonic combustion, can be directly carried over to the area of subsonic compressible flows with combustion. Here, low Mach-number subsonic, chemically reacting flows are found in most types of gas turbines, and are therefore of particular interest to the corresponding scientific community.

Simulation methods of these types of flows have reached a certain degree of maturity, offering a choice of standard spacial and time-wise discretization procedures. For details, see for example Cox et. al. [2] or Godfroy and Tissier [3]. However, several problems remain. One of the main difficulties is the treatment of turbulence, since none of the simpler models is adequate in these complex flow cases. The interaction between the turbulent effects and the chemistry, in particular the chemical rates of reaction, is difficult to predict numerically due to the uncertainties in describing this physical phenomenon theoretically. The ability of various turbulence models to predict the mixing phenomena in a Scramjet combustor was investigated by Madabhushi et. al. [1].

The present authors have been attempting to numerically simulate reacting flows in supersonic combustion chambers for some time now, [18], [19]. For laminar flows, their predictions were satisfactory; the resulting turbulent flows, however, were lacking the above realistic features. The increasing complexity of the turbulence models, from simple zero equation algebraic to one equation (k) to two equation (k - ϵ) did not bring any significant improvement. The multivariate β-pdf approach of accounting for the turbulence-chemistry interaction, described by Narayan [4], seems to result in some gain of accuracy in some cases, at a much increased cost. A modified version of the linear eddy model for subgrid combustion was developed by Chakravarthy and Menon [5]. They used it for studying the properties of turbulent premixed flames in the core region of Couette flow. Arunajatesan and Menon [6] used a simpler LES model for simulations of flow in a compact incinerator.

Therefore, in view of the difficulty to formulate an appropriate model for the turbulence-chemistry interaction, as well as the turbulence alone, the present authors decided to investigate the feasibility of large eddy simulation, applied to a case of chemically reacting supersonic flow of a air-hydrogen system.

2 Algorithm

In this work, the flow was assumed to be compressible, viscous, and a mixture of thermally perfect species. Due to the relatively low temperature and high pressure in the present configuration, the gas mixture can be treated as in vibrational equilibrium. The governing equations were in this case the compressible Navier-Stokes equations for n_s species:

$$\frac{\partial \hat{Q}}{\partial t} + \frac{\partial \hat{F}}{\partial \xi} + \frac{\partial \hat{G}}{\partial \eta} + \frac{\partial \hat{H}}{\partial \zeta} = \frac{S}{J} \tag{1}$$

where $\hat{F}$, $\hat{G}$ and $\hat{H}$ are the flux vectors in the corresponding ξ , η and ζ directions, Q is the vector of the dependent variables and J is the jacobian of the transformation of coordinates. S is the vector of the chemical source terms. The details of the governing equations are given in [18].
A simple model according to the Fick's law for the binary diffusion coefficient was used, along with the Sutherland equation for the viscous coefficient. Chemical reactions were realized with an 8-reaction model of Evans and Schexnayder [9], for the H_2-air combustion.

Three different upwind methods were extended to chemical reactions system and are optionally available in the present computer code "ACHIEVE". These methods are based on the work of Roe (flux-difference splitting) [15], van Leer (flux-vector-splitting) [12] and Liou (AUSM Advection-Upstream-Splitting-Method) [13]. However, it was found out that only the Roe Scheme worked reliably and provided physically relevant results.

2.1 Roe Scheme

The present research mostly employed a numerical scheme based on Roe's Flux Difference Splitting in finite volume form. This scheme was demonstrated to be accurate, with relatively low dissipation and dispersion. Depending on the type of limiter and MUSCL interpolation, this scheme was highly effective in providing accurate viscous results at a wide range of Mach numbers. In the present version, the reconstruction of the cell-centered variables to the cell-interface locations was done using a monotone interpolation as introduced by Grossmann and Cinella in [10]. The interpolation slope was limited by an appropriate limiter, according to the previously published MUSCL type procedure (see, for example, [18]). The details of this scheme, with the corresponding modifications, are given by Hilgenstock et. al. [19]. This includes also the positivity preserving modifications in the sence of Larrouturou [11].

2.2 Temporal Integration

Initially, the governing equations were integrated using a semi-implicit method, with different multi-stage Runge-Kutta type schemes used for the explicit operator. Only the chemical source terms were treated implicitely,

$$\left\{ I - \Delta t\, \Theta \,\frac{\partial S^n}{\partial Q^n} \right\} \frac{\partial Q^n}{\partial t\, J} = \hat{S}^n - \frac{\partial \hat{F}^n}{\partial \xi} - \frac{\partial \hat{G}^n}{\partial \eta} - \frac{\partial \hat{H}^n}{\partial \zeta} \tag{2}$$

with the relaxation parameter Θ. In most of the present computations, a two-stage Runge-Kutta procedure with $\Theta = 1$ seemed to be the best choice. The numerical effort to invert the Matrix $B = I - \Delta t\, \Theta\, \frac{\partial S^n}{\partial Q^n}$ depends on the formulation of the jacobian of the chemical source terms. Several different forms of the jacobian matrix, with increasing complexity and accuracy, were implemented and compared. The most obvious choice is to invert the full n_s x n_s matrix B. This, however, is a problem from the numerical point of view, since the inversion is CPU time consuming, and the matrix B usually illconditioned. This approach worked, but was rather inefficient. The next possibility to simplify the matrix B consists of dropping all the off-diagonal terms, while keeping only the diagonal terms. In our case of eight reactions with seven species, this turned out to be an effective means of accelerating the convergence, with stability limits given by the acoustic wave speeds. However, the highly restrictive time limit, caused by the fine spacial resolution, resulted in a prohibitively slow rates of convergence. Therefore, the fully implicit delta form of the governing equations

$$\left\{I + \Delta t \frac{\partial}{\partial \xi}A + \Delta t \frac{\partial}{\partial \eta}B + \Delta t \frac{\partial}{\partial \zeta}C - \Delta t\, \Theta \frac{\partial S^n}{\partial Q^n}\right\} \delta \hat{Q}^n = \Delta t \left(\hat{S}^n - \frac{\partial \hat{F}^n}{\partial \xi} - \frac{\partial \hat{G}^n}{\partial \eta} - \frac{\partial \hat{H}^n}{\partial \zeta}\right) \quad (3)$$

was implemented.
A point implicit Gauss-Seidel relaxation procedure with subiteration was implemented. This scheme allows large timesteps, while maintaining time accuracy on the block boundaries. Its parallelization is straight forward.
Using a multi-block grid structure resulted in a flexible code with the possibilty of working with different chemical models (nonequilibrium, equilibrium, frozen) in different blocks. Besides, some of the blocks were selectively refined, depending on the evolving results. The present geometrical treatment of the computational domain was simple, yet flexible enough.

3 Multigrid Acceleration

Previously, turbulence closure models based on statistical methods, such as the well known k-ϵ model, were used by the present authors and other investigators. The results were mixed, at the best. These methods utilize or imply a certain type of averaging of the governing equations and require modeling of the correlations of all the fluctuating quantities that represent turbulence. In the present work, the LES approach was deemed more appropriate. It separates the field variables of the turbulent flow into large-scale and small-scale quantities. The large-scale parts are computed directly by solving the filtered governing equations, while the unresolved small-scale quantities need to be modeled in terms of the resolved variables. This was accomplished by the SGS eddy viscosity model, proposed by Smagorinsky [17].The computed turbulence intensity was evaluated and used to correct the chemical rates of reactions.

4 Parallel Implementation

Ideally, in a LES, the eddies with as much energy contents as possible should be directly simulated, making resolutions of order of magnitude below y^+

necessary. Since no preferential direction is assumed, this very high resolution should be applied not only normal to the solid walls, but in all spacial directions considered in that particular simulation. Additionally, the computational grid should be uniformly distributed. Clearly, even if the above requirements are somewhat relaxed, an extremely high number of grid points (or cells in the finite volume method) have to be utilized. The corresponding computations can be carried out only on the largest computers available. A performance that is adequate for the LES is presently offered only on massively parallel computers.

Early in this work, it was decided to implement a data parallel structure, since the multiblock grid system already had data exchange between the blocks built in. The parallelization was accomplished using the MPI library using standard point to point communications. Only few global operations had to be used.

5 Results

The geometry and the boundary conditions of this configuration are decribed in detail by von Lavante et. al. [18] . Here, a hydrogen jet is injected into a square channel with parallel walls. At the inflow, the Mach number was $M = 2.97$, the static pressure was $p = 0.1 MPa$ and the temperature was $T = 1300K$. The hydrogen jet enters at sonic conditions through a hole of 1.3 mm width; its temperature was $T = 350K$. The Reynolds number was in this case $Re_h = 9\ 10^6$, based on the channel hight. The computational grid had up to 1024x1024 internal cells in two dimensions and 256x128x32 in three dimensions, allowing a reasonable resolution of the important regions. This case is of particular interest, since it has been frequently used in numerical simulations by several authors and was (and is still being) extensively experimentally investigated.

The present two-dimensional results were shown previously elsewhere, so that only the three-dimensional results will be discussed at this time. The comparison of these three-dimensional results with their two-dimensional counterpart will be given in the full version of this paper.

The contours of H_2O-concentration at the injection hole are displayed in Fig. 1. The jet, seen in the center of the picture, enters the flow channel from the top and penetrates the the supersonic air stream, generating a system of compression shocks. Noticable is also the horse-shoe vortex, forming around and behind the jet. The flow is highly unsteady, with the jet fluctuating sidewards and in the length. A comparison of measured and numerically created schlieren pictures of the flow field at and behind the injection location is offered in Fig. 2. The position of the shocks and the extent of the shear layer agree rather well. In accordance with the experimental measurements, published in [18], the H_2O is carried upstream of the injection opening by recirculating fluid in the boundary layer. Downstream of the region where most of the H_2O production occures, the flow is basically chemically frozen, with H_2O being convected. The ondulation of the H_2O contours is a clear indication of the unsteady character of the flow here. The computed velocities fall well within the limits of the experimentally determined values.

6 Conclusions

The present work was aimed at finding out the practical feasibility of LES in realistic configurations of chemically reacting supersonic flows, consisting of a hydrogen-air combustion system. A simplified model of an Scramjet combustor was analyzed numerically using various parallel computer systems. Different implicit operators were implemented and compared. The results, displayed in the physical as well as spectral domain, show the typical turbulent fluctuations in the predicted variables. The three-dimensional simulations show distinctive three-dimensional flow structures, absent from the two-dimensional data.

The fully implicit procedure was effective in accelerating the timewise computation, allowing much larger time steps than the semi-implicit formulation.

References

1. Madabhushi, R. K., Choi, D., Barber, T. J., and Orszag, S., " Computational Modeling of Mixing Process for Scramjet Combustor Applications ", AIAA Paper 97-2638.
2. Cox, C. F., Cinnella, P., and Arabshahi, A., " Multi-Block Calculations for Flows in Local Chemical Equilibrium ", AIAA Paper 93-2999.
3. Godfroy, F., and Tissier, P. Y., " CFD Analysis of Vortex Shedding Inside a Subscale Segmented Motor ", AIAA Paper 94-2781.
4. Narayan, J. R., " Prediction of Turbulent Reacting Flows Related to Hypersonic Airbreathing Propulsion Systems ", AIAA Paper 94-2948.
5. Chakravarthy, V. K., and Menon, S., " Characteristics of a Subgrid Model for Turbulent Premixed Combustion " , AIAA Paper 97-3331.
6. Arunajatesan, S., and Menon, S., "Simulation of Controlled Injection of Toxic Waste for Enhanced Destruction in a Compact Incinerator", AIAA Paper 96-3075.
7. Rausch, V. L.,McClinton, C. R. and Hicks, J. W., "Scramjets breathe new life into hypersonics", Aerospace America, July 1997, pp. 40-46.

8. "Hyper-X Production Begins In Support of 1999 Flight Test", Aviation Week & Space Technology, October 13, 1997, pp.66-67.

9. Evans, J.S., Schexnayder, C.J., "Influence of Chemical Kinetics and Unmixedness on Burning in Supersonic Hydrogen Flames", AIAA-Journal, Febr. 1980, pp. 188-193

10. Grossmann, B., and Cinella, P., "Flux-Split Algorithms for Flows with Non-equilibrium Chemistry and Vibrational Relaxation", J. Comp. Phys., vol. 88, pp. 131-168, 1990

11. Larrouturou, B., and Fezoui, L. "On the Equations of Multi-Component Perfect or Real Gas Inviscid Flow", Nonlinear Hyperbolic Problems, Lecture Notes in Mathematics, 1402, Springer Verlag, Heidelberg 1989

12. van Leer, B., "Flux-Vector-Splitting for the Euler Equations", Institute for Computer Applications in Science and Engineering, Hampton, VA, Rept. 82-30, Sept. 1982

13. Liou, M.S., "On a New Class of Flux Splitting", Proceedings , 13th International Conference on Numerical Methods in Fluid Dynamics, Rome 1992, pp. 115-119

14. von Lavante, E. and Yao, J. " Simulation of Flow in Exhaust Manifold of an Reciprocating Engine ", AIAA-93-2954

15. Roe, P.L., Pike, J., "Efficient Construction and Utilisation of Approximate Riemann Solutions", Computing Methods in Applied Sciences and Engineering, VI, pp. 499-516, INRIA, 1984

16. Shuen, Jiang-Shun, " Upwind Differencing and LU Factorization for Chemical Non-equilibrium Navier-Stokes Equations ", Journal of Comp. Phys., vol. 99, pp. 213-250, 1992

17. Smagorinsky, J., " General Circulation Experiments with the Primitive Equations ", Monthly Weather Review, Vol. 91, Num. 3, March 1963.

18. von Lavante, E., Hilgenstock, M. and Groenner, J., " Simple Numerical Method for Simulating Supersonic Combustion ", AIAA Paper 94-3179

19. Hilgenstock, M., von Lavante, E. and Groenner, J., " Efficient Computations of Navier-Stokes Equations with Nonequilibrium Chemistry ", ASME Paper 94-GT-251

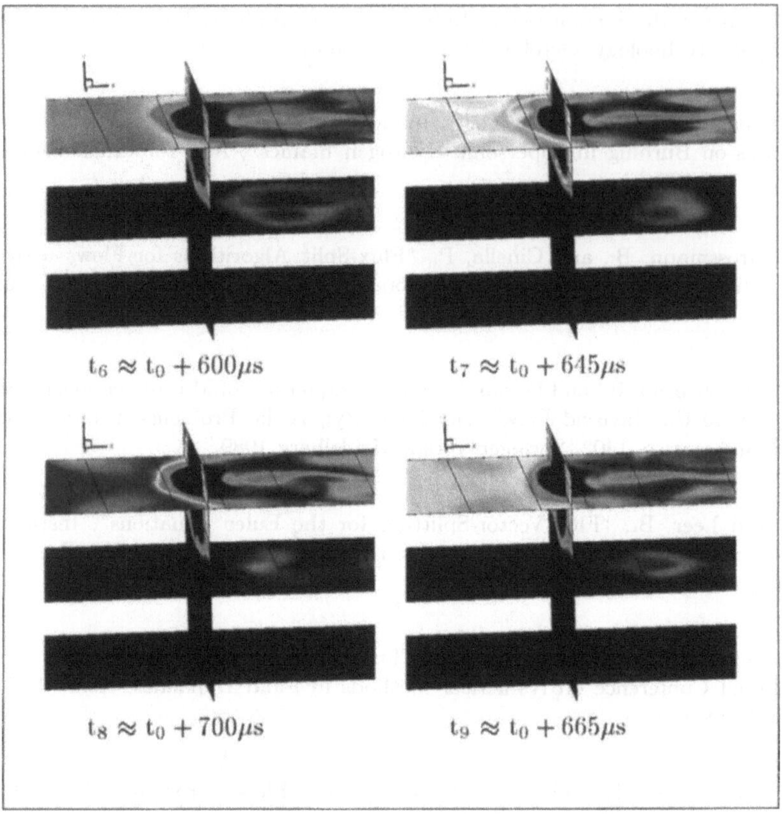

Figure 1: H_2O concentration at the injection position, showing unsteady hydrogen jet movement.

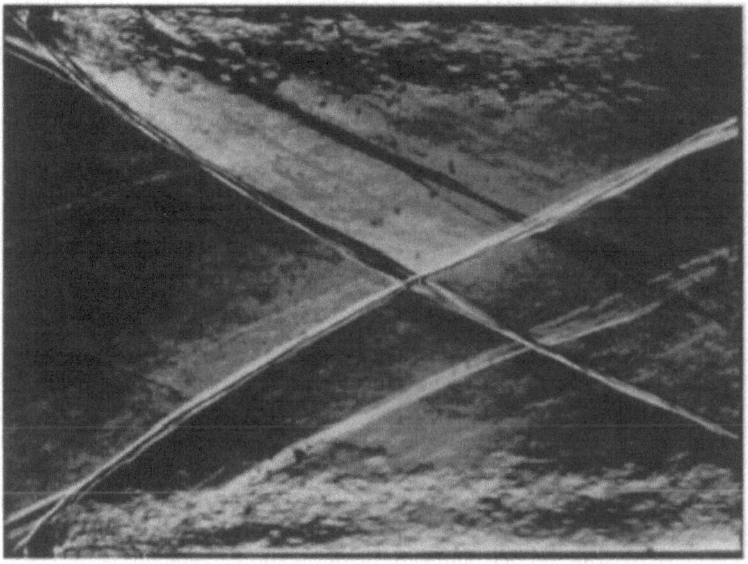

Eperimental Schlieren Picture.

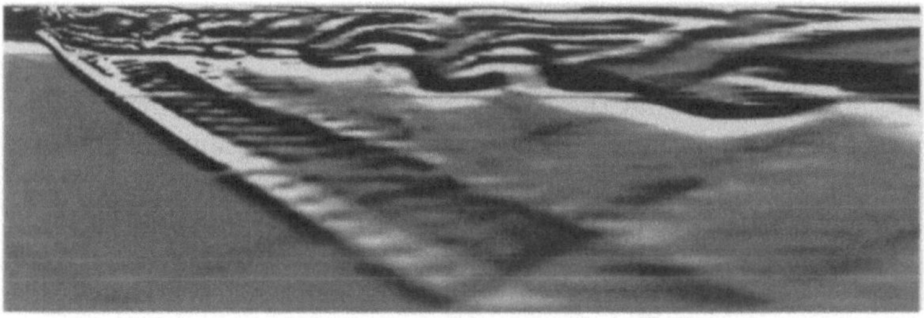

Numerically Generated Schlieren Picture

Figure 2: Comparison of the experimental and numerical schlieren pictures.

Computation of Turbulent Flows with Separation by Coherent Structure Capturing

Schmid M., Perić M.

Fluid Dynamics and Ship Theory Section,
Technical University of Hamburg-Harburg,
Lämmersieth 90, D-22 305 Hamburg

Abstract. Within the scope of technical application it is necessary to simulate flows with separation not only at "academic" Reynolds numbers by direct numerical simulation (DNS), but also at higher Reynolds numbers by means of large eddy simulation (LES) or very large eddy simulation (VLES). The flow around a sphere at Re=50 000 has been calculated on unstructured grids with the Smagorinsky subgrid model and data is compared with available measurements. It is necessary to evaluate existing models and test them on simple geometries for the future application on real problems. Results indicate that the agreement between simulation and experiment is within the range of the experimental error.

1 Introduction

It is generally accepted that the Navier-Stokes equations accurately describe the behavior of a fluid continuum. The conservative equation for the mass for an arbitrary control volume V enclosed by the surface S, here written in integral form, is known as the continuity equation:

$$\frac{\partial}{\partial t}\int_V \rho \, \mathrm{d}V + \int_S \rho \underline{v} \cdot \underline{n} \, \mathrm{d}S = 0 \, . \tag{1}$$

The momentum conservation equation in integral form reads:

$$\frac{\partial}{\partial t}\int_V \rho \underline{v} \, \mathrm{d}V + \int_S \rho \underline{v} \, \underline{v} \cdot \underline{n} \, \mathrm{d}S = \int_S \underline{\underline{T}} \cdot \underline{n} \, \mathrm{d}S + \int_V \rho \underline{b} \, \mathrm{d}V \, , \tag{2}$$

where ρ is the density, $\underline{v}$ is the velocity vector, $\underline{b}$ is the vector of the resultant body force, $\underline{\underline{T}}$ is the stress tensor and $\underline{n}$ is the unit vector normal to S and directed outwards. For flows of Newtonian incompressible fluids considered here, the stress tensor $\underline{\underline{T}}$ is defined by:

$$\underline{\underline{T}} = 2\mu \, \underline{\underline{D}} - \left(p + \frac{2}{3}\mu \, \underline{\nabla} \cdot \underline{v}\right) \underline{\underline{I}} \, , \tag{3}$$

where the strain rate tensor $\underline{\underline{D}}$ whose trace is equal zero due to the continuity equation, is:

$$\underline{\underline{D}} = \frac{1}{2}\left(\underline{\nabla} \, \underline{v} + (\underline{\nabla} \, \underline{v})^{\mathrm{T}}\right) . \tag{4}$$

For the discretization of the equation system a Finite-Volume-Method with cell-centered variable arrangement is used. Unstructured grids with local grid refinement are used to increase the accuracy in regions of large variation of the variables. The equations are discretized using approximations of second order (midpoint rule integration, linear interpolation, central differences, and fully implicit three-time-levels time integration) and linearized using Picard iteration scheme. Linear equation systems are solved sequentially using conjugate-gradient-type solvers (ICGG, CGSTAB). The equations are coupled by the well known SIMPLE method. More details about the underlying numerical method can be found in [8].

2 Filtering and LES

Filtering and Reynolds-averaging leads formally to the same equations but the operators are differently defined. In the case of filtering it is a spatial average, in the other case a time average. To filter a function $f(x)$, the filter $G(x)$ – which is defined by Leonard [5] as a convolution integral – is applied and leaves the filtered function denoted by an over-bar:

$$\overline{f}(\underline{x}) = \int_V G(\underline{x} - \underline{x}\prime) f(\underline{x}\prime) d\underline{x}\prime \,. \tag{5}$$

Applying the filter to the Navier-Stokes equations adds terms to the stress tensor, which are named subgrid stresses and need to be modelled. It is the small-scale part of the turbulent energy spectrum which is modelled, where the structure of turbulence is nearly isotropic and an eddy viscosity model can be used for all terms which are created by the filter function:

$$\overline{\underline{v}\,\underline{v}} = \underline{\overline{v}}\,\underline{\overline{v}} + \underline{\underline{T}}^{\text{sgs}} = \underline{\overline{v}}\,\underline{\overline{v}} + \left(\overline{\underline{\overline{v}}\,\underline{\overline{v}}} - \underline{\overline{v}}\,\underline{\overline{v}}\right) + \left(\overline{\underline{\overline{v}}\,\underline{v}\prime} + \overline{\underline{v}\prime\underline{\overline{v}}}\right) + \overline{\underline{v}\prime\underline{v}\prime} \,. \tag{6}$$

The three parts of $\underline{\underline{T}}^{\text{sgs}}$ in the order of appearance are named Leonard stress, cross stress, and subgrid Reynolds stress. They are modelled together by an eddy viscosity model based upon the Bousinessq approximation. For the sake of simplicity the elements of the tensor $\underline{\underline{T}}^{\text{sgs}}$ will be expressed using the Einstein summation convention:

$$T_{ij}^{\text{sgs}} = 2\mu_{\text{t}} S_{ij} = \mu_{\text{t}} \left(\frac{\partial \overline{u}_i}{\partial x_j} + \frac{\partial \overline{u}_j}{\partial x_i} \right) . \tag{7}$$

The eddy viscosity is a function of the filter size Δ, the filtered strain rate S_{ij} , and the wall distance. Close to the wall the subgrid stress must be reduced by the van-Driest damping function [1], otherwise the flow regime will be laminar at small Re. The aim of LES is to resolve as much as possible of the energy spectrum. Therefore the filter size is the cubic root of the volume over which the filtering takes place. The turbulent viscosity is computed using the Smagorinsky model [12]:

$$\mu_{\text{t}} = \rho \Delta C_{\text{s}} \left[1 - \mathrm{e}^{\left(-y^+/25\right)^3} \right]^{\frac{1}{2}} \sqrt{S_{ij} S_{ij}} \,, \tag{8}$$

where the Smagorinsky constant C_s is taken equal to 0.1, which is the value usually used in flows with strong shear [4]. The model accounts for the anisotropic part of the subgrid stress whereas the isotropic part is added to the pressure and forms a pseudo pressure:

$$p = \overline{p} + \frac{1}{3} T_{ii}^{\mathrm{sgs}} \,. \tag{9}$$

The LES thus computes the filtered velocity field and the pseudo pressure at a sequence of time steps long enough to enable computation of time-averaged quantities, which are usually obtained from experimental studies.

3 Large-Eddy Simulation of Sphere Flow at Re=50 000

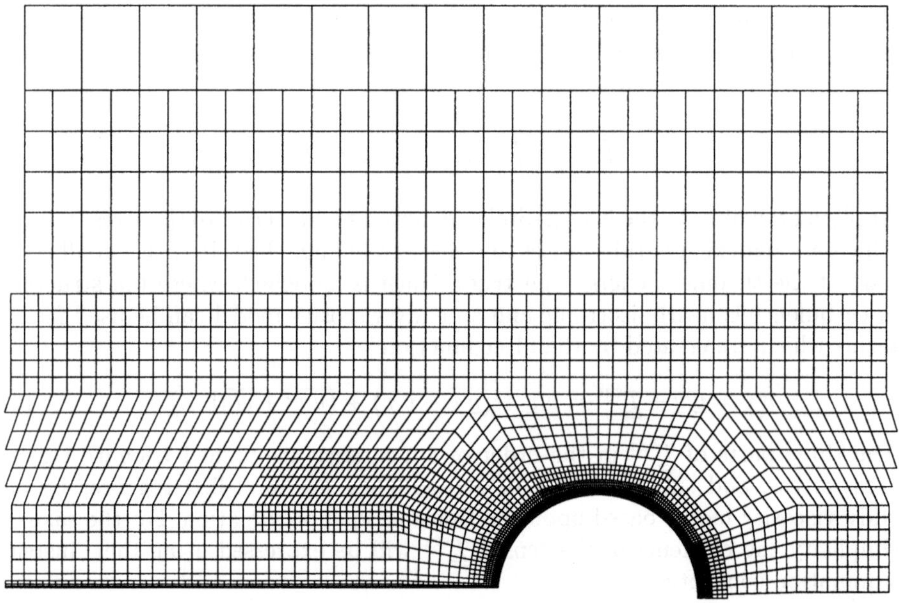

Fig. 1. Cut through the locally refined grid for the flow around the sphere (coarser grid).

The flow around a sphere held by a stick of diameter $d = 0.1D$, where D is the sphere diameter, in a uniform stream of undisturbed fluid is studied. The same configuration was the subject of an earlier study by direct numerical simulation (DNS) at Re=5 000 [11]. All results shown here are preliminary and are not fully processed yet. The dimensionless time step size $\Delta tU/D$ was 0.03 and the integration was carried out over 30 000 time steps (total integration time equal 900).

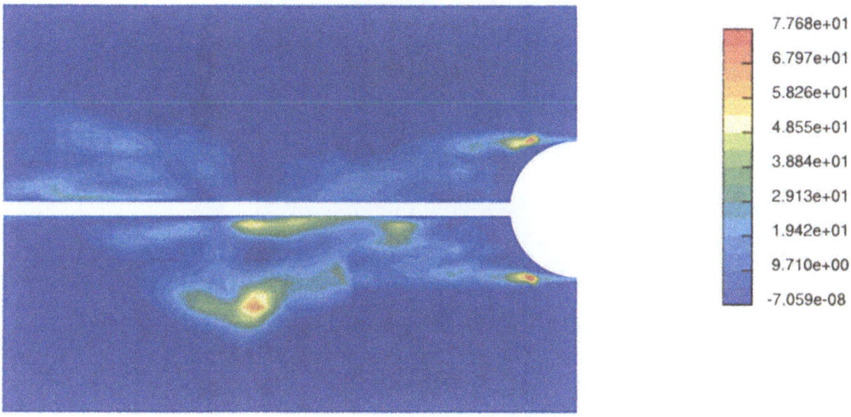

Fig. 2. Instantaneous turbulent viscosity, normalized by the dynamic viscosity, for the flow around the sphere at Re=50 000.

For the simulation the Smagorinsky model without damping function was used. It was developed and has been tested for channel flows with good results, but is considered not suitable for this case, because the transition takes place in the free shear layer and is not wall-initiated. In separated turbulent flows the law of the wall is generally not valid and it has not been shown so far that the damping function is correct for this type of flows. Therefore no wall functions were used here but a very fine grid in the normal direction has been generated near the wall in order to resolve the boundary layer. The first computational point next to the wall lies always at a normalized distance $n^+ < 1$. Simulations are performed on two grids. The coarser numerical mesh contains 840 000 CVs and is locally refined to resolve the boundary layer and to save memory and computing time in regions where the flow is unaffected by the body (see Fig. 1). A finer grid has a similar structure but is further locally refined and has around $1.9 \cdot 10^6$ CVs. The local refinement caused oscillations in regions of relatively coarse grid, which influenced the solution in regions of finer grid. This can be avoided by blending the central differencing scheme with the upwind differencing scheme, which introduces numerical diffusion and influences the results severely if it is done in the whole domain [10]. The time averaged results shown here are computed with blending 10 % of the upwind scheme together with 90 % of the central scheme. Computation with local blending (far away from sphere and wake, in regions where no turbulence occurs and where the variation in variables is small) will be done and the statistics will be compared. The mesh was divided into 64 domains and computation was done on the massively parallel computer Cray T3E at the High-Performance-Computing-Center (HLRS) in Stuttgart. The contribution of the model is small close to the wall where the mesh is fine, and large in the free

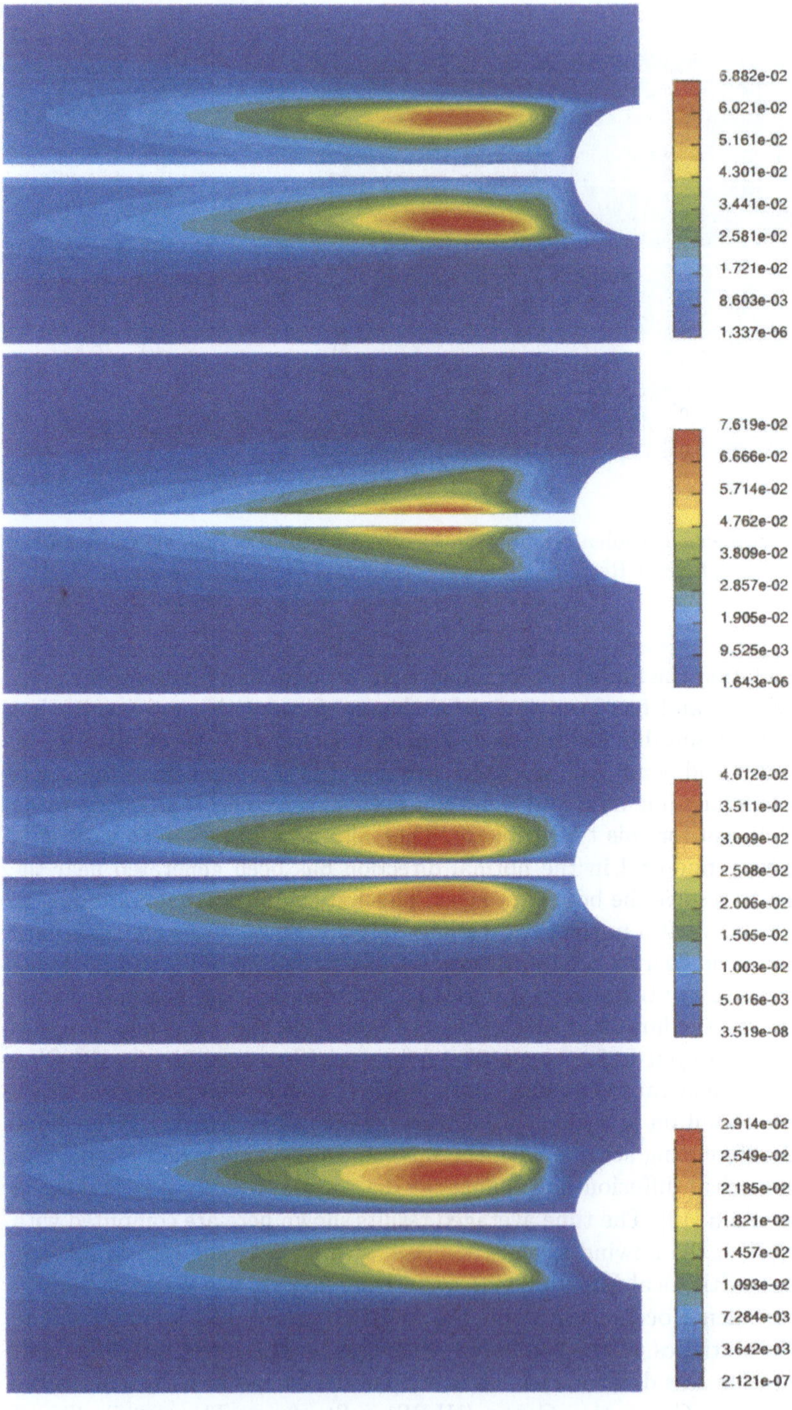

Fig. 4. Reynolds stress (from top to bottom): $R_{zz}, R_{\varphi\varphi}, R_{rz}, R_{rr}$.

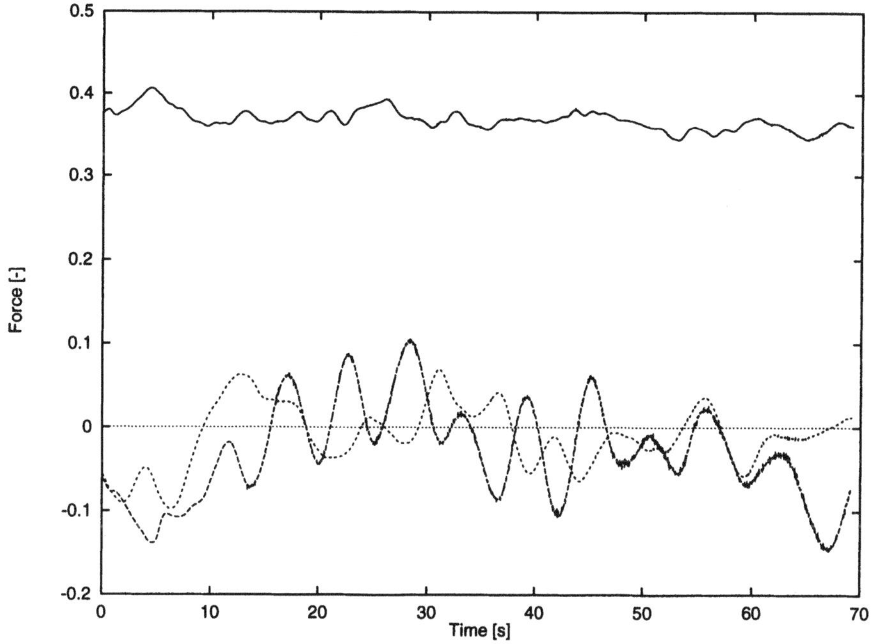

Fig. 3. Dimensionless forces acting upon the surface of the sphere.

shear layer, as required; this can be verified by checking the distribution of the eddy viscosity in Fig. 2. The Reynolds stresses are in qualitative agreement with the DNS at Re=5 000, which showed high stresses in the free shear layer [11]. The stress in circumferential direction is high close to the reattachment point at the stick and the instability starts from the streamwise direction, which can be seen from the stress R_{zz}, see Fig. 4. All stresses look symmetric, which indicates that a sufficient number of samples has been taken. The drag coefficient shown in Fig. 3 as a function of time has a mean value of approximately 0.37. In the DNS of Seidl at Re=5 000, the drag coefficient was 0.38. The curve fitted to experimental data as shown e.g. in Schlichting [9], suggests for Re=50 000 the drag coefficient of about 0.45. Preliminary results from the simulation with pure central differencing in the wake and close to the sphere show an increased drag coefficient, as do the results of the simulation on the finer grid. It is expected that the grid-independent solution will lie substantially closer to the experimental data. Figure 3 also suggests that the lateral forces oscillate at a very low frequency and an appreciable amplitude, with higher frequency oscillations superimposed. The mean lateral force should of course be zero.

4 Future Plan

Besides the Smagorinsky model, the dynamic procedure of Germano [3] is implemented (with the modification proposed by Lilly [7]) into the code and was tested for a plane channel flow at low Reynolds number. The results were improved compared to the Smagorinsky model. This will solve the problem of the constant coefficient C_s, because by the comparison between two filters the small scales will be identified and modelled whereas the large scales will not be influenced. The numerical diffusion of the used blending scheme can be removed if the blending factor is adapted to the grid size. Using the blending only in regions far from the wake, close to the outlet, in front of the sphere and at the outer boundaries – where the grid is coarse – avoids oscillations in the solution and guarantees accuracy of a second order method in the region of interest. Computations on the finer grid using 128 nodes of Cray T3E with both models

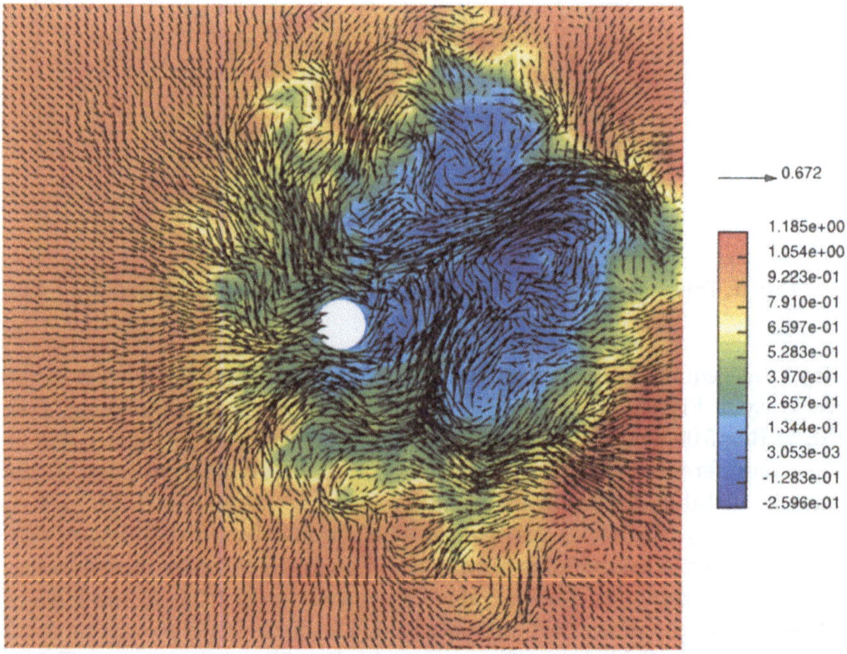

Fig. 5. Flow around the sphere at Re=50 000 in a cross section normal to the main flow direction one diameter behind the sphere (computed using the dynamic subgrid model of Germano and locally defined blending factors): contour shows the main velocity component and vectors the other two components (interpolated to a uniform presentation grid).

are on the way and statistics will be presented in the future. In Fig. 5 a cross-section through the wake normal to the main flow direction one diameter behind the sphere shows the instantaneous velocities and the structure of the flow. The comparison of solutions on the two grids will enable estimation of discretization errors (grid dependency of the solution).

5 Conclusions

The Smagorinsky subgrid-scale model was used in a LES of the flow around a sphere held by a backward stick at Re=50 000 on an unstructured grid with cell-wise local refinement. Results agree qualitatively with the available data for comparison (Reynolds stresses, drag coefficient). It has already been shown that the dynamic model [2] is capable of predicting transition and can also cope with locally laminar flow regime better than the Smagorinsky subgrid-scale model. The results expected of the large eddy simulation using the dynamic procedure will help to verify the results of the Smagorinsky model and will give more hints about the applicability of LES models to complex separated flows with highly turbulent wakes. The calculations will be compared with LDV measurements which will be done in the wind tunnel of the TUHH in order to provide detailed data for validation. Performing simulations before experiments will enable identification of regions in which measurement points should be placed and to achieve maximum reliability of model validation.

References

[1] E. R. van Driest, On turbulent flow near a wall, *J. Aero. Sci.*, **23**, 1007–1011, (1956).

[2] M. Germano, U. Piomelli, P. Moin and W. H. Cabot, A Dynamic Subgrid-Scale Eddy Viscosity Model, *Physics of Fluids A*, **3**, 1760–1765, (1992).

[3] M. Germano, The Filtering Approach, *Journal of Fluid Mechanics*, **235**, 325–336, (1992).

[4] C. Härtel, Turbulent Flows: Direct Numerical Simulation and Large-eddy Simulation, Handbook of Computational Fluid Mechanics (ed. Roger Peyret), Academic Pess, (1996).

[5] A. Leonard, On The Energy Cascade in Large-Eddy Simulations of Turbulent Fluid Flows, *Advances in Geophysics*, **18** A, 237–248, (1974).

[6] D. K. Lilly, The representation of small-scale turbulence in numerical simulation experiments, *Proceedings of IBM Scientific Computing Symposium*, IBM Form No.320-1951, 195–210, (1967).

[7] D. K. Lilly, A Proposed Modification of the Germano Subgrid Scale Closure Method, *Physics of Fluids A*, **4**, 633–635, (1992).

[8] J. H. Ferziger and M. Perić, *Computational Methods for Fluid Dynamics*, Springer, Berlin, 1996.

[9] H. Schlichting, K. Gersten, *Grenzschichttheorie*, Springer, Berlin, 1997.

[10] M. Schmid, G. Deng, V. Seidl, M. Visonneau, M. Perić, Computation of Complex Turbulent Flows, Numerical Flow Simulation I (ed. E. H. Hirschel), Vieweg Verlag, (1998).

[11] V. Seidl, Entwicklung und Anwendung eines parallelen Finite-Volumen-Verfahrens zur Strömungssimulation auf unstrukturierten Gittern mit lokaler Verfeinerung, Dissertation, Universität Hamburg, 1997.

[12] J. Smagorinsky, General Circulation Experiment With the Primitive Equations: I. The Basic Experiment, *Monthly Wheater Review*, **91**, 99–164, (1963).

Large Eddy Simulation of the Flow around a Circular Cylinder

Jochen Fröhlich and Wolfgang Rodi

Institut für Hydromechanik, Universität Karlsruhe, Kaiserstr. 12, D-76128 Karlsruhe

1 Introduction

Large–Eddy Simulation (LES) is a promissing concept for modelling turbulent flows. The principle is to solve the three–dimensional, unsteady Navier Stokes equations on a grid which is too coarse to capture all vortical motions. The unresolved part of the flow field, constituted by fine–scale motion, is accounted for by an additional so–called subgrid–scale model term in the equations. This approach allows to reduce the computational cost for the same Reynolds number or to compute the flow at a much higher Reynolds number.

The flow behind bluff bodies such as a circular cylinder is a particular candidate for LES, since the vortex shedding is difficult to model with other techniques.

2 Purpose

The present project was concerned with the simulation of the flow around a circular cylinder in uniform flow at a Reynolds number of 140000 based on the cylinder diameter in continuation of the work reported in [FRK+98]. This flow is sub–critical which means that the boundary layer along the upstream wall is laminar while the rear flow is turbulent, triggered by the transition of the separating shear layer shortly after detachment as well as the large shedding vortices resulting from the interaction of the two shear layers on either side. The code LESOCC (Large Eddy Simulation On Curvilinear Coordinates) has been developed at the Institute for Hydromechanices [BR96] and is now extensively used by this group. The purpose of the project, which started in 1997, was to perform the parallelization of the code on a shared memory architecture such as the SX4 and to assess the resulting performance that can be achieved.

3 Accomplishments

In a first phase in autumn 1997 the code was ported onto the SX4. A test case was generated. The parallelization by means of compiler directives was performed and controlled with the available tools. First tests suggested unsatisfactory performance but it turned out in the sequel that this resulted from insufficient loop length in the main solver which in turn was a consequence of choosing an almost two–dimensional setting for the test. Subsequently, in spring 1998, a full computation was made for a structured O–grid around the cylinder containing $182 \times 306 \times 96 =$ 5.3 Mio. grid points. The grid for this computation is represented in Figure 1. It resolves the near–wall structure so that a rigorous no–slip condition can be applied. For this computation the Smagorinsky subgrid–scale model was employed. Figure 2 shows an iso–pressure surface in one quadrant of the domain. The mean flow is from left to right in positive x–direction and the axis of the cylinder is at $x = y = 0$. This figure illustrates the complicated flow structure characterized on one hand by von Karman vortex shedding, i.e. vortex tubes parallel to the cylinder axis, and on the other hand by secondary structures generated during the break up of the former into threedimensional vortices.

The achieved performance during the computation was very satisfactory, as typically 2600 MFLOPS (concurrently) were obtained on 4 processors (not used exclusively during the run). During 85% of the time all 4 processors worked in parallel.

The attributed CPU time was used to perform the simulation during one shedding period. This is much shorter than estimated due to the fact that when writing the proposal the time step was estimated differently. It turned out however, that the CFL criterium was stricter than expected for the grid that had been generated and that more iterations per time step were needed. The run was then continued on another platform for organisational reasons.

4 Comparison to Message Passing

In the meantime the code has also been parallelized for distributed memory architectures using MPI. On a VPP300 with exclusive use of 4 processors 2800 MFLOPS were obtained. The algorithm is however more involved and the domain decomposition slightly increases the number of internal iterations. It also renders more cumbersome certain variants of the physical modelling such as the regularization for the dynamic subgrid–scale model as well as the pre– and postprocessing.

5 Perspectives

Running the code LESOCC on shared memory machines as the SX4 has prooven an efficient means to perform LES with large grids that require a huge amount of memory and computing time. It will in the future be considered as a valuable alternative to distributed memory machines.

References

[BR96] M. Breuer and W. Rodi. Large eddy simulation of complex turbulent flows of practical interest. In E.H. Hirschel, editor, *Flow simulation with high performance computers II*, volume 52 of *Notes on Numerical Fluid Mechanics*, pages 258–274. Vieweg, Braunschweig, 1996.

[FRK+98] J. Fröhlich, W. Rodi, Ph. Kessler, S. Parpais, J.P. Bertoglio, and D. Laurence. Large eddy simulation of flow around circular cylinders on structured and unstructured grids. In E.H. Hirschel, editor, *Notes on Numerical Fluid Mechanics*, volume 66, pages 319–338. Vieweg, 1998.

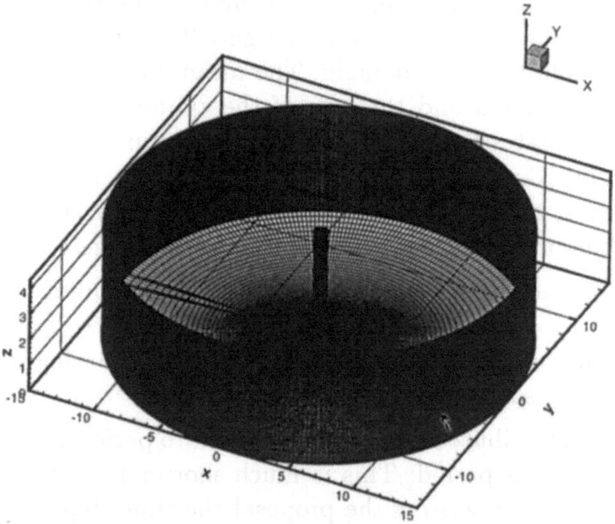

Fig. 1. O-grid used for LES of the flow around a circular cylinder.

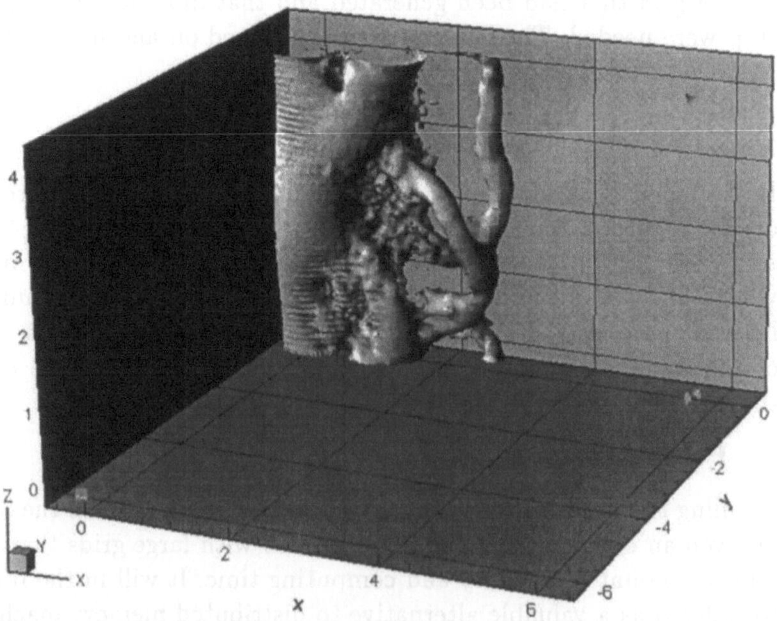

Fig. 2. Pressure iso-surface in the rear part of the cylinder.

Direct Numerical Simulations of an Adverse Pressure Gradient Turbulent Boundary Layer on High Performance Computers

Michael Manhart

Technische Universität, München , Germany

Abstract. Direct Numerical Simulations (DNS) of developing turbulent boundary layers have been performed using a Finite Volume method with a locally refined grid near the wall. In the case of a zero pressure gradient boundary layer it is shown that the use of local grid refinement saves a considerable amount of computational resources. The method is then applied to an adverse pressure gradient boundary layer. The comparison of the results with an experiment performed by Watmuff [10] shows that the method is well suited for strongly developing turbulent flows.

1 Introduction

One focus of the current turbulence research is the improvement of turbulence models for Reynolds averaged simulations (RANS) or large eddy simulations (LES). A DNS can provide highly reliable data bases for this goal, because all relevant turbulent length and time scales are resolved. Due to the limitations of computational power up to now only low or moderate Reynolds numbers and simple geometries could have been investigated by DNS. Although computational power has been continuously increasing during the last years, a DNS of a moderately complex flow, like a separated turbulent boundary layer, at a moderate Reynolds number requires the use of the fastest available computers and the most efficient algorithms currently available. A promising way to save computer resources in a DNS is using a locally refined computational grid in critical regions of the flow and a coarse grid in less critical regions of the flow.

The use of locally refined (zonal) grids in LES and DNS is a relatively new approach. Kravchenko *et al.* [3] performed a zonal grid DNS of turbulent plane channel flow using a combined B-spline/spectral method. Sullivan *et al.* [9] used zonal grids to perform a LES of a planetary boundary layer. Like Kravchenko *et al.* [3] they used spectral interpolation at the boundary between coarse and fine grid. Spectral interpolation delivers highly accurate results but has the disadvantage of being restricted to relatively simple geometries with periodic boundaries. A first step to overcome this drawback has been done by Manhart [4] and [5] in a zonal grid DNS of turbulent plane channel flow and a turbulent boundary layer, respectively.

The purpose of the current research is to extend the zonal grid approach to the DNS of turbulent boundary layers with pressure gradients and separation. The zonal grid approach has been implemented in MGLET, a well-tested finite volume code for LES and DNS of turbulent flows in complex geometries (Werner and Wengle [11,12] and Manhart and Wengle [6]). In the present paper the application to an adverse pressure gradient boundary layer is evaluated and documented.

2 Numerical Method

2.1 The code MGLET

MGLET, is based on a finite volume formulation of the Navier-Stokes equations on a staggered Cartesian non-equidistant grid. The spatial discretization is of second order (central) for the convective and diffusive terms. For the time advancement of the momentum equations an explicit second-order time step (leapfrog with time-lagged diffusion term) is used.

The Poisson equation for the pressure is solved by an iterative point-wise velocity-pressure iteration like that described in Hirt et al. [2]. It can be used as a single-grid iteration or as a smoother in a multigrid cycle. The maximum divergence is chosen in order to keep the maximal velocity error below $\Delta u_{max} \leq 10^{-5} U_\infty$ (according to the relation $div_{max} = \Delta u_{max}/\Delta x_{min} \cdot l/U_\infty$). Here, U_∞ and l are the characteristic velocity and length scales, respectively.

2.2 Zonal grid algorithm

The refinement for the local grids is done by dividing one coarse-grid cell into 8 fine-grid cells. The coarse and the fine grid are arranged in an overlapping way, so that the coarse grid is defined globally (global grid) and the fine grid is defined only locally (zonal grid). Each second cell-face of the local fine grid lies exactly on a coarse grid cell-face. The overlapping of the grids allows for flexible handling of grid refinement. The simulation can be started on a coarse grid and later be refined locally if in some region of the computation the resolution is not sufficient.

In our approach, the coarse-grid and the fine-grid solutions are fully coupled. The coupling is achieved by transferring the fine-grid solution in the overlap region to the coarse grid. This so-called restriction is done at certain steps within the solution algorithm. We use averaging over four cell faces for the velocities and averaging over 8 grid cells for the pressure restriction.

The solution on the coarse-grid level in the non-overlapping region serves as a boundary condition for the fine grid. The interpolation of the coarse-grid variables to the fine grid boundary points is done by a first order interpolation, in order to provide conservation of mass and momentum fluxes. If the interpolation were not conservative, the turbulent fluctuations would be strongly damped near the grid interface [4]. For solving the Poisson equation on both levels, we use the pressure correction on the coarse grid as a new pressure estimate for the fine grid in a multigrid cycle.

2.3 Parallelisation

For running the code on parallel computers, we employed the following strategy. The original single-grid code has been extended to a block-structured code in order to manage the multiple grids that arise from the local grid refinement, the multigrid algorithm and parallelisation. In this framework parallelisation is done over the grid blocks using the original subroutines of the single-grid code. A domain decomposition technique has been employed over the second and third direction (in the Fortran arrays) to divide each of the grids into an arbitrary number of subgrids that are treated as independent grid blocks. The communication of neighbouring grids is done using the MPI library. In order to keep the data consistent we employ a red-black algorithm in the velocity-pressure iteration. Therefore the convergence of the iterations is not dependent on the number of PE's used. The boundary conditions for the fine grid on a fine/coarse interface are stored in twodimensional buffer arrays that are updated only at two steps in the algorithm. That minimizes the communication between the local fine grid and the underlying global coarse grid that need not necessarily reside on the same processor.

2.4 Computational efficiency

Considerable efforts have been made to optimize the single processor performance as well as the parallel efficiency of the code. It achieves about 70 MFlop/sec on a single processor on an CRAY T3E-900 and nearly one GFlop/sec on a single processor on a Fujitsu VPP-700. If the problem size is large enough, parallelisation gives a linear speedup up to more than 200 PE's on the CRAY T3E-900 (see Figure 1). A detailed comparison of the performance of the code on a massively parallel computer (CRAY T3E) and a vector parallel computer (Fujitsu VPP) has been done by Manhart [5].

3 Results

3.1 Zero pressure gradient boundary layer

Configuration and boundary conditions. In order to validate the zonal grid approach for the case of a developing turbulent boundary layer, three direct numerical simulations of a zero pressure gradient boundary layer have been performed at different grid resolutions. The geometry of the flow is sketched in Figure 2. The streamwise, spanwise and wall-normal directions are denoted by x, y and z, respectively. The Reynolds number of the performed simulations based on the inlet momentum thickness is $Re_\theta = 670$. The dimensions of the computational box have been chosen as $L_X/\delta_0 = 40.96$, $L_Y/\delta_0 = 3.2$ and $L_Z/\delta_0 = 3.8$. We have run three different grid resolutions: starting from a coarse grid ("apg2") two refined runs have been done, one with a locally refined grid near the wall up to $z^+ = 60$ ("apg6") and one with a fully

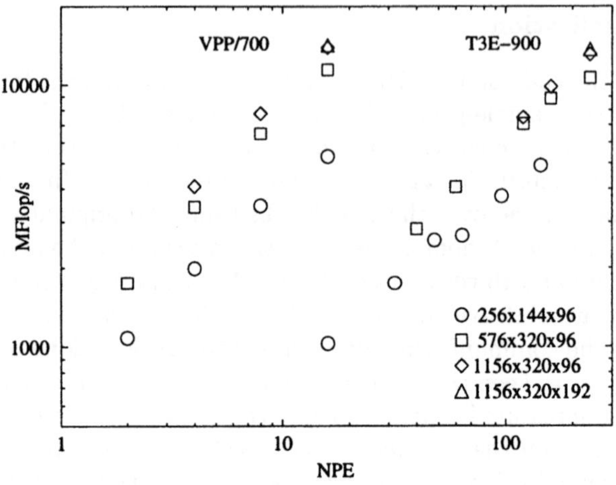

Fig. 1. Mflop-rate achieved by different test problems on a CRAY T3E-900 and a Fujitsu VPP700.

Table 1. Parameter of the grids used for evaluation of the local grid approach.

Case	"apg2"	"apg6"		"apg4"
		local	global	
N_X	512	1024	512	1024
N_Y	80	160	80	160
N_Z	96	32	96	192
N_{TOT}	$3.9 \cdot 10^6$	$3.9 \cdot 10^6$	$5.2 \cdot 10^6$	$31.4 \cdot 10^6$
Δx^+	26	26	13	13
Δy^+	13	13	6.5	6.5
Δz^+_{min}	3.2	3.2	1.6	1.6

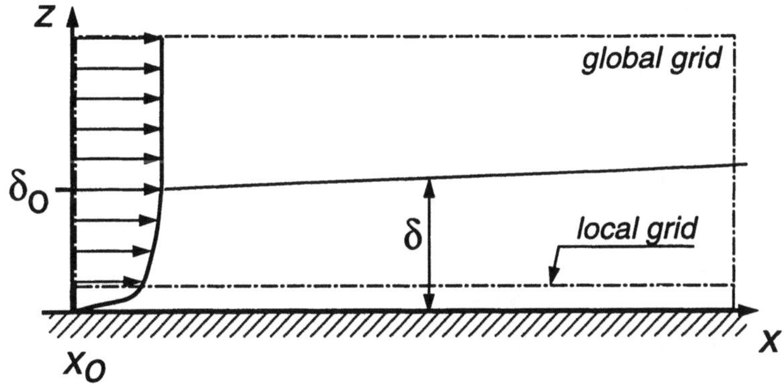

Fig. 2. Geometry of the boundary layer simulations (not to scale).

refined grid over the total domain ("apg4", see table 1). The use of the zonal grid reduced the CPU-time spent during one time step significantly from 8.0 seconds ("apg4") to 2.5 seconds ("apg6"). The coarse grid run took 1.5 seconds per time step on the machine used (a Fujitsu VPP700, 8 PE's).

At the inlet plane a time dependent boundary condition is used to prescribe turbulent fluctuations in the boundary layer downstream. It is generated by taking fluctuations from a position 10 inlet boundary layer thicknesses δ_0 downstream and superposing them onto a time mean velocity profile corresponding to the Reynolds number considered. As time mean velocity profiles we have taken the profiles from the simulations of Spalart [8]. The fluctuations are exponentially damped at a wall distance higher than δ_0 in order to prevent the boundary layer from growing in time.

On the exit plane and the top surface, the velocity derivatives normal to the boundary are set to zero (Neumann condition). For the pressure, we use a Dirichlet boundary condition that allows to introduce a pressure gradient. At the lower boundary (the wall), a noslip boundary condition is applied, and the spanwise direction is assumed to be periodic.

Results. The quality of the simulations has first been checked by an inspection of the global quantities like the skin friction coefficient c_f. The evolution of the skin friction coefficient over the momentum thickness Reynolds number is plotted in Figure 3 and compared with Coles' correlation [1]. The skin friction coefficient in the coarse grid simulation is too high, which is an indication of an insufficient grid resolution. The results of the global and the zonal fine-grid simulations approach the empirical values to within 10% accuracy. In the global fine grid simulation, a minimum occurs immediately after the inflow. This collapse of the skin friction is due to a side effect of the parallelisation (in this run, a domain decomposition was used in spanwise direction). This

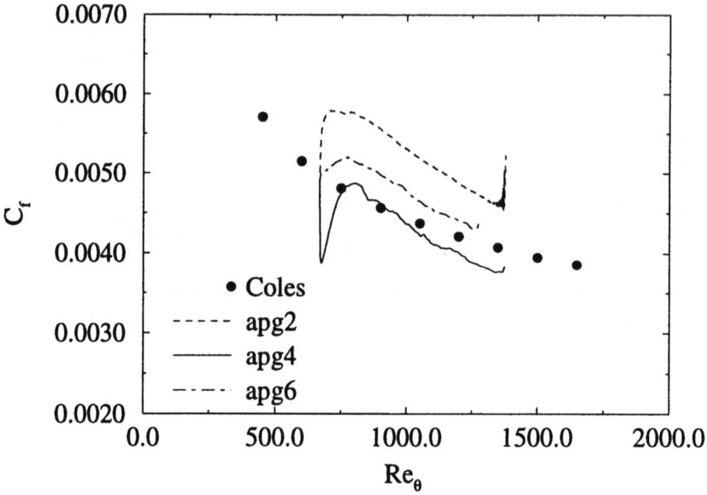

Fig. 3. Evolution of skin friction coefficient with momentum thickness Reynolds number. Symbols: Coles ([1]).

"bug" has been fixed in the zonal-grid run leading to a smoother development of the skin friction coefficient.

In Figure 4 we compare the time averaged streamwise velocity profiles at 10 boundary layer thicknesses δ_0 downstream of the inlet with the result of Spalart's DNS (spectral code). Up to a wall distance of about $z^+ = 100$ a remarkable accordance of the locally refined and the fully refined grid runs with the spectral result can be observed. The higher near-wall resolution is required to obtain the "law of the wall" $U^+ = log(z^+)/0.41 + 5.0$, and the locally refined grid clearly improves the result with respect to the coarse grid.

The comparison of the RMS-values of the streamwise velocity fluctuations (figure 5) shows that in the nearest wall region ($z^+ \leq 20$) the coarse grid values are too high, whereas in the logarithmic layer they are too low. Again, the locally refined grid improves the situation in the near wall region. But in the logarithmic region, where the grid interface is located, the RMS-values stay marginally lower than those of the spectral and the fine grid simulation. As in the averaged profiles, an effect of the grid interface is hardly visible.

It can be stated that by using a locally refined grid in the near wall region a significant improvement of the solution can be achieved without the computational cost of a fully refined grid.

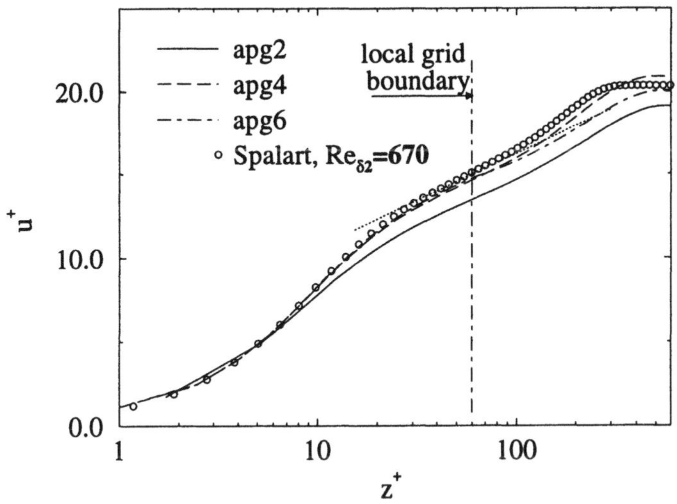

Fig. 4. Time averaged velocity profile in inner coordinates. Symbols: Spalart [8]. Dotted line: log law $U^+ = log(z^+)/0.41 + 5.0$.

3.2 Adverse Pressure gradient boundary layer

Configuration. For the simulation of the adverse pressure gradient boundary layer, we used the locally refined grid "apg6". (The adverse pressure gradient run is denoted hereafter with "apg6dp"). The only alteration compared to the zero pressure gradient boundary layer is the description of a variable pressure at the upper boundary of the computational domain. The pressure distribution has been derived from Bernoulli's equation using a u_∞ which meets the experiment of Watmuff [10]. This experiment has been selected because it is well documented and there is already a DNS available for comparison (Na and Moin [7]). In the experiment, the upper wall of the wind tunnel has been adjusted to accelerate the flow before the adverse pressure gradient (deceleration) has been applied. In our simulation, we let a zero pressure gradient boundary layer develop for 5 inlet boundary layer thicknesses δ_0 before we apply the favourable pressure gradient at the top surface at $x = 0.0$. This leads to a smooth onset of the pressure gradient at the wall (Figure 6).

The falling pressure is connected to a rising skin friction and the rising pressure to a falling skin friction (Figure 7). In the simulation, this effect seems somehow delayed as compared to the experiment. For the favourable pressure gradient region ($x/\delta_0 \leq 15.0$) this is not surprising because in the experiment it has been applied for a longer streamwise distance than in case of the simulation. For the adverse pressure gradient region this is not so obvious and for an explanation of the higher c_f in this region probably a

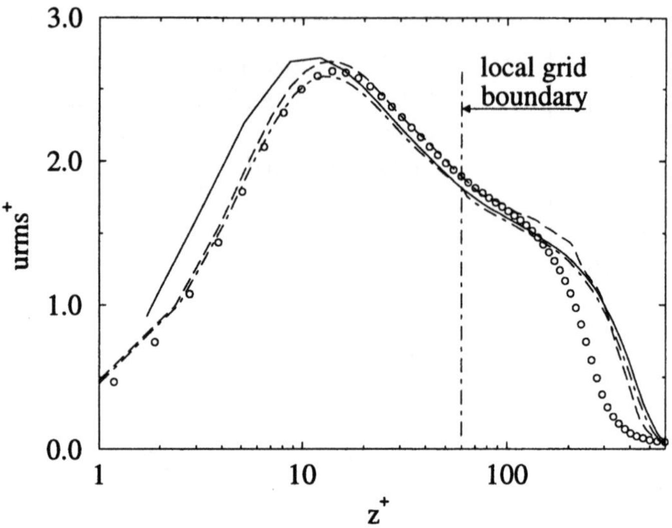

Fig. 5. RMS-values of streamwise velocity fluctuations in inner coordinates. Symbols: Spalart [8].

longer streamwise extent would be necessary in the simulation. Nevertheless, the skin friction difference between simulation and experiment is less than 10%.

The mean streamwise velocity profiles and the streamwise RMS-values obtained by the DNS are compared with the experimental values of Watmuff [10] in the figures 8 and 9. The data are normalized by the boundary layer thickness δ_0 at the position $x = 0.0$ and the freestream velocity u_0 at the same position. The behaviour of the streamwise velocity profile during acceleration ($x = 15.6$) and deceleration ($x = 31.4$) can be depicted from figure 8. Both, the acceleration and the deceleration acts stronger near the wall than at the upper edge of the boundary layer. This effect gives rise to a changing skin friction coefficient, even if it is normalized by the local freestream velocity. The deformation of the time averaged velocity profile due to this effect is reasonably well represented by the DNS, although it seems that in the buffer layer ($z/\delta_0 \approx 0.1$), the computed deceleration is smaller than that measured. This observation is consistent with the higher skin friction coefficient in the deceleration region ($x > 15.0$, figure 7). The first observation that can be made for the RMS of the streamwise velocity fluctuations (figure 9) is that throughout the whole development of the boundary layer, the DNS gives a lower peak value than the experiment. In order to clarify this observation, we included the RMS-values from the zero-pressure gradient boundary layer

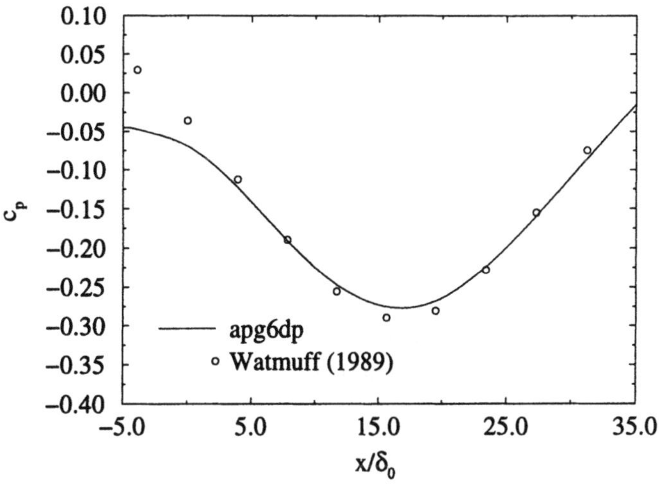

Fig. 6. Streamwise development of c_p in the experiment of Watmuff [10] and in the simulation.

(global fine grid DNS "apg4", see also figure 5) in the plot for $x = 7.8$. The DNS results for the accelerated region ($x = 7.8$) are in between the DNS results for a zero pressure gradient boundary layer and the experiment with an accelerated boundary layer. As we can see in figure 6 the acceleration that takes place for a long distance in the experiment has just started in the DNS at the position considered. We expect, that it would have taken a longer acceleration of the boundary until the DNS, that starts with a zero pressure gradient boundary layer, would have reached the structure of an accelerated turbulent boundary layer. As we can see at the position $x = 15.6$, the coincidence between DNS and experiment increases during the acceleration phase of the boundary layer. It should be noted that in the DNS of Na and Moin [7] the same effect can be observed where the peak values of the DNS are systematically lower than those of the experiment throughout the whole development of the boundary layer, even in the decelerated (adverse pressure gradient) phase. So the reason for the lower RMS values in the decelerated phase cannot be searched in the zonal grid algorithm, although in this region the grid interface at $z/\delta_0 = 0.18$ becomes slightly visible in the RMS values.

4 Conclusions

We have reported on DNS of developing turbulent boundary layers. Several features of the code used allow the efficient computation of such kind of flows. Local grid refinement saves grid points and therefore memory and CPU-time.

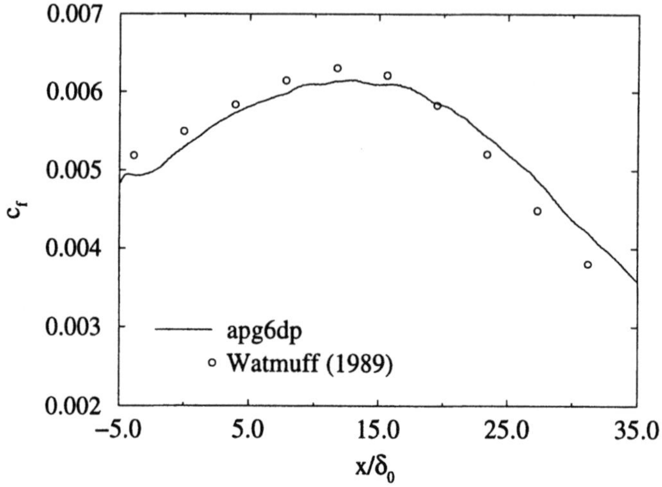

Fig. 7. Streamwise development of c_f in the experiment of Watmuff [10] and in the simulation.

The parallelisation and good vectorisation enables the use of supercomputers even with completely different architectures to reduce overall computing time.

A zero pressure gradient boundary layer served as a reference case to validate the applicability of the zonal grid approach to spatially developing turbulent flows. It has been shown that the near wall resolution is crucial to the overall quality of the results and that in the outer layer the resolution can be decreased. A zonal grid DNS of an adverse pressure gradient boundary layer has been done in accordance with an experiment of Watmuff [10]. The good agreement between experimental and DNS results shows that the use of locally refined grid is actually a means for reducing computational requirements for strongly developing turbulent flows.

Notes and Comments. We gratefully acknowledge the support of the HLRS in Stuttgart and the LRZ in Munich. The work has been supported by the DFG under grant no. FR 478/15.

References

1. D. Coles. The turbulent boundary layer in a compressible fluid. In *Report R-403-PR*. The Rand Corporation, Santa Monica, CA, 1962.
2. C.W. Hirt, B.D. Nichols, and N.C. Romero. Sola – a numerical solution algorithm for transient fluid flows. In *Los Alamos Sci. Lab.*, Los Alamos, 1975.

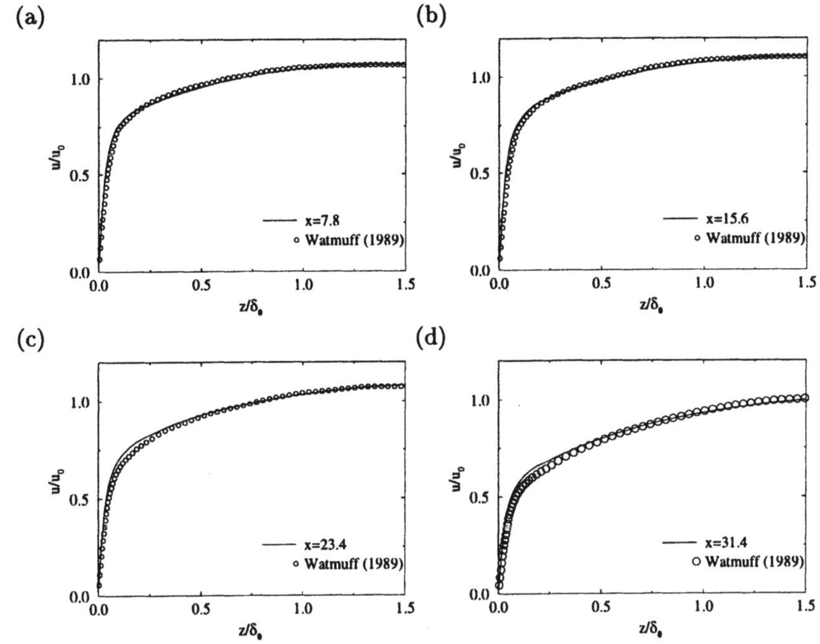

Fig. 8. Averaged streamwise velocity at different streamwise position. Symbols: Watmuff [10].

3. A.G. Kravchenko, P. Moin, and R. Moser. Zonal embedded grids for numerical simulations of wall-bounded turbulent flows. *J. Comp. Phys.*, 127:412–423, 1996.

4. M. Manhart. Zonal direct numerical simulation of turbulent plane channel flow. In R. Friedrich and P. Bontoux, editors, *Computation and visualization of three-dimensional vortical and turbulent flows. Proceedings of the Fifth CNRS/DFG Workshop on Numerical Flow Simulation*, volume 64 of *Notes on Numerical Fluid Mechanics*. Vieweg Verlag, 1998.

5. M. Manhart. Direct numerical simulation of turbulent boundary layers on high performance computers. In E. Krause and W. Jaeger, editors, *High performance Computing in Science and Engineering 1998*. Springer Verlag, 1999.

6. M. Manhart and H. Wengle. Large-eddy simulation of turbulent boundary layer flow over a hemisphere. In Voke P.R., L. Kleiser, and J-P. Chollet, editors, *Direct and Large-Eddy Simulation I*, pages 299–310, Dordrecht, March 27-30 1994. ERCOFTAC, Kluwer Academic Publishers.

7. Y. Na and P. Moin. Direct numerical simulation of turbulent boundary layers with adverse pressure gradient and separation. Report No. TF-68, Thermosciences Division, Department of mechanical engineering, Stanford University, 1996.

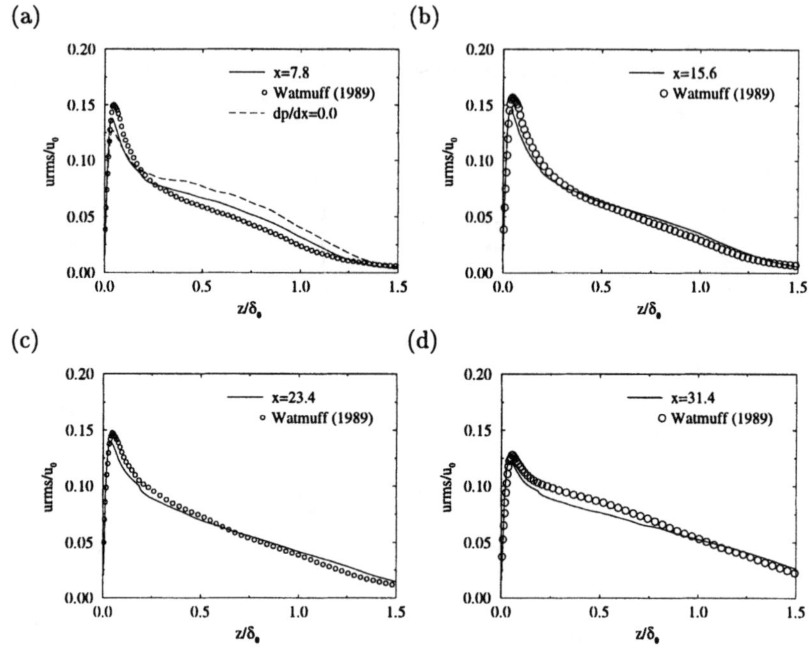

Fig. 9. RMS of streamwise velocity fluctuations at different streamwise position. Symbols: Watmuff [10].

8. P.R. Spalart. Direct simulation of a turbulent boundary layer up to $R_\theta = 1410$. *J. Fluid Mech.*, 187:61–98, 1988.
9. P.P. Sullivan, J.C. McWilliams, and C.-H. Moeng. A grid nesting method for large-eddy simulation of planetary boundary-layer flows. *Boundary-Layer Meteorology*, 80:167–202, 1996.
10. J.H. Watmuff. An experimental investigation of a low Reynolds number turbulent boundary layer subject to an adverse pressure gradient. In *Ann. Res. Briefs*, pages 37–49. Center for Turbulent Research, 1989.
11. H. Werner and H. Wengle. Large-eddy simulation of turbulent flow over a square rib in a channel. In H.H. Fernholz and H.E. Fiedler, editors, *Advances in Turbulence*, volume 2, pages 418–423. Springer-Verlag, Berlin, 1989.
12. H. Werner and H. Wengle. Large-eddy simulation of turbulent flow over and around a cube in a plate channel. In F. et al. Durst, editor, *Turbulent Shear Flows 8*, Berlin, 1993. Springer.

Aeroelastic Analysis of a Helicopter Rotor in Forward Flight

B. Buchtala, K.–H. Hierholz and S. Wagner

Institut für Aerodynamik und Gasdynamik, Universität Stuttgart Pfaffenwaldring 21, 70550 Stuttgart, Germany

Flow fields around rotary wings, especially around helicopter rotors, are extremely complex due to strong compressibility and three–dimensional effects. The solution of the governing equations for these flow problems is very difficult and can usually only be achieved with the help of numerical methods.

A sophisticated numerical analysis tool has been developed at the Institut für Aerodynamik und Gasdynamik (IAG), featuring pure aerodynamic as well as aeroelastic simulation capabilities. Fluid and structure codes constitute separate programs which are time–accurately coupled and communicate via TCP/IP socket connnections.

The aerodynamic analysis is based on the Euler equations [6–11]. Thus, typical aerodynamic phenomena which are encountered in the unsteady three–dimensional compressible flow field around a helicopter rotor can be properly modeled. The Euler equations are derived in a non–inertial, rotating frame of reference using absolute quantities for the formulation of the conservative variables. Since an aeroelastic analysis requires deformable grids, arbitrary Lagrangian–Eulerian fluxes are employed [4,5].

The governing equations are projected from physical to computational space, transforming the body–fitted structured grids into uniformly–spaced cartesian meshes.

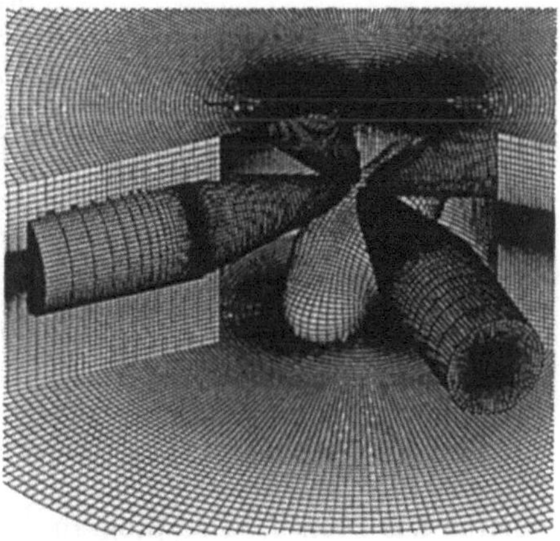

Figure1. Chimera grids

Grid generation for a helicopter model is a laborious task. Due to the complex kinematics of the system, a numerical analysis using a single grid is nearly impossible. The Chimera approach provides a possibility to circumvent these difficulties by discretizing the individual helicopter components in separate grids which are all embedded in a common background grid (Fig. 1). Thus, arbitrary motions can be achieved with a tolerable level of effort.

An efficient implicit LUSGS–algorithm drives the aerodynamic system in time, providing second–order accuracy. A cell–centered finite–volume upwind scheme based on an approximate Riemann solver is used for spatial discretization. Third–order accuracy is achieved by a uniformly–non–oscillatory (UNO) flux evaluation method.

The dynamic behavior of the rotor blades is simulated by a quasi one–dimensional finite–element method using Timoshenko's beam theory which considers strain displacement, bending, shear and torsional deformation [1,3]. The dynamic blade model comprises the calculation of coupled flap–torsion as well as coupled flap–lag motion. Blade deflections are implicitly solved in time by integrating the second–order linear system of differential equations with the generalized–α–scheme. Fluid structure coupling is achieved by partitioned procedures [2]. An offset between the time scales of fluid and structure solver conserves second–order accuracy of the global algorithm.

Figure2. Wake structure in forward flight

A typical test case for a four–bladed helicopter rotor in forward flight (Fig. 2) yields the following performance parameters on a NEC–SX4 super computing platform:

	Standard grids	Refined grids
Blade grids	$4 \times (65 \times 47 \times 18)$	$4 \times (129 \times 83 \times 31)$
Background grids	$85 \times 51 \times 49$	$193 \times 86 \times 90$
Number of grid points	432 375	2 821 488
Number of CPUs	8	8
Memory usage	270MB	1.2GB
Average MFLOPS per CPU	502	758
Overall MFLOPS	3793	5727
Average vector length	170	225
Time per revolution	19min	1h30min

Fortran has been used as programming language for both the aerodynamic and the structural program. While the dynamics code is a pure Fortran 90 program, time–critical portions of the aerodynamic solver have remained in their original FORTRAN 77 implementation due to substantial performance losses experienced with the present Fortran 90 compiler. The communication routines responsible for fluid structure coupling have been written in C.

Compiler directives contained in the program source files are used for vectorization and parallelization. The latter is achieved by domain decomposition. Fig. 3 depicts the speed–up measured for a well–balanced test case versus the ideal linear distribution.

A sophisticated three-dimensional visualization system is used for the interpretation of the complex rotor wake structures. The timewise and spatial evolution of the rotor wake can be animated in a virtual reality environment which enables the user to interact with and walk about the scene.

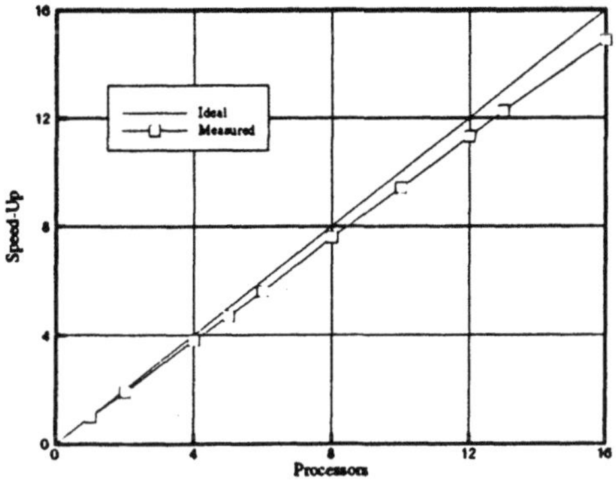

Figure3. Performance scalability in parallel mode

Figure4. Bottom side view of wake structure

References

1. ALTMIKUS, A., BUCHTALA, B., AND WAGNER, S.: **Rotary Wing Aeroelasticity in Forward Flight with Refined Structure Modelling.** Accepted for presentation at the 25th European Rotorcraft Forum, Rome, September 1999.

2. Buchtala, B., Wehr, D., and Wagner, S.: **Coupling of Aerodynamic and Dynamic Methods for the Calculation of Helicopter Rotors in Forward Flight.** 23rd European Rotorcraft Forum, pp. 5.1–5.12, Dresden, September 1997.
3. Buchtala, B., and Wagner, S.: **Rotary Wing Aeroelasticity in Forward Flight with Refined Wake Modelling.** 24th European Rotorcraft Forum, pp. 6.1–6.17, Marseilles, September 1998.
4. Hierholz, K.–H., and Wagner, S.: **Simulation of Fluid–Structure Interaction at the Helicopter Rotor.** Proceedings of the 21st ICAS Congress, Melbourne, Australia, September 13–18, 1998, ICAS–98–2.9.4, 1998.
5. Hierholz, K.–H.: **Ein numerisches Verfahren zur Simulation der Strömungs–Struktur–Interaktion am Hubschrauberrotor.** Dissertation, Institut für Aero– und Gasdynamik, Universität Stuttgart, Stuttgart, 1999.
6. Stangl, R., and Wagner, S.: **Euler Simulation of a Helicopter Configuration in Forward Flight using a Chimera Technique.** Paper presented at the AHS 52nd Annual Forum, Washington D. C., June 4–6, 1996, American Helicopter Society, Proceedings, Vol. I, pp. 453–462, 1996.
7. Wagner, S.: **Strömungsphänomene am Drehflügler und ihre Modellierung.** Hauptvortrag auf der GAMM–Tagung in Bremen, 6.–9. April 1998 (erscheint demnächst in der ZAMM).
8. Wehr, D., Zerle, L., and Wagner, S.: **Coupling Euler and Potential Methods for the Calculation of Helicopter Rotor Flows in Unsteady Forward Flight.** Paper presented at the 22nd European Rotorcraft Forum, Brighton, UK, September 16–19, 1996, Proceedings, Paper No. 59, 1996.
9. Wehr. D., Zerle, L., and Wagner, S.: **Improvement of Efficiency and Robustness of Unsteady Euler Codes for 3D Time Accurate Calculations.** In: ECARP–European Computational Aerodynamic Research Project: Validation of CFD Codes and Assessment of Turbulence Models (W. Haase, E. Chaput, E. Elsholz, M. A. Leschziner, U. R. Müller (Eds.)), Notes on Numerical Fluid Mechanics, Vol. 58, Chapter II.29, pp. 301–308, Friedr. Vieweg & Sohn, Braunschweig/Wiesbaden, 1997.
10. Wehr, D., Stangl, R., Uhl, B., and Wagner, S.: **Calculations of Rotors in Unsteady Forward Flight Using Potential and Euler Methods** In: ECARP–European Computational Aerodynamic Research Project: Validation of CFD Codes and Assessment of Turbulence Models (W. Haase, E. Chaput, E. Elsholz, M. A. Leschziner, U. R. Müller (Eds.)), Notes on Numerical Fluid Mechanics, Vol. 58, Chapter II.28, pp. 291–300, Friedr. Vieweg & Sohn, Braunschweig/Wiesbaden, 1997.
11. Wehr, D.: **Untersuchungen zum Wirbeltransport bei der Simulation der instationären Umströmung von Mehrblattrotoren mittels der Eulergleichungen.** Dissertation, Institut für Aero– und Gasdynamik, Universität Stuttgart, Stuttgart, 1999.

Flow with chemical reaction

Prof. Dr. Dietmar Kröner
Institut für Angewandte Mathematik, Universität Freiburg
Hermann-Herder-Str. 10, D-79104 Freiburg

The numerical simulation of flows with chemical reactions, in particular with turbulent combustions, is still one of the most difficult challenge in scientific computing. Because of the limitation of CPU time and memory resources the resolution of all length and time scales for realistic applications is still impossible. But due to the extensive use of parallel computers the investigations in this area are growing up rapidly and the results become more and more profitable. While some years ago the main goal was the adaption of the software to the architecture of parallel computers and to develop new software for it, we have today more reliable results which are in good agreement with measurements and which sometimes can be used already for design and development. Most of the computations have been performed in 3D. Three of the following projects have been run on the CRAY T3E, three on the NEC, and 1 of them on both machines. Most of the groups have used MPI.

In the first project of Brueggemann the simulation of a supersonic hydrogen-air diffusion flame was studied. The basic numerical method was a Monte-Carlo-Method for the PDF transport equation with a variable and adaptive number of particles. The chemical part was modeled by 7 reactions with 6 species. The results are in good agreement with measurements.

In the second project of Brueggemann multigrid methods have been applied to supersonic flames with detailed chemistry (9 species) and a low Reynolds-Number turbulence model. In order to control the influence of turbulent fluctuations to the chemical source term, PDF models have been used. The necessary damping of the defect, restringated from the fine to the coarse grid, was controlled by a sensor, which indicates the regions with strong chemical activities. The multigrid method could reduce the whole CPU time by factors between three and six. These results have been used for the improvement and design of new injectors for combustion chambers.

Werner considered an instationary internal turbulent flow with combustion in an engine. A parallelized version of the StarCD code was used to solve the compressible unsteady Navier-Stokes equations together with the $k - \epsilon$ turbulence model on unstructured moving grids. For the reaction a rough model with two species have been used. The speedup for the stationary case with a non-moving grid was nearly perfect while those for moving (i.e. moving piston) grids are less promissing. Comparisons with light-sheet measurements showed a good agreement of experiment and simulation. Using parallel computers the simulations of engines could now be reduced to one week and can be used in the product development.

The project of Bundschuh et al. deals with the numerical prediction of load changes in a coal-fired utility boiler i.e. in a real industrial-scale combustion system in 3D. In this case the numerical code is based on the implicit SIMPLEC and PISO codes. It contains submodels describing fluid flow, turbulence, combustion and radiative heat transfer which are coupled in a non-linear way. The numerical prediction is in a good agreement with full scale temperature experimental data.

The direct numerical simulation of the turbulence in chemically reacting flows was studied in the project of Lange and Warnatz. The modelling of the chemical mechanism includes 15 species and 84 elementary reactions with a modified Arrhenius law. The investigations contain studying the influence of strain and curvature to the local flame structure and flame speed (occurence of isolated pockets of cold fuel). In the numerical code dynamic grid adaption is used. Therefore if the relative discrepancies between the load of the single processors are larger than a given tolerance value, a gridpoint redistribution is performed. At the end they got a speedup of 197 for 256 processors. Furthermore they present a systematic comparison of results obtained by MPI and PVM.

The re-entry problem in 2D and 3D has been considered in the project of Frühauf et al. Large gradients and strong shock, stiffness and complex thermochemical processes are the main ingredients of the problem. The mathematical model for the high-temperature nonequilibrium flow includes advanced gas-phase (Coupled Vibration-Chemistry-Vibration) and gas-surface interaction (detailed catalysis model) and it is solved by the URANUS code. This is a coupled, implicit algorithm with an ILU preconditioning. Five different species are taken into account. For up to a number of 512 processors they got a nearly linear scale-up on the CRAY T3E. Also a 3D Parallel-Multiblock version of the code has been implemented.

Reaction-diffusion systems in 3D are capable to support localized moving structures, which can be interpreted as so called quasi paricles. This has been investigated by Schenk et al. For certain parameters the propagation speed of these quasi-particles and their behaviour in scattering processes can be predicted. Several numerical scattering experiments have been carried out and annihilation, reflection and merging of particles could by shown.

All the projects indicate, that new parallel computers can provide more computational power and thus accelerate the calculation of numerical simulation dramatically and enables the investigation of a new class of complex problems in reacting flow.

Investigation of Chemistry-Turbulence Interactions Using DNS on the Cray T3E

Marc Lange and Jürgen Warnatz

Universität Heidelberg, Interdisziplinäres Zentrum für Wissenschaftliches Rechnen
Im Neuenheimer Feld 368, D-69120 Heidelberg, Germany
E-mail: marc.lange@iwr.uni-heidelberg.de

Abstract. We use direct numerical simulation (DNS) to study the influence of turbulence in chemically reacting flows. First a short description of the governing equations, their numerical solution, and performance results on the Cray T3E are given. Then two applications are presented: The temporal evolution of a turbulent premixed methane-air flame is simulated using a chemical mechanism including 15 species and 84 elementary reactions. Isolated pockets of cold fresh gas are formed propagating into the burnt side of the flame. Examples are given for the combined influence of curvature and preferential diffusion on the chemical composition in the flame front. A correlation analysis shows that it is possible to use the formyl-radical as an indicator for the reaction intensity. The second application is the DNS of autoignition processes in turbulent hydrogen-air mixing layers. A reduction of ignition-delay time compared to the laminar case is observed and a recent result is presented concerning the locations where the first ignition spots occur.

1 Introduction

Combustion processes are important for a broad range of applications like automotive engines, electrical power generation, and heating. In most applications the reactive system is turbulent and the reaction progress is influenced by turbulent fluctuations and mixing in the flow. The optimization of combustion processes, e.g. the minimization of pollutant formation, requires accurate numerical simulations. Better and more generally applicable models for turbulent combustion are needed to be able to perform such simulations. The coupling between the chemical kinetics and fluid dynamics constitutes one central problem in turbulent combustion modeling [1].

2 Direct Numerical Simulation

Chemically reacting flows can be described by a set of coupled partial differential equations expressing the conservation of total mass, chemical species masses, momentum and energy:

$$\frac{\partial \varrho}{\partial t} + \mathrm{div}(\varrho \boldsymbol{u}) = 0 \, , \tag{1}$$

$$\frac{\partial(\varrho Y_\alpha)}{\partial t} + \mathrm{div}(\varrho Y_\alpha \boldsymbol{u}) + \mathrm{div}\boldsymbol{j}_\alpha = M_\alpha \dot{\omega}_\alpha \ , \quad \text{where } \alpha = 1, \ldots, N_\mathrm{s} \ , \tag{2}$$

$$\frac{\partial(\varrho \boldsymbol{u})}{\partial t} + \mathrm{div}(\varrho \boldsymbol{u} \otimes \boldsymbol{u}) = -\mathrm{div}\mathsf{T} - \mathrm{grad}\, p \ , \tag{3}$$

$$\frac{\partial e_\mathrm{t}}{\partial t} + \mathrm{div}((e_\mathrm{t} + p)\boldsymbol{u}) = -\mathrm{div}(\mathsf{T}\, \boldsymbol{u}) - \mathrm{div}\, \boldsymbol{q} \ , \tag{4}$$

where ϱ is the density and $\boldsymbol{u}$ the velocity; Y_α, $\boldsymbol{j}_\alpha$ and M_α are the mass fraction, diffusion velocity and molar mass of the chemical species α and N_s is the number of species; T denotes the stress tensor and p the pressure; $\boldsymbol{q}$ is the heat flux and e_t is the total energy given by

$$e_\mathrm{t} = \varrho \left(\frac{\boldsymbol{u}^2}{2} + \sum_{\alpha=1}^{N_\mathrm{s}} h_\alpha Y_\alpha \right) - p \ , \tag{5}$$

where h_α is the specific enthalpy of the species α. The source term $\dot{\omega}_\alpha$ on the right hand side of (2) is the production rate of the chemical species α, which is given as the sum over the formation rate equations for all N_er elementary reactions,

$$\dot{\omega}_\alpha = \sum_{\lambda=1}^{N_\mathrm{er}} k_\lambda (\nu_{\alpha\lambda}^{(\mathrm{p})} - \nu_{\alpha\lambda}^{(\mathrm{r})}) \prod_{\alpha=1}^{N_\mathrm{s}} c_\alpha^{\nu_{\alpha\lambda}^{(\mathrm{r})}} \ , \tag{6}$$

where $\nu_{\alpha\lambda}^{(\mathrm{r})}$ and $\nu_{\alpha\lambda}^{(\mathrm{p})}$ denote the stoichiometric coefficients of reactants and products respectively, and c_α is the concentration of the species α. The rate coefficient k_λ of an elementary reaction is given by a modified Arrhenius law

$$k_\lambda = A_\lambda T^{b_\lambda} \exp\left(-\frac{E_{\mathrm{a}\lambda}}{RT} \right) \ . \tag{7}$$

Direct numerical simulations (DNS), i.e. the time-dependent solution of the above system of equations, can provide detailed information on the interaction of chemical reactions and turbulent transport. But the required spatial and temporal resolution in conjunction with the large number of unknowns leads to extremely high computational costs [2]. This makes the DNS of most technical configurations practically impossible. Nevertheless, DNS is a valuable tool which can provide insight into the fundamental mechanisms of turbulent combustion and thus aid in the development of improved turbulent combustion models.

3 Parallel Numerical Solution Technique

We have developed a code for the direct simulation of reactive flows on parallel computers with distributed memory using message-passing communication [3–5]. Detailed models are utilized for the computation of the thermodynamical properties, the viscosity and the molecular and thermal diffusion velocities. At the moment, we restrict ourselves to the simulation of two-dimensional flow fields. As turbulence is an inherently three-dimensional phenomenon, some of its aspects cannot be reproduced precisely with such simulations [6]. On the other hand, it has been shown that the use of reduced chemistry models often leads to large errors in the obtained solution [7]. We consider the inclusion of detailed models for the chemical kinetics and molecular transport to be at least as important for the investigation of turbulent flames as an extension to three dimensions.

The spatial discretization is performed using a finite-difference scheme with sixth-order central-derivatives, avoiding numerical dissipation and leading to very high accuracy. The integration in time is carried out using a fourth-order fully explicit Runge-Kutta method with adaptive timestep control. The parallelization strategy is based on a regular two-dimensional domain decomposition. More details on the employed numerical methods, the parallel implementation, and the scaling behaviour of the code can be found in [2].

Besides simulations of flames in turbulent flows, our code is used to investigate flames interacting with isolated vortices [8], which is a valuable test case revealing many features apparent in turbulent combustion [9]. Since an adaptive gridding technique is used these simulations, a static load-distribution would lead to a poor parallel performance. Therefore, a dynamic load-balancing algorithm has been implemented, which relies on the transfer of boundary points between neighbouring processors [5]. At regular intervals during the run, the computation time needed by each node to carry out an integration step is measured. These local times are then averaged along lines and columns and gathered on one node, where the global mean value is computed. If the relative discrepancies between the measured times are larger than a given tolerance value, a grid-point redistribution is performed. In this redistribution process the nodes currently needing more computing time than the average transfer grid points to their neighbour on the side with the smallest average load per processor. The number of exchanged grid points is proportional to the additional computing time needed in comparison with the average value. This process is carried out independently in the x- and y-direction. Other cases of interest, in which this dynamic load-balancing procedure leads to good results, are simulations using adaptive models for chemical reaction kinetics and computations performed on workstation-clusters.

4 Performance of the Parallel Code

It is well known that the parallel speedup for most applications strongly depends on the problem size [10]. Typically, one is interested in the scaling behaviour for a problem size which is also usual for production runs. But for many real-world applications the amount of memory available for a one processor reference computation does not allow such a performance measurement. An indication for the scaling behaviour of such applications can be given by a scaled speedup

$$S_p^{\mathrm{s}}(N) = p \cdot \frac{T_1(N/p)}{T_p(N)} , \tag{8}$$

where the computing time $T_p(N)$ for a problem of size N using p processors is compared with the computing time for a p times smaller problem using one processor. For a typical production run – the DNS of a turbulent hydrogen-air flame with 512×512 grid points and a chemical mechanism with 9 species and 37 reactions – using 256 processors of a Cray T3E-900 we get a speedup of $S_{256}^{\mathrm{s}} = 197$. The average rate of MFlop/s per PE is 85.1 . More details on the performance of the code can be found in [2].

These measurements have been performed with a version of the code, which uses PVM for the communication. Meanwhile, we have ported our code to MPI. As a first test for a comparison between both versions we performed simulations (decaying turbulence, 9 species, multicomponent diffusion, no reaction) on a 160×160 grid using 16, 32, and 64 processors with PVM, with MPI using standard send, and with MPI using non-blocking communication. An analysis of these measurements showed that the biggest part of the differences in overall time were due to interference with other interactive jobs running at the same time. Therefore, we performed another benchmark using 384×384 grid points on 256 processors ensuring that no interference by other jobs takes place. Figure 1 shows the cpu-time (per processor) used over the physical time in this simulation. Every $t_{\mathrm{sim}} = 100\,\mu\mathrm{s}$ output files including values of the conservation variables and some other quantities needed for post-processing are saved. In this step the values of the output variables from all subdomains are gathered on one processor, which performs the I/O. This results in message-sizes scaling with the number of grid-points in the subdomains, whereas during the rest of the temporal integration the message-sizes scale with the number of grid-points along the sub-domain boundaries. The cpu-time needed for this step by the PVM-version exceeds those of the MPI-versions by more than a factor 2. The differences between the PVM-

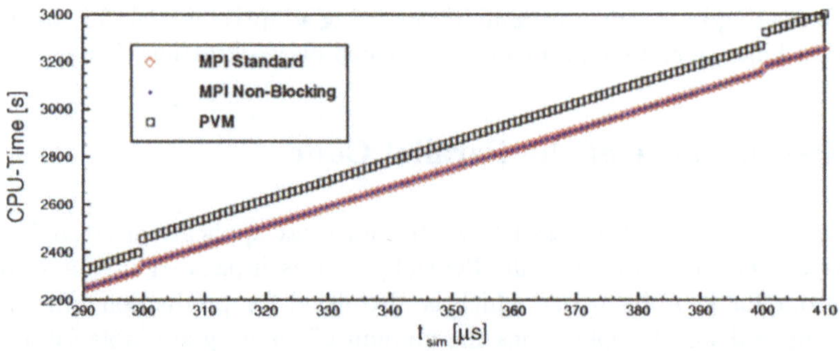

Fig. 1. CPU-time (per processor) using different communication routines

version and MPI-versions for the rest of the computation as well as between MPI standard and MPI non-blocking communication for the whole computation are less than one percent. This behaviour results from the fact, that on the Cray T3E MPI delivers a higher communication bandwidth than PVM, but has also a higher latency [11] and thus performs better if message-sizes increase. The total time required by the PVM-version for this benchmark is 5% longer than those of the MPI-versions.

5 Applications

5.1 Turbulent Premixed Methane Flame

During the last few years, DNS has been extensively used to study the interaction of flame fronts with turbulent flow fields, mainly for modeling purposes [9]. The main objectives of these investigations include studying the influence of strain and curvature on the local flame structure and flame speed. However, most of these numerical studies have been carried out using highly simplified models for molecular transport and chemical reactions. Since, for example, the response of flame temperature and upstream velocity requires the simultaneous presence of stretch and preferential diffusion [12], many important effects cannot be captured in such simulations. In addition to simulations of premixed [2,13] and non-premixed $H_2/O_2/N_2$ flames [13] interacting with decaying turbulence, we study premixed methane-air flames. The chemical reaction mechanism used includes 15 species and 84 elementary reactions [14].

The simulation is initialized by superimposing a flat laminar stoichiometric methane-air flame with a turbulent velocity field. Periodicity is assumed in the y-direction. The turbulent flow field is computed by inverse FFT from a von Kármán Pao spectrum

$$E(k) = A\frac{\left(\overline{u'^2}\right)^{5/2}}{\epsilon}\frac{(k/k_e)^4}{[1+(k/k_e)^2]^{17/6}}\cdot\exp\left(-\frac{3}{2}a(k/k_d)^{4/3}\right) \tag{9}$$

with randomly chosen phases. (In Fig. 5 the spatial distribution of turbulent kinetic energy in such a flow field can be seen.) After this initialization, the temporal evolution of the flame in the turbulent flow field is computed.

An example of such a computation is given in Fig. 2, where the temperature, the hydroxyl-radical mass fraction, and the formaldehyde mass fraction are shown after an interaction time t_i of one, two, and three large-eddy turnover times t_Λ. The integral length scale based Reynolds number $Re_\Lambda = \mathrm{r.m.s.}(u')\cdot\Lambda/\nu$ of the initial turbulence in this simulation is $Re_\Lambda = 185$ and the corresponding large eddy turnover time is $t_\Lambda = \Lambda/\mathrm{r.m.s.}(u') = 0.75\,\mathrm{ms}$. It can be seen how the flame front gets stretched and wrinkled by the interaction with turbulent eddies. Like in premixed hydrogen flames, the maximum OH concentration is located well behind the flame front and is increased

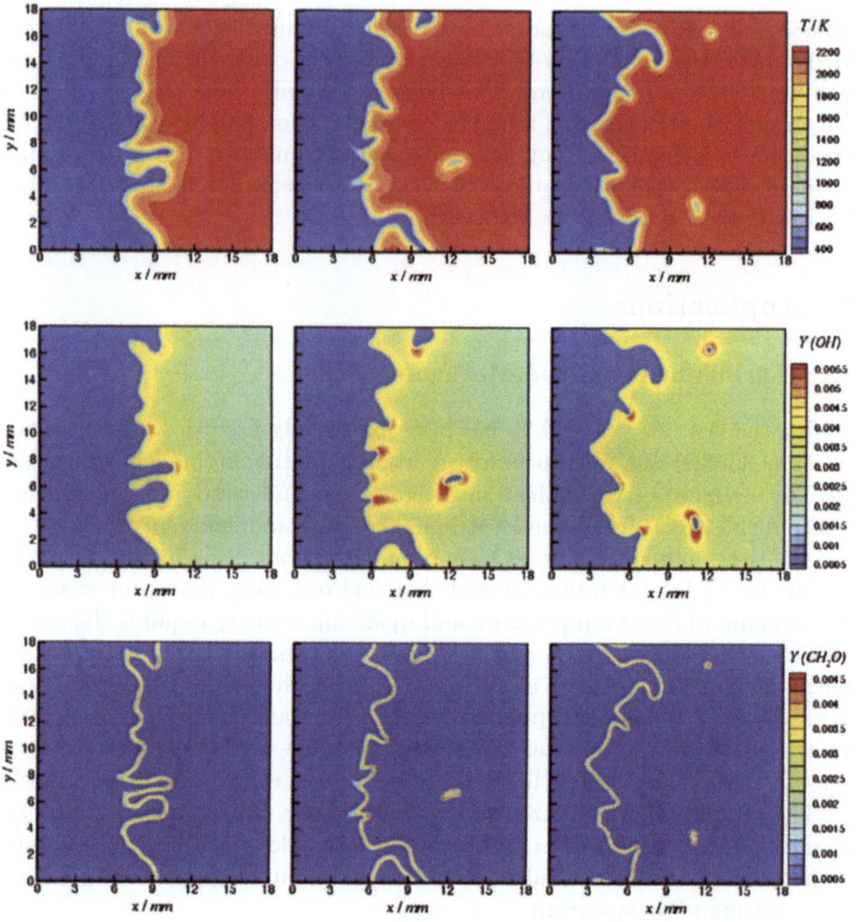

Fig. 2. Temperature, OH-, and CH_2O-mass-fraction in a turbulent premixed methane-air flame at $t_i = t_A$, $t_i = 2 \cdot t_A$, and $t_i = 3 \cdot t_A$ (from left to right).

in regions convex towards the burnt side of the flame. This effect is caused by preferential diffusion: The curvature of the flame focuses/defocuses diffusion stream lines, thus enlarging/reducing the concentration of fast diffusing species, especially of H and H_2, in the reaction zone. The changed composition alters the production to consumption ratio for other species like OH. Another interesting phenomenon which can be seen in Fig. 2 is the occurence of isolated pockets of cold fuel. The transient process of pocket formation involves mutual annihilation of parts of the flame front and is discussed and shown in more detail in [13,2] for the case of hydrogen flames. A one-point correlation analysis reveals that the heat release rate is almost perfectly cor-

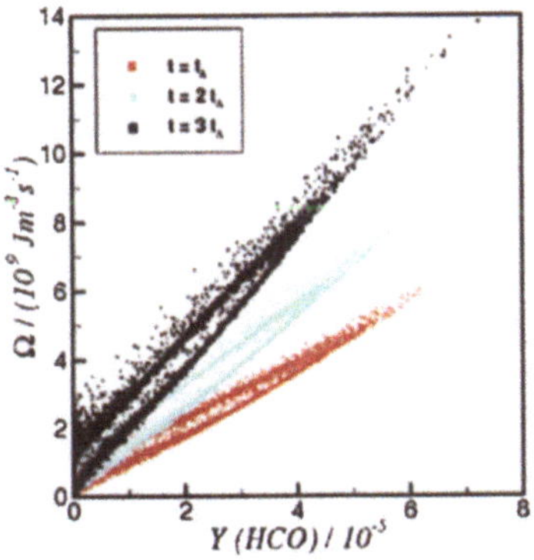

Fig. 3. Correlation of HCO mass fraction with heat release rate in a CH_4 flame interacting with turbulence at $t_i = t_\Lambda$, $t_i = 2 \cdot t_\Lambda$, and $t_i = 3 \cdot t_\Lambda$

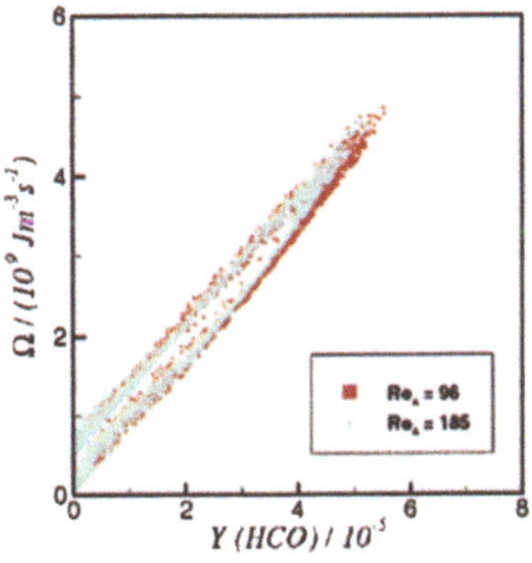

Fig. 4. Correlation of HCO mass fraction with heat release rate in two CH_4 flames interacting with turbulence fields of different intensity

related with the formyl radical concentration, as can be seen from Fig. 3 in which this correlation is shown for different interaction times. Figure 4 shows the same correlation for two simulations with different turbulence intensities. Moreover the correspondence of both simulations confirms that t_Λ serves well as a characteristic time. The result that HCO could be used as a tracer for the reaction intensity in turbulent methane flames is especially important, because high resolution measurement techniques for HCO already exist, whereas it is still common to use OH which is by far less suited for this purpose.

Computational Requirements This DNS has been performed on the Cray T3E-900 at the HighPerformance Computing Center Stuttgart (HLRS). The computing time using 256 nodes (including several restarts, up to $t = 3.3t_\Lambda$) was more than 61 hours, corresponding to an overall time of more than 21 CPU-months.

5.2 Autoignition in a Turbulent Mixing Layer

Many studies of turbulent combustion processes have been carried out considering one of the two limiting regimes corresponding to either premixed or non-premixed cases. In combustion systems like Diesel engines or aircraft jet turbines, in which fuel autoignites after being released into a turbulent oxidant of elevated temperature, chemical species can be mixed prior to reaction. Autoignition and partially premixed combustion play an important role in such configurations. Simulations of an adiabatic compression showed the

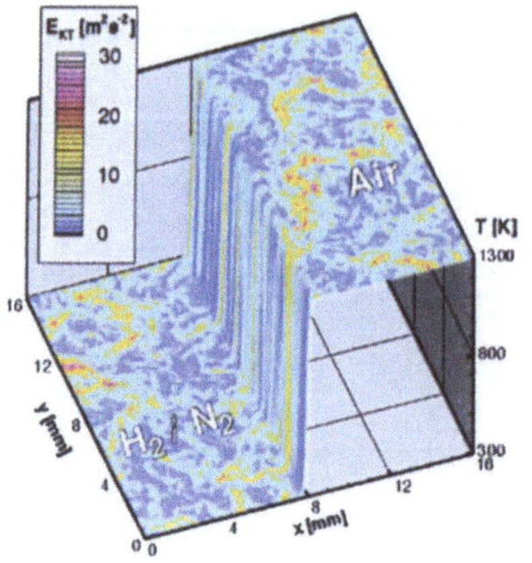

Fig. 5. Configuration for the simulation of turbulent mixing of cold fuel with hot oxydizer

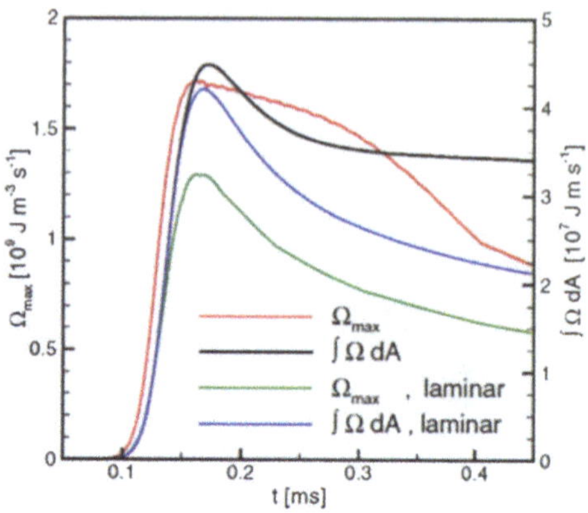

Fig. 6. Comparison of temporal evolution of heat release in an autoigniting turbulent mixing layer with the laminar case

role of edge-flames in the autoignition of such non-premixed turbulent mixtures [15]. Another interesting two-dimensional DNS-study focussed on the dependence of temporal appearance and spatial distribution of ignition spots on parameters describing the turbulent flow-field [16]. However, these studies were performed using simple one-step chemistry. By making efficient use of the computational power provided by the Cray T3E we are able to perform DNS of such a configuration using detailed models for reaction kinetics and molecular diffusion including the Soret effect. The chemical mechanism for the $H_2/O_2/N_2$ system uses 9 species and 37 elementary reactions [17].

Figure 5 shows the configuration for these simulations. A cold fuel stream consisting of 10% H_2 and 90% N_2 (mole fractions) and an air stream with an initial temperature of $T_{0,\mathrm{air}} = 1300\,\mathrm{K}$ are superimposed with a turbulence field with $Re_\Lambda = 177$. Figure 6 shows the temporal evolution of maximum heat release and heat release integrated over the computational domain for this turbulent mixing layer in comparison with the laminar case. The occurring ignition-delay is characteristic for chemical reactions which are governed by a chain-branching mechanism. During this first phase of autoignition, important radicals are produced at an exponential rate, in this case O, H, OH, and HO_2. At approximately the same time as (in contrast to the other radicals) the HO_2-concentration increase starts to slow down, the maximum heat release starts to strongly increase ($t \approx 0.11$ms). At this time, a strong spatial correlation between heat release and HO_2 mass fraction is found in the turbulent as well as in the laminar case. The main rate-limiting process is the chain-branching reaction

$$\mathrm{H} + \mathrm{O_2} \longrightarrow \mathrm{OH} + \mathrm{O}$$

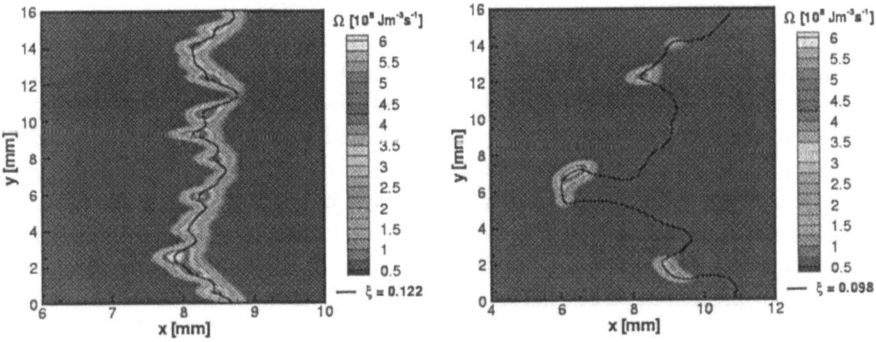

Fig. 7. Ignition spots in turbulent mixing layers with different initial temperature of the oxydizer stream (left: $T_{0,\mathrm{air}} = 1300\,\mathrm{K}$, right: $T_{0,\mathrm{air}} = 1100\,\mathrm{K}$) at the time of the maximum increase of maximum heat release t_{ign} (left: $t_{\mathrm{ign}} = 128\,\mu\mathrm{s}$, right: $t_{\mathrm{ign}} = 473\,\mu\mathrm{s}$

which competes with the terminating reaction

$$\mathrm{H} + \mathrm{O_2} + \mathrm{M} \longrightarrow \mathrm{HO_2} + \mathrm{M}\ .$$

By defining the ignition-delay time t_{ign} as the time up to the maximum increase of maximum heat release, one finds a reduction of t_{ign} of 6.1% for the turbulent mixing flow compared to the corresponding laminar case. A reduced ignition-delay time due to the influence of turbulence has also been found in experiments [18], whereas some studies based on laminar flow analysis suggested that the high strain rates caused by turbulence would delay autoignition [19,20].

An investigation of autoignition in a laminar one-dimensional mixing layer using high-energy asymptotics showed the existence of an unique most reactive mixture fraction depending only on the order of reaction, the activation energy, and the temperatures of the fuel and air streams [21]. However, the spatial location of autoignition sites in turbulent flows is not yet fully understood. Figure 7 shows the spatial distribution of heat release at $t = t_{\mathrm{ign}} = 128\,\mu\mathrm{s}$ in the described turbulent mixing layer (left) and for another simulation with $T_{0,\mathrm{air}} = 1100\,\mathrm{K}$ and $Re_{\Lambda}(t = 0) = 252$ at $t = t_{\mathrm{ign}} = 473\,\mu\mathrm{s}$ (right). The shown mixture-fraction isolines confirm that all ignition spots occur near the most reactive mixture fraction. It can also clearly be seen that ignition occurs first at locations where the mixing layer is curved convex towards the fuel side. This is caused by the focusing of fast diffusing hydrogen into the hot fuel-lean side of the mixing layer. It would not have been possible to identify this effect using simplified models for chemical kinetics or molecular transport. The inclusion of this result in ignition models should lead to more accurate simulations of autoignition in technical configurations.

6 Conclusions

A parallel code for the direct numerical simulation of turbulent reactive flows has been developed, which employs detailed models for chemical reaction kinetics and molecular transport processes in order to allow accurate simulations of the interaction between turbulence and chemical reactions. This code can be used on the Cray T3E with either PVM or MPI as message-passing library achieving high parallel speedups and performance rates. Results of the DNS of a premixed methane-air flame interacting with a turbulent flow field have been presented. The chemical composition in the wrinkled flame front is considerably altered by preferential diffusion. Pockets of cold fresh gas are formed, propagating into the burnt gas. A correlation analysis showed that HCO is ideally suited as a tracer for the reaction intensity in experiments. The influence of turbulence on the autoignition of hydrogen-air mixing layers has been studied. The ignition-delay time is reduced by the turbulence. It was shown that ignition spots first occur at those locations with the most reactive mixture fraction which are convex towards the fuel side. This effect is caused by preferential diffusion of hydrogen which is the deficient species for the most important chain branching reactions. It has been shown that the inclusion of detailed models for chemical reactions and diffusion is necessary for an accurate prediction of turbulent reacting flows. The consideration of findings from DNS studies like the presented ones in turbulent combustion models will lead to more accurate simulations of technical combustion processes.

7 Acknowledgement

We would like to thank the John von Neumann Institute for Computing at Jülich (NIC) and especially at the High Performance Computing Center at Stuttgart (HLRS) for granting us access to their Cray T3E systems. The presented DNS would not have been possible without this support.

References

1. Warnatz, J., Maas, U., Dibble, R.W.: Combustion. 2nd ed., Springer-Verlag, Berlin, Heidelberg, New York (1999)
2. Lange, M., Warnatz, J.: Detailed Simulations of Turbulent Flames Using Parallel Supercomputers. High Performance Computing in Science and Engineering '98 (Krause, E., Jäger, W., Eds.), Springer, Berlin (1999), 343–352
3. Thévenin, D., Behrendt, F., Maas, U., Przywara, B., Warnatz, J.: Development of a Parallel Direct Simulation Code to Investigate Reactive Flows. Computers and Fluids. Vol. **25**, 5 (1996) 485–496
4. Lange, M. , Warnatz, J.: Direct Numerical Simulation of Chemically Reacting Flows. Supercomputer 1997 (Meuer, H.-W., Ed.), K.G.Saur-Verlag, München (1997) 145–156

5. Lange, M., Thévenin, D., Riedel, U. , Warnatz, J.: Direct Numerical Simulation of Turbulent Reactive Flows Using Massively Parallel Computers. Parallel Computing: Fundamentals, Applications and New Directions (D'Hollander, E., Joubert, G., Peters, F. , Trottenberg, U., Eds.), Elsevier Science, Amsterdam (1998)
6. Lesieur, M.: Turbulence in Fluids. Kluwer Academic, Dordrecht (1997)
7. Sloane, T. M., Ronney, P. D.: Comparison of Ignition Phenomena Modelled with Detailed and Simple Chemistry. Proc. Fourth International Conference on Numerical Combustion. SIAM, Saint Petersburg, Florida (1991)
8. Thévenin, D., Renard, P.H., Rolon, J.C., Candel, S.: Extinction Processes During a Non-Premixed Flame-Vortex Interaction. Proc. 27th Symp. (Int.) on Combustion, The Combustion Institute, Pittsburgh (1998) 719–726
9. Poinsot, T.: Using Direct Numerical Simulations to Understand Premixed Turbulent Combustion. Proc. 26th Symp. (Int.) on Combustion, The Combustion Institute, Pittsburgh (1996) 219–232
10. Flatt, H.P., Kennedy, K.: Performance of Parallel Processors. Parallel Computing 12 (1989) 1–20
11. Anderson, E., Brooks, J., Grassl, C. Scott, S.: Performance of the Cray T3E Multiprocessor. Proc. of SC97: High Performance Networking & Computing, http://www.supercomp.org/sc97/ (1997)
12. Chung, S.H., Law, C.K.: An Integral Analysis of the Structure and Propagation of Stretched Premixed Flames. Combustion and Flame 72 (1988) 325–336
13. Lange, M., Riedel, U., Warnatz, J.: Parallel DNS of Turbulent Flames with Detailed Reaction Schemes. AIAA Paper 98-2979 (1998)
14. Smooke, M.D., Mitchell, R.E., Keyes, D.E.: Numerical Simulation of Two-Dimensional Axisymmetric Laminar Diffusion Flames. Combustion Science and Technology, 67 (1989) 85–122
15. Domingo, P., Vervisch, L.: Triple Flames and Partially Premixed Combustion in Autoignition of Non-Premixed Turbulent Mixtures. Proc. 26th Symp. (Int.) on Combustion, The Combustion Institute, Pittsburgh (1996) 233–240
16. Mastorakos, E., Baritaud, T. and Poinsot, T.: Numerical Simulations of Autoignition in Turbulent Mixing Flows. Combustion and Flame 109 (1997) 198–223
17. Maas, U., Warnatz, J.: Ignition Processes in Hydrogen-Oxygen Mixtures. Combustion and Flame 74 (1988) 53–69
18. Mizutani, Y., Kazuyoshi, N.: Effects of Turbulent Mixing on Spray Ignition. Proc. 23rd Symp. (Int.) on Combustion, The Combustion Institute, Pittsburgh (1990) 1455–1460
19. Thévenin, D., Candel, S.: Effect of Variable Strain on the Dynamics of Diffusion Flame Ignition. Combustion Science and Technology 91 (1993) 73–94
20. Liñan, A., Williams, F.A.: Ignition in an Unsteady Mixing Layer Subject to Strain and Variable Pressure. Combustion and Flame 95 (1993) 31–46
21. Liñan, A., Crespo, A.: An Asymptotic Analysis of Unsteady Diffusion Flames for Large Activation Energies. Combustion Science and Technology 14 (1976) 95–117

Multigrid Convergence Acceleration for Non-Reactive and Reactive Flows

Peter Gerlinger and Dieter Brüggemann

LTTT · Lehrstuhl für Technische Thermodynamik und Transportprozesse, Universität Bayreuth, Universitätsstr. 30, 95440 Bayreuth, Germany

Abstract. The numerical simulation of chemically reacting flows requires extremely long CPU times if finite-rate chemistry is employed. Multigrid methods belong to the most promising methods for convergence acceleration and are widely used for non-reactive subsonic and transonic flows. This paper presents the application of an implicit multigrid algorithm to non-reactive and reactive turbulent high speed flows. Special care is required for the treatment of shock waves as well as chemical and turbulent source terms to enable convergence. CPU time reductions by a factor of at least 4 are presented for a non-reactive flow with streamwise air injection and a lifted methane-air flame. All calculatione are performed on the NEC SX-4.

1 Introduction

With increasing computer power the use of more complicated turbulence closures and larger reaction mechanisms came into reach of engineering applications. Additionally more efficient numerical schemes for time integration are available to solve the fluid and chemistry equations. However, the simulation of supersonic combustion chambers still is a demanding task. Due to extremely high Reynolds numbers two equation turbulence models offer a good compromise between effort and accuracy. A grid spacing with y^+-values smaller than 1 is required for most low-Reynolds number turbulence closures. This results in grids with extremely high cell aspect ratios (up to 10000) and causes severe time step restrictions. theoretically, multigrid methods offer the possibility to obtain grid independent convergence rates [1]. Excellent results are already achieved for transonic airfoil calculations [2]. However special care is necessary if supersonic flows with strong shock waves have to be simulated [3–6] or if the governing equations include strong non-linear source terms. This is the case if low-Reynolds number turbulence models are employed [6] or if combustion is modeled by finite-rate chemistry [7,8]. However, especially these cases require some kind of convergence acceleration because of their extremely long CPU times.

2 Governing Equations and Numerical Method

The flows considered are described by the full compressible Navier-Stokes, turbulence and species transport equations. As to account for turbulence chemistry interaction an assumed pdf (probability density function) approach is used. This requires the solution of two additional conservation equations for the variance of energy and the sum of species mass fraction variances. For two-dimensional simulations the set of averaged equations is given by

$$\frac{\partial \mathbf{Q}}{\partial t} + \frac{\partial(\mathbf{F} - \mathbf{F}_\nu)}{\partial x} + \frac{\partial(\mathbf{G} - \mathbf{G}_\nu)}{\partial y} = \mathbf{S} \tag{1}$$

where the conservative variable vector is

$$\mathbf{Q} = \left[\bar{\rho}, \bar{\rho}\widetilde{u}, \bar{\rho}\widetilde{v}, \bar{\rho}\widetilde{E}, \bar{\rho}q, \bar{\rho}\omega, \bar{\rho}\sigma_e, \bar{\rho}\sigma_Y, \bar{\rho}\widetilde{Y}_i\right]^T, \quad i = 1, 2, ..., N_k - 1, \tag{2}$$

$\mathbf{F}$ and $\mathbf{G}$ are inviscid, $\mathbf{F}_\nu$ and $\mathbf{G}_\nu$ are viscous fluxes in x- and y-direction, respectively. The source vector $\mathbf{S}$ results from turbulence and chemistry. The variables in eq. (2) are density $\bar{\rho}$, velocity components $\widetilde{u}$ and $\widetilde{v}$, the total specific energy $\widetilde{E}$, the turbulence variables $q = \sqrt{k}$ (k = turbulent kinetic energy) and $\omega = \epsilon/k$ (ϵ = dissipation rate of k), the variance of energy, σ_e, the species mass fractions, $\widetilde{Y}_i$, and the sum of their variances, σ_Y. N_k is the number of different species. The simulation of methane combustion involves a 13-species (N_2, O_2, H_2, H_2O, OH, O, H, CH_3, CH_4, CH_2O, HCO, CO, and CO_2), 20-step reaction mechanism [9]. For turbulence closure a two-equation low-Reynolds-number q-ω turbulence model [10,6] is employed. Finally, the source vector appearing in eq. (1) is given by

$$\mathbf{S} = \left[0, 0, 0, 0, S_q, S_\omega, S_{\sigma_e}, S_{\sigma_Y}, S_i\right]^T, \quad i = 1, 2, ..., N_k - 1, \tag{3}$$

where S_q and S_ω are source terms of the q-ω model [10], S_{σ_e} and S_{σ_Y} are source terms within the variance conservation equations, and S_i are source terms resulting from chemistry.

Numerical time integration is performed using an implicit LU-SGS finite-volume algorithm [11,12]. Chemistry is treated in an implicit fashion and fully coupled with the fluid motion.

3 Assumed PDF Closure

A major problem in simulating turbulent reacting flows by use of averaged equations is the treatment of mean reaction rates. The instantaneous production rate of species i is given by

$$S_i = M_i \sum_{r=1}^{N_r} \left[\left(\nu_{i,r}^{''} - \nu_{i,r}^{'}\right) \left(k_{f_r} \prod_{l=1}^{N_k} c_l^{\nu_{l,r}^{'}} - k_{b_r} \prod_{l=1}^{N_k} c_l^{\nu_{l,r}^{''}} \right) \right], \tag{4}$$

where k_{f_r} and k_{b_r} are forward and backward reaction rates of reaction r and c_l are species concentrations. After Reynolds decomposition a large number of higher order correlations appear which can not be modeled. This is usually known as the chemical closure problem. Assumed probability density functions are one possibility to overcome this difficulty. In this approach the shape of the pdf may not develop freely but has a prespecified form. For temperature a Gaussian distribution is chosen [13]:

$$P_T(\hat{T}) = \frac{1}{\sqrt{2\pi\sigma_T}}\, exp\left[-\frac{(\hat{T}-\widetilde{T})^2}{2\,\sigma_T}\right], \qquad \sigma_T = \widetilde{T''^2}\,. \tag{5}$$

Due to the limited validity of the Arrhenius fuction in terms of temperature a clipped Gaussian distribution is usually employed [15]. A presumption for a multi-variate composition pdf is the possibility to obtain an analytical solution for the averaged production rate given in equation (9). Such a β-pdf has been proposed by Girimaji [14]

$$P_Y(\hat{Y}_1, \hat{Y}_2, ..., \hat{Y}_{N_k}) = \frac{\Gamma\left(\sum_{m=1}^{N_k}\beta_m\right)}{\prod_{m=1}^{N_k}\Gamma(\beta_m)}\left[\delta\left(1-\sum_{m=1}^{N_k}\hat{Y}_m\right)\prod_{m=1}^{N_k}\hat{Y}_m^{\beta_m-1}\right] \tag{6}$$

where

$$\beta_m = \widetilde{Y}_m\, B, \qquad B = \left[\frac{\sum_{m=1}^{N_k}\widetilde{Y}_m\left(1-\widetilde{Y}_m\right)}{\sigma_Y} - 1\right], \qquad \sigma_Y = \sum_{m=1}^{N_k}\widetilde{Y''^2_m}\,. \tag{7}$$

Both pdf's are completely defined by the knowledge of their first ($\widetilde{T}$ and $\widetilde{Y}_i$) and second (σ_T and σ_Y) moments. One of the basic assumptions of the used assumed pdf approach is statistical independence of temperature, gas composition, and density [13]. Thus the pdf P can be written as the product of the temperature and composition pdf's

$$P(\hat{T}, \hat{c}_1, ..., \hat{c}_{N_k}) = P_T(\hat{T})\; P_Y(\hat{Y}_1, ..., \hat{Y}_{N_k})\; \delta(\hat{\rho}-\overline{\rho})\,, \tag{8}$$

and δ is the δ-function. If the shapes of both pdf's are known, the averaged chemical production term may be calculated by integrating over all realizable values of the sample space variables $\hat{T}$ and $\hat{c}_i$:

$$\overline{S}_i = \int S_i(\hat{T}, \hat{c}_1, ..., \hat{c}_{N_k})\; P(\hat{T}, \hat{c}_1, ..., \hat{c}_{N_k})\; d\hat{T} d\hat{c}_1 ... d\hat{c}_{N_k}\,. \tag{9}$$

The second moments are obtained from additional conservation equations for the energy variance σ_e (from which the temperature variance is recovered) and species mass fraction variance σ_Y [13]. The complete analytically derived source Jacobian matrix is given in Ref. 15. For high speed flows the strongest influence of the pdf closure was found to be a strong reduction in ignition delay for lifted flames.

4 Multigrid Convergence Acceleration

The standard Full Approximation Storage (FAS) V-cycle multigrid method for non-linear problems is employed in a version for implicit approximately factored schemes [11,6]. The calculation may be started with or without a nested iteration. In case of reactive flows the nested iteration approach has the advantage of a much better flow field initialization on the finest grid. A good initial solution may be essential for the multigrid method to work in case of combustion. Convergence acceleration of multigrid methods is based on the fact that low frequency error components are damped more efficiently on coarser grid levels. This is in contrast to source terms that describe local physical phenomena. Problems arise if source terms comprise strongly non-linear contributions. If these terms are recalculated at all grid levels the connection between coarse and fine grid production terms gets lost, leading to divergence. There are two kinds of source terms that my cause problems for the multigrid methods: purly local non-linear terms and production terms, depending on spatial derivatives. Most turbulence closures incude terms of the latter kind. The main production term of the employed q-ω turbulence closure is given by

$$S = \left[2\frac{\partial u}{\partial x} - \frac{2}{3}\left(\frac{\partial u}{\partial x} + \frac{\partial v}{\partial y}\right)\right]\frac{\partial u}{\partial x} + \left(\frac{\partial u}{\partial y} + \frac{\partial v}{\partial x}\right)^2 + \left[2\frac{\partial v}{\partial y} - \frac{2}{3}\left(\frac{\partial u}{\partial x} + \frac{\partial v}{\partial y}\right)\right]\frac{\partial v}{\partial y} \,. \tag{10}$$

It is obvious that there will be strong differences in corresponding terms calculated at different grid levels that may cause divergence. A simple and very efficient method used for a great variety of different flow problems [6,8,16] is a freezing of strongly non-linear parts on coarse grid levels.
Shock waves are treated by damping the restricted residual error in their vicinity [5,6]. In addition a simple form of upwind prolongation may improve robustness of the multigrid algorithm in some cases. For details concerning the multigrid technique employed see Ref. 15 and 16.

4.1 Modifications for Reactive Flows

The greatest problem for multigrid solutions of turbulent reactive flows is the connection between strongly non-linear fine grid source terms in the species conservation equations and their representation on coarse grids. In case of finite-rate chemistry problems mainly arise due to the extreme sensitivity of the chemical source terms (see equation (4)) to changes within the flow variables. The coupling is given by the species concentrations and the exponential dependence of forward and backward reaction rates on temperature. In Ref. 15 different treatments for coarse grid source terms and source term Jacobians are described. The method used in this paper is a local damping of the restricted residual error in regions of high chemical activity. This is quite similar to the treatment of shock waves with the multigrid method in supersonic and hypersonic flows. In many cases combustion is limited to small regions. The residual errors transferred from fine to coarse grids are reduced (damped) in such regions, thus allowing the full multigrid scheme to work outside combustion zones. Note that species residual errors are damped only, resulting in a decoupling between the species and continuity equations. However, stationary conditions are unaffected by such a treatment. In order to locate regions with high chemical intensity the following sensor is calculated

$$\gamma = \left(\frac{1}{N_k} \sum_{i=1}^{N_k} \frac{|S_i|}{|S_i|_{max} + \epsilon} \right)^{\alpha} \quad . \tag{11}$$

$|S_i|_{max}$ is the maximum absolute production rate of species i within the flowfield and ϵ is a small number to avoid division by zero. It was found to be advantageous that all N_k individual production rates contribute to this sensor which is limited by $0 \leq \gamma \leq 1$. An important condition for the sensor to work is its smooth distribution. An exponent α of 0.25 worked satisfactory for different test cases. A full coarsening multigrid method is employed where coarser grid levels are obtained by eliminating every second grid line. If the standard restriction operator ($\bar{I}_{k \mapsto k+1} \mathbf{R}^k = \sum_{l=1}^{4} \mathbf{R}_l^k$) is used, the coarse grid residuals are obtained by simply adding up four fine grid residuals. In case of combustion the following modification is introduced

$$\bar{I}_{k \mapsto k+1} \mathbf{R}^k = \sum_{l=1}^{4} \mathbf{R}_l^k \; max \; \left[0, \, 1 - B^k \gamma_l^k \right] \; , \tag{12}$$

where B^k is a grid level dependent constant. In addition to the described local damping the coarse grid time step is also reduced in regions of high chemical activity. Note that the damping is only performed for grid level three and higher. While the proposed multigrid method works very well for attached flames the simulation of detached flames is still unsatisfactory. As will be shown in the following test case detached flames may be simulated. However, there still is a strong dependence on grid size and test case making further improvements necessary.

5 Results and Discussion

Two test cases without and with combustion are considered to investigate the influence of the multigrid method on the convergence behaviour. Both cases are initialized with constant inflow values.

5.1 Axial Injection into a Supersonic Stream

The first test case corresponds to problems occurring in supersonic combustion chambers where hydrogen is injected into a hot air stream. This calculation is based on an experiment [17], with air as injectant. Calculated and measured profiles of different flow variables are compared at three x-locations

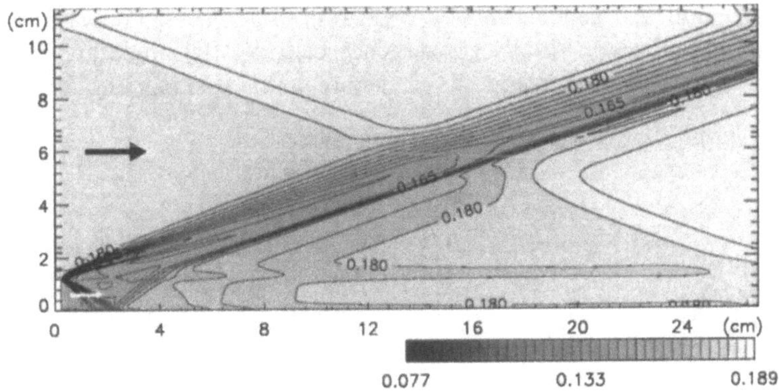

Fig. 1. Pressure contours ($\Delta p = 0.003$ bar) for a supersonic channel flow with air injection downstream of a backward facing step.

downstream of the injector. Fig. 1 shows calculated pressure contours for this test case. The upper and lower air flows are separated by a splitter plate of 0.5 *mm* thickness which is resolved by the computational grid. A 3-block grid with $160 \cdot 80$, $128 \cdot 32$ and $160 \cdot 112$ volumes is used. The inflow conditions for the mainflow and the injected jet are 100.7 and 190 K, 0.1792 and 0.1498 bar, and Mach numbers of 3 and 1.78, respectively. Calculated fully turbulent inlet profiles have been used as inflow conditions. The computational grid is strongly refined near walls and at the injector. The distance of the first cell centers to the wall is about 1×10^{-6} m, leading to cell aspect ratios of up to 450. All y^+-values for cell centers near solid walls are below 1. Fig. 2 shows a comparison with experimental data (circles are results from Ref. [17]) at different x locations. The y-coordinate is normalized with the injector height

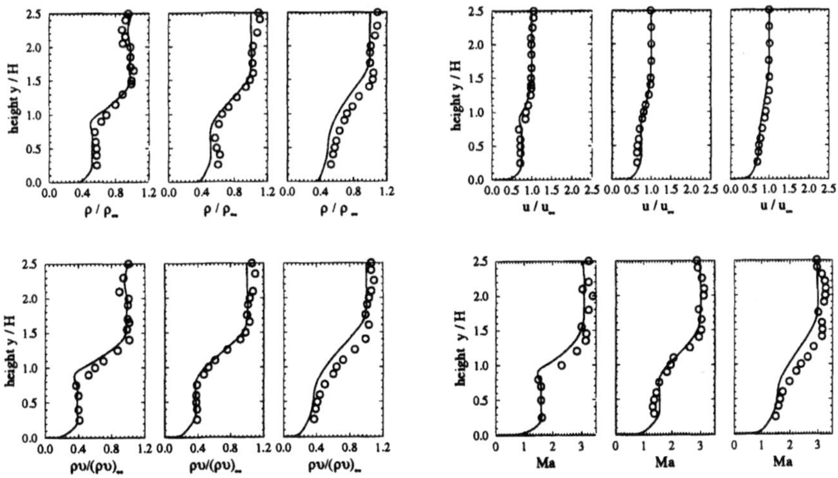

Fig. 2. Measured (o) and calculated (—) profiles of density, $\tilde{u}$-velocity, $\bar{\rho}\tilde{u}$ and Mach number at $x = 4.96$ cm, 12.7 cm, and 25.5 cm.

$H = 12.1$ mm. All profiles show a good overall agreement between calculation and experiment. Finally convergence histories for single and multigrid simulations are plotted in Fig. 3. V1 denotes a V-cycle starting at the finest

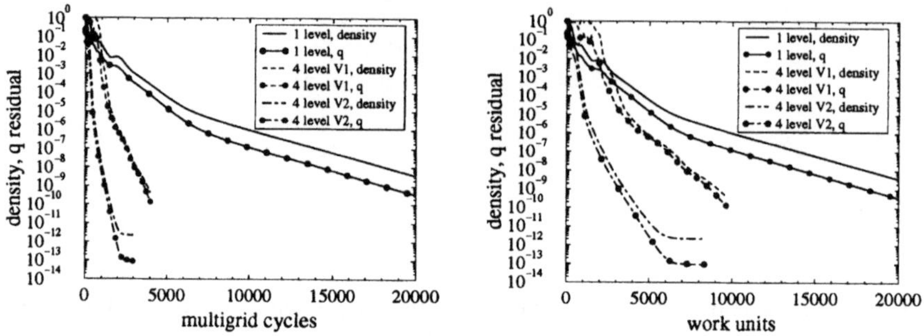

Fig. 3. Convergence histories of density and turbulence q residuals with and without the multigrid technique.

grid with one iteration performed at every grid level. V2 is a V-cycle with one iteration on the finest grid and two iterations at every coarser grid level. While in the left figure convergence histories are plotted versus the number of multigrid cycles the right figure shows the same parameters plotted over the number of work units. One work unit is the computational time necessary for one fine grid iteration. Given are the averaged normalized residuals of the density and turbulence (q).

5.2 Lifted Methane-Air Diffusion Flame

A planar supersonic shear flow over a splitter plate serves as a test case for the described multigrid modifications in case of combustion. Two fully turbulent incoming streams are separated by a splitter plate with a 4° angle. The tip of the splitter plate is located at $x = 0$ cm. While the upper flow is air, the lower

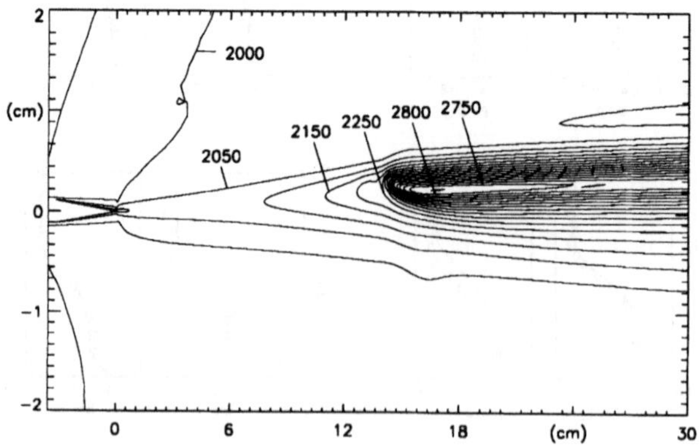

Fig. 4. Temperature contours ($\Delta T = 50$ K) for a lifted supersonic methane air diffusion flame.

flow is a nitrogen-methane mixture with a methane mass fraction of $Y_{CH_4} = 0.4$. Temperature and pressure of both streams are 2000 K and 1 bar, and the free stream velocities are 2690 m/s and 1800 m/s (air), respectively. A two block grid with $160 \cdot 64$ volumes for each block is employed and the cell aspect ratio is up to 2200. To describe methane combustion the mentioned 20 step, 13 species reaction mechanism is employed. No pdf closure is used in the present simulation. Results incuding pdf modeling may be found in Ref. [16]. However, effects of pdf modeling on the performance of the numerical algorithm are discussed in the next section. Fig. 4 shows temperature contours for the lifted flame. A significant ignition delay of about 15 cm is observed for the described inflow conditions. The convergence histories for this simulation are given in Fig. 5. Again a strong reduction in terms of multigrid cycles

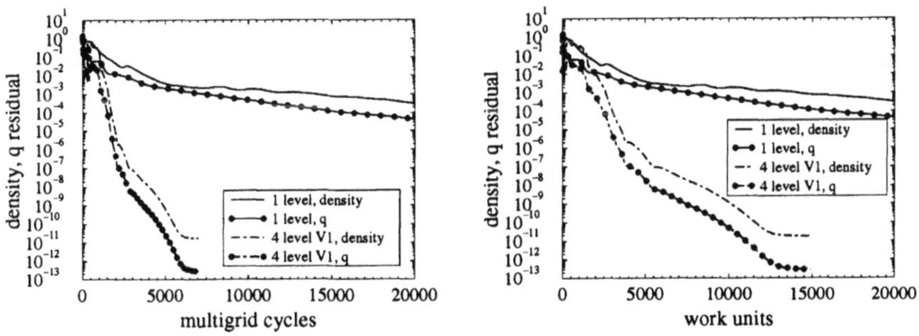

Fig. 5. Convergence histories of density and turbulence q residuals with and without the multigrid technique for a supersonic methane flame.

and work units is achieved by the proposed multigrid algorithm. However, it should be repeated, that the described damping of the transferred residual error does not enable convergence for every detached flame. While the method is quite stable for attached flames, solutions of detached flames are still test case dependent. The reason is the different ignition delay obtained at different grid levels. If the discrepancies between fine and coarse grids become to strong the solution diverges.

6 Performance

Due to the moderate memory requirements for 2D simulations all calculations are performed on a single processor of the NEC SX-4. Finite-rate chemistry is characterized by widely disperate time scales that require an implicit numerical scheme for time integration. While the right hand side (RHS) of the discretized set of equations is easy to vectorize (all subroutines reach between 500 and 1500 MFLOPS), problems usually occur for the implicit left hand side (LHS). The chosen LU-SGS algorithm has the advantage of being fully vectorizable in the implicit part, too. As to use this vectorizability, it

is necessary to perform the lower (L) and upper (U) sweep of the symmetric algorithm along diagonals of the computational domain. Due to the high numerical effort in the implicit part of the algorithm the code is programmed in this way. Without pdf closure LHS calculations require about 50% of the total CPU time. Great disadvantage of LHS calculations are short vector lengths, especially in the beginning and at the end of every sweep. The maximum vector length is defined by the smaller number of grid points of both coordinate directions. Full point block matrices have to be inverted and the Gauss algorithm employed vectorizes along diagonals. Because of the short vector length (36 in case of the analysed 3 block grid) only 260 MFLOPS are obtained in this subroutine. Other subroutines of the LHS reach about 350 MFLOPS. Thus the implicit part of the algorithm has the lowest performance due to short vector lengths caused by the algorithm. The situation is much better for three dimensional simulations where the LU-SGS algorithm sweeps along planes through the computational domain. Longer vector lengths are obtained and the increase and decrease in the beginning and at the end of every sweep is much more favourable.
The situation concerning CPU time is quite different with assumed pdf closure, for which the necessary CPU time is roughly doubled. Up to 60% is required for the calculation of averaged reaction rates and the corresponding souce Jacobians. Because these calculations are independent from neighbouring cells, they are vectorized over the total number of grid points of one computational block. 97% of the corresponding subroutine is vectorized, the average vector length is 250 and 590 MFLOPS are obtained. From the point of performance look-up tables used for averaged forward and backward reaction rates are a disadvantage as they require indirect addressing. Another time consuming part is the calculation of gas properties. For methane combustion with 13 different species about 10% of the CPU time is required and 930 MFLOPS are reached. While the LHS is the slow part of the algorithm most subroutines of the RHS have more than 1000 MFLOPS. The highest performance out of all subroutines is obtained for the calculation of species enthalpies by fourth order polynomials where 1825 MFLOPS are obtained. The avarage values of the total program is 536 MFLOPS, a vector operation ratio of 97% and an average vector length of 148. Only 300 iterations where performed with the analyser and data input still requires 5% of the necessary CPU time (at 0.5 MFLOPS). Thus the total performance may be expected to be better for more realistic calculations with several 1000 iterations.
The use of multigrid methods on vector computers is disadvantageous due to short vector lengths on coarse grid levels. For full coarsening algorithms the number of grid points decreases quadratically with increasing grid level. Thus practically reached convergence accelerations are below theoretically possible values. Nevertheless considerable speed-ups in CPU time are possible as demonstrated in the present paper.

References

1. Brandt, A.: Barriers to Achieving Textbook Multigrid Efficiency (TME) in CFD. NASA/CR-1998-207647 (1998)
2. Thomas, J. L., Bonhaus, D. L., Anderson, W. K., Rumsey, C. L., Biedron, R. T.: AN O(Nm^2) Plane Solver for the Compressible Navier-Stokes Equations. AIAA paper 99-0785 (1999)
3. Swanson, R. C., Turkel, E., White, J. A.: An Effective Multigrid Method for High Speed Flows. NASA ICASE Report 91-56 (1991)
4. Hemker, P. W., Koren, P. W.: Damped, Direction-Dependent Multigrid for Hypersonic Flow Computations. Appl. Numer. Math. **7** (1990) 309–328
5. Radespiel, R., Swanson, R. C.: Progress with Multigrid Schemes for Hypersonic Flow Problems. J. Comp. Phys. **116** (1995) 103–122
6. Gerlinger, P., Brüggemann, D.: Multigrid Convergence Acceleration for Turbulent Supersonic Flows. Int. J. Numer. Methods Fluids **24** (1997) 1019–1035
7. Edwards, J. R.: Advanced Implicit Algorithms for Finite Rate Hydrogen-Air Combustion Calculations. AIAA paper 96-3129 (1996)
8. Gerlinger, P., Stoll, P., Brüggemann, D.: An Implicit Multigrid Method for the Simulation of Chemically Reacting Flows. J. Comp. Phys. **146** (1998) 322–345
9. Westbrook, C. K.: An Analytical Study of the Shock Tube Ignition of Mixture of Methane and Ethane. Comb. Flame. **20** (1979) 5–17
10. Coakley, T. J., Huang, P. G.: Turbulence Modeling for High Speed Flows. AIAA paper 92-0436 (1992)
11. Jameson, A. , Yoon, S.: Lower-Upper Implicit Schemes with Multiple Grids for the Euler Equations. AIAA J. **25** (1987) 929–937
12. Shuen, J. S.: Upwind Differencing and LU Factorization for Chemical Non-Equilibrium Navier-Stokes Equations. J. Comp. Phys. **99** (1992) 233–250
13. Baurle, R. A., Alexopoulos, G. A., Hassan, H.A.: Modeling of Supersonic Combustion Using Assumed Probability Density Functions. J. Prop. Power **10** (1994) 777–786
14. Girimaji, S. S.: A Simple Recipe for Modeling Reaction-Rates in Flows with Turbulent Combustion. AIAA paper 91-1792 (1991)
15. Gerlinger, P., Möbus, H., Brüggemann, D.: An Implicit Numerical Scheme for Turbulent Combustion Using an Assumed PDF Approach. AIAA paper 99-3775 (1999)
16. Gerlinger, P., Brüggemann, D.: An Implicit Multigrid Scheme for the Compressible Navier-Stokes Equations with Low-Reynolds-Number Turbulence Closure J. Fluids Eng. **120** (1998) 257–262
17. Kwok, F. T., Andrew, P. L., Ng, W. F., Schetz, J. A.: Experimental Investigation of a Supersonic Shear Layer with Slot Injection of Helium AIAA J. **29** (1991) 1426–1435

Acknowledgments

We wish to thank the *Deutsche Forschungsgemeinschaft (DFG)* for financial support of this work within the Collaborative Research Center SFB 259 at the University of Stuttgart.

Quasi-Particles in a Three-Dimensional Three-Component Reaction-Diffusion System

C. P. Schenk, A. W. Liehr, M. Bode, and H.-G. Purwins

Institute of Applied Physics, Correnssstraße 2/4, 48149 Münster, Germany

Abstract. We investigate a reaction-diffusion system which consists of a set of three partial differential equations. Due to the reaction kinetics the system can be referred to as a 1-activator-2-inhibitor system. We show, that such systems are capable of supporting localized moving structures, so called quasi-particles. For certain parameters it is possible to predict the propagation speed of these solutions as well as their behaviour in scattering processes. In more general cases we have carried out simulations which reveal different scattering results depending on the parameters. We find annihilation, reflection and merging of particles.

1 Introduction

In 1952 Turing [1] introduced a two-component reaction-diffusion model for the morphogenesis of a growing organism in order to describe the formation of spatially inhomogeneous structures from an initially homogeneous state. Since then, reaction-diffusion equations have been widely used to describe pattern formation phenomena in very different biological [2,3], chemical [4,5] and physical [6–8] systems. Though a lot of phenomena, like pulse propagation in one-dimensional systems or the formation of stationary spatially inhomogeneous structures, can be explained by two-component reaction-diffusion equations, there are certain phenomena, like the destabilization of the homogeneous state by a travelling wave, that require three components. This work deals with moving, particle-like solutions of such a system described by the following set of equations

$$\dot{u} = D_u \Delta u - v - \kappa_3 w + f(u) + \kappa_1, \tag{1}$$

$$\tau \dot{v} = D_v \Delta v + u - v, \tag{2}$$

$$\theta \dot{w} = D_w \Delta w + u - w, \tag{3}$$

with $u = u(r,t)$, $v = v(r,t)$ and $w = w(r,t)$ and $r \in G \subset \mathbb{R}^n$ with $n = 2,3$ and Δ denoting the Laplacian. For analytical considerations it is useful to assume an infinite domain G. For numerical simulations periodic boundary conditions or Neumann boundary conditions on box-shaped domains G are used. The above equations can be regarded for example as a qualitative model for the description of pattern formation processes in a two-dimensional gas-discharge system [7], but similar phenomenological models can also arise in other systems [8]. The function $f(u)$ represents the only nonlinear term and is chosen such that near $(u, v, w) = (0, 0, 0)$ u can be referred to as an activator and that its nonlinear part prevents a blow-up of the solutions. The simple choice $f(u) = \lambda u - u^3$ fulfills these two conditions and will be used throughout this paper, though in principle different nonlinearities could be used. Following the structure of the reaction terms in Eqs. (1)-(3) the component u is called activator, because it acts as a source for v and w and at least in a certain amplitude range for itself, too. The components v and w are referred to as inhibitors. They suppress their own production as well as the production of the component u. Apparently, if the time constants τ and θ and the diffusion constants D_v and D_w are of the same magnitude, it is possible to omit one inhibitor equation. Thus we expect new features, as compared to a two-component model only if the inhibitors have significantly different parameters. In the following we will usually choose $\tau > \theta$, $\tau > 1$ and $D_w > D_u$. Corresponding to this choice we call v the slow inhibitor and w the fast or the diffusing inhibitor.

In order to solve the Eqs. (1)-(3) numerically a finite difference scheme is used. The domain $G = [0, L_x] \times [0, L_y] \times [0, L_z] \subset \mathbb{R}^3$ is discretised with a fixed length Δx. For the time discretisation we implemented the Crank-Nicolson scheme, due to advantages in stability and accuracy properties. The resulting system is solved iteratively using the successive-overrelaxation method which provides a reasonably fast convergence. For the task of parallelisation the domain G is divided into boxes of equal size and, as far as possible, cubic shape to reduce the ratio between boundary points and interior points. Every node on the parallel computer is assigned to one box and after each change of the components u, v and w the boundary points are updated using the Message Passing Interface (MPI). The computational effort strongly depends on the discretisation and on the parameters of the simulation. Calculations with a grid of about $35 \times 35 \times 35$ have been used to check existence, stability and speed of single solutions. Typically we have used 64 nodes for about one hour per node on the Cray T3E-900 in such cases. Simulations regarding scattering processes had to be performed on rather large domains with up to $120 \times 160 \times 120$ grid-points and with parameters leading to partially separated time-scales, which slowed down our algorithm considerably. Thus the most extensive calculations have been carried out on 256 nodes for about 8 hours per node.

2 Single Quasi-Particles

In reaction-diffusion systems a variety of different patterns can be observed. A very interesting phenomenon is the formation of stable localized current density filaments, e.g. in gas-discharge systems [9,10]. These quasi-particles can move through the system and interact with the boundaries or with each other. While stationary filaments can be described by two-component reaction-diffusion models, it was not possible to find moving localized spots in these systems until now. In two-component systems with one activator and one inhibitor there are two possible cases: If the inhibitor is fast and diffusing, stationary structures can organize. If the inhibitor is slow, dynamical structures, spirals or scroll-waves, appear. To keep a structure moving and localized in more than one dimension, two inhibitors are at least helpful [11]. Regarding the behaviour of moving quasi-particles in two-dimensional systems different scattering phenomena have already been observed and described analytically. The interaction with each other, with inhomogeneities and the system boundaries has been studied. The following work therefore deals with different phenomena in three-dimensional space.

Figure 1 a) shows a three-dimensional solution moving from the left to the right. The iso-surfaces of activator u and slow inhibitor v are drawn red and green, respectively. The direction of motion of such a solution is determined by the direction of the shift of the slow inhibitor with respect to the activator. In Fig. 1 b) the profile of components u, v and w in the direction of motion of this three-dimensional solution is plotted. The slow inhibitor is shifted with respect to the activator and gives the impression to push the activator peak. The fast inhibitor surrounds this peak and controls the size of the spot. If the spot grows or shrinks, the fast inhibitor immediately reacts and stabilizes the size of the activator distribution. For a single spot the task of controlling the size can be performed by an integral feedback as well [12–14]. However, for multiple spots this approach fails, because antisymmetric fluctuations are not noticed by such a mechanism.

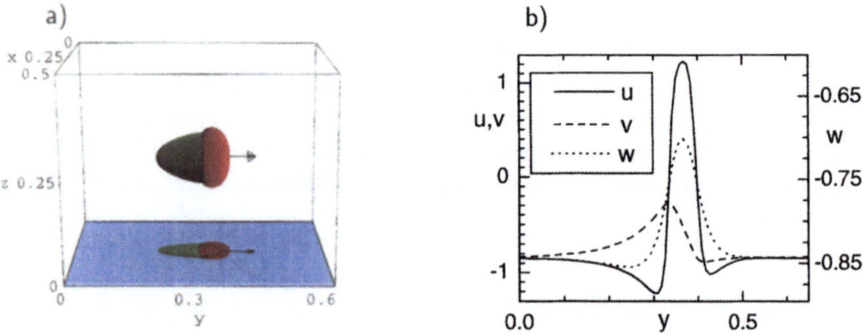

Fig. 1. a) Iso-surfaces of activator u (red) and slow inhibitor v (green) of a moving spot solution. b) The components u, v and w at $x = 0.25$ and $z = 0.25$ as a function of y of the spot of Fig. a) show the different behaviour of the two inhibitors. Parameters are $G = [0, 0.5] \times [0, 0.65] \times [0, 0.5]$, $D_u = 1.5 \times 10^{-4}$, $D_v = 1.5 \times 10^{-4}$, $D_w = 9.6 \times 10^{-2}$, $\kappa_3 = 8.5$, $\lambda = 2$, $\kappa_1 = -6.92$, $\theta = 1$, $\tau = 48$ and the discretisation in space and time was $\Delta x = 0.009$ and $\Delta t = 2 \times 10^{-3}$.

In the case $D_v = 0$ and $\theta = 0$ it is possible to derive analytical results regarding the propagation speed of these structures [15]. If there exists a stationary structure $(\bar{u}, \bar{v}, \bar{w})$ for $\tau = 0$ on an infinite domain G, in any case the onset of propagation will occur at $\tau = 1$ and the speed of the propagating solution is given by

$$c^2 = (\tau - 1)\frac{\langle \bar{u}_x^2 \rangle}{\langle \bar{u}_{xx}^2 \rangle}, \tag{4}$$

where an index x indicates the partial derivative $\partial/\partial x$ and $\langle \ldots \rangle$ denotes integration over G. It is even possible to show, that the onset of propagation is the primary bifurcation. Figure 2 b) shows numerical results for the speed of a travelling solution in three-dimensional space. The numerically estimated bifurcation point is at $\tau = 0.992$ and the slope of the speed deviates about 25 % from the theoretical result. The errors are due to the finite discretisation length and the fact, that $\theta = 0.01 \neq 0$. To reduce the dimension and thus the computational effort these computations have been performed using a cylindrical geometry. The stability of the moving spot for the parameters in Fig. 2 was checked by a three-dimensional calculation. The above description of the onset of propagation assumes, that there exists always a stationary solution which is unstable beyond the travelling bifurcation. This is not necessary, however. But this analysis holds only in the vicinity of the bifurcation point. For different parameters (very high values of τ, or $D_v \neq 0$) it is possible to find a moving stable solution, without a stationary counterpart.

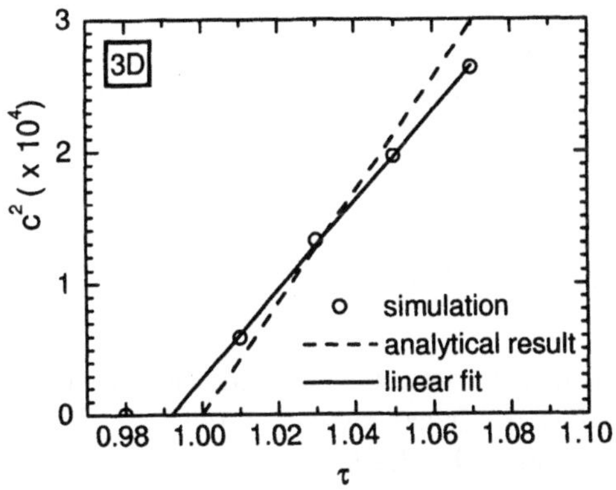

Fig. 2. Square of the speed c of a moving spot as a function of the parameter τ. The remaining parameters are $D_u = 4.67 \times 10^{-3}$, $D_v = 0$, $D_w = 10^{-2}$, $\kappa_3 = 3.33$, $\lambda = 5.67$, $\kappa_1 = -1.126$, $\theta = 0.01$. Spatial discretisation was $\Delta x = 0.02$ and for the time we used $\Delta t = 1 \times 10^{-3}$.

3 Interaction of Two Quasi-Particles

The equations (1)-(3) do not contain any global feedback terms. So, if the existence of a single moving spot is guaranteed, it is possible to construct solutions to the differential equations with two or more moving particles, at least if the individual patterns are well separated in space. As two spots approach each other, there will be an interaction process, which can lead to a variety of different scattering phenomena. In the following we will present four different cases that can result from a collision processes and discuss the relevant mechanisms.

3.1 Deflection of Quasi-Particles

The first phenomenon which we discuss are scattering processes, that lead to a mere deflection of the particles. In this case the shape of the two localized solutions does not change significantly during the process. Typically there will be a repulsive interaction between two spots as they approach. This is due to the fast inhibitor, which surrounds the particles and inhibits the propagation of the activator distribution of the other spots.

This case can be described analytically starting with the assumptions $D_v = 0$ and $\theta = 0$. According to Eq. (4) the onset of propagation occurs at $\tau > 1$. The dynamics of a quasi particle in this regime near the bifurcation point can be described by two quantities: The position $\boldsymbol{p}_i$ of a spot i, and the amplitude and direction of the shift $\boldsymbol{\alpha}_i$ of the slow inhibitor with respect to the activator. While the first quantity $\boldsymbol{p}_i$ results from the translational invariance of the equations, the second quantity describes the internal degree of freedom of a solution which is decisive for the propagation speed and direction of a particle. A value of $\boldsymbol{\alpha}_i = 0$ corresponds to a stationary, and for $\tau > 1$ unstable, solution. Using a singular perturbation approach it is possible to derive equations for the time evolution of these quantities. For analytical details refer to [16] and [17]. For a system of two particles 1 and 2 the following differential equations can be deduced for the motion of particle 1:

$$\partial_t \boldsymbol{p}_1 = \boldsymbol{\alpha}_1 - W(|\boldsymbol{p}_2 - \boldsymbol{p}_1|)(\boldsymbol{p}_2 - \boldsymbol{p}_1), \tag{5}$$

$$\partial_t \boldsymbol{\alpha}_1 = (\tau - 1)\boldsymbol{\alpha}_1 - \frac{\langle \bar{u}_{xx}^2 \rangle}{\langle \bar{u}_x^2 \rangle} |\boldsymbol{\alpha}_1|^2 \boldsymbol{\alpha}_1 - W(|\boldsymbol{p}_2 - \boldsymbol{p}_1|)(\boldsymbol{p}_2 - \boldsymbol{p}_1). \tag{6}$$

The function $W(d)$ is the interaction law between two particles with the distance d and is equal to zero for $d \to \infty$. It can be calculated numerically from the shape of a single spot solution $\bar{u}$. In the following we will restrict ourselves to the case of a repulsive interaction law. Obviously for $d = |\boldsymbol{p}_2 - \boldsymbol{p}_1| \to \infty$ the above system reduces to Eq. (4). The stationary solution $\boldsymbol{\alpha}_1 = 0$ becomes unstable for $\tau > 1$.

In Fig. 3 the trajectories of two interacting quasi-particles (spot 1 and 2) are sketched. The solid lines are from a fully three-dimensional calculation. The dashed line was calculated from (5) and (6). As initial conditions two approaching spots with an angle of about 90 degrees between their velocities have been used. As the spots approach each other the repulsive force leads to a deflection. After the collision there is a slight divergence between the three-dimensional numerical and the simplified numerical results, but the agreement is still satisfying. As long as τ is small enough the perturbation approach (5) and (6) is valid, and thus elastic scattering can be expected in these cases.

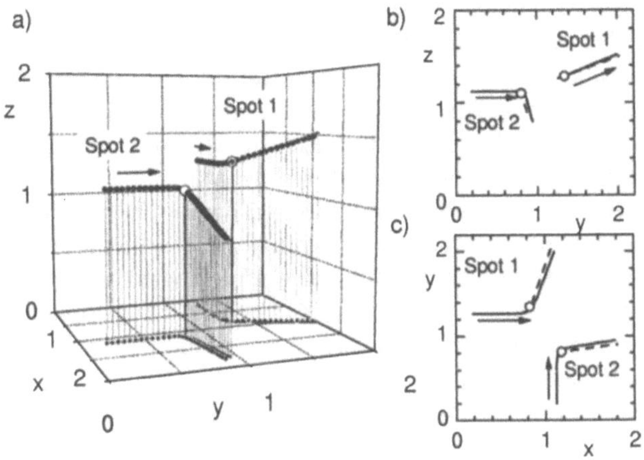

Fig. 3. Trajectories of two interacting quasi-particles. The solid lines indicate the results of the three-dimensional equations, the dashed lines are from the simplified model. a) Trajectories in three-dimensional space. b) and c) show the projections of the trajectories on the ground and on the back plane. Parameters were chosen as $D_u = 4.67 \times 10^{-3}$, $D_v = 0$, $D_w = 0.01$, $\lambda = 5.67$, $\kappa_1 = -1.126$, $\kappa_3 = 3.333$, $\kappa_4 = 3.333$, $\tau = 1.03$, $\theta = 0.01$ and $L = 2.25$. Space and time discretisation was $\Delta x = 0.0288$ and $\Delta t = 10^{-3}$.

3.2 An Overview of Typical Scattering Processes

In the previous section we have restricted the range of parameters to the case $D_v = 0$, $\theta = 0$ and $\tau \approx 1$. Thus we have been able to give an analytical description of the collision process. But since the shape and stability of the moving spots is important for this analysis certain cases are automatically excluded: The annihilation, generation or the mergence of two spots.

Before the other examples are discussed in detail Fig. 4 shall give a survey of the qualitative behaviour of two moving spots. It was not possible to define exact parameters separating different areas of qualitative behaviour in parameter space, nevertheless the basic behaviour can be described. In the following we will assume the existence of a sufficiently stable spot. Two quantities can be used to characterize the behaviour in the course of a collision process: The speed of the structure, which can be controlled by the time constant τ, and the strength of the interaction force between two particles, which can be reduced for example by decreasing the ratio of the diffusion constants D_u/D_w. Of course, changing other system parameters can equally effect the scattering phenomena, but the above mentioned can be regarded as the most important ones. The deflection process (I.) has been described already. This case can be found if the speed of the single particles is very low or if the strength of particle interaction is rather strong. Thus the particles cannot merge, but will be deflected. For slow particles, a reversal of speed is typically possible and leads to elastic scattering. For different parameters, e.g. if there exist only spot solutions with a certain minimal speed, one or both spots can be annihilated (II.) because the activator distribution cannot proceed due to the repulsive force and the slow inhibitor catches up and suppresses the activator. This type of annihilation process hasn't been examined yet in the three-dimensional case. In two-dimensional systems this mechanism has already been observed. If the repulsive forces between two quasi particles are weaker or the speed of the solutions is high, the propagation cannot be stopped and the activator of the two particles merges to an intermediate activator spot. This may happen due to very high speed of the particles (e.g. $\tau \gg 1$) or due to a very flat fast inhibitor distribution w ($D_w \gg D_u, D_v$). In this situation there are two possibilities: either the activator spot is destroyed by the inhibitor v (III.), or the activator is fast enough to escape transverse to the original direction of motion (IV.). Note that in highly excitable media the generation of new particles from remaining perturbations is possible after an annihilation process. The two cases (III.) and (IV.) are described in the next section in more detail.

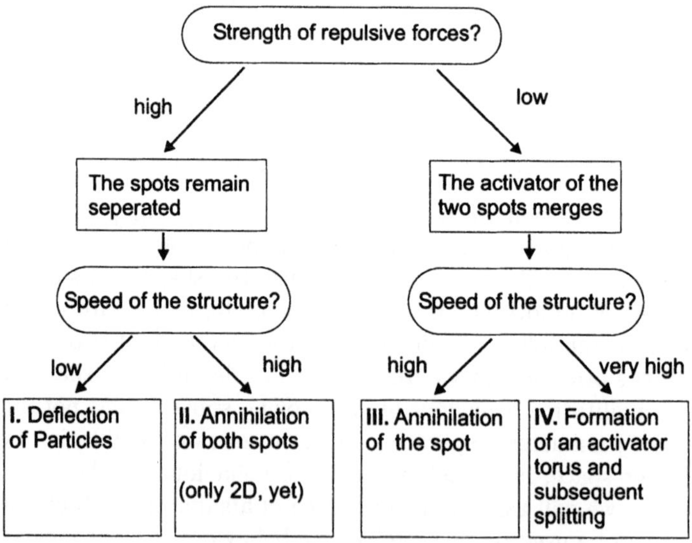

Fig. 4. Depending on the system parameters different scattering phenomena occur.

3.3 Mergence and Annihilation of Spots

In Fig. 5 the process of mergence of two spots after a head-on collision is shown. The activator u (red) and the slow inhibitor v (green) have been depicted as iso-surfaces. The spots approach each other in a head-on collision in (a) and (b). The inhibitor is clearly located behind the activator. In (c) the activator distribution merges. For the parameters given in Fig. 5 the activator is suppressed by the slow inhibitor in (d) and the spots vanish (e), corresponding to Fig. 4 (III.). Figure (f) shows the normalized activator integral $\langle u \rangle / |G|$ in the course of the process. The points in time corresponding to pictures (a)-(e) are marked with circles in this plot. The number of particles present in the system corresponds to the value of $\langle u \rangle / |G|$. At $t = 0$ there are two particles ($N = 2$), at $t > 100$ there remains only the homogenous state ($N = 0$). For $t \approx 52$ an intermediate state forms, which can be interpreted as a transient particle, formed by the activator and the fast inhibitor.

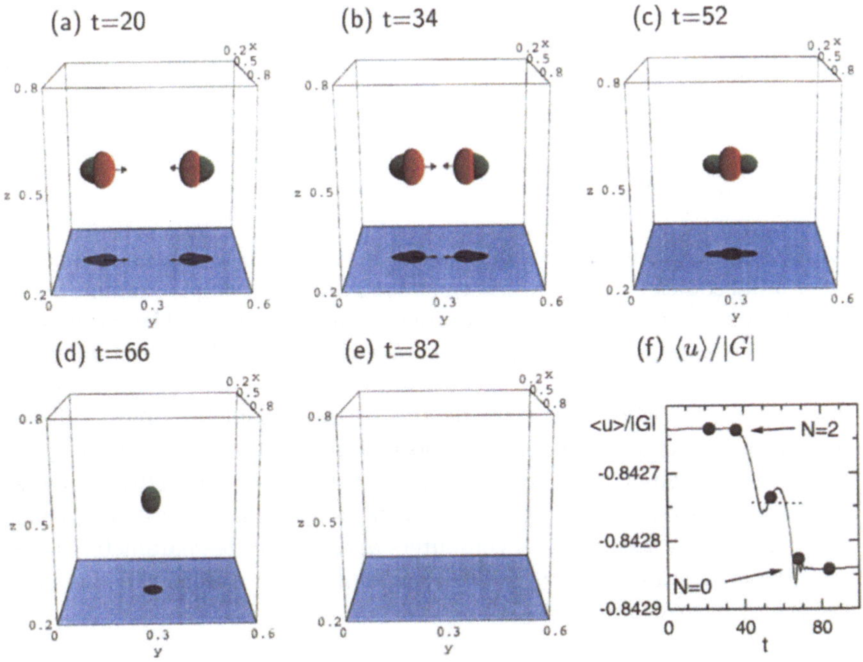

Fig. 5. Pictures (a)-(e) show an annihilation process between two particles. The iso-surfaces of the activator u are red, those of the inhibitor v are green. The normalized activator integral $\langle u \rangle / |G|$ is plotted in (f). The parameters used are $D_u = 1.5 \times 10^{-4}$, $D_v = 1.86 \times 10^{-4}$, $D_w = 9.6 \times 10^{-3}$, $\lambda = 2$, $\kappa_1 = -6.92$, $\kappa_3 = 8.5$, $\theta = 1$, $\tau = 25$. Discretisation was chosen as $\Delta x = 0.085$ and $\Delta t = 2 \times 10^{-2}$.

Starting from the annihilation process the time constant τ was increased from 25 to 48. Thus the shift of the inhibitor v with respect to the activator u is increased. This leads to a faster propagation of the spot. In Fig. 6 the results of this calculation are presented. Figures (a)-(e) show the iso-surfaces, (f) the normalized activator integral $\langle u\rangle/|G|$. Compared to Fig. 5 the inhibitor tails are longer, which reflects the higher time constant τ. After the mergence (b) the activator is fast enough to escape transverse to the direction of motion. Due to the cylindrical symmetry of the collision process a transient activator torus forms. In Fig. 6 (c) the symmetry is already broken and two new particles have been created. This case corresponds to (IV.) in Fig. 4. The plot of the normalized activator integral again shows the formation of an intermediate transient one-particle solution, before it breaks up and the two new particles form. The value of $\langle u\rangle/|G|$ at the end of the calculations was slightly higher than in the beginning, because the particles have still not reached their final shape yet. These simulations are similar in some sense to two-dimensional results of different groups [12,13], that were related to the collisions of spots in two-component models with an inhibiting global feedback . However, the asymptotic solution is unstable in the two-component case. In a three-dimensional space transversal scattering has to break the initial cylindrical symmetry, even if multi-particle solution are asymptotically stable. There even is a possibility that more than two particles emerge from the intermediate state. A further increase of the activation potential of u could lead to such a situation, but this was not observed until now.

4 Discussion

We have presented moving localized solutions of the three-dimensional partial differential equations (1)-(3), that can be regarded as a simple example for the complex dynamical behaviour in reaction-diffusion systems. The mechanisms that are necessary for the propagation of such solutions have been identified and it was possible to give an analytical description of the propagation and interaction properties in certain regions of the parameter space. Different scattering phenomena like annihilation and intermediate mergence of particles have been observed. The knowledge of these basic properties provides a basis for the study of multi-particle solutions by means of two complementary approaches: Fast parallel computers and, in suitable parameter ranges, effective interaction equations. – We gratefully acknowledge the support of the HLRS in Stuttgart and of the DFG.

References

1. A. M. Turing, The chemical basis of morphogenesis, Phil. Trans. Roy. Soc. **237**, 37 (1952).
2. J. D. Murray, *Mathematical Biology*, Springer, Berlin, 1993.

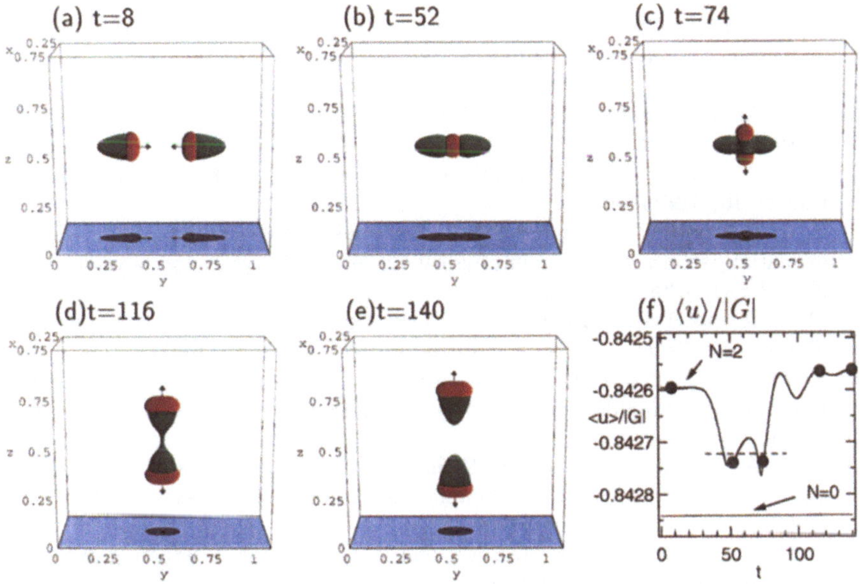

Fig. 6. After the mergence of two particles, the activator escapes transverse to the original direction, which can lead to the formation of two new particles. The parameters are chosen like in Fig. 5, but τ was increased to 48. (a)-(e) shows the iso-surfaces of the activator u (red) and inhibitor v (green). (f) is the normalized activator integral as a function of time.

3. H. Meinhardt, *The Algorithmic Beauty of Sea Shells*, Springer, Berlin, 1994.
4. Y. Kuramoto, *Chemical Oscillations, Waves, and Turbulence*, Springer, Berlin, 1984.
5. S. C. Müller, Characterization of wave front instabilities in the Belousov-Zhabotinsky reaction: An overview, Int. J. Bif. Chaos **4**, 1257 (1994).
6. G. Ertl, Ordnung und Dynamik an Festkörperoberflächen, Physikalische Blätter **46**, 339 (1990).
7. H.-G. Purwins, C. Radehaus, and J. Berkemeier, Experimental investigation of spatial pattern formation in physical systems of activator inhibitor type, Z. Naturforsch. **43a**, 17 (1988).
8. M. Bode and H.-G. Purwins, Pattern formation in reaction-diffusion systems - dissipative solitons in physical systems, Physica D **86**, 53 (1995).
9. E. Ammelt, D. Schweng, and H.-G. Purwins, Spatio-temporal pattern formation in a lateral high-frequency glow discharge system, Phys. Lett. A **179**, 348 (1993).
10. Y. A. Astrov and Y. A. Logvin, Formation of clusters of localized states in a gas discharge system via a self completion szenario, Phys. Rev. Lett. **79**, 2983 (1997).
11. C. P. Schenk, M. Or-Guil, M. Bode, and H.-G. Purwins, Interacting pulses in three-component reaction-diffusion systems on two-dimensional domains, Phys. Rev. Lett. **78**, 3781 (1997).
12. K. Krischer and A. Mikhailov, Bifurcation to traveling spots in reaction-diffusion systems, Phys. Rev. Lett. **73**, 3165 (1994).
13. L. Schimansky-Geier, M. Mieth, H. Rosé, and H. Malchow, Structure formation by active brownian particles, Phys. Lett. A **207**, 140 (1995).
14. S. Kawaguchi and M. Mimura, Collision of travelling waves in a reaction-diffusion system with global coupling effect, SIAM J. Appl. Math. **59**, 920.
15. M. Or-Guil, M. Bode, C. P. Schenk, and H.-G. Purwins, Spot bifurcations in three-component reaction-diffusion-systems: The onset of propagation, Phys. Rev. E **57**, 6432 (1998).
16. M. Bode, Front-bifurcations in reaction-diffusion systems with inhomogenous parameter distributions, Physica D **106**, 270 (1997).
17. C. P. Schenk, P. Schütz, M. Bode, and H.-G. Purwins, Interaction of self-organized quasiparticles in a two-dimensional reaction-diffusion system: The formation of molecules, Phys. Rev. E **57**, 6480 (1998).

Upwind Relaxation Algorithm for Re-entry Nonequilibrium Flows

H.-H. Frühauf[1], M. Fertig[1], F. Olawsky[1], and T. Bönisch[2]

[1] Institut für Raumfahrtsysteme, Universität Stuttgart, Pfaffenwaldring 31, 70550 Stuttgart, Germany
[2] Rechenzentrum der Universität Stuttgart, Allmandring 30, 70550 Stuttgart, Germany

Abstract. The development of the Navier-Stokes solver URANUS (Upwind Relaxation Algorithm for Nonequilibrium Flows of the University of Stuttgart) will be described.

Advanced gas-phase and gas-surface interaction modelings allow an accurate prediction of thermochemical relaxations in high-temperature nonequilibrium flows around re-entry vehicles and the associated aerothermal surface loads.

A fully coupled, fully implicit numerical algorithm, which solves the Newton-linearized equation system iteratively with arbitrary accuracy allows for large CFL numbers. Therefore, Newton-like convergence rates and a sufficient convergence grade are obtained for accurate computation of sensitive flow quantities like skin friction coefficient and Stanton number. The favorable performance of the algorithm will be demonstrated for solutions of the 2D and 3D sequential nonequilibrium codes on NEC SX-4.

Due to the large memory requirement and the large computational work of the 3D algorithm, a 3D Parallel-Multiblock version is developed which allows to compute nonequilibrium flows around re-entry vehicles with about one million mesh points on CRAY T3E to be computed.

1 Introduction

The development of a robust, efficient and accurate nonequilibrium flow solver for re-entry flows in a wide altitude and velocity range is a challenging task. Complex thermochemical relaxation processes may occur in dissociated and ionized high-temperature nonequilibrium flows. A large number of conservation equations is needed for an accurate modeling of these relaxation processes within a multiple-temperature gas-phase model. Due to complex relaxation processes, large local gradients of physical quantities may occur which require a high local spatial resolution. Furthermore, the characteristic times of elementary reaction processes may differ by many orders of magnitude leading to stiffness problems. Besides complex thermochemical relaxation processes, very strong shocks, locally large expansions and complex near surface viscous flow structures exist in nonequilibrium flows around blunt re-entry vehicles at high angles of attack. Finally, the local catalytic behaviour of the vehicle's surface strongly depends on the gas-state at the surface. Hence, there is a strong interaction between the gas-phase state and the surface conditions. Under these conditions the nonequilibrium flow solver has to predict sensitive quantities like skin friction coefficient and Stanton number with sufficient accuracy. Therefore, the residual has to drop sufficient orders of magnitude irrespective of the complexity of nonequilibrium flows.

After a description of the advanced thermophysical modeling, the theory of the fully coupled, fully implicit upwind solver will be shortly reviewed and the parallelization of the code will be described in some detail. The emphasis in this paper is on numerical aspects and parallel code development.

2 Advanced Thermophysical Modeling

The advanced thermophysical modeling has been implemented and extensively validated using the two-dimensional/axisymmetric URANUS code. Then important models which significantly influence the aerothermal surface loads of a re-entry vehicle have been transfered to the 3-D sequential URANUS code. Enhancements which have been recently implemented into the URANUS Nonequilibrium Navier-Stokes code in order to further improve the gas-phase modeling and the gas-surface interaction modeling are summarized. In order to predict the radiation cooling for high velocity re-entries accurately enough, coupled nonequilibrium flow-radiation simulations have been performed with the URANUS-PARADE-HERTA code system.

2.1 Gas Phase Modeling

The Coupled Vibration-Chemistry-Vibration (CVCV-) gas phase model [1] was recently extended to include the influence of the rotational excitation of the molecules [8]. Starting from state-selective rates, multi-temperature rates depending on the translational temperature as well as on vibrational and rotational temperatures are obtained by averaging over the distribution functions of internal energies. The model can be used for all types of reactions which are important in high-enthalpy flows like dissociation, exchange and associative ionization reactions. Consistent with the modeling of the production rates, the production rates of vibrational and rotational energies due to chemical reactions are computed. Using quasi-classical trajectory calculations, the parameters of the CVCV-model were newly calibrated for dissociation and exchange reactions [8]. With these improvements nonequilibrium flows are dissociated higher because the dissociation energy is lowered by the rotational excitation. The kinetics of the Zeldovich reactions are now in accordance with quasi-classical trajectory theory. The improved NO-chemistry modeling in the URANUS code has been validated indirectly by means of a spectrometer simulation performed with the PARADE Radiation code and the HERTA Radiation Transport code for the Bow Shock Ultraviolet Experiment II, where spectra in the wavelength range of NO have been measured. With the exception of large altitudes where weak radiation signals occur, a good agreement is obtained between measured and simulated spectra along the trajectory [8]. Furthermore, the vibration-vibration energy-exchange model has been extended for low translational temperatures, which occur in the vicinity of cold surfaces and in wakes. It has been successfully validated against Sharma/Gilmore's air nozzle flow experiment.

For significantly ionized flows the influence of the electronic excitation on the ion-chemistry of atoms is modeled via the Quasi-Steady-State theory [9], which does not assume a Boltzmann distribution function. The influence of the electronic excitation on the post-shock ionization of atomic nitrogen and oxygen has been investigated for the FIRE II experiment. A delayed ionization was demonstrated in the post-shock region.

In dissociated and ionized re-entry flows strong gradients can be observed in density, temperatures and velocities. In order to describe the exchange of mass, momentum and energy under these conditions, Capitelli's newly computed collision integrals and Chapman-Cowlings first approximations for the transport coefficients translational thermal conductivity, viscosity and mass diffusion were implemented into the URANUS code, as well as the second approximation of thermal conductivity of electrons [10]. In Fig. 1 the temperature dependence of the total heat conductivity is computed under equilibrium conditions for a pressure of 10^5 Pa. Results of the new collision integrals and mixing models are compared with those of Gupta/Yos, respectively. It can be seen that collision integrals and mixing rules significantly influence the total heat conductivity in dissociated ($T<8000$ K) and ionized ($T>8000$ K) gases.

Results of the new mixing model have been compared with the simplified model of Gupta/Yos for the high-velocity re-entry of the Stardust capsule. It has been shown that the more accurate mixing rules lead to a faster relaxation of N^+ and O^+ in the post-shock regime. More important for an accurate computation of the surface heat loads is an accurate transport coefficient modeling in the hot boundary layers. During re-entries from low earth orbit, the boundary layer edge temperature is typically 5000 K and in this case Capitelli's collision integrals typically lead to a 10% increase in the surface heat flux, while the more accurate mixing rules do not lead to significant changes of the surface heat flux. During high speed re-entries, the boundary layer edge temperature is typically 10000 K. In this case both the new collision integrals and mixing rules can lead to an increase of the surface heat flux by roughly 10%.

2.2 Gas-Surface Interaction Modeling

Gaskinetic gas-surface interaction models are implemented into the URANUS code in order to take into account near-surface rarefaction effects on the leeward side and base flows and generally in re-entry flows at high altitudes [2].

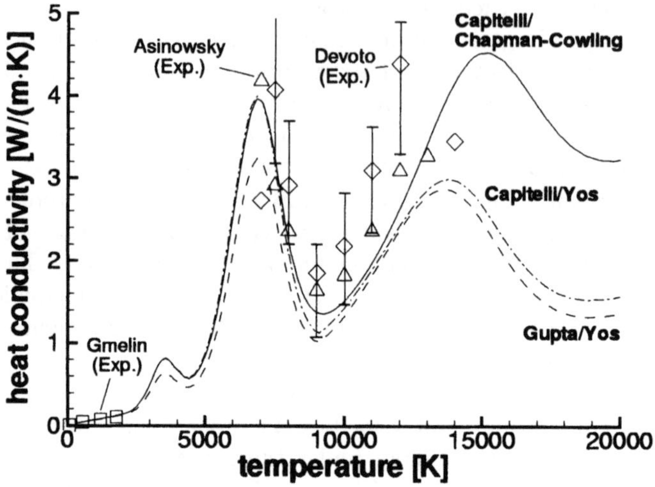

Fig. 1. Enhanced modeling of the transport coefficients through improved collision integrals and mixing rules, here: total heat conductivity

The finite-rate catalytic behaviour of a Thermal Protection System (TPS) surface significantly influences the surface heat flux to a re-entry vehicle. A detailed catalysis model for SiO_2-surfaces has been implemented into the URANUS code [2], which distinguishes adsorption, desorption and recombination reactions according to the Eley-Rideal and the Langmuir-Hinshelwood mechanisms as well as dissociative adsorption reactions. Futhermore, an expression for the chemical energy accommodation coefficient of the Eley-Rideal mechanism has been derived. This detailed model which consists of elementary reactions is able to account for the important non-Arrhenius behaviour of the recombination coefficients at high surface temperatures and has been successfully validated for the Shuttle re-entry. The finite-rate catalytic behaviour can also be predicted by a phenomenological model, which assumes complete chemical energy accommodation at the surface. The surface temperature dependence of the overall recombination coefficients of atomic species is determined in a large temperature range in high-enthalpy facilities. Reasonably accurate finite-rate surface heat flux predictions can be performed with this phenomenological model for the MIRKA capsule experiment. The outer heat shield is made out of C/SiC. Its catalytic behaviour is modeled by overall recombination coefficients for SiC which were measured by Stewart in a large surface temperature range.

Figure 2 shows the nonequilibrium flow around MIRKA under peak heating conditions and demonstrates the large thermal nonequilibrium between translational temperature and vibrational temperature of NO which is predicted by the accurate multi-temperature gas-phase model described in section 2.1. Once the gas-phase state at the surface is predicted by the gas-phase model, the gas-surface interaction model accounts for the catalytic behaviour of the TPS-surface which strongly influences surface heat flux and temperature. Figure 3 shows that the time history of surface heat flux density can be computed along the flight trajectory in good agreement with the experiment.

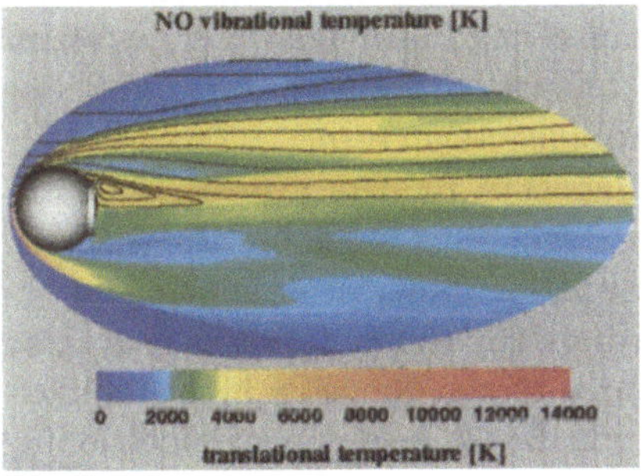

Fig. 2. Peak heating nonequilibrium flow around MIRKA (Freestream conditions: $p_\infty = 22$ Pa, $v_\infty = 6875\ \frac{\text{m}}{\text{s}}$, $\rho_\infty = 3.07 \cdot 10^{-4} \frac{\text{kg}}{\text{m}^3}$, $H = 60.1$ km)

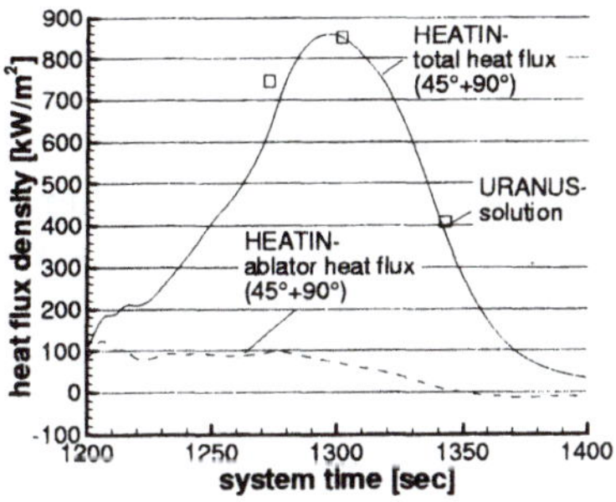

Fig. 3. Time evolution of measured and predicted heat flux densities along MIRKA flight trajectory (solid line: measurements, open squares: computations)

3 Numerical Scheme

The Navier-Stokes solver URANUS has been developed for the accurate simulation of nonequilibrium flows around re-entry vehicles in a wide altitude-velocity range. The unsteady, compressible Navier-Stokes equations in the integral form are discretized in space using the cell-centered finite volume approach. The inviscid fluxes are formulated in the physical coordinate system and calculated with Roe/Abgrall's approximative Riemann solver. Second order accuracy is achieved by a linear extrapolation of the characteristic variables from the cell centers to the cell faces. TVD limiter functions applied on forward, backward and central differences for non-equidistant meshes are used to determine the corresponding slops inside the cells, thus characterising the direction of information propagation and preventing oscillations at discontinuities. The viscous fluxes are discretized in the computational domain using classical central and one-sided difference formulas of second order accuracy.

Time integration is accomplished by the Euler backward scheme and the resulting implicit system of equations is solved iteratively by Newton's method, which theoretically provides the possibility of quadratic convergence for initial guesses close to the final solution. The time step is computed locally in each cell from a given CFL number. To gain full advantage of the convergence behaviour of Newton's method, the exact Jacobians of the flux terms and the source term have to be determined. While the source term Jacobians are implemented exactly, the Jacobians of the inviscid flux terms are only computed approximately following Barth by assuming a locally constant Roe-matrix. The viscous Jacobians are obtained from the Thin-Layer approximation of the viscous fluxes. The resulting linear system of equations is iteratively solved by the Jacobi line relaxation method. An underlying subiteration scheme is used to minimize the inversion error.

As can be shown, the source term Jacobian renders a positive contribution to the diagonal of the LHS-matrix and thus enhances the diagonal dominance of the matrix. Therefore, a superior diagonal dominance is obtained for nonequilibrium flows with nonvanishing source terms as compared with perfect gas flows with vanishing source terms. Hence, a higher convergence rate can be obtained for nonequilibrium flows. A simple preconditioning technique is implemented which improves the condition of the linear system and simplifies the LU-decomposition of the block-tridiagonal matrices to be solved in every line relaxation step.

The boundary conditions are implemented implicitly to preserve the large convergence rate of the Newton-like method. Slip and no-slip conditions are available. Different models for the surface catalysis are implemented, covering non, partial and fully catalytic surfaces. Global and detailed heterogeneous reaction models can be used. The surface temperature can be either prescribed or computed with the assumption of a radiation equilibrium surface. Note that the slip condition is derived from the gaskinetic approach and can be used in the slip and continuum flow regimes as well, allowing for thermal nonequilibrium at the surface. The slip condition reduces the spatial resolution requirements in the surface vicinity because unphysical temperature gradients do not have to be resolved.

4 2D/3D Sequential Nonequilibrium Codes

The same numerical scheme is applied for the 2D/axisymmetric and 3D codes. Very large spatial gradients in temperatures and concentrations may occur in nonequilibrium flows especially in the vicinity of shocks and surfaces due to complex local thermochemical relaxation processes. Therefore, in order to resolve these steep gradients, the spatial resolution can be locally as small as the local meanfree path. A sufficient convergence grade which is important for an accurate prediction of the surface heat flux, is obtained with the Newton-like method. Even for complex nonequilibrium flows CFL numbers up to 1000 can be obtained and result in favorable convergence properties in many applications. A problem is still the rather long initial iteration phase which is needed in high-velocity flows to get a sufficiently accurate initial solution from which the Newton iterations proceed.

4.1 2D/Axisymmetric Code

In order to predict the aerothermal loads during re-entry in a wide altitude and velocity range with sufficient accuracy, the axisymmetric URANUS nonequilibrium Navier-Stokes code consists of 18 conservation equations. Ten species mass conservation equations, 2 momentum equations and 6 energy equations are required to determine not only the typical flow field properties but also the partial densities of N_2, O_2, NO, N, O, N_2^+, O_2^+, NO^+, N^+, O^+ and e^-, as well as the vibrational energies of the neutral molecules, the rotational energies and the translational energies of the heavy particles and of the electrons.

The 2D/axisymmetric code was validated in numerous applications ranging from perfect gas flows to significantly ionized high-enthalpy flows [7].

A sufficient convergence grade and favorable Newton-like convergence rate were obtained in many applications [3].

The computation time per mesh point and iteration is 0.7 ms and 1 ms on NEC SX-4 in the case of the effective diffusion model (Yos/Gupta) and polynary diffusion model (Chapman-Cowling), respectively. The serial code reaches a peak performance of 720 MFLOPS. The memory requirement per mesh point is 57 KByte.

4.2 3D Code

Because of memory restrictions of available supercomputers, the gas-phase modeling of the 3D code is limited to non-ionized cases. This gas-phase modeling is sufficiently accurate for typical earth re-entries of space vehicles like Shuttle, HERMES and X-38 Crew Rescue Vehicle. Therefore, 5 species (N_2, O_2, NO, N, O) and 4 temperatures (T_{tr}, T_{vib,N_2}, T_{vib,O_2}, $T_{vib,NO}$) are used in the gas-phase modeling of the 3D code. The system of 12 conservation equations consists of 5 species equations, 3 momentum equations and 4 energy equations. The advanced CVCV thermo-chemical relaxation model is used as in the 2D/axisymmetric code. The transport coefficients are computed with the new collision integrals of Capitelli and the mixing rule of Yos/Gupta which is accurate enough for the non-ionized flows.

Up to now the 3D nonequilibrium code has been validated for different forebody flows. A URANUS computation is described for the flow around the COLIBRI capsule. The semi-ballistic COLIBRI capsule is a sphere-cone configuration with a scarfed bottom surface. Total length and maximum diameter are 1.5m and 1.0m, respectively. Nonequilibrium flow computations have been performed on the NEC SX-4 computer for the angle of attack $\alpha = 22.4°$, Mach number Ma$= 18.2$ and Reynolds number $Re_{L,\infty} = 8.07 \cdot 10^4$.

A mesh with 55×26×18 cells and a metric singularity has been used. The singularity is discretisized explicitly. No bow shock adaption of the mesh was performed and the laminar boundary layer is resolved with roughly 16 points. The surface is assumed to be noncatalytic with a constant temperature of T_s=1800 K. As in the 2D/axisymmetric-computations, CFL numbers up to 1000 can be obtained. Favorable convergence grade and rate can be derived from Fig. 4. The computed nonequilibrium flow around the COLIBRI capsule can be seen in Fig. 5, where vibrational temperature of oxygen and the near surface stream lines are plotted.

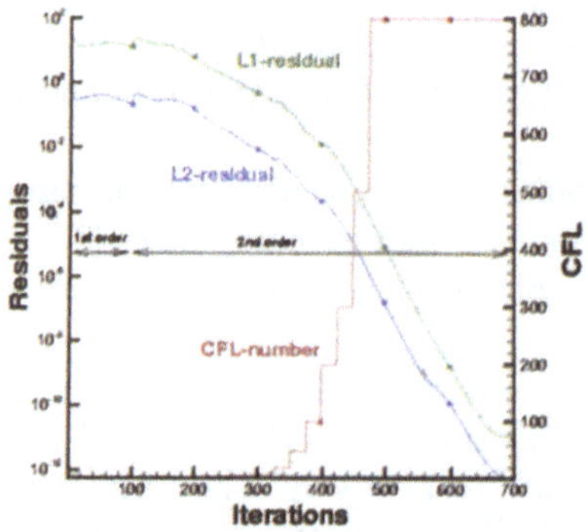

Fig. 4. Convergence of the sequential 3D nonequilibrium code, initial condition: uniform flow

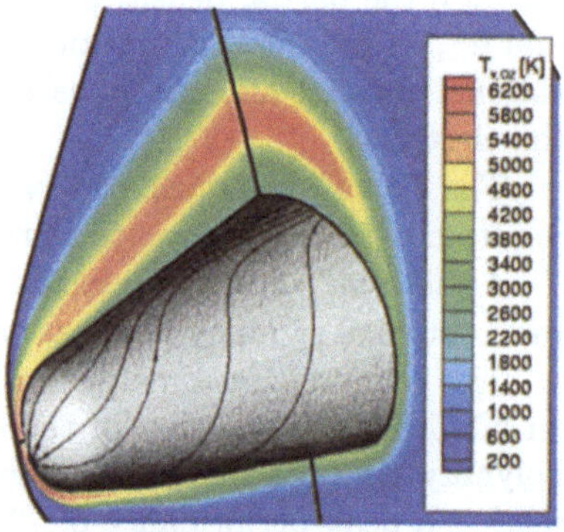

Fig. 5. Nonequilibrium flow around the COLIBRI Capsule: T_{vib,O_2}-contours and near surface stream lines

The computation time per mesh point and iteration is 0.4 ms on NEC SX-4 and the memory requirement per mesh point is 20.4 Kbyte. The total computation time (500 iterations) is 1.43 h.

The 3D implementation of the URANUS code is nearly identical to the 2D implementation. Therefore, the convergence behaviour is very similar.

In order to increase the numerical efficiency further, the Jacobi line relaxation method will be exchanged for a Krylov subspace method like GMRES(n) with an ILU preconditioning. In addition, the solver will be implemented matrixfree such that the Jacobian is computed consistently with second order accuracy. Thereby a convergence closer to the quadratic convergence of the Newton method is expected.

5 3D Parallel Codes

Due to the large memory requirement and the large computational work of the 3D Newton-like method, a powerful parallel computer is needed to compute nonequilibrium flows around re-entry vehicles on meshes with one million and more points. Therefore, in a first step the URANUS code has been parallelized.

5.1 Parallel-Code

The program was parallelized by domain decomposition. The computational domain is split into several subdomains, one for each available processor. At the inner boundaries of each of these subdomains a two-cell overlapping region is used. The values of cells from neighbouring domains are recomputed. Therefore, nearly no communication is needed while setting up the equation system. In parallelizing the solver the coupling between the subdomains is reduced. The equation matrix is split into two parts, one part which is the local heptadiagonal matrix and one part which consists only of the coupling elements. In this way the solution can be computed locally on each processor and then can be updated with the coupling matrix. Therefore, after each subiteration communication, it is nessesary in the solver to update the solution at the boundaries for the next subiteration. This method works fine on current massively parallel processors like the Cray T3E because of their good communication network.

To ensure portability, the message-passing paradigm MPI was used in the implementation. The portability of the parallel code was achieved. The code was tested on severals platforms, e.g. NEC SX-4, Paragon, Hitachi SR 2201 and Cray T3E.

A flexible parallel code is obtained by introducing dynamic data structures. These avoid the recompilation of the code for every grid size used. Fortran 90 pointers were used to implement these dynamic data structures because in this way the program structure of the sequential code did not need to be changed. Just a few statements for memory management were added.

Note, that the parallel code maintains the sequential code convergence rate. Speedup and scaleup were measured on Cray T3E with up to 512 processors. The measured computation times in Fig. 6 show a good speedup on the Cray T3E for various problem (mesh) sizes. The measurements were done for 200 iterations. Much more important than the speedup is the scaleup because it shows the behaviour of the code when problem size and number of processors are increased simultaneously. In Fig. 7 the scaleup on the Cray T3E is shown starting from the problem size of 27648 computed on 16 processors up to the problem size of 884736 computed on 512 processors. Problem size and number of processors are doubled step by step. The obtained scaleup which is larger then 95% is very good and ensures that computations with more than 1000 processors for a large scale computation are efficiently possible with this program.

The perfect gas flow around the HERMES 1.0 model has been computed for ONERA S4MA wind tunnel conditions: angle of attack $\alpha = 30°$, upstream Mach and Reynolds number $\mathrm{Ma}_\infty = 10$ and $\mathrm{Re}_{L,\infty} = 3.28 \cdot 10^6$, respectively.

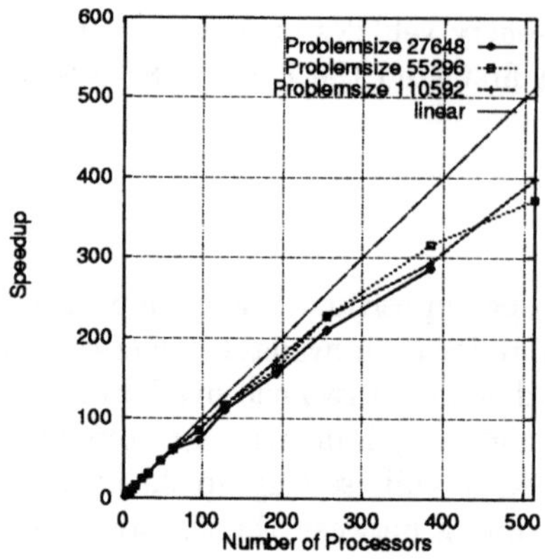

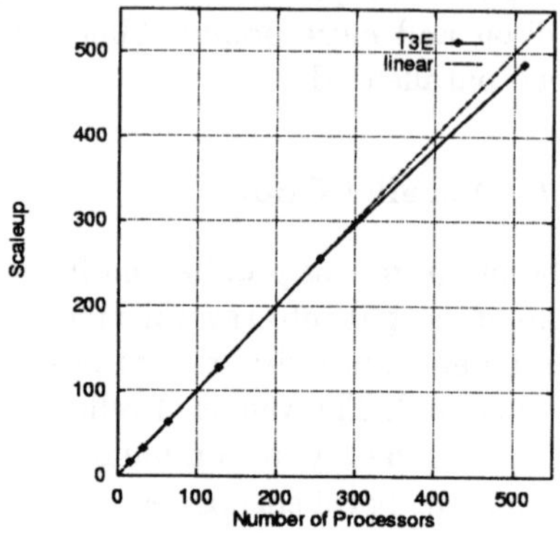

Fig. 6. Speedup for three problem sizes, 200 iterations

Fig. 7. Scaleup (from 16 up to 512 processors, from 27648 up to 884736 cells)

The surface temperature was assumed to be 300K. Laminar flow was assumed and zero body-flap, elevon and rudder deflection. The upstream stagnation conditions are: Total pressure and temperature $P_0 = 85$ bar and $T_0 = 1100$ K, respectively.

The intention was to perform a Parallel-URANUS computation on a roughly one million cell mesh. The parallel computation was performed using an available 884736 cell mesh with a metric singularity at the nose. A converged first order solution was obtained on this mesh using 128 processors of the CRAY T3E computer. In Fig. 8 the first order Mach number contours are shown for this Parallel-URANUS computation on a typical fine mesh. The total computation time on 128 processors of the CRAY T3E is 1.6 and 55 minutes for 110592 and 884736 cell meshes, respectively.

5.2 Parallel-Multiblock-Code

Based on the described parallel flow solver a multiblock version was developed. Multiblock meshes consist of several structured mesh blocks. They are used to mesh complicated topologies trying to combine the advantages of unstructured meshes in capturing the topology and structured meshes in their computational efficiency and accurate boundary layer computation.

Due to the topology of multiblock meshes, a certain block may have an orientation different from its neighbours. This means that the local coordinate systems of the blocks and therefore their representation in the memory will differ. When exchanging data between the blocks this must be taken into account in order to provide the cells with the correct data. The halo cells used at the inner subdomain boundaries contain the values of the corresponding block transformed to the local coordinate system. The connectivity file from

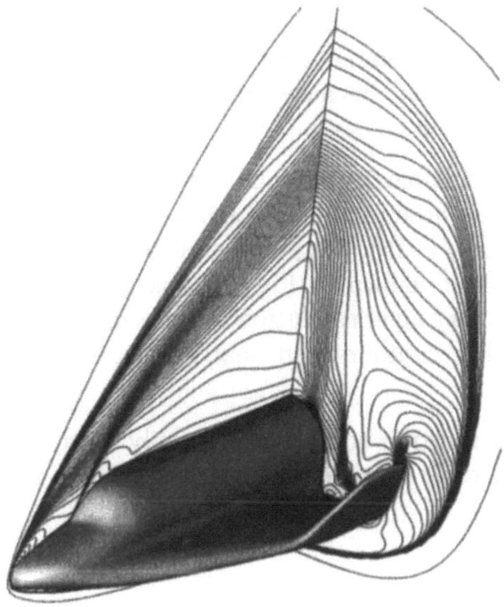

Fig. 8. HERMES 1.0 vehicle flow for ONERA S4MA wind tunnel conditions ($\alpha = 30°$, $Ma_\infty = 10$, $Re_\infty/m = 3.28 \cdot 10^6$, $T_w = 300$ K, $T_0/T_w=3.66$)

the grid generator provides the envisaged block with the orientation of its six neighbours. These data are used as an additional input for the data exchange routines. In multiblock meshes boundary types are not bound to one block side as in the single block version, but can appear on each of its faces due to the different local coordination systems. Therefore, each boundary type routine was extended to work not only on one special index but on all sides of the block. Furthermore, the handling of all the different edge types was implemented.

The parallel line relaxation solver used in the single block code exchanges the boundary data between the subdomains after each line relaxation step. This leads to its good numerical efficiency. In the multiblock code this solver cannot be used because the directions of the lines computed on the different blocks in parallel in each line relaxation step need not to be equal. Therefore, another version of the parallel solver was used, which begins with a complete solving step, then exchanges the resulting data at the subdomain boundaries and finally does an additional solving step to reach the necessary accuracy. This solving method is independent of the line directions, but needs an additional computing effort of about 20%.

Presently the multiblock solver is in a stage where it is possible to compute every block on more than one processor with the precondition that all blocks are divided in the same way. This is because the decomposed blocks are each computed by their own process and are not allowed to have more than one neighbour on each of its six sides so that no communication problems can occur at the subdomain boundaries. Additionally, the blocks should have more or less the same size, because the program is not able to do any load balancing at the moment.

Up to now the multiblock code has been validated for perfect gas flows. The perfect gas flow around the COLIBRI capsule was computed for the upstream conditions $\alpha = 15°$, $\mathrm{Ma}_\infty = 4$ and $\mathrm{Re}_{L,\infty} = 7.3 \cdot 10^6$. The 10-block mesh with 80000 cells, shown in Fig. 9a was generated with the GridPro program. The boundary layer resolution is performed in separate near surface blocks. Mach number contours and near surface streamlines are shown in Fig. 9b.

The multiblock solution compares favorably with the solution obtained with the sequential code on a similar fine mesh with a singularity.

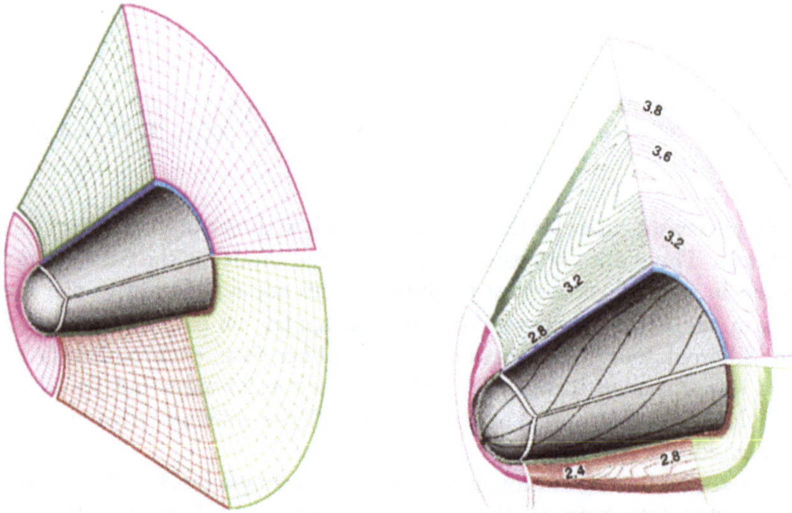

a) 10-block mesh with 80000 cells

b) Mach number contours and near surface streamlines

Fig. 9. Perfect gas flow around the COLBRI capsule, ($\alpha = 15°$, Ma_∞=4, $\mathrm{Re}_{L,\infty} = 7.3 \cdot 10^6$)

The work done at the moment aims to obtain a complete flexible parallel multiblock code which can deal with different block sizes, different neighbour numbers on each block side and different block numbers on the processors. As soon as this parallel 3-D multiblock code is finished and tested, components of the sequential 3D nonequilibrium code will be implemented.

6 Conclusions

- Advanced gas-phase and gas-surface interaction models have been implemented in the 2D/axisymmetric-URANUS code which are able to accurately predict thermochemical relaxations in the gas-phase and aerothermal loads at the surface of re-entry vehicles in a large density and velocity range.
- A reduced thermophysical modeling has been implemented in the sequential 3D URANUS code for non-ionized nonequilibrium flows.
- No stiffness problems are encountered using the fully coupled, fully implicit algorithm which solves the Newton linearized system iteratively with arbitrary accuracy. Large CFL numbers allow for Newton-like iterations. Therefore, favorable convergence rate and convergence grade have been obtained with the 2D and 3D sequential codes for many complex high-temperature nonequilibrium flows.
- Due to the large memory requirement and the large computational work of the 3D Newton-like algorithm, the URANUS code has been parallelized in order to compute nonequilibrium flows around re-entry vehicles on meshes with about one million points.
- The 3D Parallel-URANUS code, which has been developed so far for perfect gas flows, maintains the convergence rate of the sequential code. A nearly linear scale-up was obtained for processor numbers as large as 512 on CRAY T3E. Hence, efficient parallel computations with more than 1000 processors seem to be possible.
- The 3D Parallel-Multiblock URANUS code has been implemented and tested up to now for nearly equally sized blocks and is presently extended to deal with different block sizes, neighbour numbers and block numbers on the processors.

7 Acknowledgements

This research work has been performed within the Special Research Center "Sonderforschungsbereich 259".

The authors would like to thank Deutsche Forschungsgemeinschaft (DFG) for their support.

References

1. KNAB O., FRÜHAUF H.-H., MESSERSCHMID E.: Theory and Validation of the Physically Consistent Coupled Vibration-Chemistry-Vibration Model, *AIAA Journal of Thermophysics and Heat Transfer, Vol.9, No.2, April-June 1995.*
2. DAISS, A., FRÜHAUF, H.-H., MESSERSCHMID, E.W.: Modeling of Catalytic Reactions on Silica Surfaces with Consideration of Slip Effects, *AIAA Journal of Thermophysics and Heat Transfer, Vol.11, No.3, July-September 1997.*

3. FRÜHAUF, H.-H.: Efficient Prediction of Reentry Nonequilibrium Flows, *Proceedings of the 15th IMACS World Congress on Scientific Computation, Modelling and Applied Mathematics, Berlin, August 24-29, 1997.*
4. MPI Forum: MPI: A Message-Passing Interface Standard, University of Tennessee, Knoxville, Tennessee, USA, 1995.
5. BÖNISCH, T., GEIGER, A.: Implementierung eines 3-D Strömungscodes auf Parallelrechnern, Forschungs- und Entwicklungsberichte RUS-35, Computing Center University of Stuttgart, Germany, 1996.
6. BÖNISCH, T., RUEHLE, R.: Portable Parallelization of a 3-D Flow-Solver in Parallel Computational Fluid Dynamics, Proceedings of the Parallel CFD '97 Conference, Elsevier, Amsterdam, 1998.
7. FRÜHAUF, H.-H., FERTIG, M., KANNE, S.: Validation of the Enhanced URANUS Nonequilibrium Navier-Stokes Code, *AIAA-Paper 99-3683, 33rd Thermophysics Conference, Norfolk, Virginia, USA, 28 June - 1 July, 1999.*
8. KANNE, S.: Thermochemical Relaxations through Collisions and Radiation, Paper offered to the AIAA Journal of Thermophysics and Heat Transfer.
9. PARK, C.: *Nonequilibrium Hypersonic Aerothermodynamics*, Wiley, New York, 1990.
10. FERTIG, M., DOHR, A., FRÜHAUF, H.-H.: Transport Coefficients for High-Temperature Nonequilibrium Air Flows, *AIAA-Paper 98-2937*, 1998 and accepted contribution to the AIAA Journal of Thermophysics and Heat Transfer, to be published.

3D Simulation of instationary turbulent flow and combustion in internal combustion engines

A. Werner

Rechenzentrum Universität Stuttgart
70550 Stuttgart. Germany

Abstract. This paper describes the advances of internal combustion engine simulations achieved in the ***parice3d*** project. The aim of the project was to accelerate ICE simulation technology by using massively parallel computers. The paper will only mention the advances in pre- and post-processing and focus on the computing part of the simulation.

1 Introduction

The flow inside an internal combustion engine cylinder is a highly complex 3D instationary coupled with turbulent combustion. In the past, Computer simulations of the combustion process were too time-consuming to be used in the development process. Time-to-market for a car is usually about five years in Europe and will have to be reduced to less than three years to stay competitive in global markets. With one to two weeks of grid generation, up to months of calculation time and weeks of post-processing, computer simulations of internal combustion engines were used for basic research, but not in the development cycle.

New massively parallel computers can provide much more computational power and thus accelerate the calculation of numerical simulations dramatically. To use these systems, the existing simulation codes have to be parallelised. This project extended a simulation code parallelised in the Esprit Europort-1 project for the needs of combustion simulation. The main goal of the project was to perform the complete simulation cycle in one week. To make this possible, not only the simulation, but also pre- and postprocessing had to be accelerated.

2 Pre-processing

The mesh generation for internal combustion engines was one of the major problems limiting the usefulness of ICE simulations for the engineering development cycle. During the project, a semi-automatic grid generation system was developed, which allows to generate moving grids including all cell activation and de-activation events automatically, removing a long and error-intensive part of the engine simulation.

3 The simulation code

The simulation code StarCD solves the fully compressible unsteady Navier-Stokes equations together with the conservation equation for thermal energy. In all simulations a standard k-ε turbulence model was used, although other models are available as well. The solution is calculated using a CG solver for an implicit finite volume calculation on unstructured moving grids,. The topology of the grid changes dynamically to maintain grid quality. This grid movement is performed by an external program, which is not yet parallelised.

Combustion is simulated with either a 1-equation or a 2-equation Weller combustion model [1]. This model has been developed in a European research project and reflects the physics of premixed combustion and the heat release rather accurately. The 1-equation Weller model solves a transport equation for one additional variable c, which represents the progress of the reaction between 0 (unburned) and 1 (fully burned).

The 2-equation model solves a second transport equation for the local wrinkled flame area. In general, this model is more robust, requires less tuning and gives more accurate results than the 1-equation model. Extended versions of both models were implemented into StarCD via user-defined subroutines [4] and comparisons of the numerical results with light-sheet measurements by Daimler-Chrysler in experimental engines showed a good agreement of experiment and simulation.

4 Parallelisation

The parallelisation of StarCD was based on domain decomposition.

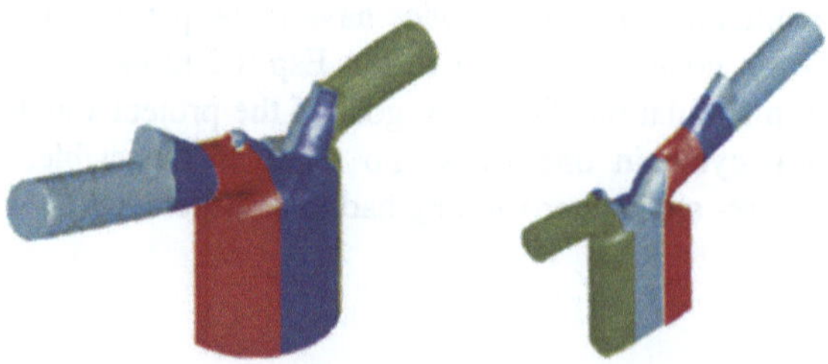

Fig. 1. 4-proc decomposition of engine mesh

Each processor calculates a part of the computational domain and communicates a one- or two-layer set of cells to its neighbours. The basis of the parallelisation was distributed-memory, using MPI as the communication layer.

As mentioned above, the mesh movement is performed by an external program, either ProSTAR or ProICE.

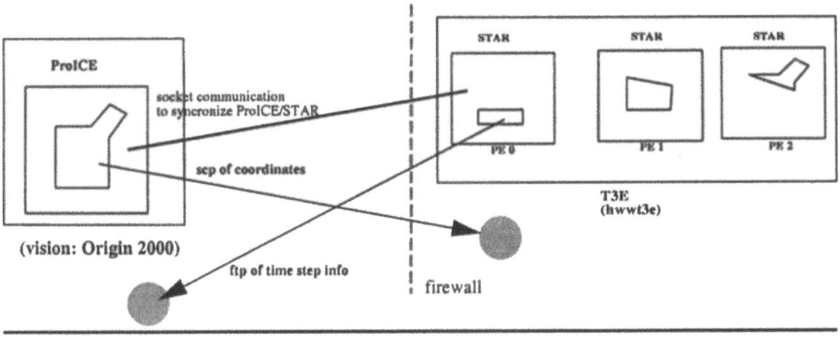

Fig. 2. Simulation set-up with mesh movement and simulation

The mesh movement program communicates with Star by creating files with the new vertex positions, which are then read in by one node of the MPP and broadcasted to all other nodes. The mesh movement calculations need a significant amount of memory for larger cases. It also must react interactively on the events from the simulation while having long idle-times between the events. Since none of the machines behind the HWW Firewall could supply this, an Onyx-2 outside of the Firewall was used, leading to the configuration shown in Fig. 2. This configuration is obviously not optimal, but since the mesh movement is not integrated into the simulation code and not yet parallelised, a T3E node could not be used for memory limitation reasons.

5 Load Balancing

Although planned in the beginning of the project, no dynamic load-balancing was integrated into the code. Instead, the domain decomposition was performed in a way that cell activation and de-activation was automatically evenly distributed across the nodes. This was done by splitting into cylindrical slices along the moving axis in the cylinder and along the valves. Even for distributions to large node numbers this leaded to a very equal distribution of the load, although the interfaces between the node domains were unnecessarily large. This is partly compensated by latency hiding inside the code, partly by the extremely low latency of the T3E communication.

Experiments with the distributed load-balancer Jostle were performed, but did not lead to better partitioning both due to constraints introduced by the mesh movement algorithm and problems in Jostle with the test cases

6 Post-Processing

For the post-processing, the visualisation system Covise, which was developed by the University of Stuttgart Computing Centre, was parallelised both for SMP and MPP systems. Covise is a modular system, which allows network-distributed and collaborative visualisation and simulation steering in a supercomputing environment. The parallel modules now allow interactive handling of large 3D transient data sets, with interaction times allowing even Virtual Reality usage.

7 Results

For non-moving grids, the Europort result code shows excellent speedup behaviour, especially for large model sizes:

# of CPUs	Time [s]	Speedup	Par. Efficiency
1	11181	1.0	1.0
4	2668	4.2	1.05
8	1334	8.4	1.05
16	632	17.7	1.1
32	385	29.0	0.9

Fig. 3. Benchmark of large stationary test-case

These results were achieved on an Origin-2000 server with a 3.1 million cell stationary flow case. The results show that the code itself parallelised nearly perfectly, achieving even superlinear speedup due to caching effects. The first results achieved with the moving-mesh code on the distributed-memory T3E-900 looked quite less promising: (300.000 cell moving grid, 100 steps)

# of CPUs	Time [s]	Speedup	Par. Efficiency
16	1283.65	1	1
32	879.49	1.46	0.7

Fig. 4. First benchmark of moving-grid case

Obviously the speedup was now far less optimal. The analysis of the benchmarking resulted in the fact, that the simulation code itself was well-balanced, but had long waiting-times for the grid movement performed by the external program ProSTAR whenever the mesh connectivity changed. These so-called 'events' needed significantly longer than the calculation of meshes where only simple movements in the mesh were performed.

To achieve further speedup, multiple ProSTAR sessions were run in parallel, looking ahead several steps to prevent waiting times for the parallel simulation job. First, the additional sessions were applied to the next timesteps to be calculated. This method did not help the problem, since the delays on the 'event' time-steps was still significant. The solution to the problem was to have one ProSTAR session calculating the mesh for the next time-step and multiple others to pre-calculate the next 'event' time-steps.

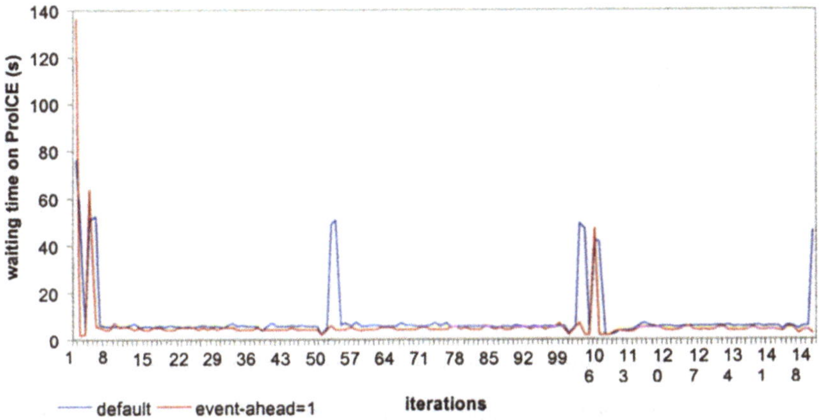

Fig. 5. Comparison with and without event look-ahead

The blue graph shows the idle time of the parallel solver waiting for the mesh movement. After setting up the first grid, a single event at step 55, two events at steps 103 and 106 and a single event at step 150 cause significant delay to the application. Having a second ProSTAR application running one event in advance removes all peaks except the 'double-peak' where the second ProSTAR cannot create the new 'event' fast enough and the parallel application is still held back by the mesh movement generation. The obvious method is to add more ProSTAR processes to make sure the next grid is always available when the solver finishes.

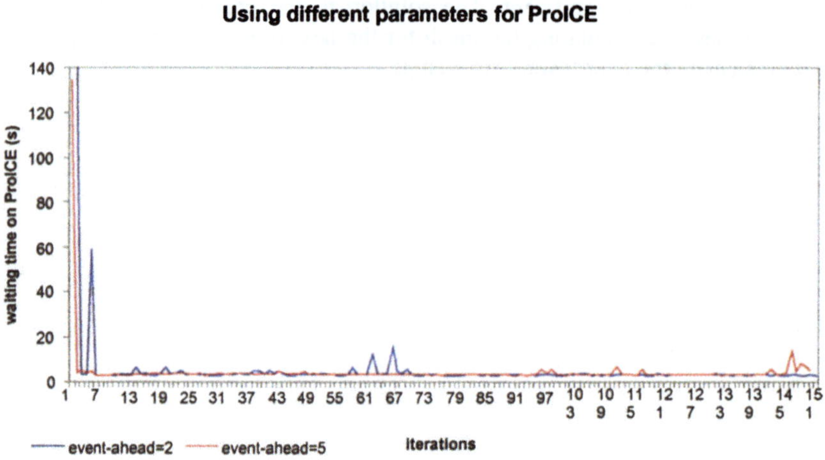

Fig. 6. Comparison with ‚more event look-ahead steps

This look-ahead reduced the computing time significantly. As the final result, the scalability of the moving-grid case was now improved even for higher numbers of processors: (600.000 cells, 250 Time-steps, 16 events, T3E-900 + Onyx-2)

# of CPUs	Time [s]	Speedup	Par. Eff.	ProSTAR wait
64	8063 sec	1		750 sec
96	6376 sec	1.265	0.84	740 sec

Fig. 7. Final benchmark of moving-grid case

The waiting times on ProSTAR are now mainly I/O related and not real wait periods. To achieve higher speedups, these times should be removed by parallelising the cell movement and integrating it into the simulation code.

8 Discussion

The results achieved in the project now allow to calculate a full engine simulation including pre- and post-processing within one week, which was the projects primary goal. This result opens two perspectives for the future of simulation in engine development:

- The simulation of engines is now a tool, which can be used in the product development cycle and not only in research. This opens new possibilities in engine development, since the experimental possibilities in engine test-beds are limited to only a few measurable species, while the simulation allows direct insight into the complete field distributions inside the cylinder.

- The development of new simulation models, especially in combustion and turbulence is improved, since it is now possible to run series of experiments. Half-empirical models can be refined by testing model parameter changes against test-bed experiment results.

Still, some issues are still open, especially concerning grid movement, which should be integrated into the simulation to prevent I/O, and load balancing, which is not yet implemented at all. Both issues will become even more important when scaling to larger numbers of nodes to gain further acceleration. For stationary cases, the code has proven scalability to much larger node numbers, so the industrial end-users will soon request even higher performance for larger grids with more complex physical models for moving-grid simulations.

Automatic cell refinement based on error estimation could also increase performance by specifically using high grid resolution only at places with high gradients. Especially for transient problems, hand-made grids will usually not be able to 'track' the flow effects as an automatic refinement tool will, enforcing high mesh density for all steps wherever the gradients become high even once during the complete simulation.

9 Publications

During the project, the following publications were done:

Harald Echtle, D. Liang, Monika Wierse: Transient Simulation of Fluid Flows in Internal Combustion Engines on Parallel Computers. Paper presented at the ECCOMAS Conference in Athens 1998.

Andreas Werner, Harald Echtle, Monika Wierse: High Performance Simulation of Internal Combustion Engines. Paper presented at the Supercomputing Conference in Orlando 1998.

H. Jasak, J. Luo, B. Kaludercic, A. D. Gosman, H. Echtle, Z. Liang, F. Wirbeleit, G. Fernstrom, A. Karlsson, S. Rips, A. Werner, M. Wierse: Rapid Simulation of Internal Combustion Engines. Paper presented at the SAE International Congress, 1999, Detroit.

10 Literature

H.G. Weller, S. Uslu, A.D. Gosman, R.R. Maly, R. Herweg, B. Heel: Prediction of Combustion in Homogeneous-Charge Spark-Ignition Engines. 3rd Int. Symp. COMODIA94, pp. 163-169, 1994

Numerical prediction of load changes in a coal-fired utility boiler

Alexander Bundschuh, Benedetto Risio, Uwe Schnell, Klaus R.G. Hein

Institute of Process Engineering and Power Plant Technology (IVD)
University of Stuttgart, Pfaffenwaldring 23, D-70550 Stuttgart, Germany
Fax: 0049-711-685 3491, E-mail: bundschuh@ivd.uni-stuttgart.de

Abstract. The present work deals with the numerical transient simulation of load changes in a pulverised coal-fired utility boiler. The boiler is a tangential firing system with an electrical output of 750 MW. The code used for the calculations is the in-house developed 3D combustion code AIOLOS. The implemented solution procedure is based on either the well-known SIMPLEC method (*Semi Implicit Method for Pressure Linked Equations Consistent*) or the PISO approach (*Pressure Implicit with Splitting of Operators*). The stability of the procedure is independent of the time step, since an implicit temporal discretisation scheme (*Euler Implicit*) is used. A validation exercise is presented by comparing the numerical results with temperature measurements available from LSE (Lehrstuhl für Systemtheorie in der Elektrotechnik), Saarland University, Saarbrücken.

1 Introduction

Besides empirical methods, numerical simulation has become a useful and supplementary tool in optimising the operating regime of industrial-scale, coal-fired furnaces. The objectives are manifold: enhancement of total efficiency, reduction of pollutant emissions, stabilisation of burner operation, investigation of the effects of off-axis flow distribution, avoiding of erosion, corrosion, fouling and slagging etc. As an example, Antifora et al. (1997) design a NO_x reduction retrofit of a 320 MW_e opposite wall fired boiler using experimental tests as well as mathematical modelling simultaneously. Risio et al. (1998) compare numerical results with in-furnace measurements of temperature and oxygen. The experimental activities are performed for the purposes of an optimisation of the burner aerodynamics in a 150 MW_e brown coal fired utility boiler. Specifically, concerning the coupled dynamical processes of the steam-water cycle and the furnace system, a potential for optimisation is hold. Such simulations require a time-dependent formulation in order to describe e.g. temperature fluctuations and peaks.

For a huge number of problems in science and engineering time-dependent solutions are demanded. Concerning computational fluid dynamics, for many applications the time-dependent differential equations must be adopted. However, there is no or only little attempt to perform the simulation of transient processes in industrial-scale combustion systems. This may be due to the fact that these large scale numerical simulations require high computational resources which can not be provided by conventional computers. However, the recent development in high performance computing (vector and parallel supercomputers) delivering both computational power and memory resources enables the solution of such complex problems.

The objective of the present work is to investigate the reliability of numerical predictions of time-dependent processes in full-scale combustion systems. In order to assess the mathematical models, the numerical transient simulation of load changes in a pulverised coal-fired utility boiler is demonstrated. The boiler is a tangential firing system with an electrical output of 750 MW. The calculations are performed using the in-house developed 3D combustion code AIOLOS. The code contains submodels describing fluid flow, turbulence, combustion and radiative heat transfer which are coupled in a strongly non-linear way. In order to prove the reliability of the tool, the numerical predictions are compared with experimental data. In recent years, a variety of validation exercises at full scale have been carried out, showing that for the steady-state case AIOLOS is a reliable tool in predicting carbon in ash, wall heat fluxes, temperatures and oxygen concentrations (Knaus et al., 1997a, 1997b, Risio et al., 1997, 1998, Schneider et al., 1995, Schnell et al., 1995). In the present work the calculated results are compared with measurements of mean path temperatures obtained by acoustic pyrometry. In order to meet the high computational effort, the calculations are carried out on a parallel vector computer NEC SX-4/32 using up to 8 processors.

The remainder of the paper is organised as follows. Section 2 introduces the mathematical models and numerical methods involved in the AIOLOS code. Section 3 presents the results of the numerical simulation and a comparison with full scale temperature measurements. Finally, section 4 gives a summary and conclusions with recommendations for future work.

2 Mathematical models and numerical methods

The numerical simulation of turbulent reactive flows within industrial utility boilers involves submodels describing fluid flow, turbulence, combustion and radiative heat transfer which are coupled in a strongly non-linear way. In the following, the several parts are briefly introduced. For a detailed description of the models, the reader is referred to Schneider et al. (1995), and Schnell et al. (1995). The system of differential equations, describing turbulent reactive flows, is derived from the conservation of momentum, mass, species concentrations and energy in a non-stationary formulation. Each equation can be expressed in form of a general transport equation

$$\frac{\partial}{\partial t}(\rho\Phi)+\frac{\partial}{\partial x_j}(\rho u_j\Phi)=\frac{\partial}{\partial x_j}\left(\frac{\mu}{\sigma}\frac{\partial\Phi}{\partial x_j}\right)+S_\Phi \quad , \tag{1}$$

by substituting suitable quantities for the variable Φ. Equation (1) describes the local changes of the Favre-averaged quantity Φ due to temporal change, convective and diffusive transport, and a source term contribution. The closure of the Favre-averaged momentum equations is done by either the standard k,ε-model or the differential Reynolds stress model (Leschziner, 1994). Coal combustion is described by a four step reaction scheme containing two heterogeneous and two gas phase reactions, respectively, while the heterogeneous reactions are subdivided into the pyrolysis and the burn-out of the char particles. The interaction between turbulence and reaction kinetics is modelled using the Eddy Dissipation Concept (EDC), proposed by Magnussen (1989). Heat transfer within industrial furnaces is mainly dominated by thermal radiation. Within the AIOLOS code several radiation models are implemented which are distinguished between ray tracking methods and differential methods. The former class is represented by the Discrete Transfer method, proposed by Lockwood and Shah (1980). The latter one involves the six-flux model of de Marco and Lockwood (1975), and the Discrete Ordinates method, introduced by Fiveland (1984). The discretised form of the general transport equation (1) arises from a conservative finite-volume approach with collocated grid definition. The convective and diffusive cell face fluxes are approximated by applying a suitable discretisation scheme leading to the linearised equation

$$\frac{(\rho\Phi)^{n+1}-(\rho\Phi)^{n}}{\Delta t}V + a_P^{n+1}\Phi_P^{n+1} = \sum_{nb} a_{nb}^{n+1}\Phi_{nb}^{n+1} + S_\Phi^{n+1}V \tag{2}$$

for each control volume. Here the subscript nb denotes the neighbours of the grid point P in the centre of the cell. The superscripts n+1 and n represent the actual and previous time level, respectively. The diffusive fluxes are discretised using the *Central Differencing Scheme*. To avoid numerical diffusion, in addition to the *Upwind Differencing Scheme* of first order, higher-order discretisation schemes (MLU, *Monotonised Linear-Upwind*) are applied to the calculation of the convective fluxes, introduced by Noll (1992). The discretisation in time is done by the first order *Euler Implicit* scheme, since it is unconditionally stable. The chosen non-staggered grid arrangement can lead to a decoupling of the velocity and pressure fields, resulting in non-physical solutions. In order to avoid odd-even oscillations of the solution, the pressure-weighted interpolation method of Rhie and Chow (1983) is applied. The coupling between the conservation of mass and momentum is achieved by either the pressure correction method SIMPLEC, proposed by van Doormal and Raithby (1984), or the PISO approach (Issa, 1985).

3 Full scale validation

The purpose of the present chapter is to investigate the predictive capability of the AIOLOS code. In order to assess the numerical simulation, a comparison with full-scale temperature measurements is carried out. In the following, a brief description of the boiler facility, the process data as well as the measurement system is given, before the numerical results are presented.

3.1 Description of the facility

The boiler is a tangential corner firing system with an electrical output of 750 MW. The size of the furnace is 76 m in height and a cross section of about 18 m squared. It is subdivided into 4 burner planes each consisting of 4 jet burners which are fired by pulverised, bituminous coal. The burner consists of 2 coal nozzles, an intermediate air nozzle, and 2 ignitors which deliver air during operation. In order to reduce NO_x emissions, additional overfire air nozzles are installed above the upper burner plane. The furnace is discretised using approximately 310.000 grid cells, with a minimal grid resolution of 8 cm in the burner near field.

3.2 Process data

In figure 1, the control signal of the feeder speed and the electrical output are plotted during a time period of 8 hours. A control signal of 75 % corresponds to full load operation. The diagram is restricted to feeder 1 and 2, since feeder 3 and 4 run equivalent to feeder 2. It indicates a load reduction step by step from full load to part load operation. At the first two steps, the supply of coal at the lowest burner plane (feeder 1) is reduced until it gets out of operation. In the sequel, a stable operation mode can be found until the remaining 3 burner planes are reduced simultaneously, step by step.

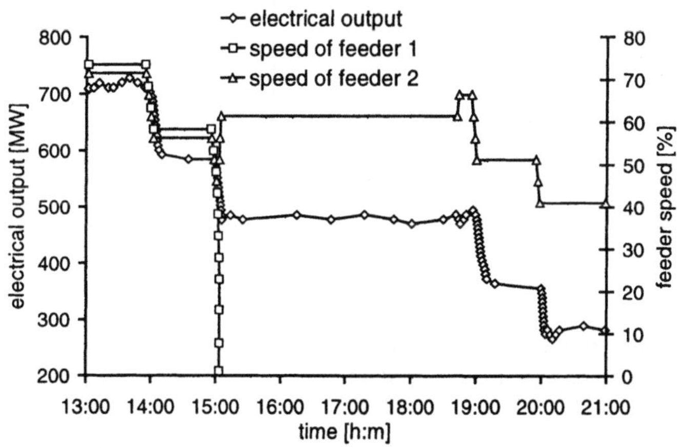

Fig. 1: Electrical output and control signal (corresponding to feeder speed) over a time period of 8h

Since the fuel mass flow is not available, it is indirectly calculated by means of the produced electrical output and the speed of the feeders. In addition, the process data are not collected continuously, but only if the data are changed by a constant incremental. Therefore, fluctuations and even absolute quantities can not be exactly captured.

3.3 Measurements

The temperature measurements are carried out by LSE (Lehrstuhl für Systemtheorie in der Elektrotechnik), Saarland University, Saarbrücken. The mean path temperatures of the flue gas are measured by acoustic pyrometry. The measuring system uses the travel time of an acoustic signal between transmitting and receiving units to determine the mean temperature along an acoustic path. A cycle, i.e. the period in which all paths deliver a measured value, lasts about two minutes. In figure 2, the vertical arrangement of the measuring system as well as the horizontal location of the acoustic paths are shown. For more details of the measuring method and their application in combustion systems, the reader is referred to Spliethoff (1996).

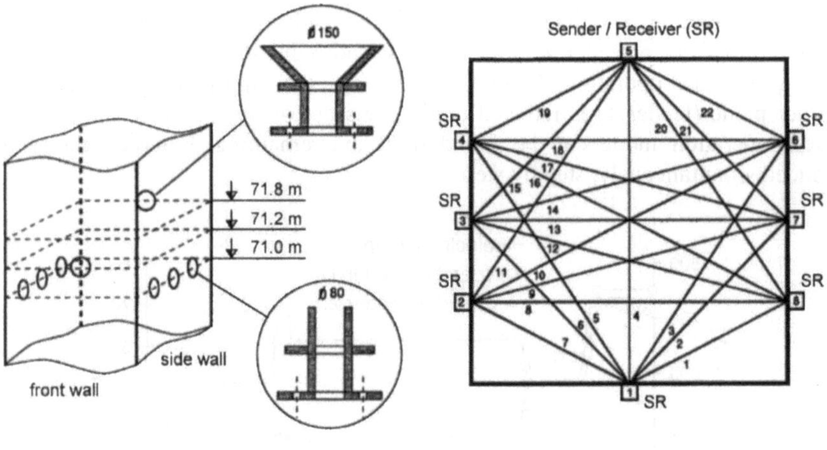

Fig. 2: Vertical arrangement of measuring system (left), and horizontal location of the acoustic paths (right)

Unfavourable to the measuring campaign was that the installed measuring system had to manage with the available openings at the boiler. It was observed that due to the small openings at the side walls some measurements failed, because fly ash covered the openings. Hence, some temperature measurements had to be rejected, since their values were beyond any physical range. In order to reduce the cycle time without significant loss of information, the measurements are restricted to only a few number of paths, while paths of comparable characteristics are combined into path groups by calculating their arithmetic mean value (see Spliethoff, 1996). Classifications are defined by combining all paths passing through the interior of the furnace as well as paths representing the region near corners (path numbers are seen from figure 2, right):

$$\bar{\vartheta}_{inner} = (\vartheta_4 + \vartheta_{10} + \vartheta_{13} + \vartheta_{16})/4$$
$$\bar{\vartheta}_{corners} = (\vartheta_2 + \vartheta_6 + \vartheta_{15} + \vartheta_{21})/4 \,. \quad (3)$$

Furthermore, the measurements sometimes were affected by the noise which was produced by sootblower operation. However, it is ensured that the data considered in the present work are not concerned by these effects.

3.4 Numerical and computational aspects of the simulations

Concerning the numerical simulations, exact analysis of accuracy is not possible for non-linear problems. The temporal discretisation error (also named as truncation error) can be estimated by combining the solutions of the numerical simulations obtained by systematically reduced time steps. For more details, the reader is referred to Ferziger and Peric (1996) where the approximation of the spatial discretisation error is described. Using the SIMPLEC approach and a time step of 0.01 s, the truncation error ranges between 0.01 % and 1 %. Therefore, this solution can be considered as reference. For the SIMPLEC approach and a time step up to 1 s as well as for the PISO approach using time steps up to 0.1 s, it is found that the average deviation between the obtained solutions and the reference solution is less than 1 %. Since no noticeable differences between the various solutions are observed, it is likely that the temporal discretisation error is insignificant compared to the spatial discretisation error. However, an analogous investigation of the spatial discretisation error is not conducted, since only one grid is considered.

The calculations are performed using the parallel vector computer NEC SX-4/32 of the system-management company HWW GmbH (Höchstleistungsrechner für Wissenschaft und Wirtschaft), founded by the High-Performance Computing-Centre of the University of Stuttgart, and industrial partners. The SX-4 is a shared memory computer consisting of 32 vector processors and a main memory of 8 GB. Each vector unit provides a theoretical peak-performance of 2 GFLOP/s leading to an aggregate peak-performance of 64 GFLOP/s. For the present work, the shared memory programming model, supported by the NEC-specific Microtasking concept, is applied. It is based on DO-loop parallelisation. For a comprehensive investigation of vector as well as parallel performance of the AIOLOS code, the reader is referred to Risio et al. (1998).

Access to the SX-4 is generally managed in batch mode. Since the machine mostly works at full capacity, jobs demanding a high number of processors are often queued for days, and even running jobs are often suspended and replaced by jobs of higher priority. Hence, the calculations are carried out using not more than 8 processors.

What follows is a performance analysis of the two numerical schemes SIMPLEC and PISO using various time steps and solution procedures. Within the SIMPLEC approach the transport equations are solved using SOR, while for the solution of the pressure correction equation SIP is applied. This combination of solvers has become an efficient choice, since only a few inner iterations are necessary to get overall convergence. In contrast to this, within the PISO algorithm the linearised equations have to be solved more accurately. Comprehensive investigations of the solvers, implemented in the AIOLOS code, reveal that the convergence of the SOR method is poor, while the CG algorithm (ILU BiCGSTAB, *Bi-Conjugate Gradient Stabilised with Incomplete LU-Factorisation*) outperforms both SOR and SIP. Therefore, ILU BiCGSTAB is applied to the PISO approach.

Table 1 summarises the computational effort spent on the SX-4 for simulating the combustion process during two different time periods, both taking 10 minutes (from 14:04:38 till 14:14:38, in the sequel referred to as time period i; from 15:05:41 till 15:15:41, referred to as time period ii). As mentioned above, during regular operation of the SX-4 no dedicated access is possible. Nevertheless, the performance measurements were carried out ensuring that the demanded number of processors as well as the memory were available, so that the calculations could be considered not to be affected by other jobs running on the machine simultaneously.

Table 1: Performance statistics of the numerical schemes SIMPLEC and PISO for various time steps

	SIMPLEC			PISO	
Time step [s]	0.01	0.10	1.00	0.01	0.10
Time period i [CPUh]	1286	198	25	389	55
Time period ii [CPUh]	4822	627	87	490	58

As can be seen from the measurements in table 1, the convergence of the SIMPLEC algorithm strongly depends on boundary conditions and initial values, respectively. Due to the constant number of outer iterations in case of the PISO method, the calculation times of time period ii are only slightly increased compared to time period i. Using the same time step, the PISO algorithm by far outperforms the SIMPLEC method. Nevertheless, as mentioned above, there is no significant difference between the various solutions obtained with different time steps. A comparison between both approaches, using in each case the time step which leads to a minimal computational effort (SIMPLEC with a time step of 1 s; PISO with a time step of 0.1 s), indicates superiority of SIMPLEC for time period i, but for time period ii just the opposite. Since for the PISO approach the computational effort within a time step is approximately constant, a rough estimation of the total computational effort of the simulations can be a priori obtained, which is not valid for the SIMPLEC method. Therefore, the numerical simulations are carried out using the PISO method with a time step of 0.1 s.

3.5 Numerical results versus temperature measurements

Concerning the comparison of the temperature measurements with the numerical results, the time period from 14:04:38 till 19:17:39 is considered. As mentioned above, the load changes step by step. It is observed that a few minutes after the load reduction is completed a steady operation is found. Hence, the whole time period of about 5 hours is not fully calculated. Instead, right before the load changes, a steady-state solution, serving as initial solution for the transient simulations, is calculated by means of the stationary solver. The transient calculations are performed until the boundary conditions do not change and a steady-state solution is obtained. Referring to table 2, the total time period is subdivided into six parts. They are classified into time periods where the solution is dependent on time, and periods where a steady-state solution is found. As can be further seen in table 2, the total computational effort on the SX-4, including the calculations of the stationary solutions, is about 390 CPUh. Only about 6.6 % of the total computational effort is spent in the stationary solution procedure. Since some jobs were unfortunately suspended during operation, it was not possible to get reliable real time performance statistics. The following reflections are therefore undertaken neglecting loss of speed-up. Assuming optimal speed-up, the whole calculations would last about 2 days using 8 processors, and a half day using 32 processors. Compared with this, the calculations performed on a typical workstation available at the institute (Digital Personal Workstation 433au with a clock cycle of 433 MHz) would even last about 130 days. For smaller time steps the total computational effort would increase excessively.

Table 2: Definition of time periods and computational effort spent on the NEC SX-4

	Period	Classified	Computational effort [CPUh]
I	Till 14:04:38	Steady	10.4
II	14:04:38 – 14:19:38	Unsteady	82.5
III	14:19:38 – 15:05:41	Steady	7.9
IV	15:05:41 – 15:30:41	Unsteady	145.0
V	15:30:41 – 18:43:39	Steady	7.8
VI	18:43:39 – 19:17:39	Unsteady	140.1

Concerning the comparison of measurements and simulation results, the path groups, defined in equation (3), are discussed next. In figures 3 and 4, the numerical predictions of mean path temperatures of the path groups compared with temperature measurements are illustrated. The path group representing the inner region of the furnace is considered for the whole period of about 5 hours, while the path group representing the region near corners is focused on a time interval containing the time periods II and IV.

In each case, the calculated temperatures reflect the similar shape of the load changes and feeder speed, respectively (seen from figure 1). For lack of measurements during nearly the whole time period II as well as for time period IV (see table 2 for the definition of the time periods), the predictions can not be verified. Nevertheless, for those measurements which are available within both time periods II and IV a good agreement with the predicted values is found, e.g. at the beginning and the end of time period IV, seen in figure 4. Fortunately, as can be seen from figure 3, for time period VI a sufficient number of measured values is available. Referring to both figures 3 and 4, the general trend of the measurements is maintained in the calculations. Taking into account an accuracy of the measurements which is estimated to ±40 K (see Breuer et al., 1992), the measured temperatures are sufficiently well reproduced by the calculated values.

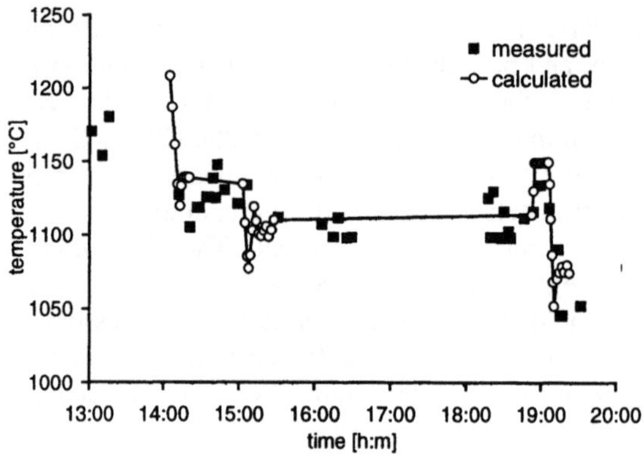

Fig. 3: Comparison of measured and calculated path temperatures: path group defining inner region of furnace

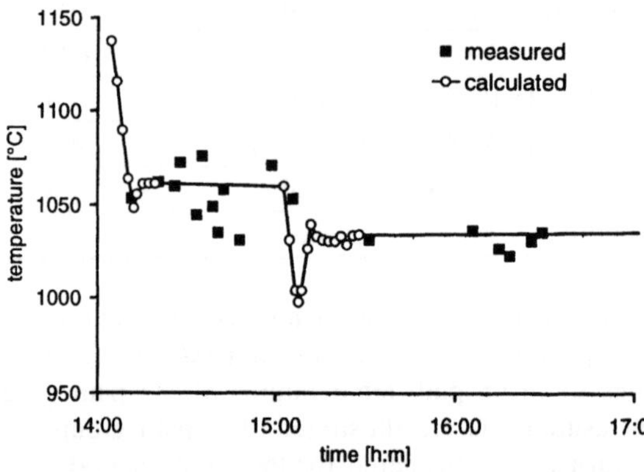

Fig. 4: Comparison of measured and calculated path temperatures: path group defining region near corners

The calculations of the temperatures are further evaluated by a comparison with paths of different characteristics. The comparison is restricted to time period VI, where, as mentioned above, a sufficient number of measured values are available. In figures 5 and 6, the mean path temperatures of paths 9 and 13, respectively, are shown, both representing paths passing through the interior of the furnace. In figures 7 and 8, the temperatures in two different corners are considered, represented by paths 2 and 19, respectively.

Despite the above mentioned uncertainties in the process data, and therefore in setting the boundary conditions, the results indicate (similar to the path groups above) a sufficiently good agreement between the measured and calculated values. In general, the major trends of measured temperatures are reproduced by the predicted values. However, a slight shift of the profiles of a few minutes is observed. This may be attributed to the process data not continuously available.

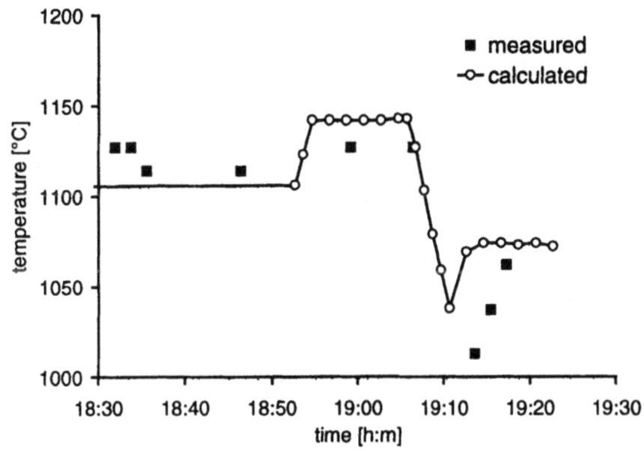

Fig. 5: Comparison of measured and calculated path temperatures: path 9, inner region

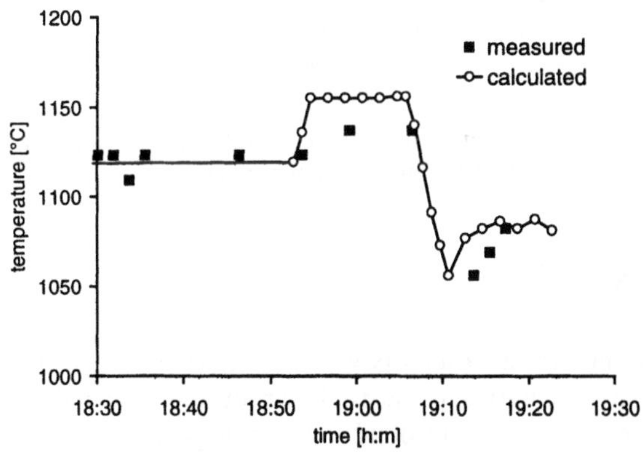

Fig. 6: Comparison of measured and calculated path temperatures: path 13, inner region

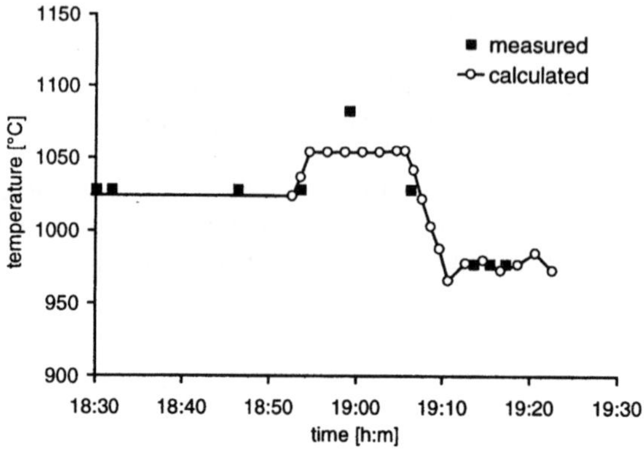

Fig. 7: Comparison of measured and calculated path temperatures: path 2, lower right corner region

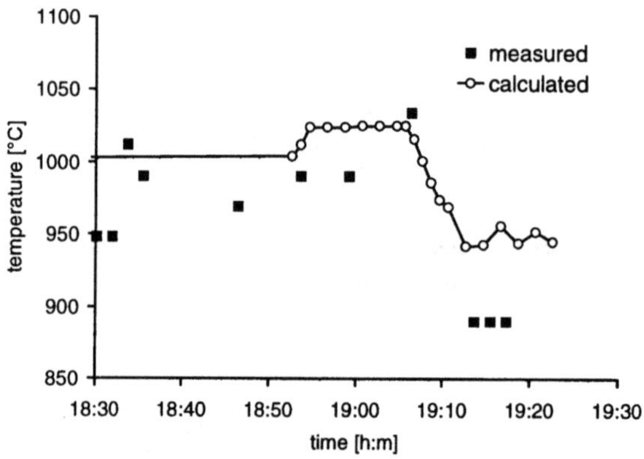

Fig. 8: Comparison of measured and calculated path temperatures: path 19, upper left corner region

3.6 Example of results

An example of the numerical results is given in figure 9, where the temperature distribution at the lowest burner plane is shown during load reduction. The first three pictures in the sequence correspond to the time period II (see table 2), while the remaining pictures describe time period IV. Due to the jet burners which are directed tangentially to an imaginary cylinder in the furnace centre, the flame stream is forced to take on a rotational motion leading to an even temperature profile across the flame axis. Load reduction causes a decrease of the average temperature with local fields of higher temperature giving the impression of a disrupting flame vortex.

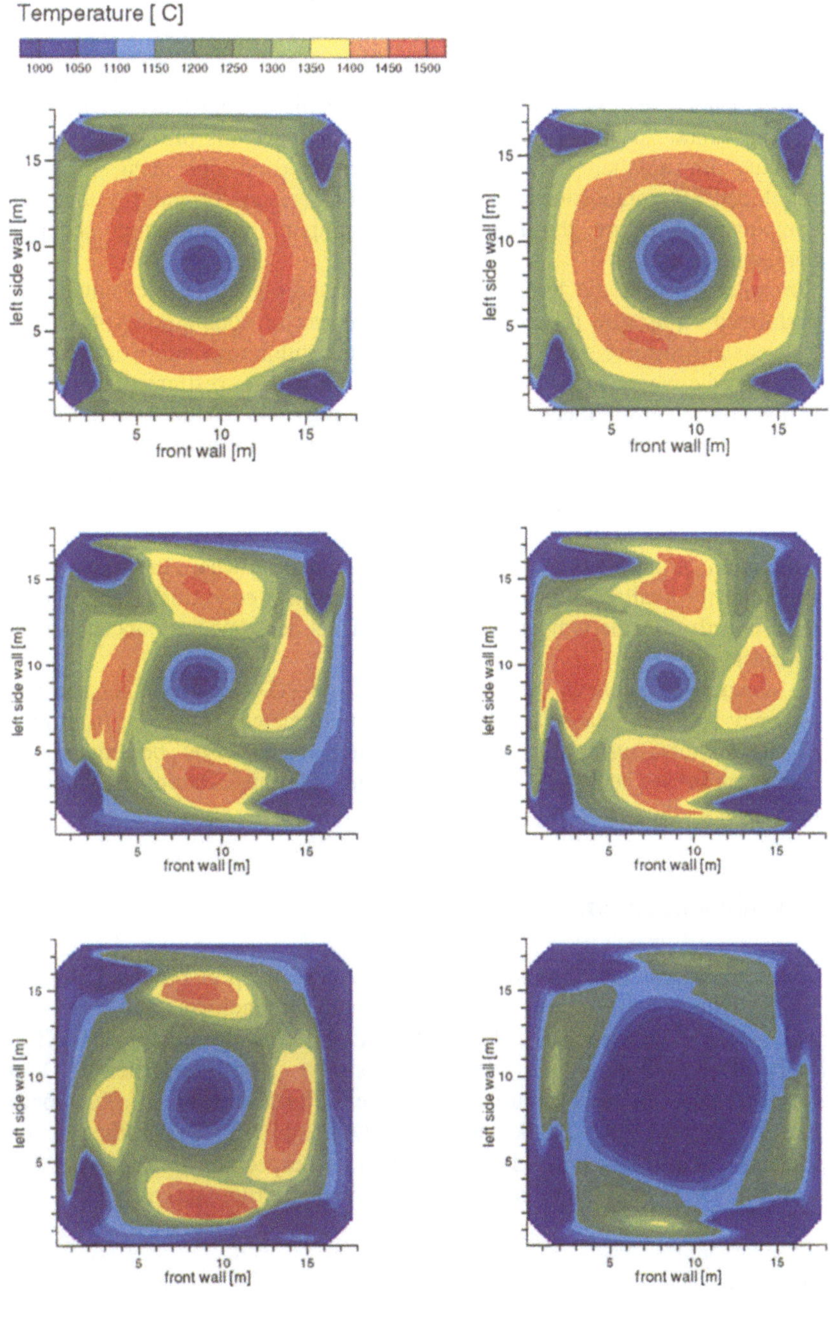

Fig. 9: Sequence of temperature distribution at lowest burner level at 14:05:38 (upper left), 14:08:38 (upper right), 14:12:38 (middle left), 15:06:41 (middle right), 15:10:41 (lower left), 15:13:41 (lower right)

4 Conclusions and further remarks

A validation exercise has been presented based on the numerical simulation of transient reactive flow in a coal-fired utility boiler. From the knowledge of the authors, such 3-dimensional transient simulations of full-scale combustion systems are unique. The calculations were carried out using the in-house developed 3D combustion code AIOLOS. For reasons of stability and minimal computational effort, the *Euler Implicit* temporal discretisation scheme was applied.

In order to meet the high computational requirements, the calculations were performed on up to 8 processors of a parallel vector computer SX-4/32 of NEC. This enables to solve such large-scale problems within reasonable time frames.

The predictive quality of the AIOLOS code was investigated by comparing temperature measurements, carried out during load changes, with numerical results. The comparison revealed a good agreement between calculated and measured values. A parallel shift of the temperature profiles of a few minutes was observed. This may be a consequence of the uncertainties of the process data. However, the major trends of measured temperature values were maintained in the calculations. Future work is focused on an application where the process data are continuously available, ensuring that the boundary conditions can be captured more exactly. Beside temperature measurements, also species concentrations will be considered.

A comparison between the SIMPLEC and the PISO approach, on the basis of accuracy and computational effort, has been attempted. However, from the view of the authors, universally valid conclusions cannot yet be drawn. Further investigation is necessary, also examining the application of sophisticated numerical methods like multigrid solver.

Acknowledgements

The authors gratefully acknowledge the financial support provided through the BMWi (Bundesministerium für Wirtschaft und Technologie) and RWE Energie AG within the project No. 0327066A: "Verbundprojekt KOMET 650: Teilvorhaben Optimierung der Betriebsweise von modernen CO_2-armen Kohlekraftwerken durch detaillierte gekoppelte Prozeßsimulation". We would also like to express our thanks to LSE (Lehrstuhl für Systemtheorie in der Elektrotechnik, Universität des Saarlandes) and especially to Dipl.-Ing. K. Blug and Prof. Dr. H. Spliethoff for the contribution of their measurements.

References

Antifora, A., Coelho, L.M.R., Hepburn, P.W., Hesselmann, G., Irons, R., Martins Neves, L.F., Pasini, S., Perrini, A., Sala, M., Schnell, U., Spliethoff, H.

(1997): Integrating Numerical Modelling and Test Rig Data into the Design of a Coal-Over-Coal Reburn Process at Vado Ligure Unit #4 (320 MW_e), *Int. Joint Power Generation Conference*, Denver, USA

Breuer, N., Derichs, W., Hess, F. (1992): Einsatz akustischer Temperaturmeßsysteme an einem braunkohle-befeuerten Kraftwerkskessel, VGB Kraftwerkstechnik 72, Heft 1

van Doormal, J.P., Raithby, G.D. (1984): Enhancements of the SIMPLE Method for Predicting Incompressible Fluid Flows, *Numerical Heat Transfer*, Vol. 7, pp. 147-163

Ferziger, J.H., Peric, M. (1996): *Computational Methods for Fluid Dynamics*, Springer Verlag, Berlin

Fiveland, W.A. (1984): Discrete-Ordinates Solution of the Radiative Transport Equation for Rectangular Enclosures, *J. of Heat Transfer*, vol. 106, pp. 699-706

Issa, R.I. (1985): Solution of the Implicitly Discretised Fluid Flow Equations by Operator-Splitting, *Journal of Computational Physics*, Vol. 62, pp. 40-65

Knaus, H., Schneider, R., Han, X., Ströhle, J., Schnell, U., Hein, K.R.G. (1997a): Comparison of Different Radiation Heat Transfer Models in Coal-Fired Utility Boiler Simulations Using Boundary-Fitted and Cartesian Grids, *4th Int. Conference on Technologies and Combustion for a Clean Environment*, Lisbon, Portugal

Knaus, H., Maier, J., Schnell, U., Hein, K.R.G. (1997b): Advantages of Applying Boundary-Fitted Grids to the Simulation of Pulverised Coal-Fired Utility Boilers with Mixed Staging Burners, *4th European Conference Industrial Furnaces and Boilers*, Porto, Portugal

Leschziner, M.A. (1994). Refined Turbulence Modeling for Engineering Flows, *Invited Lecture, Proc. 2nd European Comp. Fluid Dyn. Conf.*, pp. 33-46

Lockwood, F.C., Shah, N.G. (1980): A New Radiation Solution Method for Incorporation in General Combustion Prediction Procedures, *Proc. 18th Symp. (Int.) on Comb.*, The Comb. Inst., pp. 1405-1414

Magnussen, B.F. (1989): The Eddy Dissipation Concept, *XI Task Leaders Meeting - Energy Conservation in Combustion*, IEA

de Marco, A.G., Lockwood, F.C. (1975). A New Flux Model for the Calculation of Radiation in Furnaces, *La rivista dei Combustibili*, vol. 29, pp. 184-196

Noll, B. (1992): Evaluation of a Bounded High-Resolution Scheme for Combustor Flow Computations, *AIAA Journal*, vol. 30, pp. 64-69

Rhie, C.M., Chow, W.L. (1983): Numerical Study of the Turbulent Flow Past an Airfoil with Trailing Edge Separation, *AIAA Journal*, vol. 21, pp. 1525-1532

Risio, B., Förtsch, D., Schnell, U., Hein, K.R.G. (1997): Prediction of Pulverized Coal Fired Utility Boiler Performance on Individually Discretized, Non-Matching Grids, *4th Int. Conference on Technologies and Combustion for a Clean Environment*, Lisbon, Portugal

Risio, B., Bundschuh, A., Förtsch, D., Schnell, U., Hein, K.R.G., Klinge, Th., Derichs, W. (1998): Industrial-Scale Validation of the 3D-Furnace Simulation Code AIOLOS, *Int. Symposium on Computational Technologies for Fluid / Thermal / Chemical Systems with Industrial Applications*, San Diego, USA.

Risio, B., Schnell, U., Hein, K.R.G. (1998): Towards a Reliable and Efficient Furnace Simulation Tool for Coal Fired Utility Boiler, *High Performance Computing in Science and Engineering*, Springer Verlag, Berlin

Schneider, R., Risio, B., Schnell, U., Hein, K.R.G. (1995): Application of a Differential Reynolds-Stress Turbulence Model to the Numerical Simulation of Coal-Fired Utility Boilers, *Proc. 3rd International Symposium on Coal Combustion*, Beijing, China, pp. 336-343

Schnell, U., Schneider, R., Magel, H.-C., Risio, B., Lepper, J., Hein, K.R.G. (1995): Numerical Simulation of Advanced Coal-Fired Combustion Systems with In-Furnace NO_x Control Technologies, *Proc. 3rd Int. Conf. Comb. Tech.*, Lisbon, Portugal

Spliethoff, H. (1996): Untersuchungen und Entwicklungen schallpyrometrischer Temperaturmeßverfahren an zwei Kohlenstaubfeuerungen, *VGB Kraftwerkstechnik 76*, Heft 7

Structural Mechanics and Electrical Engineering

Prof. Dr. Wolfgang E. Nagel, Institut für Wissenschaftliches Rechnen
Technische Universität Dresden
Zellescher Weg 12-14, D-01069 Dresden

Prof. Dr.-Ing. Dr.-Ing.E.h. Dr.h.c. mult. Erwin Stein,
Institut für Baumechanik und Numerische Mechanik
Universität Hannover

In this chapter, a selection of engineering applications from structural mechanics and electrical engineering are presented which currently get resources on the HPC platforms at HLRS. While in the past high-end workstations were in most cases sufficient to perform the necessary computations in an appropriate time, today's problems are becoming more and more complex, taking several weeks or even months of CPU time on a single processor. Both engineering areas are not traditional fields for supercomputer usage. Even so, they have evolved as very interesting fields for the use of high performance computers over the last couple of years, and some research groups have made significant effort to improve their codes. Algorithm development as well as programming methods have been targeted to parallel computing, and today some encouraging results from that work can be presented here.

An interesting aspect to observe is the fairly rapid move from old-style Fortran77 codes to more advanced and modern software engineering techniques and implementation languages. The use of object oriented concepts is growing, and today quite a few large production applications that are in regular use have been written in C++. The template concept has enabled very complex structures, and it would have been a major burden on the programmer to implement these codes without such techniques. As one of the examples will show, even latency-hiding techniques for a metacomputer application running MPI over a long-distance transatlantic ATM internet communication line between two parallel CRAY computers (Pittsburgh and Stuttgart) can be integrated easily in such a production code. On the other hand we have to point out that, beyond strongly required competence in the algorithmic and methodological field, the analysis of runtimes and the optimization steps in the diverse phases of a program still is very important. Experience has shown that tools for detailed performance analysis, complemented by a qualified knowledge of the underlying computer architecture, are an absolutely necessity to get the performance expected on parallel architectures.

Design and Application of Object Oriented Parallel Data Structures in Particle and Continuous Systems

Kai Höfler, Matthias Müller, and Stefan Schwarzer

Institut für Computeranwendungen I
Pfaffenwaldring 27,
70569 Stuttgart, Germany

Abstract. The complexity of parallel computing hardware calls for software that eases the development of numerical models for researchers that do not have a deep knowledge of the platforms' communication strategies and that do not have the time to write efficient code themselves. We show in three examples how two "*container*" data structures, one for particles, the other an array for arbitrary data types, can address the needs of many potential users of parallel machines that have so far been deterred by the complexity of parallelizing code. These examples comprise (i) parallel solutions of partial differential equations, (ii) a molecular dynamics simulation of more than 10^9 particles and (iii) a simulation of particles sedimenting in a viscous fluid.

1 Introduction

According to Moore's laws the capabilities of computing hardware increase exponentially in time. At the same time more and more scientific and engineering problems become amenable to numerical treatment and ever more complex and efficient algorithms are being developed to make even better use of the existing computing power. However, in order to enable researchers to make efficient use of both, it is necessary to provide software that hides the complexity of the computing platform and of the algorithm alike.

In this paper we describe the implementation in C++ of two collections of data structures and algorithms *(containers)* that are designed to support the modeling of (i) particle systems in Lagrangian formulation and (ii) the numerics arising from finite differencing of partial differential equations (PDEs) or that of discrete automaton models (spin models of statistical physics, discrete growth models, etc.). These containers employ the C++ template mechanism so that the user can freely specify new data structures on which they will operate *after* the containers have already been written and tested. We will demonstrate how these containers aid in molecular dynamics simulations, the solution of hydrodynamic equations and how they can be made to work together in a large-scale simulation of two types of rigid particles sedimenting in a fluid.

In these containers we have used the locality inherent in many physical problems to our advantage. In the case of particle systems we will require that the interaction between particles vanishes beyond a certain cutoff distance, as is the case for example in granular systems [1]. Likewise, discretization of differential operators requires only knowledge of the variables in local neighborhoods. Thus, we formulate parallel update algorithms such that they require only communication between neighboring processors and in a regular fashion. Moreover, by splitting the loops over particles or array indices in a part involving the update of degrees of freedom at the inter-processor boundary and in a part only involving 'inner' variables, latency hiding can be employed to efficiently overlap computation and communication, details are explained later in the text. The communication is realized by calls to the MPI library that guarantees portability.

The complexity of parallelism is hidden behind an interface that tries to stick as close as possible to the guiding example of the C++ standard template library, making the code easy to learn. The data structures are provided in source code form as independent pieces of software, resulting in comparatively straightforward maintainance. At the same time this concept allows to develop programs for parallel machines on workstations or workstation clusters, later moving the tested program onto a parallel machine for production runs.

To aid debugging, we employ a multilevel safety check and information message concept, providing array bounds checking and "safe" versions for the most important data types of the standard library, tests for memory leaks, etc., all configurable freely at compile time. The containers are supplemented by a graphics library with several graphical output devices, including VRML, OpenGL, X11 and postscript, which is another independent component of the software. Another important part is formed by statistical evaluation tools for histogramming, (auto)correlating, and averaging of quantities, with the same code in serial and parallel environments. A checkpointing concept applies to all components of the source library and permits long lasting runs to be restarted independent on the availability of checkpointing support from the computing environment. The current version of the source code is freely available on request to `p3t@ica1.uni-stuttgart.de`.

2 Particles with short-range interactions

The principle of a Molecular Dynamics (MD) program is: find the interaction partners for every particle, calculate the forces between the particles and integrate the equations of motion. A real MD program can be rather complex, i.e. by using optimization techniques like linked cell algorithms for short range interaction. Different MD programs need different types of particles with different degrees of freedom and forces. Parallel computers with distributed memory add another level of complexity: particles have to be distributed among the processors and interactions across processor boundaries have to be calculated.

With this in mind we developed a MD package for short range interaction. It allows the easy modification of particle type, interaction force, and integration scheme and allows to switch between the serial and parallel implementation without changing the main program. We decided to use C++ and the Message

Passing Interface (MPI) for implementation because both are available on most supercomputers as well as on the workstations where the code is developed. In addition C++ was designed to have no or only a minimal runtime overhead while still offering the advantages of an object oriented language.

2.1 Design

To free the ordinary programmer from the details of bookkeeping and still provide him with the freedom to implement his own particle types, forces and integration scheme we decided to provide a template container for particles similar to the containers that are provided by the standard template library (STL). In addition to the normal members provided by a STL container a particle container provides a template member **for_each_pair** to allow for iterations over pairs of particles. Here the function object concept of STL is applied, the force class has to supply an operator() for two particles:

```
template<class PARTICLE>
class Force{
public:
  inline void operator()(PARTICLE & p1 , PARTICLE & p2){
     Vector force;
     // calculation of f goes here
     p1.force += force;
     p2.force -= force;  // actio = reactio
  }
};
```

This example also demonstrates the possibility to formulate the algorithm in a natural, dimension independent way. To avoid possible performance problems due to the many small vectors used, we employ the template expression technique [2] to perform loop unrolling, inlining and other optimizations at compile time.

The main loop of a MD simulation will look similar to this example.

```
double dt=0.1;                                  // time step
MD_int_vv_start<Particle> int_start(dt);        // integration
MD_int_vv_finish<Particle> int_finish(dt);
while( t<maxT ){
  // velocity verlet integration first part:
  for_each_particle(box.begin(),box.end(),int_start);
  // force calculation:
  box.update();                                 // communication
  box.for_each_pair(myForce);
  // velocity verlet integration second part:
  for_each_particle(box.begin(),box.end(),int_finish);
  // advance time:
  t+=dt;
}
```

2.2 Performance

While recent benchmarks have shown that C++ can compete with C or Fortran [3] it is not clear whether this holds for a particular application. Because performance depends on the abstraction techniques used it is a non trivial task to find a suitable balance between certain techniques and performance. Because the force calculation is the most time consuming part in an MD simulation, the danger of an abstraction penalty at this point is high.

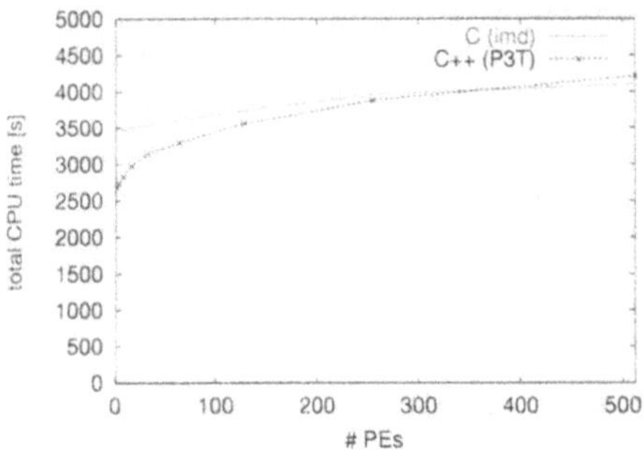

Fig. 1. Total runtime of two different Molecular Dynamics programs depending on the number of PEs.

To get an estimate of the performance penalty of using C++ and its abstraction we made a comparison with a program (imd) written in C that was developed independently [4]. The basic linked cell algorithm is the same. For a simulation of a Lennard-Jones fcc crystal with 442368 atoms the C++ version is between 3 percent slower and 20 percent faster. There are several reasons why the speed up of the C++ program is worse. First, latency hiding is applied and its advantage is lost rapidly with decreasing numbers of particles. Second, Newton's third law is not applied across PE boundaries. This increases the workload on a single PE and results in larger overall runtimes when the cost of the additional communication for exchanging forces is negligible. For large particle numbers the overlapping of communication and computation will also overcome performance problems due to possible congestion of the T3Es network [5].

3 Application: Metacomputing

Metacomputing employs several computers where the capability of one alone is not enough. We participated in a testbed that coupled the T3E of the HLRS and a similar T3E at PSC Pittsburgh resulting in 1020 nodes available to the application.

Metacomputing puts more stringent constraints on the techniques for latency hiding and efficient use of communication bandwidth within the program.

The container addresses the bandwidth problem by providing ways to restrict the amount of communicated particle data if necessary and the PACX-MPI metacomputing library uses data compression schemes to further reduce the amount of data exchanged between the connected machines.

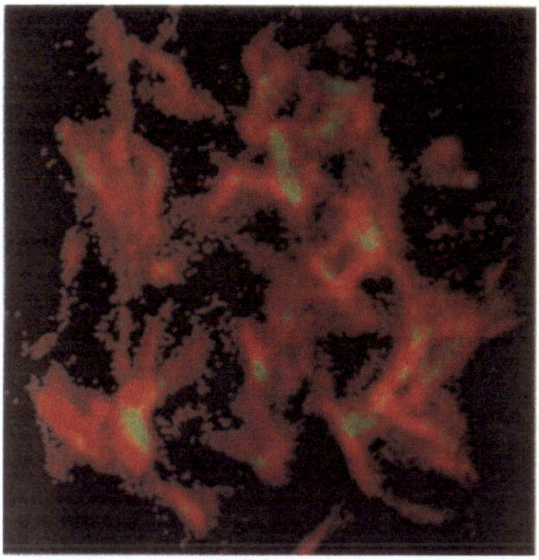

Fig. 2. Cluster formation in a three dimensional granular gas.

Latency on a metacomputer depends on the interconnection between the computers and the protocol used. For PACX-MPI using TCP/IP via the external I/O interface on a T3E the latency between a a T3E in Pittsburgh and its counterpart in Stuttgart adds up to 70ms. To achieve latency hiding it was decided to group the communication in two sections. First the communication for the first dimension is initiated by calls to `MPI_Isend`. In a second stage this communication is completed and the calls for all remaining dimensions are performed, allowing a partial overlap of communication with computation. In the basic version of the template container all communication is done in `update`. Now this call is split into `update_init` and `update_wait` representing the two stages of communication. In this application the force calculation for particles that interact only with particles of the core domain is performed between the first and the second stage. The optimized container provides the calls `for_each_inner_pair` and `for_each_bound_pair` to calculate the pairwise interaction for the corresponding domains.

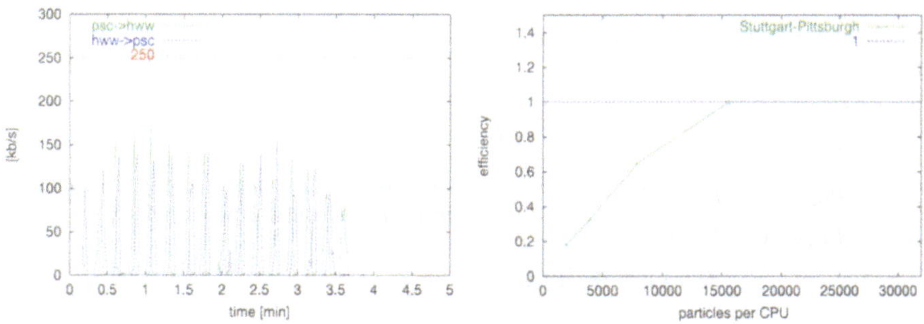

Fig. 3. Communication pattern of a typical DSMC run on a metacomputer (left). Efficiency of the metacomputing run compared with the same programm running on one single computer with the same number of nodes.

With these optimizations we proved that a metacomputer can be as efficient as a single huge computer with a Direct Simulation Monte Carlo (DSMC) simulation that used the described particle container (see Fig. 3). We were also able to demonstrate the power of metacomputing by simulating 1.759.165.695 particles in a transatlantic DSMC simulation of a granular gas (see Fig. 2).

4 Quadrilateral arrays, regular grids

Our second example is the implementation of regular quadrilateral arrays of arbitrary data structures. We use the concept of inheritance to create class hierarchy which represents the levels of abstraction. The classes are templates in order to work for any user specified data type. The base class `PT_Array<TYPE>`[1] mimicks the behavior of dynamically allocated C arrays in d dimensions, and permits arbitrary index ranges and optional index checks for testing. We have found in benchmarks that, depending on compiler and architecture, there is no or only a small overhead in comparison to built-in C arrays.

For array data that needs to be distributed over several PE's (e.g. the discretized physical quantity in a PDE), we provide `PT_PArray<TYPE>`, which automatically arranges for a balanced distribution of the index range over the available processing nodes and provides shadow rows that contain copies of the data on neighboring processors and consistent with the specified boundary conditions (e.g., periodic or open).

As an example we describe some details of an implementation of a 3D Poisson solver, employed as a part of a multigrid scheme for the Navier-Stokes equations

$$\nabla^2 p = f(\mathbf{x}). \qquad (1)$$

which can be discretized at each lattice point (i, j, k) by

$$\begin{aligned} \frac{1}{h^2}(\; & p_{i-1,j,k} + p_{i+1,j,k} \qquad (2) \\ & + p_{i,j-1,k} + p_{i,j+1,k} \\ & + p_{i,j,k-1} + p_{i,j,k+1} \\ & \qquad - 6p_{i,j,k} \;) = f_{i,j,k}. \end{aligned}$$

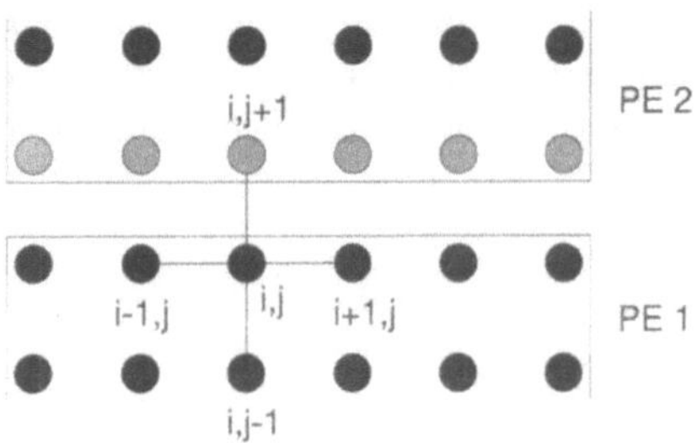

Fig. 4. 2D slice of the computational grid decomposed onto 2 processors.

In order to compute the left hand side of Eq. 2 at (i, j, k) we need the values of p also on the neighboring grid points. To do that next to the inter-processor boundary we must provide one shadow row in each direction. In Fig. 4 these are shown in grey to demonstrate that they only contain copies of the current data kept actually on PE 2.

Apart from the data distribution across PE's one also needs a generic way to handle the necessary communication between the PE's and to apply arbitrary

[1] In the following, we will restrict ourselves to the 3 dimensional case, but the general approach is not limited.

boundary conditions. To this end, we use (predefined) function objects that inherit a common interface from a class PT_PArray_update and that are registered with PT_PArray after construction.

4.1 Equation solvers

Due to the fact that the class interface is the same for serial and parallel versions, the same source code can be used in both cases. This is demonstrated for instance in the example below where a Gauss-Seidel smoothing is performed. The method update_bc() ensures consistency of the shadow values with boundary conditions and across processors. The single processor version of this subroutine does not perform communication, but uses copy operations on shadow data elements to implement boundary conditions. The operation to be performed on the data elements of the array is specified in the object GaussSeidel in the same fashion as in the Force object of Sec. 2.1. The two members begin() and end() return the index values of the PE local part of the parallel array. Please note that all, if any, parallelism of the program fragment is hidden within the PT_Parallel object and its member functions.

```
// node for hydrodynamics computation
struct Node { Vector v; double p; };
// .. distributed over processors
PT_PArray<Node>   fluid(...);

while( <not sufficiently smoothed> ){
  // update shadow rows consistent with boundary conditions
  fluid.update_bc();
  // apply GaussSeidel operator alternatingly linewise
  apply_alw(fluid,fluid.begin(),fluid.end(),GaussSeidel);
}
```

More intelligence is built into the arrays by providing specific classes for solution schemes. In the case above, a full multigrid hierarchy is erected on top of the array holding the variables of the PDE to be solved and the Gauss-Seidel smoother is part of it. Further technical details are described in [6].

4.2 Code Example

We now show a more complete code example which demonstrates the usage of the classes mentioned above.

After the declaration of the sizes (line 1-5) we create an object describing the layout of the PE's (line 7) and two double arrays (line 8,9) which hold the values of the two physical fields of the Poisson equation (1). After computing the right hand side of the Poisson equation (line 9) and assigning it to the Poisson equation class PT_Poisson (line 11) which provides the full multigrid solver for the Poisson equation in the method solve. Like the particle containers, also PT_?Array provides input-output operators to save and restore its content to and from files (line 18) either in ASCII or binary format.

```
const IntVector cpu_layout(2,4,2);
const IntVector boundary_cd(1,0,1);
const int       dimension=3;
const int       n_shadow=1;
const IntVector array_size(257,513,257);

PT_Parallel      PE_layout(argc,argv,Dimension,cpu_layout,boundary_cd);
PT_PArray<double> pressure(PE_layout,field_size,n_shadows);
PT_PArray<double> div_f(PE_layout,field_size,n_shadows);
PT_Poisson       poisson(PE_layout,field_size,n_shadows);

// ... compute rhs (div_f) and solve Poisson equation
poisson.rhs() = div_f;
poisson.solve();

pressure       = poisson.x();
ofstream  save("restart.dat");
save << pressure;
```

4.3 Performance

As already mentioned above, the use of inlining, template mechanisms and aggressively optimizing compilers produces executables which experience no or only minor performance penalties due to the employed abstraction mechanisms as compared to programming languages optimized for procedural design. In Figure 5 we show the parallel efficiency, i.e., the ratio of execution time of a scaled algorithm on N nodes T_N to execution time of the 1 processor subproblem T_1, of the class `PT_PArray<double>` when executing the loop shown in section 4.1. Only at specific sizes of the array we observe a decrease of the parallel efficiency which might be due to a large number of cache misses in this cases.

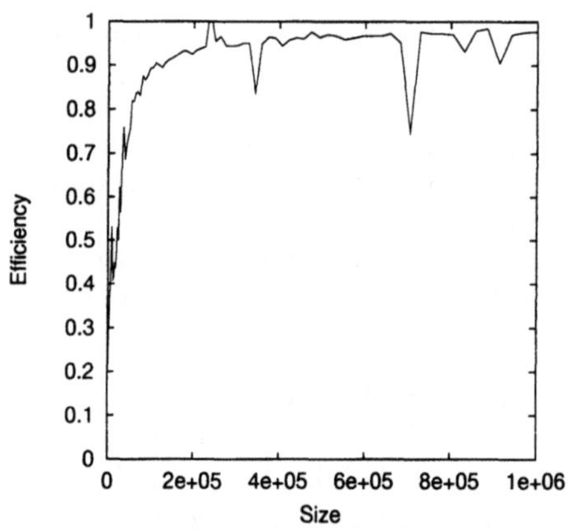

Fig. 5. Parallel efficiency of the Gauss-Seidel iteration for $N = 3 \times 3 \times 3 = 27$ PE's versus the total number of data elements in a cubic array.

5 Application: Multi-Phase Flow

As one application of the software described above, we consider the direct solution of the Navier-Stokes equation in three dimensions in the presence of moving, rigid particles. This problem has outstanding technological importance as it arises, e.g., in the physics of fluidized beds or catalytic chemical reactors and at the same time some fundamental questions concerning the container shape and size dependence of the fluctuating physical quantities in this system still remain open.

This type of multiphase flow problem can quite straightforwardly be addressed with the software components described above. First, we model the particle phase using the Lagrangian particle container described in 2.1, complementing the data structures to account for translation and rotation. Second, the fluid is modeled by the Navier-Stokes equations, discretized on a staggered grid. The fluid-solver uses `PT_PArray`'s of 3 dimensional vectors for the degrees of freedom of the Navier-Stokes equation and a `PT_Poisson` class to solve the pressure equation arising in the employed time-explicit pressure-implicit solution scheme. The no-slip boundary conditions on the particle surfaces are imposed by Lagrangian multipliers appearing as force density in the fluid code. Their values are computed from the trajectory of massless tracer particles advected by the fluid at the location of the particle boundary [7, 6, 8]. These tracers are likewise stored in a particle container.

Using this model, we have performed simulations of the rheological behavior of monodisperse suspensions [9], of up to about 10^6 particles, typically a few ten thousand. Simulations of bidisperse suspensions (Fig. 6) are under way. Only in direct simulations of systems of this size one can observe the collective phenomena like vortex formation and breakup that control the physics [10]. The goal of our work is to finally obtain a more precise continuum formulation of two-phase flow at low and moderate Reynolds numbers $Re \approx 20$ on the particle scale.

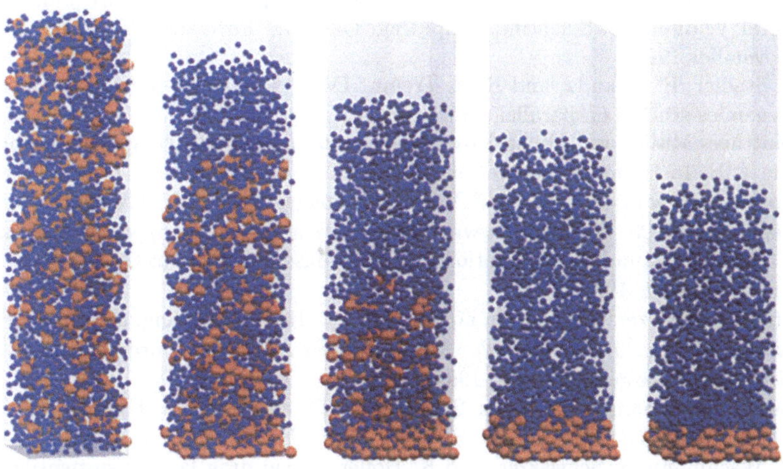

Fig. 6. Timeseries of the sedimentation process of a bidisperse suspension of 875 large (red) particles of radius twice that of the 7000 small (blue) particles. The fluid is discretized on $65 \times 257 \times 65 \approx 10^6$ grid points.

6 Conclusions

We have shown in demanding applications that object-oriented concepts help to design flexible, portable and easy-to-use tools for important problem classes frequent in science and engineering. In particular, judicious application of the template and inlining mechanisms of C++ *can* lead to program performance at par or only slighlty worse than that of classical procedural languages. In several cases, the container abstractions described above have proved to yield programs easily portable to parallel platforms and between 2 and 3 dimensions.

As an technical aside we see that the use of optimizing C++ compilers is often severely hampered by the available memory and file transfer rates. In addition, on the software side, compiler vendors only recently support the important new features of C++ specified in the ANSI/ISO standard.

We acknowledge financial support of the SFB 382, the SFB 404 and our "Bundesprojekt" at the HLRS for the research into granular systems and suspensions. We would also like to thank the PACX-MPI development team at the HLRS for their effort and help and several colleagues at the ICA1 who take an active part in developing and improving the described software, among others Oliver Kitt, Christian Manwart, Reinmar Mück, Gerd Sauermann.

References

1. H.J. Herrmann and M. Müller. Simulations of granular media. In N. Attig and R. Esser, editors, *Molecular Dynamics on Parallel Computers*. World Scientific, 1999.
2. Todd Veldhuizen. Expression templates. *C++ Report*, pages 26–31, June 1995.
3. Todd Veldhuizen. Scientific computing: C++ vs. Fortran. *Dr. Dobb's Journal*, November 1997.
4. J. Stadler, R. Mikulla, and H.-R. Trebin. IMD: A software package for molecular dynamics studies on parallel computers. *Int. J. Mod. Phys. C*, 8:1131–1140, 1997.
5. Matthias Müller and Michael Resch. Pe mapping and the congestion problem on the T3E. In Hermann Lederer and Friedrich Hertweck, editors, *Proceedings of the Fourth European Cray-SGI MPP Workshop*, pages 20–28. IPP, Garching, Germany, September 1998. see http://www.rzg.mpg.de/mpp-workshop/proceedings.html.
6. Kai Höfler. Räumliche Simulation von Zweiphasenflüssen. Master's thesis, Universität Stuttgart, 1997.
7. Stefan Schwarzer. Particles in liquids. In J.-P. Hovi, S. Luding, and H. Herrmann, editors, *Physics of Dry Granular Media*. Kluwer Academic, Dordrecht, 1998. [Proc. NATO ASI Cargèse, Corsica, 1997].
8. Bernd Wachmann and Stefan Schwarzer. Three-dimensional massively parallel computing of suspensions. *Int. J. Mod. Phys. C*, 9(5):759–775, 1998.
9. B. Wachmann, S. Schwarzer, and K. Höfler. Local drag law for suspensions from particle-scale simulations. *Int. J. Mod. Phys. C*, 1998. [Refereed proceedings of the 7th International Conference on the Discrete Simulation of Fluids, Oxford, UK, 1998].
10. P. N. Segrè, E. Herbolzheimer, and P. M. Chaikin. Long-range correlations in sedimentation. *Phys. Rev. Lett.*, 79(13):2574–2577, 1997.

Computation of Electromagnetic Fields by the Method of Moments on the CRAY T3E: Iterative Solution Techniques and Large Scale Applications

Ulrich Jakobus

Institut für Hochfrequenztechnik, University of Stuttgart, Pfaffenwaldring 47, D-70550 Stuttgart, Germany

Abstract. For certain classes of electromagnetic radiation and scattering problems, fast and efficient techniques have been developed, which enable the solution of these problems on relatively small desktop computers or workstations. However, for validation purposes it is indispensable to have accurate reference data available, which we gain from an application of the method of moments (MoM) as implemented in the computer code FEKO on the CRAY T3E. Iterative solution techniques are presented as a means of reducing run-time and improving parallel scalability, and examples for large scale applications are given.

1 Introduction

The classical method of moments (MoM) [1] is a very accurate technique for the solution of electromagnetic radiation and scattering problems for metallic or homogeneous dielectric bodies.

The MPI-based parallel implementation of the various phases of the MoM in the computer code FEKO has already been described in detail in [2,3]. The main bottleneck for the analysis of electrically large problems (i.e. problems where the geometrical dimensions exceed several wavelengths) is the solution of a dense system of linear equations. For surface modelling, the number of unknowns N grows proportional to the square of the frequency f, resulting in a computational complexity proportional to f^6. So far, a direct LU decomposition based on ScaLAPACK [4] has been used in FEKO. In Section 2 we investigate iterative solution methods as an alternative with better scalability.

In order to reduce memory and CPU-time requirements of the MoM, several fast techniques have been developed in recent years, e.g. the IML (impedance matrix localisation) [5], the AIM (adaptive integral method) [6], wavelet-based methods [7–10] or the FMM (fast multipole method) [11–13], just to name a few. We have also implemented a sequential version of the FMM, and some first results of this method executed on a Pentium II Linux system are compared to a full MoM solution on the CRAY T3E in Section 3.

Apart from the mentioned fast techniques, which are quite general and therefore applicable to a wide class of problems, it is possible to develop even more efficient and faster methods for specific problems. For the analysis of electromagnetic sources (e.g. a portable telephone) radiating in the close vicinity of a human person with an implanted pacemaker, we have developed such a computational technique by hybridising the MoM with the GMT (generalised multipole technique) [14,15]. Choosing the correct order of the multipole expansion and especially the optimal spatial positioning of the multipoles is very critical with respect to the solution accuracy, and reference values obtained through the traditional MoM on the CRAY T3E proved to be indispensable for validation purposes, see Section 4.

2 Iterative solution methods

As already mentioned in the introduction, the most time consuming step in the MoM solution of problems with a large number of unknowns is the solution of the dense complex system of linear equations. On the CRAY T3E using e.g. 256 nodes, the available main memory allows us to store a matrix with about 40,000 unknowns (not using out-of-core techniques), and the CPU–time for the LU decomposition with ScaLAPACK then dominates over the other phases of the solution process, such as geometry setup, matrix fill, or the near– and far–field computations.

A variety of different iterative solution methods exist [16], and also their application to electromagnetic problems can be found in literature, e.g. [17–20]. The main drawback is that the convergence behaviour is rather discouraging for most examples, and also by applying preconditioning techniques, convergence does not improve dramatically in general.

By trying different iterative techniques in combination with different preconditioners and by varying some of the parameters (e.g. block–size of a block Jacobi preconditioner), it is most often possible to find an iterative technique which shows a very satisfying convergence behaviour. But this selection can hardly be generalised. Applying the same solver to a different structure (e.g. dielectric problem versus metallic problem or just a different geometry or frequency range) may fail or take much longer than the LU–decomposition. After having compared different iterative techniques for the same problem, one can in most cases claim that one specific choice outperforms the LU decomposition. But of course we want to know in advance which method should be selected for which problem, and then convergence should be guaranteed and the required CPU–time should preferably not exceed the LU decomposition and back–substitution time.

Different iterative techniques were compared (see e.g. [21]), including CG, BiCG, GMRES, CGR, CGNE, CGS, QMR, TF–QMR and others. These were taken from packages such as PIM [22], QMRPACK [23], or also implemented from scratch. To illustrate these comparisons, two representative results are depicted in Figs. 1 and 2.

In Fig. 1 the scattering of a thin dielectric plate is investigated with $N = 3969$ unknowns. For this specific problem, the volume equivalence principle is

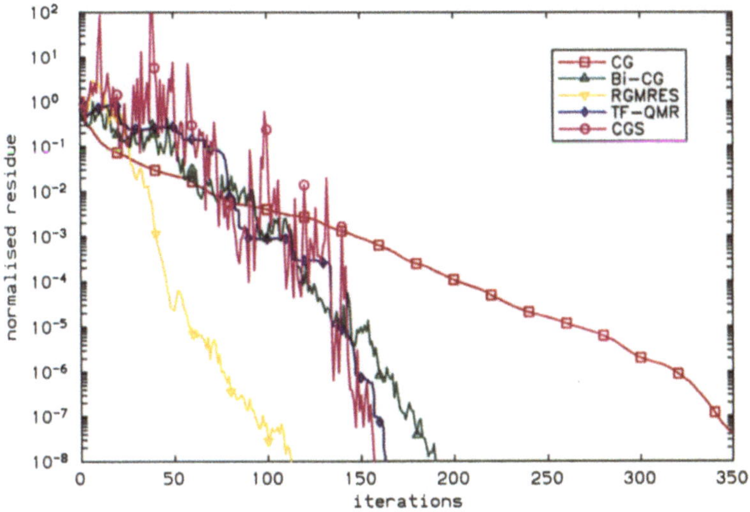

Fig. 1. Convergence of the iterative solution for a dielectric scattering problem (volume equivalence principle) with $N = 3969$ unknowns. Comparison of different iterative techniques.

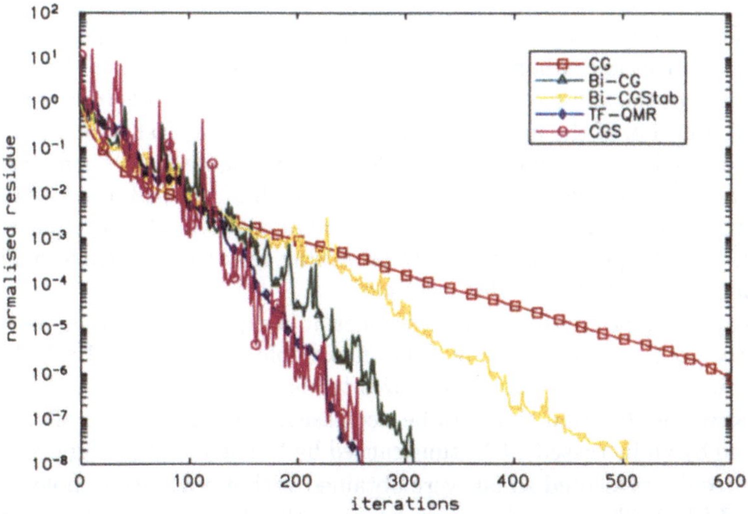

Fig. 2. Convergence of the iterative solution for a metallic scattering problem with $N = 1068$ unknowns. Comparison of different iterative techniques.

applied, resulting in a Fredholm integral equation of the second kind, which explains the excellent convergence behaviour for the various methods (for most of the iterative methods under consideration, $\frac{N}{6} \approx 662$ iterations denote the break–even point when the number of floating point operations is the same as for a LU decomposition).

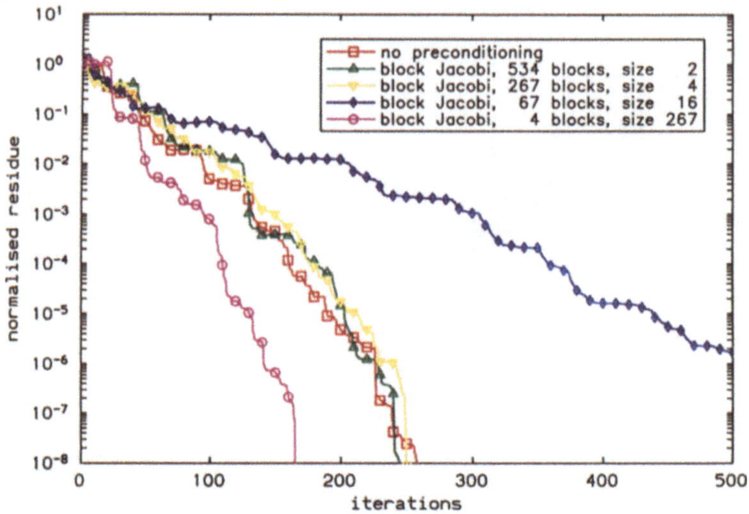

Fig. 3. Convergence of the iterative solution for a metallic scattering problem with $N = 1068$ unknowns. TF–QMR is used with different preconditioners.

Unfortunately, most of the practical problems we aim to solve are either metallic or dielectric, when the surface equivalence principle must be applied as opposed to the volume equivalence principle, and this leads to a Fredholm integral equation of the first kind. Convergence is not as fast then, see Fig. 2 for an example. The underlying problem in this figure is the computation of the radar cross section of a metallic cube with $N = 1068$ unknowns. For instance, for a stopping criterion of 10^{-6} none of the methods requires less than $\frac{N}{6} = 178$ operations. The very promising restarted GMRES in Fig. 1 does not even converge for this example.

By applying preconditioning techniques, it is expected to reduce the number of required iterations, see Fig. 3 for an example of applying a Block–Jacobi preconditioner with different block sizes to the example of Fig. 2. For large block sizes the iteration time can be decreased, but this effect is often compensated by an increased CPU–time caused by the preconditioning overhead.

All results presented so far were obtained with a sequential implementation of FEKO. They can, however, be generalised for a parallel version, as far as the convergence behaviour (e.g. graphs in Figs. 1 to 3) and the number of required iterations are concerned. We have not presented any absolute CPU–times for the sequential version, since we found that the behaviour on parallel machines is somewhat different.

For various reasons (e.g. hybridisation of the MoM with asymptotic high–frequency techniques, see [2]), we are using a one–dimensional block cyclic row distribution scheme for the matrix of the system of linear equations in a parallel distributed memory environment. The consequence is, that a parallel matrix–vector product can be performed quite efficiently with the

Table 1. Performance of CGS with a block–Jacobi preconditioner (block size 2) as compared to ScaLAPACK for the analysis of a metallic cube on different parallel machines with a varying number of processors. Run–times include preconditioning (CGS) and back–substitution (ScaLAPACK).

number of unknowns	machine	number of nodes	CGS iterations	CGS run–time in sec	ScaLAPACK run–time in sec
5724	Linux–Cluster	2	542	2107.4	3425.6
	Pentium II 266 MHz	4	567	1483.2	2157.9
9096	CRAY T3E–900	16	754	442.6	317.2
		32	740	204.6	190.3
		64	771	108.3	168.3
		128	725	57.5	145.7
18468	CRAY T3E–900	64	1546	833.9	—
		128	1486	429.1	1055.2
		256	1506	250.8	836.6

PZGEMV subroutine from PBLAS as opposed to the product transposed matrix times a vector. Some of the iterative Krylov subspace techniques, e.g. CGS (conjugate gradient squared) and TF–QMR (the transpose free variant of quasi minimum residual) require only the product matrix times a vector and avoid the computation of the product transposed matrix times a vector. These two methods CGS and TF–QMR also showed the most promising and robust convergence behaviour during the sequential tests, albeit in a sequential implementation the CPU–time cannot compete in general with a direct LU solution (for the LU decomposition a highly efficient BLAS level 3 implementation of ZGEMM for Pentium II processors was used).

The situation is different for a parallel implementation. Table 1 summarises some of the benchmark computations that were performed. It can be noted that for a large number of nodes the CGS solution requires considerably less CPU–time than the ScaLAPACK application. Also ScaLAPACK requires some additional internal buffers, and this prohibits for instance the solution of the problem with $N = 18468$ unknowns on the CRAY T3E using 64 nodes. It was found that the CPU–time for ScaLAPACK is very sensitive concerning the block size N_B for the block cyclic row distribution scheme (not related to the block size for the block Jacobi preconditioner), see e.g. [3, Fig. 6]. As a consequence, for some examples ScaLAPACK scales quite well (e.g. [3, Table 1]), but for others it does not. The run–time for CGS is not so sensitive on N_B, but of course the effect of the preconditioner block size is significant. For instance for the $N = 5724$ problem on the Linux cluster with 4 processes, the run–time of 1483.2 sec can be reduced to 1053.2 sec by applying a simple symmetrical diagonal preconditioner. The same technique, however leads to divergence for the larger problems. For all examples consid-

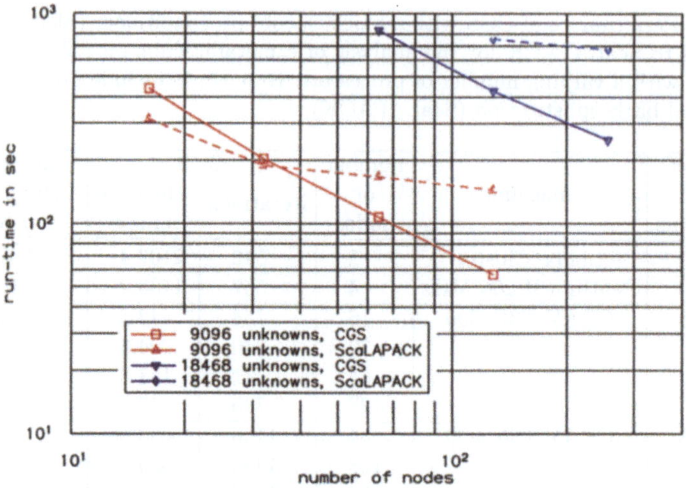

Fig. 4. Run–time for CGS and ScaLAPACK on the CRAY T3E as a function of the number of nodes for the two metallic scattering problems in Table 1.

ered, a block Jacobi preconditioner with a block size of 2 leads to satisfactory convergence, even though it is not the optimal choice for some examples.

In order to demonstrate the scaling behaviour of CGS as compared to ScaLAPACK, Fig. 4 graphically presents the run–times from Table 1 for the two CRAY T3E–900 examples. It is obvious that the slope of the two CGS results is much closer to the optimum curve than that of ScaLAPACK.

3 Fast multipole method (FMM)

A brief overview of fast techniques (e.g. AIM, FMM, IML) was given in the introduction. The main motivation is the reduction of both memory requirement and CPU–time as compared to a full MoM solution. For instance, for the FMM [11–13] this goal is achieved by grouping basis functions and by considering the interaction between distant groups through a multipole expansion. We have implemented a 2–level FMM [24], and for the validation of the results and for the optimisation of parameters (e.g. the order L of a so-called transfer function, see cited references for more details) the availability of a full MoM reference solution obtained on the CRAY T3E turned out to be extremely useful. The FMM itself has not yet been implemented in parallel, but due to the much lower memory requirement, we are able to execute it on a standalone PC or workstation.

One example shall be discussed: When construction works in an airport area are planned and a crane must be erected, it is important to address the possible effect of the metallic structure on the navigational radar system. A simplified solid model of the crane jib is shown in Fig. 5, its length is

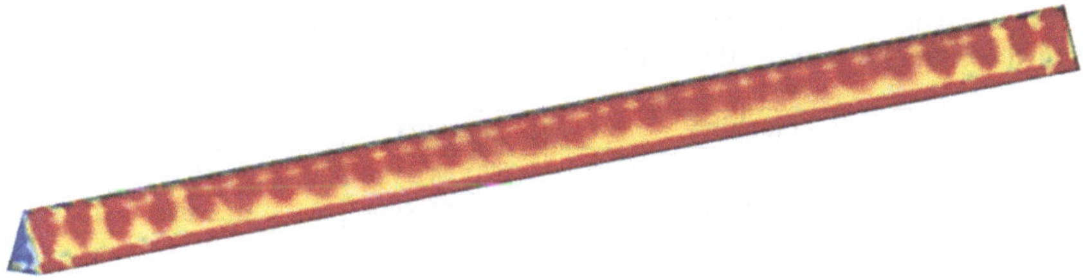

Fig. 5. Surface current density on the solid model of a crane jib illuminated by a plane wave at a frequency of 170 MHz (horizontally polarised).

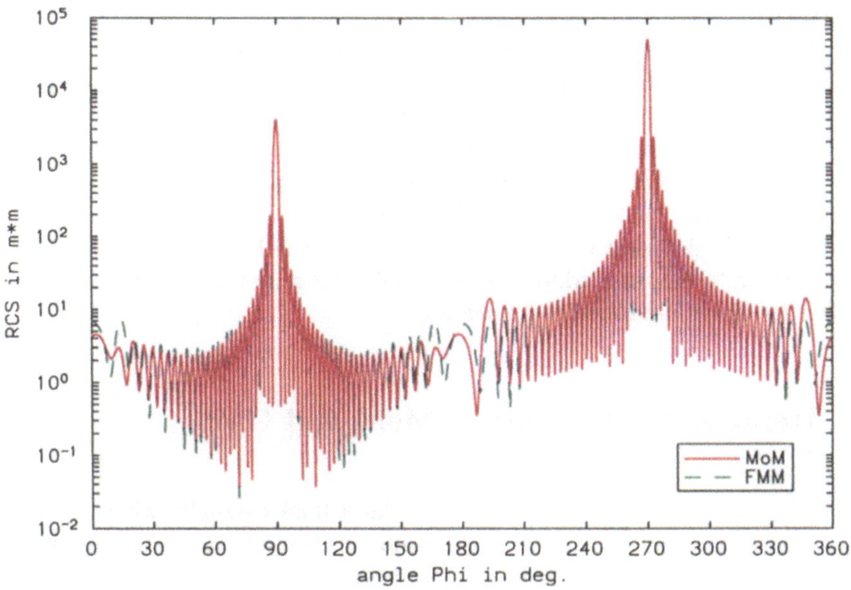

Fig. 6. Bistatic radar cross section of the crane jib for the MoM and the FMM.

50 m. The FMM is very well suited for such a long and small structure, since then the separation distance of many basis and weighting functions, which are defined on the metallic surface, is larger than a certain threshold. As a consequence, the near-field interaction matrix will be highly sparse.

The magnitude of the surface current density for the FMM solution is depicted in Fig. 5. Optically it cannot be distinguished from the MoM solution (not shown here). Some small differences can be realised in the computed bistatic radar cross section, which is plotted in Fig. 6 as a function of the azimuth angle φ in the horizontal plane. The plane wave is incident from $\varphi = 90°$, and in both angular ranges close to forward scattering ($\varphi = 270°$) and back scattering ($\varphi = 90°$) there is an excellent agreement between the FMM result and the MoM reference solution. Small differences can be observed

Table 2. Memory requirement and total CPU–time for the analysis of the scattering form the crane jib with FMM and MoM.

method	machine	memory	CPU–time
FMM	Pentium II 400 MHz (seq.)	523 MByte	15.4 h
MoM	CRAY T3E–900, 64 nodes	5164 MByte	$64 \times 18.1\,\text{min} = 19.3\,\text{h}$

only in the perpendicular directions, but the accuracy can be improved by increasing the order of the transfer-function for the FMM (here only $L = 8$ was used).

The geometrical model is constructed of 11696 triangular patches, leading to 17544 basis functions. Memory and CPU–time requirements for the FMM and MoM can be found in Table 2. Memory can be reduced by almost a factor of 10. The CPU–time is difficult to compare on the two different machines, and also for the FMM we were not yet able to fully exploit optimisation possibilities (e.g. using sparse BLAS subroutines).

Future development includes the extension of the FMM to the treatment of dielectric bodies and addressing the challenging problem of the parallelisation for distributed memory machines.

4 Iterative combination of MoM and GMT

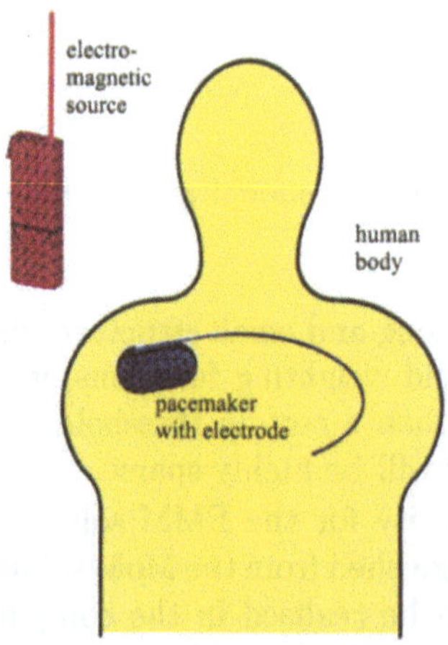

Fig. 7. Model of an electromagnetic source in the vicinity of a human body with implanted pacemaker.

As a final example, where CRAY results of the full MoM turned out to be very useful for validation purposes, a study conducted recently concerning the coupling of electromagnetic fields into the electrode of implanted pacemakers shall be mentioned, see Fig. 7 for a typical configuration. Since for the worst case search regarding the induced input voltage a huge number of analyses had to be performed (e.g. different electrode types, different pacemaker positions in the human body, different position, distance and orientation of the electromagnetic radiator in front of the human body etc.), an efficient computational method based on the hybrid combination of MoM and GMT (generalised multipole technique, sometimes also referred to as

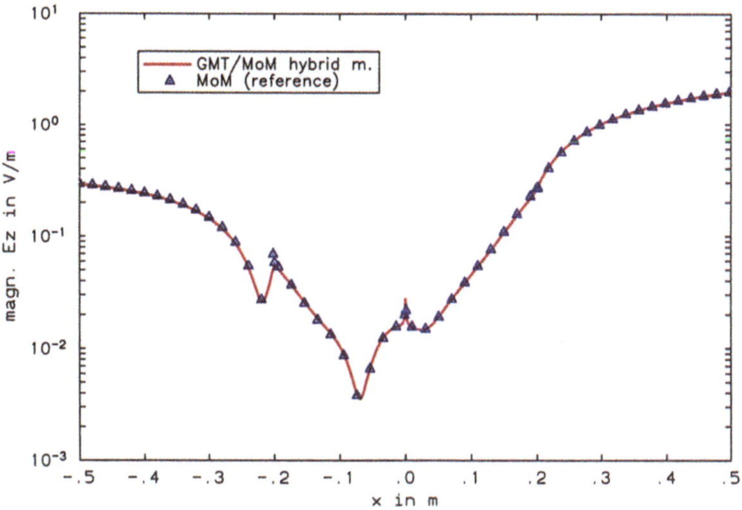

Fig. 8. Comparison of the electric field strength along a cut line through the human body for a dipole antenna in front of a spherical body model.

MMP, multiple multipole method) has been developed, see [14,15,25] for details.

One result is shown in Fig. 8 where the GMT/MoM hybrid method executed on a Linux PC is compared to the full MoM executed on the CRAY T3E. In order to get such a good agreement, the placement of the GMT multipoles is quite critical, thus the availability of full MoM validation results is essential.

5 Conclusions

It has been shown that by applying iterative techniques for the solution of a dense system of linear equations, a better scalability can be achieved. For large scale applications with many nodes, the required CPU–time can be reduced significantly as compared to a LU decomposition with ScaLAPACK.

Furthermore, the solution of practical electromagnetic field problems on the CRAY T3E has been addressed. Even though faster or hybrid techniques have been developed, which often enable the solution of electromagnetic radiation and scattering problems on a PC or workstation, the full MoM results on the CRAY T3E are important for validation purposes.

References

1. R. F. Harrington, *Field Computation by Moment Methods.* New York: Macmillan Company, 1968.
2. U. Jakobus, "Application of integral equation and hybrid techniques to the parallel computation of electromagnetic fields in a distributed memory environment," *Applied Computational Electromagnetics Society Journal*, vol. 13, pp. 87–98, July 1998. Special Issue on Computational Electromagnetics and High Performance Computing.
3. U. Jakobus, "Parallel computation of electromagnetic fields based on integral equations," in *High–Performance Computing in Science and Engineering: Transactions of the High Performance Computing Center Stuttgart (HLRS) 1998* (E. Krause and W. Jäger, eds.), pp. 377–386, Berlin: Springer Verlag, 1999.
4. L. S. Blackford, J. Choi, A. Cleary, E. D'Azevedo, J. Demmel, I. Dhillon, J. Dongarra, S. Hammarling, G. Henry, A. Petitet, K. Stanley, D. Walker, and R. C. Whaley, *ScaLAPACK User's Guide*, 1997. URL `http://www.netlib.org/scalapack/scalapack-home.html`.
5. F. X. Canning, "Solution of impedance matrix localization form of moment method problems in five iterations," *Radio Science*, vol. 30, pp. 1371–1384, Sept. 1995.
6. E. Bleszynski, M. Bleszynski, and T. Jaroszewicz, "AIM: Adaptive integral method for solving large–scale electromagnetic scattering and radiation problems," *Radio Science*, vol. 31, pp. 1225–1251, Sept. 1996.
7. B. Z. Steinberg and Y. Leviatan, "On the use of wavelet expansions in the method of moments," *IEEE Transactions on Antennas and Propagation*, vol. 41, pp. 610–617, May 1993.
8. F. X. Canning and J. F. Scholl, "Diagonal preconditioners for the EFIE using a wavelet basis," *IEEE Transactions on Antennas and Propagation*, vol. 44, pp. 1239–1246, Sept. 1996.
9. H. Deng and H. Ling, "Moment matrix sparsification using adaptive wavelet packet transform," *Electronics Letters*, vol. 33, pp. 1127–1128, June 1997.
10. Z. Baharav and Y. Leviatan, "Impedance matrix compression (IMC) using iteratively selected wavelet basis," *IEEE Transactions on Antennas and Propagation*, vol. 46, pp. 226–233, Feb. 1998.
11. R. Coifman, V. Rokhlin, and S. Wandzura, "The fast multipole method for the wave equation: A pedestrian prescription," *IEEE Antennas and Propagation Magazine*, vol. 35, pp. 7–12, June 1993.
12. J. M. Song and W. C. Chew, "Multilevel fast–multipole algorithm for solving combined field integral equations of electromagnetic scattering," *Microwave and Optical Technology Letters*, vol. 10, pp. 14–19, Sept. 1995.
13. X. Q. Sheng, J.-M. Jin, J. Song, W. C. Chew, and C.-C. Lu, "Solution of combined–field integral equation using multilevel fast multipole algorithm for scattering by homogeneous bodies," *IEEE Transactions on Antennas and Propagation*, vol. 46, pp. 1718–1726, Nov. 1998.
14. U. Jakobus, H.-O. Ruoß, and F. M. Landstorfer, "Analysis of electromagnetic scattering problems by an iterative combination of MoM with GMT using MPI for the communication," *Microwave and Optical Technology Letters*, vol. 19, pp. 1–4, Sept. 1998.

15. H.-O. Ruoß, U. Jakobus, and F. M. Landstorfer, "Iterative coupling of MoM and MMP for the analysis of metallic structures radiating in the presence of dielectric bodies," in *Conference Proceedings of the 14th Annual Review of Progress in Applied Computational Electromagnetics, Monterey*, vol. II, pp. 936–943, Applied Computational Electromagnetics Society, Mar. 1998.
16. R. Barrett, M. Berry, T. F. Chan, J. Demmel, J. Donato, J. Dongarra, V. Eijkhout, R. Pozo, C. Romine, and H. V. der Vorst, *Templates for the Solution of Linear Systems: Building Blocks for Iterative Methods.* Philadelphia: SIAM Society for Industrial and Applied Mathematics, 1994.
17. J. Shen, T. Hybler, and A. Kost, "Preconditioned iterative solvers for complex and unsymmetric systems of equations resulting from the hybrid FE–BE method," *IEEE Transactions on Magnetics*, vol. 33, pp. 1764–1767, Mar. 1997.
18. G. Bürger, H.-D. Brüns, and H. Singer, "Advanced method of moments based on iterative equation system solver," in *IEEE 1997 International Symposium on Electromagnetic Compatibility, Austin, Texas*, pp. 236–241, Aug. 1997.
19. J. Rahola, "Iterative solution of dense linear systems in electromagnetic scattering calculations," in *Conference Proceedings of the 14th Annual Review of Progress in Computational Electromagnetics, Monterey*, vol. II, pp. 1126 1133, Mar. 1998.
20. J. v. Hagen, R. Mittra, and D. Werner, "MMSNEC — multiple matrix solver NEC," in *Conference Proceedings of the 15th Annual Review of Progress in Applied Computational Electromagnetics*, pp. 428–435, Mar. 1999.
21. S. Großmann, "Parallele Verfahren zur direkten und iterativen Lösung dichtbesetzter linearer Gleichungssysteme," Diplomarbeit, Institut für Hochfrequenztechnik, Universität Stuttgart, 1998.
22. R. D. da Cunha and T. Hopkins, *PIM 2.2: The Parallel Iterative Methods package for Systems of Linear Equations User's Guide (Fortran 77 version)*, 1997. URL `http://www.mat.ufrgs.br/pim-e.html`.
23. R. W. Freund, "Conjugate gradient–type methods for linear systems with complex symmetric coefficient matrices," *SIAM Journal on Scientific and Statistical Computing*, vol. 13, pp. 425–448, Jan. 1992.
24. M. Murso, "Schnelle Lösungsverfahren zur numerischen Berechnung elektrisch großer Streuprobleme: Die Schnelle Multipolmethode FMM," Diplomarbeit, Institut für Hochfrequenztechnik, Universität Stuttgart, 1999.
25. U. Jakobus, H.-O. Ruoß, L. Geisbusch, and F. M. Landstorfer, "Hybridisation of MoM and GMT for the numerical analysis of electromagnetic sources radiating in the vicinity of persons with implanted cardiac pacemakers," in *IEEE AFRICON'99, 5th Africon Conference in Africa, Cape Town*, Sept. 1999. Accepted for publication.

Numerical Treatment of Time Varying Magnetic Fields in Power Transformers by Using the Boundary Element Method (BEM)

Dipl.-Phys. Christian Bollmann

Fachgebiet für Elektrische Energieversorung und Hochspannungstechnik, Universität der Bundeswehr Hamburg, Holstenhofweg 85, 22043 Hamburg

Abstract It is possible to calculate the leakage inductances of power transformers from their blueprints by using the boundary element method. Empirical values are not needed. The resulting dense systems of linear equations with up to 100000 unknown variables can be solved by using block matrices. Since it is possible to describe the greatest part of all windings of power transformers, the calculated leakage inductances differ only less than two and a half percent from measured values by the manufacturer.

1 Introduction

For calculation of leakage inductances of power transformers a formula found by Rogowski [1] in 1905 is still being used today as far as the author know. The dimensions of the windings are the most important factors of this formula. But the assumptions made by Rogowski like constant width or constant length of all turns in all windings are not fulfilled by modern power transformers.

To remove this fault, empirical values are introduced which are revealed in test departments. It is necessary to verify these empirical values on every new transformer. The so calculated leakage inductances have only little errors, if similar transformers have been tested before. A direct calculation throughout the real magnetic fields in the transformer does not need such empirical values. By using the Boundary Element Method the magnetic fields can be determined.

Furthermore, transformers with asymmetrical windings or asymmetrical feeding of the windings, i.e. transformers with stepping switches, can be treated as well. For this kind of transformers, the method of Rogowski does not apply. Other possibilities to calculate the leakage inductance do not exist.

2 Principals of the Boundary Element Method

To compute the magnetic fields the Boundary Element Method is used as shown in [2]. This method was specialized in [3] for the calculation of leakage inductances. It is assumed to have conductors of infinite length along the

z-direction. Their magnetic fields must be independent of the z-coordinates because of the symmetry. As a result, one gets a twodimensional field problem in the xy-plain. This assumptions describe especially longish objects like long conductors, but transformers too. In this case the magnetic fields are focussed into the plain in which the iron core of the transformer is lying. The concentric winded turns are described as unwinded coils (see Fig. 1).

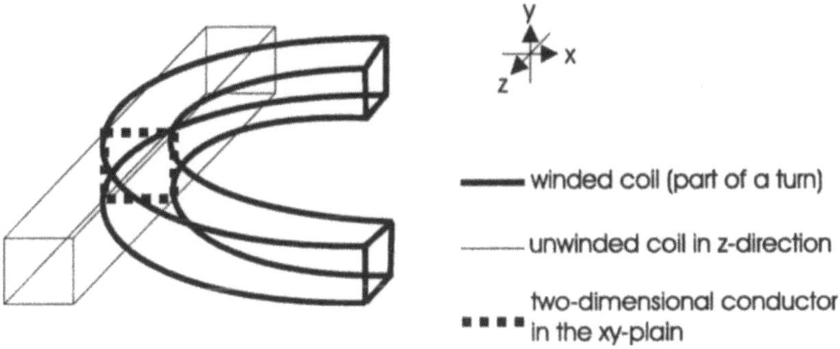

Figure1. Concentric winded turn modelled as unwinded coil

In the xy-plain, there are M areas of conductors (see Fig. 2). Each conductor m of them has a conductivity σ_m and a relative permeability μ_m which are time constant. Non-conductors have a conductivity $\sigma_m = 0$ and then can be treated as conductors. So there is no further difference between them.

Conductors are allowed to have hollow spaces like the windows of transformer iron cores. Other conductors can take place in these areas. Every conductor can be feed with an harmonic current I_m. This currents frequency is f with $\omega = 2\pi f$. As an integral value I_m makes no statement on the local current density. Due to eddy currents the exact current density is going to be calculated. As a result of the currents in z-direction there are only components of the magnetic field in x- and y-direction. $\underline{\boldsymbol{B}} = \text{rot } \underline{\boldsymbol{A}}$ (underlining means complex values) yields to a vector potential $\underline{\boldsymbol{A}} = (0, 0, \underline{A}_z(x, y))$. Deriving from the Maxwell equations as described in [3] one can get a Helmholtz differential equation for the z-component of the vector potential in the interior of a conductor m:

$$(\Delta - j\omega\mu_m\sigma_m)\left(\underline{A}_z(x, y) + \underline{K}_m\right) = 0. \tag{1}$$

$\underline{K}_m$ is another constant of the conductor m. It can be seen as voltage drop along the conductor in z-direction and is still unknown, but it can be determined by the Boundary Element Method. Due to $\sigma_m = 0$ in the exterior of the conductors the Laplace differential equation

$$\Delta\underline{A}_z(x, y) = 0 \tag{2}$$

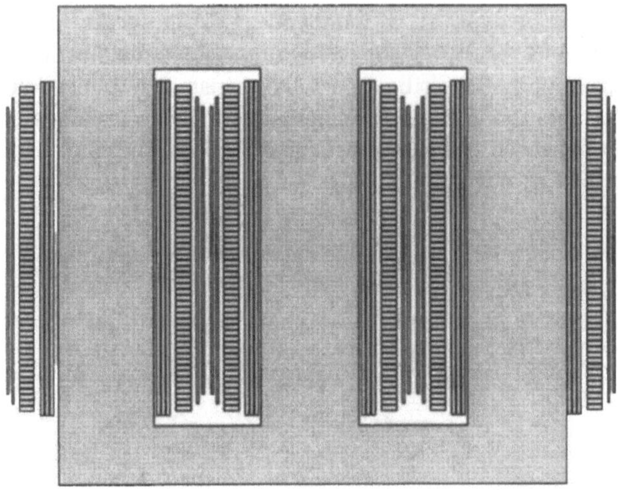

Figure2. Model of a 110 kV/10 kV power transformer with the boundary element method

is valid. At the surfaces of the conductors the well known continuity conditions for the perpendicular component of the magnetic field in both bordering media $\underline{B}_{n1} = \underline{B}_{n2}$ and the tangential component of the electrical field $\underline{E}_{t1} = \underline{E}_{t2}$ are valid. This leads to a continuos $\underline{A}_z$ at the border, while the perpendicular derivative of $\underline{A}_z$ differs àccording to the permeabilities of the both media.

To solve the field problem, $\underline{A}_z(x, y)$ has to be calculated for every point in the xy-plain. By using Greens Theorem it is possible to build a connection between the values of $\underline{A}_z(x, y)$ at every point in the xy-plain and $\underline{A}_z$ and $\frac{\partial}{\partial n}\underline{A}_z$ on the boundaries of the conductors by integrating around the boundaries. In the same way one gets an integral equation for $\frac{\partial}{\partial n}\underline{A}_z(x, y)$. For the numerical computation the model has to be made discrete. The boundaries of the conductors are divided into small parts, the so called boundary elements. If these boundary elements are small enough, the terms $\underline{A}_z$ and $\frac{\partial}{\partial n}\underline{A}_z$ would be rather constant on each boundary element. Now the assumption is made that the values of $\underline{A}_z$ and $\frac{\partial}{\partial n}\underline{A}_z$ *are* constant on each boundary element. Due to this measure the integral equations derived from Greens Theorem become sum equations.

Knowing $\underline{A}_z(x, y)$ and $\frac{\partial}{\partial n}\underline{A}_z(x, y)$ in every point (x, y) of the xy-plain means knowing them in a point (x_n, y_n) on the boundary element n, too. If there are N boundary elements, alltogether one can get $2N$ equations which determinate $\underline{A}_z(x_n, y_n)$ and $\frac{\partial}{\partial n}\underline{A}_z(x_n, y_n)$ for all $1 \leq n \leq N$ - as points of the plain - from all $\underline{A}_z(x_n, y_n)$ and $\frac{\partial}{\partial n}\underline{A}_z(x_n, y_n)$ - as constant values on the boundaries. Since these equations are linear in $\underline{A}_z$ and $\frac{\partial}{\partial n}\underline{A}_z$, the result is a system of linear equations. Because of the M still unknown voltage drops

$\underline{K}_m$ in this system, there are $2N + M$ unknown values at $2N$ equations. To get a unique solution of this system of linear equations one can make use of the Amperes Theorem which connects the vector potential on the boundaries of the conductors with the current fed in. With M conductors there are M more linear equations so there are at last the needed $2N+M$ linear equations to solve.

These M voltage drops per unit of length $\underline{K}_m$ are extremely important for the computation of the leakage inductances. By knowing this voltage drop and the current in a conductor the impedance can be calculated. $Z' = R' + \mathrm{j}X'$ may be the impedance per unit of length. This impedance includes all inductive couplings since it is computed out of the voltage drops $\underline{K}_m$ which results from a complete solution of the field problem including these inductive couplings. In this way eddy currents are included, too.

Building a winding one needs two conductors for each turn in this two-dimensional model. Each time the winding crosses the xy-plain in positive z-direction there is a conductor with a positive current. Each time the winding crosses the xy-plain in negative z-direction, there is a conductor with a negative current. The absolute value of all currents in the winding must be the same of course. Two matching conductors build a pair of conductors. These pairs are the elementary units which are used in this model to describe a transformer. According to the opposite direction of the current in each conductor of a pair $\underline{K}_p = \underline{K}_{m1} - \underline{K}_{m2}$ is the resulting voltage drop along the pair.

By feeding only one pair of conductors with a single unit of current, while all other pairs are without currents at all, the inductive couplings between the one pair and all others can be calculated from the voltage drops. The inductive part of the voltage drops per unit of length yields to the self-inductance per unit of length of the pair with the current fed in and to the mutual inductances per unit of length between this pair and all other pairs.

Treating every pair of conductors in this way one gets all inductive couplings of the system of pairs as a matrix of inductances. This matrix can be taken into a matrix of inductances per unit of length of windings and not only of pairs if all pairs belonging to a winding are connected. Mathematically, the inductances of this pairs are summed up. The resulting self- and mutual inductances per unit of length are quite exact, because no integration over magnetic fields was necessary.

For a single phased transformer with two windings this computation leads to a inductance matrix per unit of length

$$\begin{pmatrix} L'_{11} & M'_{12} \\ M'_{21} & L'_{22} \end{pmatrix} . \tag{3}$$

By multiplication with the length of the conductors this matrix of inductances per unit of length becomes a matrix of inductances. But which length must be taken? In the model all conductors have the same length infinity, while in a real transformer each turn has its own finite length. A simple arithmetic mean of lengths of all the real turns depends on how fine each turn is modelled. Rather indepent of this is a mean weigthing the magnetic energy of each turn. The magnetic energy of a pair of conductors i with a real length l_i is

$$w_{m_i} = l_i L_i' I_i. \tag{4}$$

I_i is the current fed into this turn. L_i' is the self-inductance per unit of length as computated from the field problem. The energy-weigthed length results as

$$l_{\mathrm{EW}} = \frac{\sum_{i=1}^{w} l_i L_i' I_i}{\sum_{i=1}^{w} L_i' I_i}. \tag{5}$$

Using this length l_{EW} the matrix of inductances can be calculated. In addition to that the primary and secondary leakage inductances and the total leakage inductance can be gained:

$$L_{\sigma 1} = l_{\mathrm{EW}} \left(L_{11}' - M_{12}' \right) \tag{6}$$

$$L_{\sigma 2} = l_{\mathrm{EW}} \left(L_{22}' - M_{21}' \right) \tag{7}$$

$$L_{\sigma} = L_{\sigma 1} + L_{\sigma 2} \tag{8}$$

Finally the relative short-circuit voltage u_{k} becomes with the leakage inductance L_σ

$$u_{\mathrm{k}} = \frac{\omega L_\sigma S_{\mathrm{rT}}}{U_{\mathrm{rT}}^2} \cdot 100\%. \tag{9}$$

In this formula U_{rT} and S_{rT} are the rated voltage and the rated power of the transformer. After discussing the theoretical fundamentals of this method to compute these important values of power transformers only from their blueprints one have to take a look on the way to solve great systems of linear equations by computer.

3 Solution of Large Systems of Linear Equations

While modelling power transformers with the Boundary Element Method (see Fig. 2) the size of the system of linear equations to solve has to be considered. An typical 110kV/10kV-transformer might have, for example, 950 turns in the high-end voltage winding and about 150 turns in the low-end voltage winding. Both windings may consist of two parallel interlaced windings, and the low-end windings can have up to 20 parallel turns. Altogether there are 7900 turns per phase or 23700 turns in the entire transformer. There are two conductors to be modelled per turn, so there are 47400 conductors to consider. With only one boundary element on each of the four sides, at least 189600 boundary elements exist. This leads to 426600 equations. The small number of equations belonging to the iron core can be neglected.

An the other hand, if the leftmost and rightmost phase are ignored and only the phase in the middle of the transformer remains, there are only 15800 conductors or 71100 equations. In this case, the symmetry of geometry and the currents fed in can be used. But the system of linear equations is filled up to 50 % in this case. Therefore, sparse matrix techniques are not available.

To solve these system of linear equations efficiently and with as less memory requirements as possible, a Gaussian algorithm for block matrices is used. The system matrix divides into blocks of numbers not equal zero and other blocks equal zero according to the geometry of the problem. Knowing which blocks are zero leads to advantages during the calculation. Lots of operations with zero blocks can be skipped. This results in less used computation time and memory. Using dynamically assigned memory for the block matrices it was possible to reduce the memory requirements to 25 % to 30 % of the size of the system matrix. This means that with 4GB main memory and some 35 GB harddisc space up to 100000 equations of variables using 16 Bytes per value can be solved.

One have to take care of the numerical errors which must appear solving such large linear systems. Furthermore, the two-dimensional model may lead to errors, too. In spite of this the results are quite accurate as the following comparison shows.

4 Comparision of the Numerical Results with Measured Values

There are three different cases of models which are examinated. At first, there is only the phase in the middle and the whole iron core described. In this way the most conductors can be calculated since there is a symmetry to use. The second model is one single phase of the both outer ones. Now there is no more symmetry, and the number of possible conductors is roughly half the number in the first model. Only the sixth part of conductors of the first model can be computed in the third model. Here are all three phases considered.

Comparing the results of these three kinds of models shows that the calculation of a single outer phase leads to a leakage inductance of 1 % to 2 % less than a single medium phase. The results of a three phase calculation lie between them, mostly near the mean value. Alltogether it is possible to use the single phase models to describe more turns because the error of a calculation with to few conductors is greater then 2 %.

The first examined transformer was a real build 110 kV/10 kV-transformer. The medium phase could be described with all parallel turns. On the other hand the outer phases can be modelled only with eight of 20 parallel turns in the low-end winding. The high-end winding was complete. Three positions of the stepping switch has been taken into account: The neutral position and the both extreme positions. The numerical results for these three positions for the relative short-circuit voltage as a measure for the leakage inductances are u_k=15,80 %, u_k=17,23 % and u_k=14,64 %. These results differ from the measured values of the manufacturer with a relative error from -1,7 % to +0,3 %.

Another produced 110 kV/20 kV-transformer was examined, too. With its lower number of parallel turns the outer phases could be modelled complete. There are also the three position of the stepping switch like before. The computed results are u_k=12,77 %, u_k=13,63 % and u_k=12,08 % for the relative short-circuit voltages. Here, the numerical results are 1,6 % to 2,0 % above the measure values. But they are lying between the measured values and the results gained from the formula of Rogowski.

At last there was a seldom 110 kV/10 kV/10 kV-transformer with each two low-end and high-end windings. This transformer contains much more turns so that the outer phase could be modelled without any parallel turn only. Because of the two low-end windings, there are much more possible short-circuit voltages that could be examined. The manufacturer has measured ten of them. I.e. the calculated relative short-circuit voltage for a normal case use of this transformer is u_k=14,69 %, which is 2,80 % below the measured value. The other numerical values lie between 0,7 % and 3,3 % below the measured values. A short-circuit experiment between both low-end windings could not be calculated with single phase models in sufficient way. The errors are up to 20 %. With three phase models the error could be as low as 3 % but only with rather low numbers of turns to be described. This was the only short-circuit making problems to calculate and especially it is not a typical way to use the transformer. Therefore, the result is of academic interest only.

The numerical computations have shown that the measured values for the relative short-circuit voltage respectively the leakage inductance from measurements in testing departments could be reproduced with only a small error between 1 % and 3 %. Due to the fact that even this measurements are not exact this method is an alternative way to calculate leakage inductances of power transformers. Its disadvantage is the great amount of computational power needed, but as advantage compared with Rogowskis formula it needs no empirical values - only the blueprints.

5 Evaluation of the Used Method

Since there is a good correspondence between the measured values and the numerical results it can be said that the two-dimensional model is no undue simplification. The transformers iron core focusses the magnetic field in the xy-plain. The third dimension is rather unimportant. A different situation araises if synchronous inductances should be calculated. In this case, the magnetic fields not further close in the interior of the iron core but far in the exterior, too. To this effect three-dimensional magnetic fields are needed. The calculated synchronous inductances could be only estimations.

The variety of examined models shows that with rather crude descriptions of the transformer the leakage inductances can be calculated quite well. This supports the use of Rogowskis formula which also describes modern power transformers rather crude. But in addition to the possibility of Rogowski, the use of the boundary element method allows to compute leakage inductances of transformers with stepping switches or unusual construction types. Furthermore, the method of the boundary elements allows to consider alternate currents. Using Rogowski formula only direct current are taken into account.

Knowing the exact fields in the transformer yields to the exact current distribution in the conductors and to the forces having an effect on them. Furthermore, the model can be enlarged to take capacitive couplings into account. This leads to frequency responses of gate impedance matrices of the transformers to find out the internal resonant frequencies. This topics should be examined closer later.

6 The Program REM on the NEC-SX4 at HLRS

The calculations described before are done with a program REM (from the German notation *R*and-*E*lement-*M*ethode) written in Fortran 90 by the author. Due to the problem, the solution of a large system of linear equations, a parallel computer with a vector unit like the NEC-SX4 fits best for this problem. The program uses shared-memory parallel programming.

The block matrix algorithm splits the used computation time to 90 % to the multiplication of matrices. Further 8 % are matrix inversions. The rest of time is needed to build up the matrices, for IO and global administration. Because of missing shared-memory linear algebra packages the matrix operations are done with parallel routine written by the author. These simple routines are - for not so large matrices and on one CPU - comparable with BLAS routines. But according to tests on other platforms there might be a loss of 20 % of used time compared with what would be possible on this computer.

In the program REM the used matrices have typically dimension 1000 to 5000, sometimes even larger. Jobs with such large matrices are queued in the NP8CPU2GB- or even better in the NP16CPU4GB-queue. They ran 100 to 150 hours. At vectorlength of some 252 the vector operation ratio is 99,5 %. The achieved peakperformance was 20 GFLOPS on 16 CPU. The difference to the nominal 32 GFLOPS of 16 CPU may have its reason in bank conflicts which total to a third part on the ellapsed time (according to F_PROGINF). Furthermore, these large jobs put up to 400 GB of data temporary on the harddisc and a small part of this into XMU. These transfers reduce the peakperformance. By using smaller amounts of main memory the performances drops dramatically. There is only one reason to use the queues with smaller memory: the smaller waiting period until the job starts.

Two problems appearing during the port of the program REM to the NEC-SX4 should be mentioned here, too. At first, all arrays which are given to a subroutine as parameters are copied to the stack. Running a subroutine all parameter arrays are located twice in the memory which costs a lot of time and even more memory. Using much memory this behaviour can cause serious problems. As solution not the whole array but its first element is stated as parameter in the calling routine. The called routine can have the whole array as parameter without problems. This uses the fact that the memory address of the array and its first element are the same and that numbers are not copied to the stack. Tests have shown that this copying into the stack costs up to 35 % of the computation time.

The second problem belongs to the allocation of file buffers. At the first use of a file handle, a buffer is allocated which is not deallocated if the file is closed. The program REM opens the first file when there is no more memory to allocate a new matrix to put same data temporary on harddisc. This is a rather bad moment to put a permanent buffer somewhere into the memory. In bad cases two buffer can divide the memory in three nearly equal parts. Instead of five equal sized matrices filling the memory than only three of them take place. This means an effective memory reduction of 40 %. A solution to this problem was the opening of dummy files at programs start before data matrices are allocated.

Both problems leads to sudden program ends due to a lack of memory. In both cases, it was not clear from the beginning what the reason was. After a number of trials and with significant help from the staff of the HLRS and NEC, the problems could be identified and solved, but it took some month instead of weeks as planned to migrate REM to the NEC-SX4.

References

1. W.Rogowski,Über das Streufeld und den Streuinduktionskoeffizienten eines Transformators mit Scheibenwicklung und geteilten Erdspulen, Mitteilungen über Forschungsarbeiten auf dem Gebiete des Ingenieurwesens, herausgegeben vom Verein Deutscher Ingenieure, Band 71 (1905), Seite 1-35
2. A.Kost, Numerische Methoden in der Berechnung elektromagnetischer Felder, Springer Verlag 1994, Berlin Heidelberg New York
3. T.Meyer, Zweidimensionale feldtheoretische Berechnung von Streuinduktivitäten sowie der Frequenzgänge von verlustbehafteten induktiven Systemen, Dissertation 1996, Universität der Bundeswehr Hamburg.

Direct and Inverse Electromagnetic Scattering

Wolfgang Rieger, André Buchau, Günther Lehner and Wolfgang M. Rucker

Institut für Theorie der Elektrotechnik, Universität Stuttgart, Pfaffenwaldring 47, 70565 Stuttgart

Abstract. The direct and inverse time-harmonic electromagnetic scattering from inhomogeneous media is considered. The physical problem of electromagnetic scattering from known objects is mathematically described by volume integral equations. Being able to master the direct problem is an absolute prerequisite to solving the corresponding inverse problem, which is naturally closely connected. When solving inverse scattering problems one tries to retrieve information about the unknown scatterer from the knowledge of incident probing waves and measured scattering data. We especially investigate methods to reconstruct the geometry and the material properties of inhomogeneous media from scattering data. The objects considered in this context are either isotropic or anisotropic lossy dielectrics. The objects are assumed to be nonmagnetic. The inverse scattering problem can be formulated as a nonlinear optimization problem which is solved by means of iterative optimization schemes. Numerical examples demonstrate the efficiency of the proposed methods.

1 Introduction

This paper deals with the numerical solution of the inverse problem of reconstructing an unknown object from the knowledge of its scattering data and the incident probing waves. With the fast development of computer technology during the last few years there was also an increasing interest in numerical solutions to inverse problems. A precise definition of the terms *direct* and *inverse* problems can be derived if a mathematical model with uniquely defined input and output quantities is taken as a basis [20].

In the following, the direct problem is to compute the scattered electromagnetic fields from an known object which is irradiated by time-harmonic plane waves. The geometry and the material properties are given. We assume the objects to be linear, nonmagnetic, isotropic or anisotropic, inhomogeneous dielectrics.

The inverse electromagnetic scattering problem is posed in the following way: The unknown object is irradiated by a certain number of time-harmonic plane waves (the frequency of the waves is known) and the corresponding scattered field outside the object is measured. From the knowledge of the scattered field and the incident probing waves, the geometry and the material properties are determined. This kind of inverse problem can be called microwave tomography.

In practice there are a lot of inverse problems. The task in medical imaging is to provide information in a non-invasive way about the internal structure of the human body. There are several different imaging techniques. Some methods provide imformation about the density of the tissue such as x-ray computer tomography and ultrasound computer tomography. A distinction between properties of the tissues is possible with magnetic resonance imaging and impedance computer tomography [11]. In this context microwave tomography seems to be very promising [4,5], as in comparison to x-ray computer tomography and ultrasound computer tomography, different quantities like permittivity and conductivity are involved in the reconstruction process and this could give supplementary information in diagnostics. A further application is the nondestructive testing of materials.

The aim is to develop efficient reconstruction algorithms for the solution of inverse electromagnetic scattering problems. Being able to master the direct problem is an absolute prerequisite to solving the corresponding inverse problem. On the one hand, the direct problem constitutes the basis of the inverse problem. On the other hand, scattering data should be provided very precisely for the investigation and the judgement of the developed reconstruction algorithms. In section 2, the direct scattering problem is described and the fundamental volume integral equations are presented.

Section 3 deals with the inverse scattering problem. Most of the inverse problems are *ill-posed* or *improperly posed*. This does not mean that we chose a bad mathematical model for the considered physical problem, this means that the mathematical model has an inherent property, the ill-posedness. There are three requirements to have an *well-posed* problem: existence of a solution, uniqueness of the solution and continuous dependence of the solution on the data. A problem satisfying all three requirements is called well-posed, otherwise it is called ill-posed. The inverse scattering problem of reconstructing the geometry and the material properties is ill-posed because the solution of the problem does not depend continuously on the given scattered data. This is an inherent property of the problem. During the last few years there was an increasing interest in inverse scattering problems, where several different reconstruction algorithms were developed [9,10,15,19]. In our previous papers we presented a variable metric method to reconstruct isotropic materials from synthetic [14,16] or measured, i.e. experimental, scattering data [13,17]. We also showed the improvement of reconstructions by ensuring physical meaningful quantities [12,18]. The storage and CPU time requirements are moderate in this cases, so that we used a DEC ALPHA Station 600 or DEC ALPHA SERVER 1200 which is available at our institute. In section 3 we describe a method to reconstruct anisotropic materials from scattered electromagnetic field data, based on an iterative optimization procedure, the variable metric method. As in the anisotropic case the number of unknowns is three times larger than in the isotropic case, also the CPU time increases which makes it necessary to have fast machines available like the NEC SX4.

In section 4 we give some numerical results and section 5 concludes with the main statements of the paper.

2 Direct Electromagnetic Scattering

In this section we describe the two-dimensional direct scattering problem. The object, located entirely within a test domain Ω and embedded in free space, is successively irradiated at a single frequency by N_I time-harmonic TE-polarized electromagnetic plane waves $\mathbf{E}_l^{\mathrm{inc}}(\mathbf{r}) = \mathbf{E}_0 \exp(j\mathbf{k}_l \cdot \mathbf{r})$, where $\mathbf{k}_l$ specifies the wave vector with absolute value $|\mathbf{k}_l| = k_0 = \omega\sqrt{\varepsilon_0\mu_0}$, $l = 1, \dots, N_\mathrm{I}$ and angular frequency ω. The scattered electric field is measured by J receivers on a circle that encloses the test domain Ω. The geometrical configuration is shown in Fig. 1.

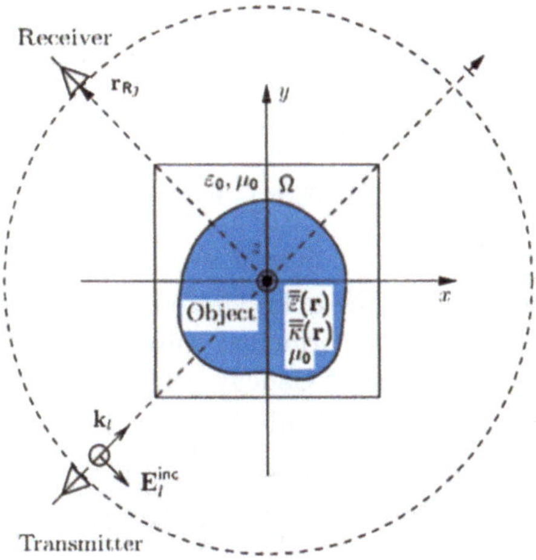

Fig. 1. Geometrical configuration of the problem.

The irradiated nonmagnetic material is described by the complex permittivity tensor

$$\overline{\overline{\varepsilon}}_\mathrm{c}(\mathbf{r}) = \overline{\overline{\varepsilon}}(\mathbf{r}) + j\frac{\overline{\overline{\kappa}}(\mathbf{r})}{\omega}, \quad \mathbf{r} \in \Omega \ , \tag{1}$$

where $\overline{\overline{\varepsilon}}(\mathbf{r})$ is the symmetric permittivity tensor and $\overline{\overline{\kappa}}(\mathbf{r})$ is the symmetric conductivity tensor. As we consider crystals of higher symmetry classes the principal axes of the permittivity and the conductivity tensor coincide in direction [1]. In the following, we consider two separate cases. First, we assume that the axes of our coordinate system coincide with the principal axes of the crystal. We call these materials *biaxial materials*. Therefore, the complex permittivity tensor reads

$$\overline{\overline{\varepsilon}}_\mathrm{c}(\mathbf{r}) = \begin{bmatrix} \varepsilon_{xx}(\mathbf{r}) + j\kappa_{xx}(\mathbf{r})/\omega & 0 \\ 0 & \varepsilon_{yy}(\mathbf{r}) + j\kappa_{yy}(\mathbf{r})/\omega \end{bmatrix} . \tag{2}$$

On the other hand, in our second case, the complex permittivity tensor is given in the following way

$$\overline{\overline{\varepsilon}}_{\mathrm{c}}(\mathbf{r}) = \begin{bmatrix} \varepsilon_{xx}(\mathbf{r}) + j\kappa_{xx}(\mathbf{r})/\omega & \varepsilon_{xy}(\mathbf{r}) + j\kappa_{xy}(\mathbf{r})/\omega \\ \varepsilon_{xy}(\mathbf{r}) + j\kappa_{xy}(\mathbf{r})/\omega & \varepsilon_{yy}(\mathbf{r}) + j\kappa_{yy}(\mathbf{r})/\omega \end{bmatrix} . \tag{3}$$

In this case the axes of the coordinate system don't conincide with the principal axes of the crystal. We call the materials in this case anisotropic materials.

The contrast tensor $\overline{\overline{m}}$ is defined as

$$\overline{\overline{m}}(\mathbf{r}) = \overline{\overline{\mathrm{I}}} - \frac{1}{\varepsilon_0}\overline{\overline{\varepsilon}}_{\mathrm{c}}(\mathbf{r}) \tag{4}$$

and vanishes not only outside of Ω but also at points in Ω exterior to the scatterer. Then the second kind integral equation with $\mathbf{r} \in \Omega$

$$\begin{aligned} &\mathbf{E}_l(\mathbf{r}) - \overline{\overline{\mathrm{L}}} \cdot \overline{\overline{m}}(\mathbf{r}) \cdot \mathbf{E}_l(\mathbf{r}) \\ &+ k_0^2 \; P.V. \int_{\Omega} \overline{\overline{G}}(\mathbf{r},\mathbf{r}') \cdot \overline{\overline{m}}(\mathbf{r}') \cdot \mathbf{E}_l(\mathbf{r}') \, \mathrm{d}\Omega' = \mathbf{E}_l^{\mathrm{inc}}(\mathbf{r}) \ , \end{aligned} \tag{5}$$

holds for the total field $\mathbf{E}_l(\mathbf{r})$. The abbreviation *P.V.* (principal value) stands for the shape dependent Cauchy principal value of the area integral [2]. As our numerical method to calculate Cauchy principal value integrals is based on a circular exclusive area, the dyade $\overline{\overline{\mathrm{L}}}$ reads $\overline{\overline{\mathrm{L}}} = \frac{1}{2}\overline{\overline{\mathrm{I}}}$, where $\overline{\overline{\mathrm{I}}}$ specifies the identity matrix [7,21]. The expression $\overline{\overline{G}}(\mathbf{r},\mathbf{r}')$ denotes dyadic Green's function and is given by $\overline{\overline{G}}(\mathbf{r},\mathbf{r}') = \left(\overline{\overline{\mathrm{I}}} + \frac{1}{k_0^2}\nabla\nabla\right) G(\mathbf{r},\mathbf{r}')$, where $G(\mathbf{r},\mathbf{r}') = \frac{j}{4}\mathrm{H}_0^{(1)}(k_0|\mathbf{r}-\mathbf{r}'|)$. The scattered electric field at the J receiver locations $\mathbf{r}_{\mathrm{R}j}$, $j = 1, \dots, J$ is determined by

$$\mathbf{E}_l^{\mathrm{sca}}(\mathbf{r}_{\mathrm{R}j}) = -k_0^2 \int_{\Omega} \overline{\overline{G}}(\mathbf{r}_{\mathrm{R}j},\mathbf{r}') \cdot \overline{\overline{m}}(\mathbf{r}') \cdot \mathbf{E}_l(\mathbf{r}') \, \mathrm{d}\Omega' \ . \tag{6}$$

3 Inverse Electromagnetic Scattering

To numerically solve the integral equations (5) and (6), the point-matching method is used. The electric field and the material properties are assumed to be constant within each finite element. Discretizing (5) and (6) by P finite elements yields

$$\{E_l\} = (\,[I] + [G_{\mathrm{D}}][M]\,)^{-1}\,\{E_l^{\mathrm{inc}}\} \ , \tag{7}$$

$$\{E_l^{\mathrm{s}}\} = [G_{\mathrm{S}}][M]\{E_l\} \ , \tag{8}$$

and with (7) in (8)

$$\{E_l^{\mathrm{s}}\} = [G_{\mathrm{S}}][M]\,(\,[I] + [G_{\mathrm{D}}][M]\,)^{-1}\,\{E_l^{\mathrm{inc}}\} \ , \tag{9}$$

where the complex $2P \times 2P$ matrix $[M]$ is, in the case of biaxial objects, a diagonal matrix,

$$[M] = \begin{bmatrix} m_{xx,1} & 0 & 0 & 0 & \cdots \\ 0 & m_{yy,1} & 0 & 0 & \cdots \\ 0 & 0 & m_{xx,2} & 0 & \cdots \\ 0 & 0 & 0 & m_{yy,2} & \cdots \\ \cdots & \cdots & \cdots & \cdots & \cdots \end{bmatrix} , \tag{10}$$

whose elements $m_{kk,p}$, $kk = [xx, yy]$, are the complex values of $m_{kk}(\mathbf{r})$ at the center of the pth finite element, $p = 1, \ldots, P$. In the case of anisotropic materials, the matrix $[M]$ reads

$$[M] = \begin{bmatrix} m_{xx,1} & m_{xy,1} & 0 & 0 & \cdots \\ m_{xy,1} & m_{yy,1} & 0 & 0 & \cdots \\ 0 & 0 & m_{xx,2} & m_{xy,2} & \cdots \\ 0 & 0 & m_{xy,2} & m_{yy,2} & \cdots \\ \cdots & \cdots & \cdots & \cdots & \cdots \end{bmatrix} , \tag{11}$$

whose elements $m_{jk,p}$, $jk = [xx, xy, yy]$, are the complex values of $m_{jk}(\mathbf{r})$ at the center of the pth finite element, $p = 1, \ldots, P$. The vector $\{E_l^{\mathrm{inc}}\}$ is the lth incident field vector consisting of $2P$ complex numbers, while $\{E_l^{\mathrm{s}}\}$ is the lth scattered field vector with $2J$ complex numbers. $[G_{\mathrm{D}}]$, a complex $2P \times 2P$ matrix, and $[G_{\mathrm{S}}]$, a complex $2J \times 2P$ matrix, are composed of the elements

$$\overline{\overline{\mathrm{g}}}_{\mathrm{D}}^{k,l} = k_0^2 \; P.V. \int_{\Omega_l} \overline{\overline{G}}(\mathbf{r}_k, \mathbf{r}')\, \mathrm{d}\Omega' - \frac{1}{2}\delta_{k,l}\overline{\overline{\mathrm{I}}} \; , \tag{12}$$

$$\overline{\overline{\mathrm{g}}}_{\mathrm{S}}^{j,l} = -k_0^2 \int_{\Omega_l} \overline{\overline{G}}(\mathbf{r}_{\mathrm{R}j}, \mathbf{r}')\, \mathrm{d}\Omega' \; , \tag{13}$$

where $k, l = 1, \ldots, P$ and $j = 1, \ldots, J$.

In view of equation (9), the nonlinear inverse scattering problem is posed in the following way: given the scattered field vectors $\{E_l^{\mathrm{s}}\}$ and the incident field vectors $\{E_l^{\mathrm{inc}}\}$, the material coefficients $m_{jk,p}$ of the contrast matrix $[M]$ have to be found. In order to formulate the inverse scattering problem as an unconstrained, nonlinear optimization problem, a solution vector $\{s\}$ has to be defined. In the following we consider the real and imaginary part of the contrast function $m_{jk}(\mathbf{r})$ as independent functions. This leads to a real solution vector $\{s\}$ of size $6 \cdot P$ in the case of anisotropic objects and to a real solution vector $\{s\}$ of size $4 \cdot P$ in the case of biaxial objects.

In order to constrain the domain of values for the tensor $\overline{\overline{m}}(\mathbf{r})$ we rewrite its components according to

$$m_{jk}(\mathbf{r}) = -\left(\eta_{jk}(\mathbf{r})\eta_{jk}(\mathbf{r}) + j\,\xi_{jk}(\mathbf{r})\xi_{jk}(\mathbf{r})\right) \; . \tag{14}$$

This ensures the constraints

$$\frac{\bar{\bar{\varepsilon}}(\mathbf{r})}{\varepsilon_0} - \bar{\bar{\mathrm{I}}} \geq \bar{\bar{0}} \quad \text{and} \quad \frac{\bar{\bar{\kappa}}(\mathbf{r})}{\varepsilon_0\,\omega} \geq \bar{\bar{0}} \ . \tag{15}$$

If we accumulate the P values $\eta_{jk,p}$ and $\xi_{jk,p}$ of the functions $\eta_{jk}(\mathbf{r})$ and $\xi_{jk}(\mathbf{r})$ in the vectors $\{\eta_{jk}\}$ and $\{\xi_{jk}\}$ respectively, the solution vector $\{s\}$ can be written as

$$\{s\} = \begin{bmatrix} \{\eta\} \\ \{\xi\} \end{bmatrix} \ , \tag{16}$$

where the vectors $\{\eta\}$ and $\{\xi\}$ are constructed according to

$$\{\eta\} = \sum_{jk} [T_{njk}]\{\eta_{jk}\} \quad \text{and} \quad \{\xi\} = \sum_{jk} [T_{xjk}]\{\xi_{jk}\} \ . \tag{17}$$

The matrices $[T_{njk}]$ and $[T_{xjk}]$ ensure that the components of the vectors $\{\eta_{jk}\}$ and $\{\xi_{jk}\}$ are appropriately assigned to the vectors $\{\eta\}$ and $\{\xi\}$. The solution vector $\{s\}$ is built in the following way

$$\{s\} = [\eta_{xx,1}, \eta_{xy,1}, \cdots, \eta_{xy,P}, \eta_{yy,P}, \xi_{xx,1}, \xi_{xy,1}, \cdots, \xi_{xy,P}, \xi_{yy,P}]^{\mathrm{T}} \ . \tag{18}$$

The inverse scattering problem formulated as an unconstrained, nonlinear optimization problem is solved by minimizing the cost function F

$$\begin{aligned} F(\{s\}) &= F_{\mathrm{D}}(\{s\}) + F_{\mathrm{REG}}(\{s\}) \\ &= \sum_{l=1}^{N_{\mathrm{I}}} \|\{\hat{E}_l^{\mathrm{s}}\} - \{E_l^{\mathrm{s}}\}\|_{\mathrm{S}}^2 + F_{\mathrm{REG}}(\{s\}) \ , \end{aligned} \tag{19}$$

where $\{f_l\} = \{\hat{E}_l^{\mathrm{s}}\} - \{E_l^{\mathrm{s}}\}$ denotes the error between measured and computed scattered electric field for the lth incident field. Since the inverse scattering problem is severely *ill-posed* [3], the regularization term F_{REG} is added to ensure stability and uniqueness of the solution. As in our previous papers [14,16] we apply the total variation as regularization functional to both the real part of $m_{jk}(\mathbf{r})$ and the imaginary part of $m_{jk}(\mathbf{r})$. Thus, we obtain the penalty term

$$\begin{aligned} F_{\mathrm{REG}} = &\sum_{jk} \alpha'_{jk,\mathrm{R}} \int_{\Omega} \sqrt{|\nabla \mathrm{Re}\,(m_{jk}(\mathbf{r}'))\,|^2 + \beta_{jk,\mathrm{R}}^2}\,\mathrm{d}\Omega' \\ &+ \sum_{jk} \alpha'_{jk,\mathrm{I}} \int_{\Omega} \sqrt{|\nabla \mathrm{Im}\,(m_{jk}(\mathbf{r}'))\,|^2 + \beta_{jk,\mathrm{I}}^2}\,\mathrm{d}\Omega' \ . \end{aligned} \tag{20}$$

To solve the unconstrained optimization problem we use a variable metric method [16]. During the optimization process, iterative solutions of the form $\{s^{i+1}\} = \{s^i\} + \gamma^i\{d^i\}$ are constructed, where $\gamma^i \in \mathbb{R}^+$ is the descent steplength and $\{d^i\}$ denotes the current descent direction. To build up the descent directions $\{d^i\}$ we use the Davidon-Fletcher-Powell [8] method which involves exact first-order gradients of F. The gradients are determined by calculating the directional derivative of F at the point $\{s^i\}$ in the direction $\{d^i\}$ [16,18] and then identifying the gradient

$$\left.\frac{\partial F(\{s^i\} + \gamma\{d^i\})}{\partial\gamma}\right|_{\gamma=0} = \nabla F(s^i)^{\mathrm{T}}\{d^i\} \ . \tag{21}$$

4 Numerical Results

We consider an anisotropic quadratic cylinder (1 m × 1 m) with the following symmetric material properties:

$$m_{xx}(\mathbf{r}) = 1 - \frac{\varepsilon_{xx}(\mathbf{r})}{\varepsilon_0} - j\frac{\kappa_{xx}(\mathbf{r})}{\varepsilon_0\omega} = -(0.6 + j\,0.63) \ , \tag{22}$$

$$m_{xy}(\mathbf{r}) = 0 - \frac{\varepsilon_{xy}(\mathbf{r})}{\varepsilon_0} - j\frac{\kappa_{xy}(\mathbf{r})}{\varepsilon_0\omega} = -(0.3 + j\,0.4) \ , \tag{23}$$

$$m_{yx}(\mathbf{r}) = 0 - \frac{\varepsilon_{yx}(\mathbf{r})}{\varepsilon_0} - j\frac{\kappa_{yx}(\mathbf{r})}{\varepsilon_0\omega} = -(0.3 + j\,0.4) \ , \tag{24}$$

$$m_{yy}(\mathbf{r}) = 1 - \frac{\varepsilon_{yy}(\mathbf{r})}{\varepsilon_0} - j\frac{\kappa_{yy}(\mathbf{r})}{\varepsilon_0\omega} = -(0.5 + j\,0.5) \ . \tag{25}$$

The principal axes of the permittivity tensor and the conductivity tensor coincide in direction. The object was probed by $N_{\mathrm{I}} = 40$ plane waves at a frequency $f_0 = 450$ MHz whose angles of incidence were equally spaced over a circle. Our near field data were simulated by numerically solving the corresponding forward problem using a mesh (1 m × 1 m) with 40 × 40 constant finite elements. The scattered field was generated at $J = 40$ receiver locations on a circle of radius $R = 5$ m. As the contrast tensor $\overline{\overline{m}}(x,y)$ is symmetric, only the coefficient functions $m_{xx}(x,y)$, $m_{xy}(x,y)$ and $m_{yy}(x,y)$ of the original material properties are shown in the figures 2 to 4.

The test domain Ω $(3\lambda_0 \times 3\lambda_0)$ was discretized by 29 × 29 finite elements. Therefore 5046 real unknowns had to be determined. The initial guess was $\eta_{jk}(\mathbf{r}) = 0.1$ and $\xi_{jk}(\mathbf{r}) = 0.1$. Assuming equal sized finite elements Ω_p, we chose $\alpha_{jk,\mathrm{R}} = \alpha'_{jk,\mathrm{R}} \int_{\Omega_p} \mathrm{d}\Omega' = 10^{-5}$, $\alpha_{jk,\mathrm{I}} = \alpha'_{jk,\mathrm{I}} \int_{\Omega_p} \mathrm{d}\Omega' = 10^{-5}$ and $\beta_{jk,\mathrm{R}} = \beta_{jk,\mathrm{I}} = 0.01$. The reconstruction from 'exact' near field data after 500 iteration steps is given in the figures 5, 6 and 7. The error of reconstruction after 500 iteration steps is $e_{\mathrm{rek},500} = 11.2\%$.

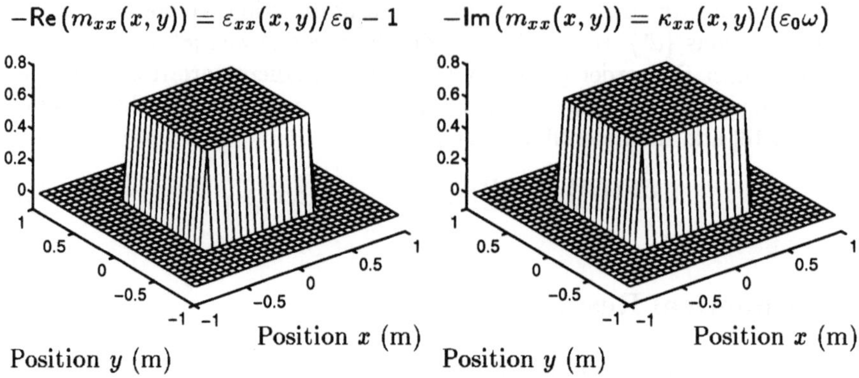

Fig. 2. Original real and imaginary part of $-m_{xx}(x,y)$

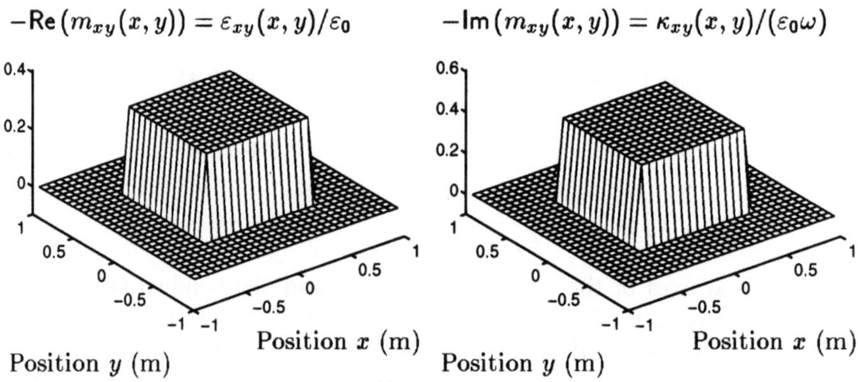

Fig. 3. Original real and imaginary part of $-m_{xy}(x,y)$

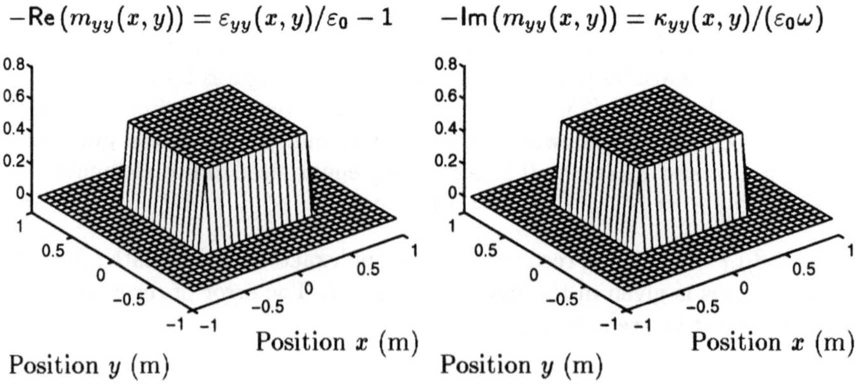

Fig. 4. Original real and imaginary part of $-m_{yy}(x,y)$

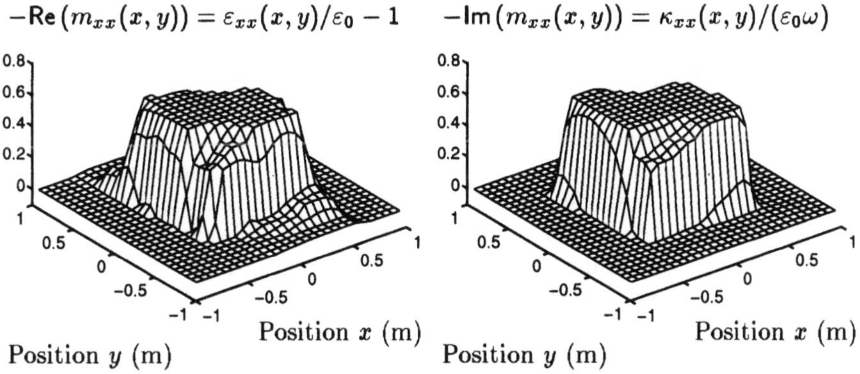

Fig. 5. Reconstructed real and imaginary part of $-m_{xx}(x,y)$; $e_{\mathrm{rek},500} = 11.2\%$

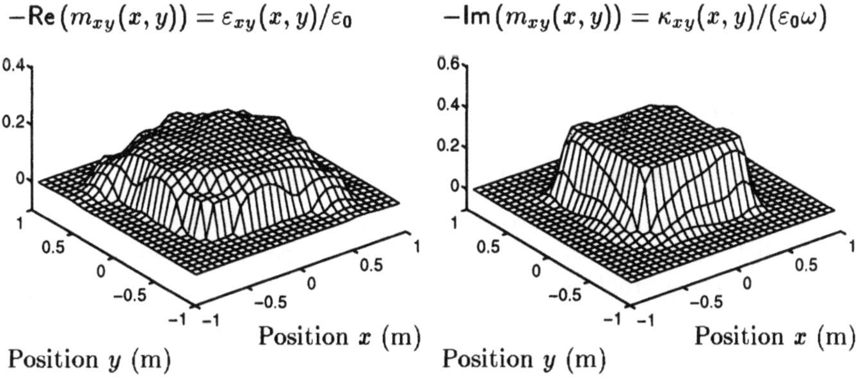

Fig. 6. Reconstructed real and imaginary part of $-m_{xy}(x,y)$; $e_{\mathrm{rek},500} = 11.2\%$

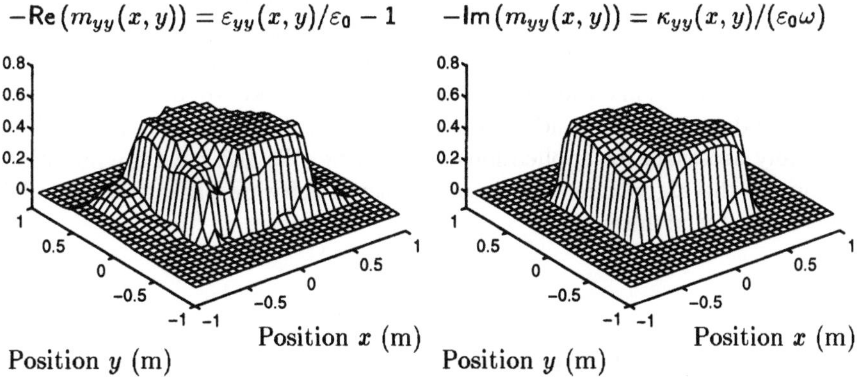

Fig. 7. Reconstructed real and imaginary part of $-m_{yy}(x,y)$; $e_{\mathrm{rek},500} = 11.2\%$

While the reconstructed functions $m_{xx}(x,y)$ and $m_{yy}(x,y)$ in Fig. 5 and 7 show clearly the object's boundary, problems occur in the reconstruction of the function $m_{xy}(x,y)$ (see Fig. 6). In [6] difficulties are also reported with the function $m_{xy}(x,y)$. It seems that the scattered near field does not contain the information necessary to reconstruct the function $m_{xy}(x,y)$.

It took approximately 4 hours to calculate the 500 iteration steps for the reconstruction on a single processor of the NEC SX4 32, whereas about 30 hours were needed on our DEC ALPHA Server 1200. This example shows very well the necessity of using a machine like the NEC SX4. The reconstruction algorithm, described in the previous sections, depends on a lot of parameters which influence the numerical solution of the inverse problem. To obtain an idea on how to choose these parameters, several tests have to be conducted. The main parameters are listed below:

- Wave length of the incident plane waves in relation to the dimensions of the object.
- Polarisation of the incident waves.
- Number of incident waves.
- Number of measurement points.
- Discretization of the test domain Ω.
- Type of regularization and regularization parameters.

The current example was a two-dimensional problem. So the computational costs (CPU time and storage requirements) for solving three-dimensional inverse scattering problems will dramatically increase.

5 Conclusion

We proposed a variable metric method to solve the inverse electromagnetic scattering problem of reconstructing the material properties of anisotropic objects. According to the increasing problem size with anisotropic objects, the CPU time is a critical factor which makes it necessary to use high performance machines like the NEC SX4. In the future, we would like to extend these two-dimensional considerations to the three-dimensional case, which is especially necessary for applications in medicine. We also have the intention to convert the algorithms of the current serial software package to parallel supercomputer architectures.

References

1. Born M. and Wolf E. *Principles of Optics*. Pergamon Press, Oxford, 1964.
2. Chew, W. C. *Waves and Fields in Inhomogeneous Media*. Van Nostrand Reinhold, New York, 1990.
3. Colton, D. and Kress, R. *Inverse Acoustic and Electromagnetic Scattering Theory*. Applied Mathematical Sciences 93. Springer, Berlin Heidelberg New York, 1992.

4. Colton, D. and Monk, P. The detection and monitoring of leukemia using electromagnetic waves: mathematical theory. *Inverse Problems*, Vol. 10: 1235–1251, 1994.
5. Colton, D. and Monk, P. The detection and monitoring of leukemia using electromagnetic waves: numerical analysis. *Inverse Problems*, Vol. 11: 329–342, 1995.
6. Fadoulourahmane, S. *An Inverse Problem for Time Harmonic Electromagnetic Waves In An Inhomogeneous Orthotropic Medium*. Dissertation, University of Oulu, Oulu University Press, 1997.
7. Huber, C. J., Rieger, W., Haas, M. and Rucker, W. M. The numerical treatment of singular integrals in boundary element calculations. *ACES Journal*, Vol. 12, No. 2: 121–126, 1997.
8. Jacoby, S. L. S., Kowalik, J. S. and Pizzo, J. T. *Iterative Methods for Nonlinear Optimization Problems*. Prentice-Hall, Inc., Englewood Cliffs, New Jersey, 1972.
9. Kleinman, R. E. and van den Berg, P. M. Two-dimensional location and shape reconstruction. *Radio Science*, Vol. 29: 1157–1169, 1994.
10. Lobel, P., Kleinman, R. E., Pichot, Ch., Blanc-Feraud, L. and Barlaud M. Conjugate-gradient method for solving inverse scattering with experimental data. *IEEE Antennas and Propagation Magazine*, Vol. 38: 48–51, 1996.
11. Louis, A. K. *Inverse und schlecht gestellte Probleme*. Teubner, Stuttgart, 1989.
12. Rieger, W., Buchau, A., Haas, M., Huber, C., Lehner G. and Rucker W. M. 2D-TE Inverse Medium Scattering: An Improved Variable Metric Method. *IEEE AP-S International Symposium Digest*, Vol. 3: 2140–2143, 1999.
13. Rieger, W., Buchau, A., Haas, M., Huber, C., Lehner G. and Rucker W. M. Image Reconstruction from Real Scattering Data. *COMPEL-The International Journal for Computation and Mathematics in Electrical and Electronic Engineering*, Vol. 18, No. 3: 382–394, 1999.
14. Rieger, W., Haas, M., Huber, C., Lehner G. and Rucker W. M. Inverse Electromagnetic Medium Scattering Using a Variable Metric Method. *IEEE AP-S International Symposium Digest*, Vol. 4: 2621–2624, 1997.
15. Rieger, W., Haas, M., Huber, C., Lehner G. and Rucker W. M. Reconstruction of Inhomogeneous Lossy Dielectric Objects in One Dimension. *ACES Journal*, Vol. 12, No. 2: 54–59, 1997.
16. Rieger, W., Haas, M., Huber, C., Lehner G. and Rucker W. M. A New Approach to the 2D-TE Inverse Electromagnetic Medium Scattering. *IEEE AP-S International Symposium Digest*, Vol. 2: 706–709, 1998.
17. Rieger, W., Haas, M., Huber, C., Lehner G. and Rucker W. M. An Improved Iterative Scheme with Incorporated a priori Information applied to Real Scattering data. *USNC/URSI National Radio Science Meeting Digest*, p. 17, 1998.
18. Rieger, W., Haas, M., Huber, C., Lehner G. and Rucker W. M. A Novel Approach to the 2D-TM Inverse Electromagnetic Medium Scattering. *IEEE Transactions on Magnetics*, Vol. 35, No. 3: 1566–1569, 1999.
19. Wang, Y. M. and Chew, W. C. An iterative solution of the two-dimensional electromagnetic inverse scattering problem. *International Journal of Imaging Systems and Technology*, Vol. 1: 100–108, 1989.
20. Wolfersdorf von, L. *Inverse und schlecht gestellte Probleme; Eine Einführung*. Akademie Verlag, Berlin, 1994.
21. Yaghjian, A. D. Electric dyadic green's functions in the source region. *Proceedings of the IEEE*, Vol. 68: 248–263, 1980.

Computer Science

Prof. Dr. Christoph Zenger, Institut für Informatik, FORTWIHR,
Technische Universität München
Arcisstr. 21, D-80333 München

Strictly speaking, program development for high performance computing is confronted in every case with many areas of computer science. Usually these aspects are not a matter of research but more a question of effective application of methods, tools, and results from computer science.

In this section only those projects are collected where research in computer science is in the center of interest or where computer science aspects are of predominant importance for the project.

Typical questions are related to software engineering aspects for large program packages (these aspects are frequently disregarded to the disadvantage of those projects), parallelization and efficient load balancing (reflecting the fact that high performance computers today are parallel computers), the intelligent decomposition of large and complicated problems in smaller and simpler problems, and the development of tools supporting programming for high perfomance computing. But it should be mentioned that research in this direction does not necessarily need high performance computers, these are only needed for testing and proving the effectiveness of concepts for complex applications. Therefore, the broad research in this direction is covered only to a very small part in this volume.

As already observed in the previous volume, projects in typical research areas of computer science related to high perfomance computing – like data mining, combinatorial optimization like scheduling, control of large distributed systems etc. – are still not present. Future work in this direction is encouraged. High performance computing is not restricted to numerical simulation, and the investigation of new fields of applications is an interesting and challenging area of research.

Fine-Grained Multithreading on the Cray T3E

Andreas Grävinghoff and Jörg Keller

FernUniversität-GHS Hagen, FB Informatik, D-58084 Hagen, Germany
Andreas.Graevinghoff|Joerg.Keller@FernUni-Hagen.de

Abstract. Fine-grained multithreading can be used to hide long-latency operations encountered in parallel computers during remote memory access. Instead of using special processor hardware, the emulation of fine-grained multithreading on standard processor hardware is investigated. While emulation of coarse-grained multithreading is common in modern operating systems, in the fine-grained case research on emulation has been limited and design of multithreaded processors has been favored. A set of tools was developed to support emulation of multithreading on the Cray T3E parallel computer. Several experiments based on parallel matrix multiplication were performed.

1 Introduction

An important problem faced by parallel computers is the latency of accessing remote memory. As this latency usually increases with the number of processors, massively parallel computers are especially affected. While the latency of remote stores can be ignored since they return no result, remote loads have to be completed before computation can proceed (at least beyond a certain point).

A popular approach to attack this problem is to avoid latency by the use of coherent caches. However, this approach causes widely varying memory access times, therefore these machines are called ccNUMA (cache coherent non-uniform memory access) architectures. The non-uniform access time complicates application development, especially for irregular applications. Instead of avoiding latency, one can try to hide latency by *multithreading*, which is explained in the next paragraph.

A multithreaded processor switches context between different *threads* in order to perform useful computations while other threads are waiting for completion of outstanding operations, thus hiding the latency of those operations from the user. This is only possible if several threads per processor are available, hence the application must possess more parallelism than in a ccNUMA machine of comparable size. However, many applications possess this amount of parallelism if the number of threads per processor is moderate.

We distinguish between *fine-grained* (switches context after one or a few instructions) and *coarse-grained* (switches context after a block of instructions) multithreading. In the context of parallel computers, fine-grained multithreading is more interesting (e.g. to achieve higher degrees of parallelization). Multithreading can be implemented by special processor hardware or by emulation in software on off-the-shelf hardware. Special processor hardware is expensive, time-consuming to design, error-prone and often slower than commercial processors. Examples for multithreaded processors are Sparcle [AKK+93], Anaconda [Moo96], SB-PRAM [KPS94] and the Tera MTA [BH95]. Emulation in software is state-of-the-art in coarse-grained multithreading: it is called *multitasking* and is used by almost every modern operating system. Several new operating systems (e.g. Solaris) support threads in the form of *lightweight processes*. The purpose of these lightweight processes is not latency hiding but reduced overhead compared to normal processes and abstraction from the number of processors available. However, the overhead is still too high for our purposes, hence these lightweight processes are not considered further.

We will describe the basic concept in the following section. The third section covers the implementation on a Cray T3E in detail, along with a description of the developed tools. Experimental results based on a parallel matrix multiplication algorithm are presented in the fourth section. An outlook with special emphasis of ongoing and future work concludes the paper.

2 Basic Concept

Emulating fine-grained multithreading on off-the-shelf processors can be done as follows: For each thread, the executed program as well as the *context* (processor state of the thread) are stored in memory. The data structure that contains the context of a thread (and some management information for the emulation program), is called *frame*. To execute an instruction from a given thread, the emulation program restores the processor state, fetches and executes the instruction and updates the processor state in memory. Afterwards, the next thread is executed. To decrease the time to switch contexts, only the part of the context used or modified by the fetched instruction is restored or saved. Note that we assume the emulating and emulated instruction sets to be identical. If this is not the case, then the emulated instructions have to be replaced by one or more instructions of the emulating instruction set.

```
                              ldq  r1, #rs1(frame)
                              ldq  r2, #rs2(frame)
<op> rs1, rs2, rs3   ----->   <op> r1, r2, r3
                              stq  r3, #rd(frame)
                              ret
```

Fig. 1. Example for Subroutine Generation

To implement emulation of mutlithreading, we modify the program code to replace every instruction block `I` by a subroutine. These subroutines restore every register that will be read or modified by the instructions in `I`, execute the instructions in `I`, save every register that has been written or modified, and return. An example of the modification process for an instruction block containing a single instruction op is shown in Fig. 1. The left side contains the original instruction op, which uses two registers **rs1, rs2** as operands and stores the result in register **rs3**. The corresponding subroutine is shown on the right side: First, the registers **r1, r2** are loaded with the values of **rs1, rs2** from the current frame. Afterwards the instruction op is executed, storing the result in a third register **r3**. This register is stored to the current frame before the subroutine returns. Note that the registers **r1**, **r2**, **r3** are not necessarily identical to registers **rs1**, **rs2**, **rs3**, but can be any of the architected registers.

The size of the instruction block can be as small as a single instruction, thus enabling very fine-grained multithreading. The overhead associated with emulation of multithreading decreases with increasing block size since fewer context switches are required, but large block sizes may complicate latency hiding. Note that the size of the instruction block is variable, i.e. it is not necessary to use the same blocksize throughout the whole program. This fact makes further optimizations possible, e.g. identifying sequences of instructions that operate on the same registers and omitting unneccessary accesses to the frame.

The number and location of registers to be read, written or modified by instructions in I can be determined from the instruction set architecture and the instruction itself. This information is used during generation of the corresponding subroutine. The emulation program basically consists of a single loop that calls these subroutines from all threads in a round-robin manner:

1. Initialization
2. Load program counter (PC) from current frame into register *threadPC*.
3. Execute subroutine by jump to value stored in *threadPC*, saving old program counter in register *mainPC*.
4. Return from subroutine to program counter stored in *mainPC*, saving old program counter in register *threadPC*.
5. Store program counter from register *threadPC* to current frame.
6. Load pointer to next frame into register *framePtr*.
7. If the next thread is not the last one, go to step 2.

If synchronous operation of multiple processors is desired, a barrier synchronisation can be performed after each round, i.e. after each subroutine from every thread on a given processor has been executed. The main loop, which uses three different registers (*threadPC*, *mainPC*, *framePtr*), should be compact and fast, since execution is required for every subroutine. We will cover these details in the following section.

We assume that the high-level language source code of the programs to be emulated is available and that the program already uses threads. The thread-related system calls (e.g. creation, deletion) are then replaced by our own routines during recompilation. After modification of the assembler source the program is linked with a small library containing our routines (e.g. main loop). We further assume that all threads operate in user mode. Thus only the registers accessible in user mode are shared between different threads and form the context of a thread. By confining threads to user mode, the switching time between threads is significantly reduced. Calls to the operating system are not emulated, but are executed as usual by system calls and traps, i.e. there are no changes to the operating system. Obviously, we can not handle self-modifying code, since we perform all code modifications during compilation. However, this is not a serious restriction since self-modifying code is generally considered as unfavorable.

3 Implementation

Emulation of Multithreading was first implemented on the Alpha architecture. Alpha is a 64 bit architecture, i.e. all registers are 64 bits wide, instructions are 32 bits wide. There are 32 integer as well as 32 floating-point registers, one register in each class is a dedicated zero source/sink register. Apart from the floating-point control register FPCR, which is only used for some rounding modes, there are no special registers. Alpha is load/store architecture, i.e. all operations are performed between registers. Supported data types include longword (32 bit) and quadword (64 bit) integers as well as five different floating-point formats (2 VAX, 3 IEEE). Recently, support for byte (8 bit) and word (16 bit) integers as well as motional video instructions were added to the Alpha architecture. Implementations of the Alpha architecture, e.g. the 21064, 21164 and 21264 chips, continue to score top ratings on the SPEC benchmarks since the introduction of the architecture in 1992.

The Alpha architecture was chosen for several reasons: Since there are no special registers or condition codes, the user moce context contains only data registers, the program counter (PC) and the floating-point control register (FPCR). The FPCR has to be saved/restored only for some floating-point instructions, otherwise only the general-purpose registers specified in the instruction have to be saved/restored, which eases the work of the assembler converter. Manipulation of the program counter can be done easily with the supported jump-type instructions, which makes implementation of the main loop straight-forward. Last but not least, massively parallel computers based on the Alpha architecture are available, e.g. the Cray T3E, which is described in the next subsection.

3.1 Platform

The Cray T3E supports between 2 and 2048 processing nodes interconnected by a bidirectional 3D torus. Each processing node contains a 21164 Alpha processor running at 300, 450 or 600 MHz. The 21164 is a superscalar processor featuring two integer as well as two floating-point function units and a sustained (in-order) issue rate of four instructions per clock cycle. On-chip caches include two 8 KB direct-mapped data and instruction caches as well as a 96 KB, 3-way set-associative unified second level cache. Support for an external third-level cache with up to 64 MB is included. Virtual address space is 43 bit large, while only 40 bit are implemented physically. Apart from the processor, each node contains up to 2 GB of local memory, a router and other supporting circuitry. Instead of an external third-level cache, stream buffers are used to speed up access to local memory via prefetching on previously detected access patterns.

Access to remote memory, which is non-cacheable, is performed via a large number (512 user + 128 system) of E-registers. A remote load is performed by specifying the target address and an E-register for the result. The result can then be collected from the E-register by a load instruction, which will stall if the result is not yet available. The E-registers significantly increase the number of outstanding loads, since the processor alone can only sustain two outstanding loads. The E-registers therefore provide the necessary support for hiding the latency of remote memory. Unfortunately, the Cray shared memory programming environment ShMem uses synchronous loads, which requires us to implement our own set of communication routines in order to fully utilize the E-registers capabilities.

For a 2048 processor system, the average and maximum network latency is approximately 1500 and 2500 ns, respectively. At a processor speed of 600 MHz, this translates to 900 and 1500 clock cycles, respectively [ST96]. These times exclude memory access and processing at the processing nodes and are therefore quite optimistic. We performed our implementation on the T3E installed at HLRS (Höchstleistungsrechenzentrum Stuttgart) in Germany. This machine has 512 processors running at 450 MHZ and 64 GB of memory (128 MB per processing node).

3.2 Tools

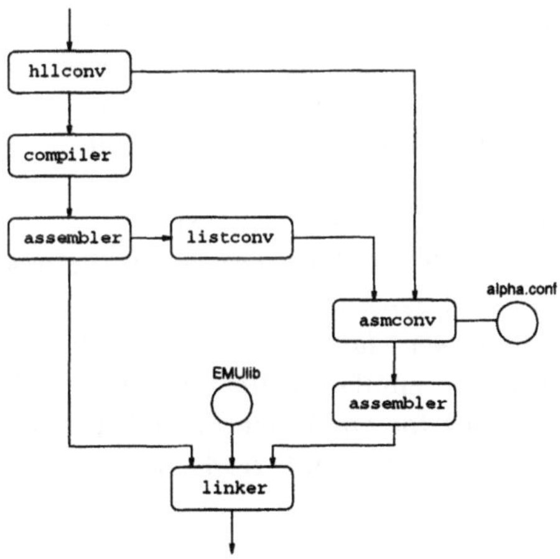

Fig. 2. Design Flow

Several tools were developed to facilitate emulation of multithreading. The design flow and interaction between these tools and the standard programming environment (compilers, assembler, linker) is shown in Fig. 2. The program source code is first modified by the high-level language converter *hllconv*, which searches the source file for user-supplied function names and modifies only those functions. This allows emulation of multithreading to be applied on a function-by-function basis, i.e. only on those functions that benefit from the emulation. The modification process works recursively through all of the given functions. All modified functions are duplicated before the thread-related function calls are substituted with the corresponding calls to the emulation library *EMUlib*. This duplication is necessary since functions may be emulated or executed depending on the caller. The modified source code is forwarded to the compiler, while the list of all modified functions is forwarded to the assembler converter *asmconv*.

The modified source file is compiled using one of the aforementioned compilers, the resulting assembler source is forwarded to the standard assembler *cam*. If there are no functions to be modified, an object file is generated for later use of the standard linker *cld*. Otherwise the assembler listing output is converted to a valid assembler source by *listconv* before it is passed to the assembler converter. The use of the assembler listing instead of the assembler source allows us to ommit support for the advanced *cam* features (e.g. macros) in the assembler converter.

The converted assembler source is processed by the assembler converter, modifiying only those functions given by *hllconv*. If one of those functions is found, the necessary instruction blocks are generated based on the assembler source, the selected grainsize/optimization level as well as the configuration file. This configuration file contains information about instruction styles, modification rules for instructions as well as characteristics of external calls. Since these largely machine-dependent informations are contained within the configuration file, porting of the assembler converter to a different architecture is quite simple.

The assembler converter supports fixed and variable grain-sizes up to several hundred instructions per block. Therefore the number of context switches can be tailored to the application's requirements. In addition, several optimizations that improve instruction scheduling within blocks are available. Calls to external functions are recognized and treated accordingly.

After the whole source file has been processed, the assembler generates an object file from the modified assembler source. After all required object file are generated, the standard linker *cld* combines these object files with the standard libraries as well as the emulation library *EMUlib*. This library contains the thread-related functions for emulation of multithreading, i.e managment, communication and synchronization of threads. The resulting executable can be executed in the usual way.

3.3 Portability

Emulation of Multithreading currently only supports the Alpha architecture. As we already mentioned, this architecture provides a good match for our approach. However, emulation of multithreading is not restricted to the Alpha architecture and can be ported to other architectures. These architectures should meet the following requirements: First of all, it should be possible to load/store the program counter from/to the general-purpose registers. A jalr (jump and link register) type of instruction is sufficient, i.e. is is not required that the program counter is one of the general-purpose registers. For performance reasons, the architecture should support hints to branch prediction logic for this type of instruction, i.e. allow the program counter to be pushed/popped to/from the return-address stack depending on the opcode. The Alpha architecture solves this problem by providing 4 different jalr-type instructions that differ only in the hints to the branch-prediction logic.

Second, the architecture has to provide a minimum number of 6 registers. As we already mentioned above, the main loop uses 3 different registers (framePC, threadPC, returnPC). These registers should be callee-save, otherwise they must be saved/restored upon entry/return to/from external calls, e.g. calls to the operating system. Up to 3 additional registers are required for instruction execution, i.e. to store the operands and results of a single instruction. A larger number of available registers is beneficial for grain-sizes larger than one, since the number of register load/stores from/to the frame can be reduced.

Last but not least, the execution of instructions should depend only on the general-purpose register values, i.e. not on special registers or condition codes. If the architecture has special registers, the load store of these registers from/to the general-purpose registers must be supported. The number of special registers in connection with the number of instructions that depend on these registers affect the size of the frame as well as the average length of the generated subroutines.

In the case of parallel architectures, the hardware must support asynchronous communication between different nodes as well as multiple outstanding communications per node. Since massively parallel computers tend to have large communication latencies, some form of these features is likely to be supported. For performance reasons, the time required to initialize a communication should be small compared to the communication latency.

4 Measurements

The functionality of the assembler converter and the emulation library was tested via several small programs. These programs cover important milestones in the design process, e.g. recursion, external calls and parallelism. After these programs could be successfully modified and executed, we started implementation of a parallel matrix multiplication algorithm.

The program first checks the command-line parameters t (number of threads) and n (matrix size), which must be a multiple of t times p (numer of processors), before shared memory for the three matrices A, B and C is allocated via *shmalloc()*. Matrices A and C are stored in row-order, while matrix B is stored in column-order. The rows and columns are mapped in blocks of t rows resp. columns to the processors in a round-robin manner. Each processor stores random integer values to the corresponding elements of matrices A, B after the random number generator was initialized with the processor's id. After the matrices have been allocated and initialized, a barrier synchronization is performed among all processors before the matrix multiplication begins.

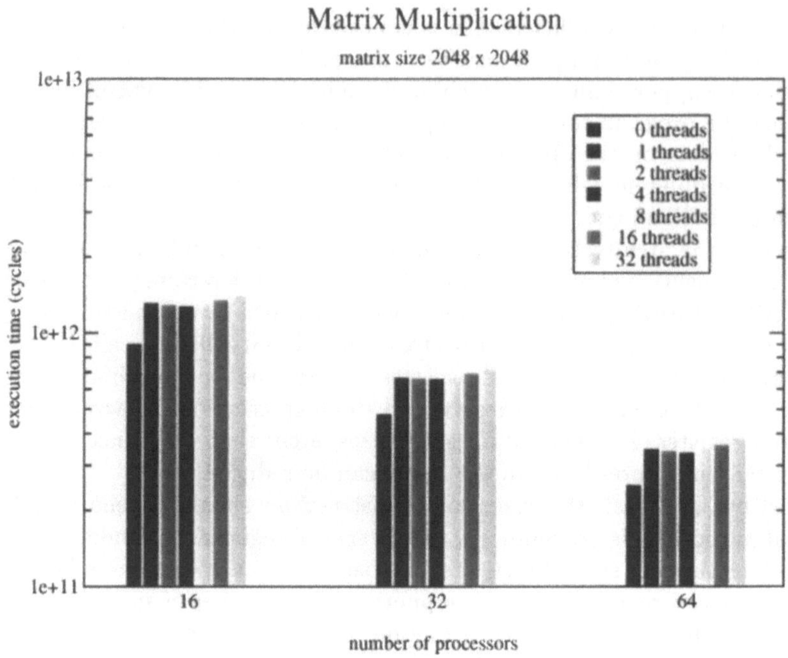

Fig. 3. Experimental Results

The matrix multiplication uses emulation of multithreading, therefore some initialization is required before the multiplication routine can be called. This routine uses the standard matrix multiplication algorithm and operates on all elements mapped to the given thread/processor combination. The elements are read from matrices A, B via *shmem_get()*, while the result is written to matrix C via *shmem_put()*. As we mentioned before, *shmem_get()* works synchronously, i.e. the routine returns only after the desired data has been placed in the target location. Therefore no latency hiding is possible between different invocations of *shmem_get*, i.e. there is no benefit to the emulation of multithreading in this case.

In order to maximize the overhead associated with emulation of multithreading, a fixed grain-size of 1 was used, i.e. context was switched after every instruction. The values reported below therefore provide an upper bound on the overhead associated with emulation of overhead on this application. The execution time of the conventional (unmodified) program was measured as well and provides the baseline for the comparison below. In both cases, the execution time of the matrix multiplication was measured and averaged among all processors. In the case of emulation, this time includes the overhead associated with initialization of threads as well as the matrix multiplication itself.

We performed several experiments with a matrix size of 2048 x 2048 elements and configurations of 16, 32 and 64 processors. For each configuration, the conventional program as well as the multithreaded version using 1, 2, 4, 8, 16 or 32 threads was executed. The results are depicted in Fig. 3. The black bar labeled "0 threads" represents the conventional program and has the smallest execution time. As we explained earlier, this comes as no surprise, since there are no benefits by the emulation of multithreading without latency hiding.

The execution times of the multithreaded programs are quite similar, but increase slightly with the number of threads. Overall, the multithreaded programs run about 1.4 times slower than the conventional version. This is an encouraging result, since the overhead of emulation is smaller than expected. After the emulation library has been extended with asynchronous communication in order to support latency hiding, we are going to repeat these experiments using latency hiding. As the latency to remote memory increases with machine size, we will extend our investigations to larger ($128 \leq 512$ PEs) configurations, since the benfits of emulation increase with latency and number of threads [GK98].

5 Conclusions

Commercial off-the-shelf microprocessors support neither fine- nor coarse-grain multithreading in hardware. As corresponding features (e.g. multiple contexts) have not been announced for any of the mainstream microprocessor families, this situation will not change in the near future. Therefore implementations of multithreading traditionally favored the design of custom microprocessors, but the results were seldom comparable to commercial products. The concept described in this paper utilizes the capabilities of modern microprocessors to enable fine-grained multithreading without hardware support. After development of the basic concept, we exemplified an implementation on the Cray T3E platform. This paper-and-pencil implementation provided the basis for a first analysis of emulating multithreading [GK98]. A grant of computing time from the HLRS allowed us to develop a real implementation.

We successfully implemented the critical tools, i.e. the assembler converter and the emulation library. In both cases development is complicated by the low-level nature of these programs. The assembler converter is quite mature and lacks only instruction scheduling optimizations in the case of larger grainsizes. The converter has been ported to the SimpleScalar architecture [BA97] to facilitate microarchitectural investigations, while the emulation library is currently complemented with asynchronous shared-memory communication and synchronisation primitives. The extended library will allow the porting of whole applications like *fview*, a graph distortion program and *rayo*, a ray tracing program. Afterwards we plan to port some of the SPLASH2 [WOT+95] applications and kernels. A detailed investigation using the aforementioned programs will provide a better view about the capabilities of emulation of multithreading. These upcoming developments will require significantly more resources, since our main focus are large-scale (512 processing elements) configurations.

References

[AKK+93] Anant Agarwal, John Kubiatowicz, David Kranz, Beng-Hong Lim, Donald Yeung, Godfrey D'Souza, and Mike Parkin. Sparcle: An evolutionary processor design for large-scale multiprocessors. *IEEE Micro*, 13(3):48–61, June 1993.

[BA97] Doug Burger and Todd M. Austin. The SimpleScalar Tool Set, Version 2.0. Technical report, University of Wisconsin-Madison, Computer Sciences Department Technical Report no. 1342, June 1997.

[BH95] Gregory T. Byrd and Mark A. Holliday. Multithreaded processor architectures. *IEEE Spectrum*, 32(8):38–46, August 1995.

[GK98] Andreas Grävinghoff and Jörg Keller. How to Emulate Fine-Grained Multithreading. In *Proceedings of the 2nd IASTED Conference on Parallel and Distributed Computing and Networks*, pages 584–589. ACTA press, 1998.

[KPS94] J. Keller, W.J. Paul, and D Scheerer. Realization of PRAMs: Processor design. In *Proceedings of the 8th International Workshop on Distributed Algorithms (LNCS 857)*, pages 17–27, September 1994.

[Moo96] Simon W. Moore. *Multithreaded Processor Design*. Kluwer Academic Publishers, Norwell, MA, 1996.

[ST96] Steven L. Scott and Gregory M. Thorson. The Cray T3E network: Adaptive routing in a high performance 3D torus. In *Proceedings of HOT Interconnects IV*, August 1996.

[WOT+95] Steven Cameron Woo, Moriyoshi Ohara, Evan Torrie, Jaswinder Pal Singh, and Anoop Gupta. The SPLASH-2 Programs: Characterization and Methodological Consideration. In *Proceedings of the 22nd International Symposium on Computer Architecture*, pages 24–36. ACM, 1995.

ParGrad System: Dynamical Adaptation of the Parallelism Degree of Programs on Cray T3E

Otilia Werner-Kytölä

Karlsruhe University, Department of Computer Science
Am Fasanengarten 5, 76128 Karlsruhe, Germany
owerner@ira.uka.de
Phone: +49 721 608-7358 Fax: +49 721 608-7343

Abstract. Typically, parallel programs are scheduled to use all available processors. However, communication costs can overwhelm the gain through parallelism, causing the program to run slower than with a smaller number of processors. Moreover, allocating many processors to a parallel program may produce poor machine utilization.
We show that a parallel program can adapt its parallelism degree automatically to achieve an optimum with respect to criteria such as minimal execution time, achievement of a efficiency threshold or minimal cost-benefit ratio. The adjustment is based on execution time measurements. It only requires programmers to identify the loop, whose parallelism degree is to be adapted, and to set some parameters for the adaptation.
The technique has been implemented and tested on the Cray T3E with real applications in the fields of Mathematics, Electrical Engineering and Geophysics. The results concerning four of these applications are presented here. Besides of the extra time needed for the adaptation, the optima were found in all analyzed cases.

1 Introduction and Motivation

A well-known phenomenon in parallel computing is that using more processors may slow down a parallel program. This is caused by a high communication/computation ratio or an insufficient parallelism in the application. A second phenomenon is that the addition of processors does speed up the application, but not proportional to the number of processors used, thus leading to poor system utilization. While most users are only concerned with absolute speed, there are situations where efficiency is important. One of them is when users are charged for the processors used. The second one is that a high allocation of processors prevents other applications from running, even though the processors are largely underutilized. Utilization in this case is more interesting for resource management, but a better utilization can also be of interest to the users, because it may allow them to run other applications simultaneously, thus reducing the overall runtime.

This paper presents an automatic adaptation of the parallelism degree of programs written in C++ and running on parallel computers with distributed memory. The automatic adaptation of the parallelism degree has three goals:

- to minimize the execution time of parallel programs
- to achieve an efficiency threshold
- to minimize the cost-benefit ratio of parallel programs

2 Related Work

We do not address the problem of processor allocation on a multiprocessor which is a well known problem with adequate solutions, e.g. [6], [9], and [3]. This work assumes that allocation services are available. It optimizes the number of processors used by a parallel program. We build on the ideas of Reimer [11], who first demonstrated the feasibility of automatic adaptation of the parallelism degree. Reimer considered a shared memory environment, while we study distributed memory. In such an environment, one of the additional problems faced is the redistribution of data when the degree of parallelism changes. We also added the cost-benefit ratio and the execution time prediction strategies in order to find the optimal parallelism degree. Mary Hall and Margaret Martonosi [5] proposed a run-time mechanism for shared memory machines which dynamically adjusts the number of processors used by an application employing threads. Besides the target architecture and the way of exploiting parallelism, another dissimilarity to the present work is that the only criterion they used to adapt the number of allocated processors was the speed-up behaviour of parallel loops.

3 Adaptation of the Parallelism Degree

We make the following assumptions: The program of interest has been parallelized for a distributed memory parallel computer. Furthermore, the processors and communication channels have identical speeds and capacities.

3.1 Strategies for the Adaptation of the Parallelism Degree

The proposed strategies are classified in local and global strategies. Strategies from the first group are able to find a local optimum. The strategies in the second group find a global optimum but cause higher overhead, because they analyze all possible parallelism degrees. The later will not be considered in this paper.

In the following, the local strategies RISS, PeSS, EISS (based on [11]) and CBISS are introduced. RISS and PeSS take program execution time into account, EISS and CBISS consider machine utilization criteria.

- **R***untime* **I***ncremental* **S***earch* **S***trategy* (RISS) - Here, the minimization of the execution time is the criterion for the adaptation of the degree of parallelism. The number of allocated processors is reduced as long as the execution time gets lower. The advantage of this strategy is that it does not need an one-processor run in order to calculate the speed-up and efficiency. So, the optimal degree of parallelism is found more quickly and with lower costs, especially when the optimal degree happens to be far from 1.

- **P***rediction-driven* **S***earch* **S***trategy* (PeSS) - In this strategy, the program execution time for the next possible degree of parallelism is predicted. For a parallel program, the execution time sinks along a given number of processors and begins to rise again afterwards. PeSS finds the point in the execution time curve for which the lowest value is delivered.
  The prediction of the execution time for the next possible degree of parallelism is based on two hypotheses:
  1. The execution of a given program with p processors is p times lower than with one processor. Furthermore, communication costs must be added. So, for the calculation of the execution time with p processors (T_p) we get:

$$\boxed{T_p = \frac{T_1}{p} + Communication\ Costs} \tag{1}$$

     where p is the number of processors and T_1 the execution time with one processor. The prediction of the execution time with p processors is then the sum of two curves, namely the execution time and the communication costs curves.
  2. The communication costs can be approximated by a polynomial:

$$Communication\ Costs = c_2 \times p^2 + c_1 \times p + c_0 \tag{2}$$

  Measurements are taken in four points, i.e. for the first four degree of parallelism starting at the highest values. The resulting linear equation system is solved and the unknown constants are calculated. After this, the execution time for the fifth point is not measured, but approximated through equation (1). If its value is lower than the later one, the degree of parallelism is reduced. In this case, the predicted value for the fifth point is replaced by the measured value and the prediction can proceed.
  This strategy has two advantages. First, the one-processor measurement must not be carried out. Second, the adaptation of the parallelism degree and the corresponding data redistribution happen only if it is feasible, i.e. just after discovering through the time prediction that the execution time is going to sink with fewer processors, the number of processors is changed. The costs for the data redistribution are saved, when no improvement in the execution time was predicted.
- **E***fficiency-driven* **I***ncremental* **S***earch* **S***trategy* (EISS) - The speed-up and efficiency are calculated for the parallel loop in which the adaptation of the degree of parallelism takes place. When the efficiency is lower than a predetermined efficiency threshold, the number of allocated processors decreases. Otherwise, it remains the same.
- **C***ost*-**B***enefit-driven* **I***ncremental* **S***earch* **S***trategy* (CBISS) - The adaptation criterion is the minimization of the cost-benefit relation. The degree of parallelism will be reduced as long as this relation gets lower. For this strategy, cost is represented by

$$C = T_p \times p$$

where p is the number of allocated processors and T_p the execution time with p processors. The benefit is represented by the speed-up. The cost-benefit ratio is then

$$R_{CB} = \frac{T_p^2 \times p}{T_1}$$

CBISS searches for the minimum of R_{CB}.

3.2 Adaptation Method

The adaptation of the parallelism degree is carried out for an already parallelized program. Two types of adaptation are proposed: the intrarun and the interrun adaptation. The information which is needed by both adaptation types is obtained dynamically during execution time.

Intrarun Adaptation In intrarun adaptation, the degree of parallelism is adjusted according to a adaptation strategy during one execution of the program, more specifically loopwise. The user instruments the code by specifying parameters (e.g. adaptation strategy, structures which should be redistributed when the number of allocated processors is changed or the data distribution mode) for each adaptation environment and by defining the places in program where the parallelism degree should be adapted.[1] Figure 1 shows the execution sequence of a program when applying the adaptation of the parallelism degree.

Interrun Adaptation In this adaptation mode, the degree of parallelism is adjusted between different executions of the program. This kind of adaptation is necessary for programs, for which the overhead of an intrarun adaptation would be prohibitive, or the entire runtime too short for adaptation. Also for this kind of adaptation, parameters like the adaptation strategy or the number of program executions should be specified (see figure 1).

The optimal degree of parallelism is found by applying a search method, namely a simplified binary search. According to this method, the parallelism degree is halved until the adaptation strategy criterion is fulfilled.

[1] To each loop in which a degree adaptation takes place belongs an *adaptation environment* which contains the parameters for the adaptation of the parallelism degree within that loop.

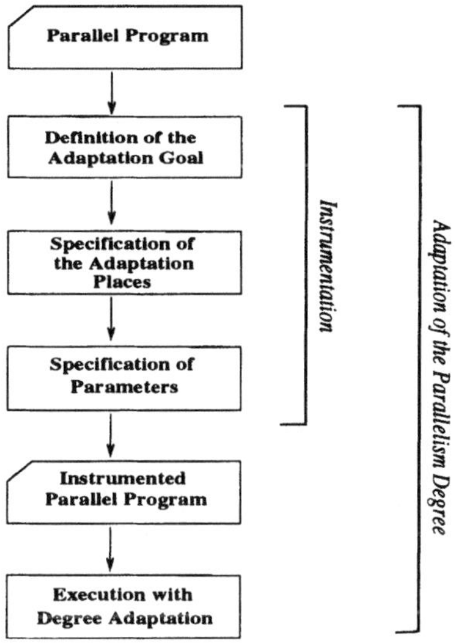

Fig. 1. Execution sequence with adaptation of the parallelism degree

3.3 Data Redistribution

Every time the number of allocated processors for the execution of a program changes, the data must be redistributed among the memory units of the distributed memory machine.

The redistribution of data structures such as arrays, matrices and cubes is already implemented. Also a set of different structures or objects with variations in their fields can be redistributed.[2] Arrays, matrices and cubes can be redistributed blockwise or cyclic. Moreover, matrices and cubes can be redistributed rowwise or columnwise. The redistribution of each built-in data structure is specified in a subclass of the *Distributer* class (see figure 2).

Additionally, when the data to be redistributed is stored in a structure which is different from the ones described above, it is possible to extend the model in order to support it. The model can be easily extended by specifying a new subclass (e.g. *My_Data_Struc* in figure 2) of the *Distributer* and writing the data importation and exportation methods for it.

3.4 ParGrad System

A system called *ParGrad* was developed to adapt the parallelism degree of programs using the presented strategies. *ParGrad* is written in C++ and runs on a Cray T3E/900 [10] using the E-registers and the *shmem*-library for the interprocessor communication.

[2] Structure fields are the different parts of a *struct* data type in the programming language C or C++.

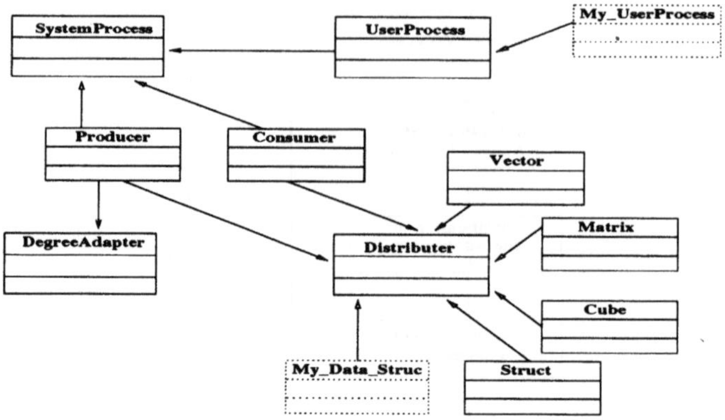

Fig. 2. Partial class diagram of the system

The user of the *ParGrad* system decides in which loop or loops the adaptation of the parallelism degree will take place as well as which kind of improvement is desired (see section 3.1). According to the expected improvement, the user will set the adaptation strategy.

To adapt the parallelism degree of a program which has been parallelized manually or automatically, the user instruments the program by adding *ParGrad* library calls and initializing parameters for the adaptation. The following partial codes show examples of the instrumentation. In line 9, a *ParGrad* library call has been included by the user. This call is responsible for invoking the measurements, which the adaptation is based on, the changing in the parallelism degree and the consequential data redistribution.

```
1    for (pass = 0 ; pass < DIGITS_PER_KEY ; pass++)
2    {
3     ...
4     all_fill_buckets(pass);
5     all_scan_buckets(pass);
6     all_coalesce(pass);
7     ...
8     ParGrad();
9    }
```

The parameters for the adaptation are initialized by specifying a class constructor for a subclass of the *UserModule* class (see the subclass *My_UserModule* in figure 2). Inside this constructor, the user sets parameters such as the adaptation strategy to be employed and the data redistribution mode. The following partial code shows the definition of some ParGrad parameters in a class constructor.

```
My_UserModule::My_UserModule()
{
 set(DATA_DISTRIBUTION, BLOCK);
```

```
  set(DISTRIBUTION_MODE, VECTOR);
  set(STRATEGY, EISS);
  set(EFFTHRESHOLD, 0.50);
}
```

Programs have to be only slightly modified when using *ParGrad*. Table 1 shows the number of added lines and the rate of the added lines when comparing an instrumented to a non-instrumented program. In each one of the four considered programs, a single parallel loop is instrumented for degree adaptation.

Program	Added lines	Overhead (%)
Radix Sort	20	3.1
PDE Solver	32	2.3
Veltran Operator	25	3.3
TLM-Method	33	2.6

Table 1. Overhead caused by the instrumentation considering four programs

4 Experimental Results

The following results are based on measurements which were carried out on a Cray T3E/900 at HLRS of the Stuttgart University, Germany. This implementation of the Cray T3E has 512 DEC Alpha processors running at 450 MHz. The communication between processors was implemented using the E-registers directly and the *shmem*-library.

4.1 Application Characteristics

Radix Sort The radix sort algorithm [13, 12] is an iterative sort method, for which the key exchange between processors corresponds to an all-to-all communication pattern. The implementation of this kernel is based on the Split-C [2] benchmark.

Veltran Operator The veltran operator is a geophysical operator which is used for the analysis of the earth's subsurface layers. The input data is stored in a three dimensional matrix and is distributed columnwise among the processors. Each processor needs data in columns which are located in other processors. The implementation of this algorithm is based on the Fortran version in [7].

Transmission Line Matrix (TLM) Method The TLM method is an electrical engineering application which is used in the analysis of the eletromagnetical compatibility between electronical equipments. The interprocessor communication implied by the parallel version of this method is of neighbor-neighbor type. The implementation of this application is based on [4].

Partial Differential Equation (PDE) Solver The PDE solver from the NAS benchmark suite [1] numerically solves a partial differential equation using three dimensional forward and inverse FFT's. The processors communicate with each other at the transpose phase. The implementation used in this work is based on [8].

4.2 Results

Some of the results achieved by applying the intrarun as well as the interrun adaptation (see section 3.2) are presented below. The left hand side of the figures in this section correspond to program executions without the adaptation of the parallelism degree. The x axis represents the number of processors which were used during the whole execution of a program. The y axis shows the execution time, the efficiency or the benefit-cost relation. Figures on the right hand side correspond to executions with adaptation. In this case, the x axis represents the maximal number of processors available for the execution of a program.

Each bar in the figures corresponds to the time of one program execution, when considering an intrarun adaptation, and of several executions when applying an interrun adaptation. A bar on the right side contains several segments, each segment represents the execution time of a part of the program with the corresponding number of processors for an intrarun adaptation and the execution time of the whole program in the case of an interrun adaptation. In these segments, the numbers in parentheses represent the number of processors which were actually used during the execution of a certain part of the program or during one whole execution of the program, respectively for an intrarun and an interrun adaptation. The numbers above the bars appearing in the figures on the right represent the variation of the degree of parallelism along the time. The last number is the degree of parallelism which was chosen by the corresponding strategy.

For each application, the adaptation takes place for only one parallel loop.

Radix Sort The data structure which must be redistributed because of a change in the parallelism degree is the vector containing the elements to be sorted.

The measurement as well as the data redistribution costs for all analyzed strategies were lower than 1% of the whole execution time for the radix program.

Here, the goal of the adaptation was the minimization of the execution time. Figure 3 shows a case, whose problem size is 128K elements. The correct decision was taken under the PeSS strategy for three cases with a maximal processor number which varies between 32 and 128, namely the parallelism degree has not been reduced. A reduction would have increased the execution time. For these three cases, the PeSS strategy needed two adaptation steps less than the RISS strategy. For a maximal processor number equal to e.g. 128, the parallelism degree was reduced to 64, 32, 16 (see figure 3b). On the other hand, the RISS strategy adapted the parallelism degree stepwise to 64, 32, 16, 8, and finally to 16 processors.

The program execution time is up to a factor of 3,5 lower under the PeSS strategy than without the degree adaptation. This can be observed in figure 3b for a maximal processor number equal to 128 compared to the corresponding bar in figure 3a.

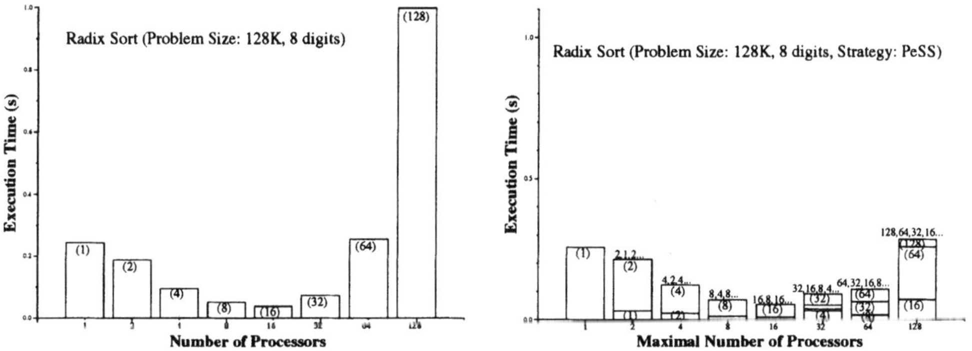

(a) Without adaptation of the parallelism degree (b) Under the PeSS strategy

Fig. 3. Radix Sort under the PeSS strategy with intrarun Adaptation

This example shows a case where the optimum is not the maximum number of processors. The adaptation worked successfully when a initial processor number much higher than the optimal one was chosen.

PDE Solver The data structures which must be redistributed because of a change in the parallelism degree are four arrays which contain intermediary and final results.

Here, the goal of the degree adaptation was the reduction in the execution time of the program. The result of an interrun adaptation of the parallelism degree over 6 iterations under the RISS strategy can be seen in figure 4b. As expected from observing figure 4a, the parallelism degree was adapted to 16.

This example shows that the parallelism degree was properly adapted in order to deliver the minimal execution time.

Veltran Operator The data structures which must be redistributed because of a change in the parallelism degree are two 3-dimensional matrices which contain input data and partial results.

The measurement as well as the data redistribution costs for all analyzed strategies were lower than 10% of the whole execution time for the veltran operator.

The goal of the adaptation here is to improve the program efficiency in order to augment the utilization rate of the whole system. Therefore, the EISS strategy was applied. The efficiency of the veltran operator with an intrarun adaptation of the parallelism degree for e.g. 16 processors was up to 12% higher than without

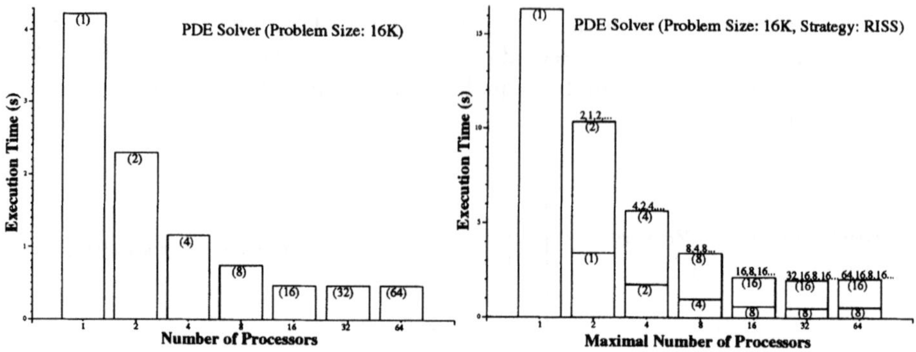

(a) Without adaptation of the parallelism degree (b) Under the RISS strategy

Fig. 4. PDE Solver in an interrun Adaptation under the RISS Strategy

adaptation, while the execution time was up to four times higher than without adaptation. For problem sizes of 6400 and 12800, the degree of parallelism was adapted to 2, when the predetermined efficiency bound was set to 90%.

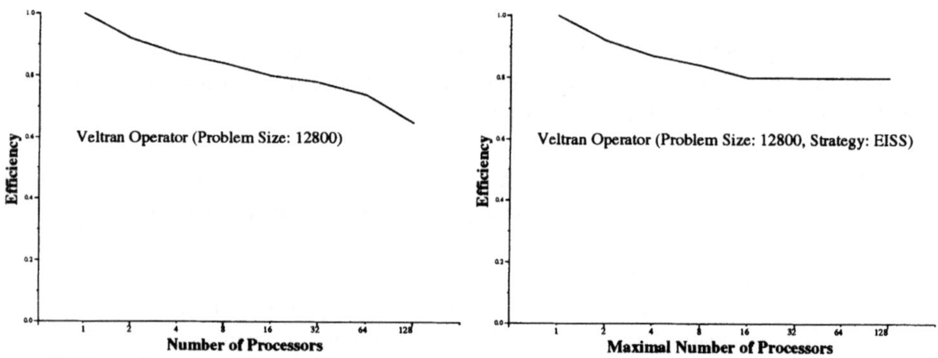

(a) Without adaptation of the parallelism degree (b) With adaptation under the EISS strategy

Fig. 5. Efficiency of the Veltran Operator before and after the interrun Adaptation under the EISS Strategy

The parallelism degree was adapted to 16 for a problem size equal to 12800 elements and an efficiency bound of 80% under the EISS strategy, when applying an interrun adaptation. The program efficiency without the degree adaptation behaves as shown in figure 5a. The efficiency is equal or higher than 80% for parallelism degrees varying from 1 to 16. For a maximal number of processors equal to e.g. 128, the program execution time with the optimal parallelism degree was in a factor of 6 slower, but the efficiency was ca. 24% higher than without the degree adaptation, i.e. with 128 processors.

This example shows that ParGrad properly adjusts the parallelism degree to maintain a minimal efficiency.

TLM Method The data which must be redistributed because of a change in the parallelism degree are in four different structures.

The goal of the adaptation was the maximization of the benefit-cost relation. Therefore, the CBISS strategy was applied.

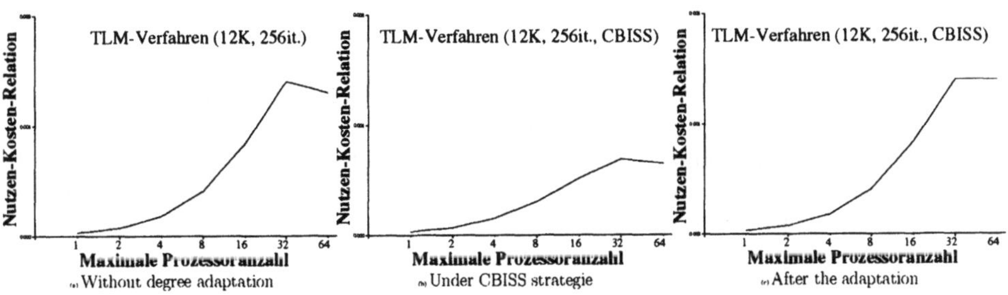

Fig. 6. Benefit-Cost Relation of the TLM Method under the CBISS Strategy

Figure 6 shows that the parallelism degree was adapted to the maximum when using 32 or less processors. When 64 processors were available, the parallelism degree was adapted to 32. Through this adaptation the benefit-cost relation reached its maximum (ca. 23% higher than without adaptation, the execution time was ca. 27% longer).

This example demonstrates that ParGrad is able to find the optimal degree of parallelism according to a criterion which takes the minimal cost-benefit ratio into account.

5 Conclusion

This paper presented the automatic adaptation of the parallelism degree of programs written in C++ and running on parallel computers with distributed memory. The parallelism degree of a program is adapted according to a strategy which is chosen by the user and which defines the parameter (e.g. execution time, efficiency, or cost-benefit relation) to be optimized.

The adaptation of the parallelism degree normally takes place during a single execution of the program (*intrarun*). The degree adaptation can also occur in a *interrun* manner for programs, for which the *intrarun* adaptation would have prohibitive costs because of the data redistribution.

Information which is needed for the adaptation is gained dynamically through measurements during execution time, i.e. no compilation time support is required. The parallelism degree is automatically adapted and the user needs to specify only a few parameters for the adaption through a class constructor. The added code for the adaptation and data distribution is between 2.3 and 3.3% of the total application.

The advantage of employing the adaptation of the parallelism degree is the improvement in the execution time and the speed-up, as well as in the efficiency or cost-benefit relation. An improvement was observed for the radix sort algorithm whose execution time was up to a factor of 3,5 lower. As another example, the cost-benefit relation of the TLM method has been improved up to a factor of 2 when compared to the execution without adaptation of the parallelism degree.

References

1. D. Bailey, J. Barton, T. Lasinski, and H. Simon. The NAS parallel benchmarks. Technical Report 103863, NASA, July 1993.
2. David E. Culler, Andrea Dusseau, Seth C. Goldstein, Arvind Krishnamurthy, Steven Lumetta, Thorsten von Eicken, and Katherine Yelick. Parallel programming in Split-C. In *Proceedings of Supercomputing '93*, pages 262–273, Los Alamitos, CA, November 1993. IEEE Computer Society Press.
3. M.S. Squillante et al. An analysis of gang scheduling for multiprogrammed parallel computing environments. In *Proceedings of the 8th ACM Symposium on Parallel Algorithms and Architectures*, pages 89–98, New York, NY, June 1996. ACM.
4. Martin Gebhardt. Parallelisierung des 3D-TLM-Algorithmus mittels MPI. Master's thesis, Department of Electrical Engineering, Karlsruhe University, 1998.
5. Mary Hall and Margaret Martonosi. Adaptive parallelism in compiler-parallelized code. In *Proceedings of the 2nd SUIF Compiler Workshop*, Stanford University, August 1997. USC Information Sciences Institute.
6. Hans-Ulrich Heiss. *Prozessorzuteilung in Parallelrechnern*. BI-Wissenschaftsverlag, 1994.
7. Matthias Jacob. Implementing large-scale parallel geophysical algorithms using the Java programming language - a feasibility study. Master's thesis, Department of Informatics, Karlsruhe University, 1998.
8. Honghui Lu, Sandhya Dwarkadas, Alan L. Cox, and Willy Zwaenepoel. Message passing versus distributed shared memory on network workstations. In *Proceedings of the Supercomputing'95*, pages 64–65, New York, NY, December 1995. ACM.
9. Cathy McCann, Raj Vaswani, and John Zahorjan. A dynamic processor allocation policy for multiprogrammed shared-memory multiprocessors. *ACM Transactions on Computer Systems*, 11(2):146–178, May 1993.
10. Wilfried Oed. Massiv-paralleles Prozessorsystem Cray T3E. Technische Dokumentation, Cray Research GmbH, München, November 1996.
11. Niels Reimer, Stefan U. Hänßgen, and Walter F. Tichy. Dynamically adapting the degree of parallelism with reflexive programs. In *Proceedings of the Third International Workshop on Parallel Algorithms for Irregularly Structured Problems (IRREGULAR)*, pages 313–318, Santa Barbara, CA, USA, August 1996. Springer LNCS 1117.
12. Steven C. Woo, Moriyoshi Ohara, Evan Torrie, Jaswinder P. Singh, and Anoop Gupta. The SPLASH-2 programs: Characterization and methodological considerations. In *Proceedings of the 22nd Annual International Symposium on Computer Architecture*, New York, NY, June 1995. IEEE Computer Society Press.
13. Marco Zagha and Guy E. Blelloch. Radix Sort for Vector Multiprocessors. In *Proceedings of the Supercomputing'91*, pages 712–721, Los Alamitos, CA, November 1991. IEEE Computer Society Press.

Comparative Measurements of the Solution of PDE's on the PARAGON and the SB-PRAM

Andreas Paul, ap@cs.uni-sb.de

Universität des Saarlandes, FB Informatik, LS Prof. Wolfgang Paul
66123 Saarbrücken, Germany

Abstract. In the project PARDEAL we want to compare the applicability of two approaches to solve partial differential equations: the SB-PRAM, an experimental shared memory machine and the grid partitioning method on distributed memory architectures. This report refers to speedup measurements on the Paragon of the HLRS.

1 Project description

This is a report about project number 11727
"PARDEAL – Solving PDE's on the SB-PRAM".

1.1 Background

The project PARDEAL is part of the SB-PRAM project: The SB-PRAM is an experimental shared memory computer, wich is currently under construction at our institute. It is an advancement of the the theoretical *FLUENT MACHINE* by RANADE ([RBJ88]). Both machines are based on the *PRAM* model ([FW78]) well known in theoretical computer science:

Definition 1. *A PRAM (Parallel Random Access Machine) consists of p independent processors, which have access to a shared memory M. The structure of a single processor corresponds to that of a RAM ([AHU74]). In each step every processor either performs an internal instruction or an access to the shared memory (LOAD/STORE). The duration of each step is independent of p.*

If we want to realize a PRAM we have to overcome two major difficulties:

- Constant access time to the shared memory: Current memory modules have only a limited number of I/O ports leading to a serialization if a large number of processors is accessing the memory cells on one single module.
- Routing time: Processors and memory have to be connected by a network. The time to send requests to the memory usually increases with the number of processors p.

The SB-PRAM ([AKP91], [Kel92]) physically emulates a PRAM and provides from the programmer's point of view a uniformly fast access time to the shared memory. This property is ensured by the following concepts:

Hashing: We use a linear hash function to distribute addresses in a pseudo-random manner to the physical memory: The memory address a (seen by the programmer) is mapped to the physical address address $h(a)$ with

$$h : \{0, \ldots, 2^{32} - 1\} \to \{0, \ldots, 2^{32} - 1\};\ a \mapsto \lambda \cdot a \pmod{2^{32}} \tag{1}$$

λ is an odd number $\neq 1$.

Combining: We use an intelligent network, which combines simultaneous requests to the same memory cell. This technqiue avoids bottlenecks and serialization.

Multithreading: To hide the latency of the interconnection network, each processor executes several (32) independent threads. Switching to another thread is done automatically in hardware after execution of *one* instruction of the current thread. Since the programmer only sees the threads and has no control over the physical processor, the threads are also called *virtual processors*.

Delayed load: The result of a load operation I on the shared memory is expected at the beginning of the *second* operation after I. Delayed load and the concept of virtual processors proved to be sufficient to hide the latency of the network, if the hash function is chosen suitably.

In addition to the memory operations LOAD and STORE the SB-PRAM supports parallel *multiprefix computation*:

Definition 2. *Let $\odot$ be an associative operator. The processors $p_1, \ldots, p_n$ hold the values $d_1, \ldots, d_n$, and the memory cell with address a holds the value d. Each processor executes the multiprefix operation $r_i = mp \odot (a, d_i)$.*

After the operations p_i holds $r_i = d \odot \bigodot_{j=1}^{i-1} d_j$ and cell a holds $d \odot \bigodot_{j=1}^{n} d_j$.

The SB-PRAM supports the operators $\vee$, $\wedge$, $+$ and max (on signed 32-bit integers). The computations are performed in parallel by the network and are completely invisible to the programmer. Parallel prefix computation and hiding the memory access time enable us to use very efficient data structures (barriers, locks, queues, lists etc.) to control the program flow and distribute work to the processors ([Röh96], [PR98]).

1.2 Goal

The SB-PRAM has already showed excellent performance on the SPLASH benchmark suite ([SWG92]). However, the size and importance of the considered problems was quite limited. We therefore want to show that the SB-PRAM also performs well on two extremely important applications: A database server and the solution of partial differential equations. The latter problem class is studied in the project PARDEAL.

To achieve meaningful results, we want to compare the SB-PRAM using a suitable parallelization strategy with the most common approach: grid partitioning on distributed memory machines. We started our comparative measurements on the Paragon, since it offered a high degree of parallelism. In a second step, we plan to use a Cray T3E. To ensure high numerical efficiency, we employ the approved methods of finite elements and multigrid solvers with adaptive grid refinement. The parallel solver on the SB-PRAM is an adaption of the library DEAL written at IWR Heidelberg ([RBS95]).

1.3 Strategy

Among various software environments for the parallel solver on the Paragon, the system UG written at ICA3 Stuttgart ([ugp]) promised to have the most advanced features with an easy access to already implemented solvers. As a first benchmark we choosed the well known problem of the Poisson equation with dirichlet boundary conditions. Since UG can already handle this problem, we could minimze the effort to adapt the benchmark for our demands.

The use of the Paragon can be divided into two stages:

1. Generating the library, verification of its correctness and generating the benchmark.
2. Running the benchmark with modification of parameters (number of processors, load balancing strategy,...)

2 Results

2.1 Resources

The division of our work into two stages is also reflected by the consumption of resources: In stage one we only needed a very limited number of processors; in stage two the number of processors was intended to vary between 1 and the maximum of 228 (cf. section "Problems"). Due to the nature of the problem, typical execution times were in the range of approximately one hour.

The following table shows an excerpt of the resource consumption report. All values are in DM.

Period	Computing time	Disk space	
01.12.98–31.12.98	23.06	3.73	≈ stage 1
01.01.99–31.01.99	134.10	3.38	≈ stage 2
01.02.99–28.03.99	5.45	3.42	

We did not need special preinstalled libraries for our benchmarks.

2.2 Problems

Unfortunately, during stage 2 we encountered three problems which inhibited very meaningful results and prevented us from observing other differential equations or boundary conditions:

- Apparently, the memory management under UG had some little bugs which led to undeterministic program crashes with increasing grid size.
- Because the load balancing algorithm for adaptively refined grids is very memory intense, it could only be used for rather small problem sizes.
- During runs using more than 64 processors, the whole machine (not only the program) regularly crashed during times of heavy communication. According to the manufacturer Intel, this is a known bug of the Paragon, for which there exists no satisfying bugfix.

2.3 Measurements and Analysis

Due to the problems described above, we could only use up to 64 processors and had to widely abandon load balancing (only possible for small grids, i.e. during the first steps of adaptive grid refinement). We were interested in comparing speedups of the SB-PRAM and the Paragon for different grid sizes and numbers of processors. The following figure shows their typical progressions (linear solver with multigrid preconditioning, on Paragon two load balancing strategies lb1, lb2):

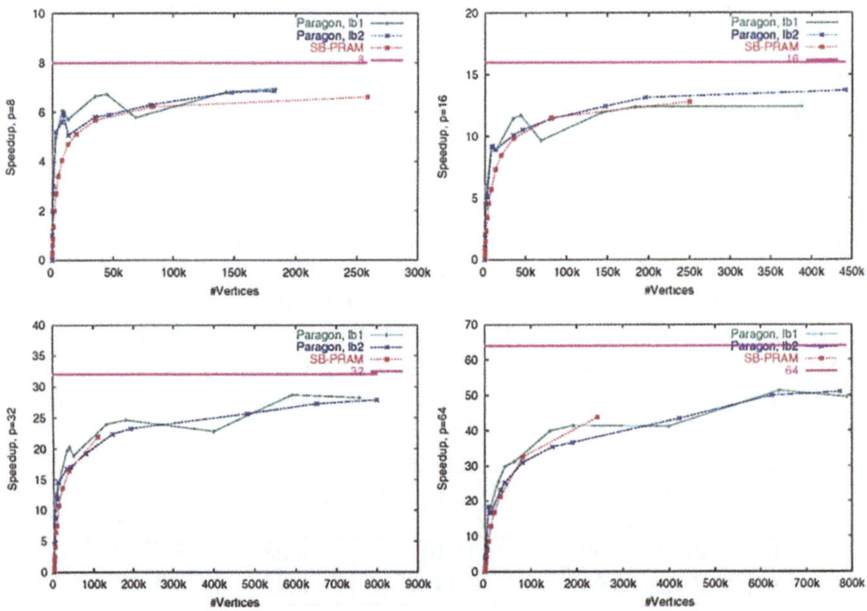

We see that the speedups of PARDEAL on the SB-PRAM and UG on the Paragon are approximately identical for this benchmark (efficieny $\approx 80\%$). Further analysis ([Pau99]) shows, that the main problem on the SB-PRAM is not the large number of threads we have to keep busy ($\approx$ 2000 for $p = 64$) but the overhead of parallel data structures and control mechanisms (parallel task queues, barriers etc.).

The parallelization overhead on the SB-PRAM only weakly depends on the problem we have to solve. Therefore we expect advantages for the SB-PRAM on more complex problems. E.g. it has been showed in [BEL96] that the efficiency of a Cray T3D using UG for linear elasticity calculations is approximately 65% for $p = 64$ and 700000 nodes. The performance of the SB-PRAM such problems has to be examined in the future.

References

[AHU74] A.V. Aho, J.E. Hopcroft, and J.D. Ullman. *The Design and Analysis of Computer Algorithms.* Addison-Wessley Publishing Company, Reading, Massachusetts, 1974.

[AKP91] Ferri Abolhassan, Jörg Keller, and Wolfgang J. Paul. On the Cost–Effectiveness and Realization of the Theoretical PRAM Model. FB 14 Informatik, SFB-Report 09/1991, Universität des Saarlandes, May 1991.

[BEL96] Peter Bastian, Knut Eckstein, and Stefan Lang. Parallel adaptive multigrid methods in plane linear elasticity problems. *Numerical Linear Algebra with Applications*, 1996.

[FW78] Steven Fortune and James Wyllie. Parallelism in Random Access Machines. In *Proceedings of the 10th ACM Annual Symposium on Theory of Computing*, pages 114–118, 1978.

[Kel92] Jörg Keller. *Zur Realisierbarkeit des PRAM Modells.* Dissertation, Universität des Saarlandes, FB Informatik, 1992.

[Pau99] Andreas Paul. *Lösen partieller Differentialgleichungen auf der SB–PRAM.* Dissertation, Universität des Saarlandes, FB Informatik, vor. 1999.

[PR98] Andreas Paul and Jochen Röhrig. Implementing a parallel list on the SB–PRAM. In *Proceedings of the 5th International Conference on High Performance Ccomputing*, pages 52–59, Los Alamitos, California, December 1998. IEEE Computer Society.

[RBJ88] Abhiram G. Ranade, Sandeep N. Bhatt, and S. Lennart Johnson. The Fluent Abstract Machine. In *Proceedings of the 5th MIT Conference on Advanced Research in VLSI*, pages 71–93, Cambridge, MA, 1988. MIT Press.

[RBS95] G. Kanschat R. Becker and F.-T. Suttmeier. DEAL - Differential Equations Analysis Library. Available via http://gaia.iwr.uni-heidelberg.de, IWR Heidelberg, Germany, 1995.

[Röh96] Jochen Röhrig. Implementierung der P4-Laufzeitbibliothek auf der SB–PRAM. Diplomarbeit, Universität des Saarlandes, FB Informatik, 1996.

[SWG92] Jaswinder Pal Singh, Wolf-Dietrich Weber, and Anoop Gupta. SPLASH: Stanford Parallel Applications for Shared-Memory. *Computer Architecture News*, 20(1):5–44, 1992. Technical report: CA 94305, Computer Systems Laboratory Stanford University.

[ugp] UG home page. Available under http://www.ica3.uni-stuttgart.de.

KaHPF: Compiler generated Data Prefetching for HPF

Matthias M. Müller*

Institute for Program Structures and Data Organization,
University of Karlsruhe, Germany.
muellerm@ira.uka.de

Abstract. In many parallel applications, network latency causes a dramatic loss in processor utilization. This paper presents KaHPF, an optimizing HPF compiler, that uses software controlled access pipelining (SCAP) to hide communication latency. Various benchmarks of common, but complex communication patterns demonstrate KaHPF's performance improvement on a Cray T3E. KaHPF's compiled code is faster than code that uses highly optimized shared-memory system functions by a factor of 1.2 to 3.2. In comparison to PGI's HPF compiler KaHPF shows an improvement of 3.9 to 30.4.

1 Introduction

As microprocessors get faster and the gap between computation and communication speeds widens, network latency becomes the dominant factor of the execution time of fine-grained parallel programs. Given a 450 MHz clock, a $1.5\mu s$ latency corresponds to 675 clock cycles on the Cray-T3E/900. Thus, instead of a single communication operation one could perform 675 arithmetic instructions. This situation becomes worse by a factor of up to 10 once the software overhead of communication libraries is taken into account.

Software controlled access pipelining (SCAP) and its vector counterpart (VSCAP) overcome these shortcomings by means of prefetching. The potential of SCAP has been demonstrated by analytic models, simulations, and hand coded benchmarks [War97, MWT98]. Now, this paper presents results of compiler generated VSCAP code. Our KaHPF compiler is a source-to-source compiler from HPF to Fortran 90. KaHPF reads HPF's data distribution directives and focuses on forall-statements for parallelization. Remote data accesses are recognized and fetched into local memory using communication pipelines only consisting of prefetch and access instructions. These communication pipelines are inserted into the final Fortran 90 code replacing any calls to communication libraries.

Prefetching is not new, most research addresses it in the context of prefetching cache lines, non-blocking loads, scheduling techniques, and speculative execution on uniprocessors or small-scale multiprocessors [CB92, RL92, MLG92,

* Supported by the Graduiertenkolleg 'Beherrschbarkeit komplexer Systeme'.

CKP91, GGV90]. Prefetching is also used by software distributed shared memory systems to prefetch whole memory pages [LCD+97, BPA98].

But little is known about the effects of latency hiding applied to communication networks in massively parallel computers with distributed memory. And to our knowledge, KaHPF is the first work targeting compiler directed prefetching for these architectures.

The remainder of this paper is organized as follows: Section 2 gives an overview about SCAP while Section 3 presents the architecture of KaHPF. Benchmarks and results are discussed in Section 4. Conclusions are drawn in Section 5.

2 SCAP

In fine-grained parallel applications as in most other parallel applications latency prevents fast access to non-local memory. Software controlled access pipelining (SCAP) targets latency hiding through both overlapping computation and communication by splitting non-local memory access into low overhead prefetch and access. This Section explains in brevity the basic concepts of SCAP. A detailed explanation can be found in [MWT98, War97].

2.1 Basic Idea of SCAP

The aim of SCAP is a run-time improvement achieved with overlapping several communication requests leading to a communication pipeline in fine grained data parallel applications. For a better understanding of the basic principle of SCAP, we explain first how communication is usually done.

The processor issues a request to the network and waits until the network replies. Only then, the processor continues its execution and issues a new request. This is done as long as the processor requires remote data elements to perform its local part of computation. As the processor blocks after each data request, we call this kind of communication the ***blocking mode***, see the upper half of Figure 1. This is our base model.

Now, let us assume the processor could issue all its communication requests and the network would be able to process them overlapped. This would lead to a shorter waiting period for the processor accessing the first and all other successive remote data elements. Finally, communication could be performed faster compared to the above mentioned blocking execution. We call this kind of communication ***overlapping communication***, see the lower half of Figure 1.

To enable overlapping communication, the network interface has to provide a ***prefetch buffer*** that decouples the processor from the network execution, see Figure 2. Its size must be large enough to hold all outstanding remote data requests issued by the processor.

We use the Cray T3E as our target platform for our implementations since it is the only machine architecture providing overlapping communication by means of E-registers that can be used as prefetch buffers.

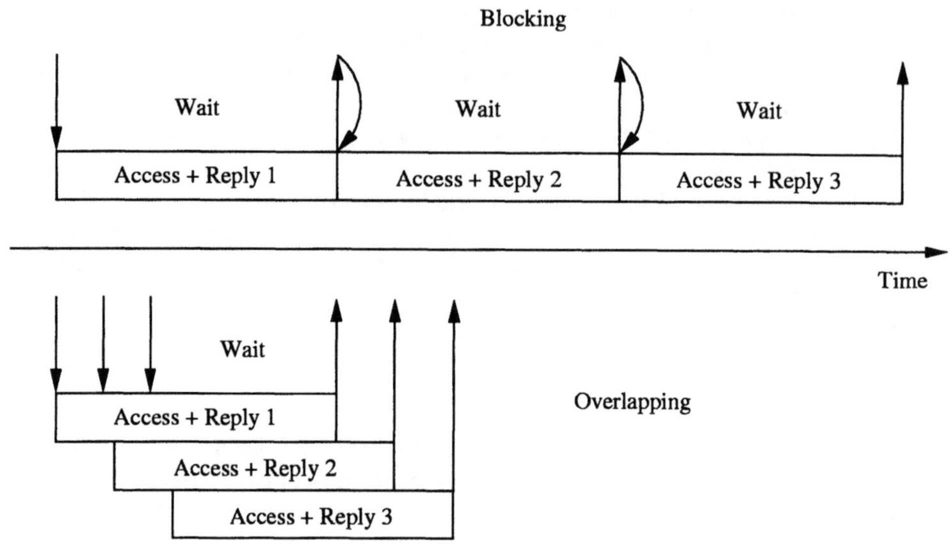

Fig. 1. Basic idea of SCAP

2.2 SCAP Code Transformation

The basic transformation rule of SCAP is illustrated using the following simple forall-statement:

```
FORALL  ( I = 0:N-1 )
   A(I) = B(q(i))
END FORALL
```

The program fragment updates array A in parallel, indexing array B with permutation q. A parallelizing compiler maps the problem size N onto P real processors. This technique is called ***virtualization***. Assuming a `CYCLIC(k)` distribution, each processor emulates $V = \lceil \frac{N}{P} \rceil$ virtual processors within a virtualization loop. Both A and B are distributed over the P processors. Since the value of $q(i)$ can not be determined at compile time the compiler has to insert remote memory accesses. The virtualization of the program fragment is as follows:

```
FORALL  ( J = 0:P-1 )                    ! Forall processors in parallel
   DO  k=j*V,(j+1)*V-1                   ! Simulate V virtual processors
      a1 = calculate_address(B(q(k)))    ! Calculate remote address
      A(k) = remote_read (a1)            ! Read remote data element
   END  DO
END FORALL
```

In the worst case, every processor issues V non-local memory accesses. These stall the processor if the network can not serve the desired values fast enough.

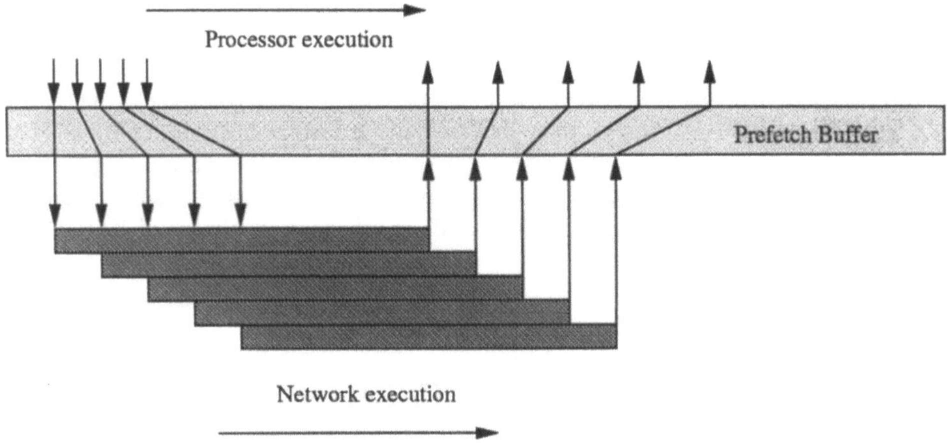

Fig. 2. Usage of Prefetch Buffer

Hence, execution time of this loop is at least V times the network latency. This would be the code for blocking execution.

The following transformation of the loop shows how communication and computation can be overlapped:

```
FORALL  ( J = 0:P-1 )
   DO  k=j*V,(j+1)*V-1
      a1 = calculate_address ( B(q(k)) )  ! Calculate remote address
      prefetch(a1)                        ! Start read request
   END DO
   DO  k=j*V,(j+1)*V-1
      a1 = calculate_address ( B(q(k)) )
      A(k) = access(a1)                   ! Access data element
   END DO
END FORALL
```

In this transformation, the main loop is split into two instances: a prefetch and an access (or calculation) loop. Instead of stalling on a remote memory access as in blocking mode, the processor issues remote memory requests. After the prefetch loop is executed the calculation loop accesses non-local memory without waiting time (if the data is already present). If the number of non-local memory accesses is too large to fit in the prefetch buffer, SCAP changes from a two to a three loop execution where the middle loop alternates between access and prefetch instructions.

In the best case, program speed-up is about $(V-1)$ times the network latency because there is at most one waiting period (arrival of first data item) compared to V waiting times in a blocking network.

VSCAP uses additionally vector-prefetch and/or -access operations for communication reducing run time overhead and decreasing communication time further.

3 KaHPF

This section presents an overview of the prototype HPF compiler of Karlsruhe KaHPF. KaHPF is a source-to-source compiler transforming a data parallel HPF program into an executable Fortran 90 node program that uses E-register operations for communication. Its program transformations concentrate on the forall-statement. The single analysis and transformation phases of KaHPF are shown in fig. 3.

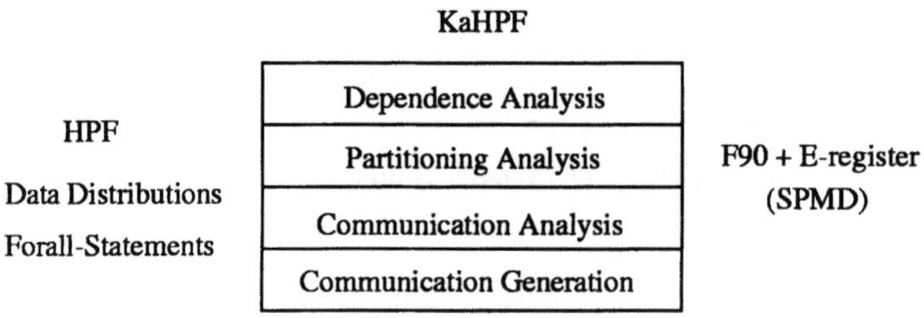

Fig. 3. Architectural Overview of KaHPF

KaHPF is based on the ADAPTOR [Bra94] front-end developed by Brandes at the German National Center for Computer Science (GMD). All subsequent analysis and transformation phases operate on the abstract syntax tree built by the front-end. The dependence and partitioning analysis phases use common techniques to perform their tasks.

3.1 Communication Analysis

Communication analysis determines which variable references cause non-local data accesses. The *rhs* references to distributed arrays are examined. According to the data distribution and the subscript function of each distributed dimension, non-local data references are determined, and the forall-statement is splitted into a local loop accessing only local variables and a remote loop accessing only remote data. If static analysis can not detect non-local data sets, e.g. in irregular access patterns, the whole loop is marked as remote. This causes all elements to be treated as remote and to be accessed using prefetch operations. We call this ***speculative prefetch.*** The advantage of speculative prefetch is its less runtime overhead compared to the ***inspector-executor*** model [KM90] because no additional data structures for communication schedules have to be built. Instead of performing speculative prefetching, it would be also possible to insert a runtime test to determine locality of data elements and to access local elements directly, but this is subject to further studies.

3.2 Communication Generation

The task of the communication generation phase is two fold: (1) It determines an appropriate communication pipeline. The selection does not only depend on the communication pattern determined by the communication analysis but also on the communication mode selected on the command line of KaHPF. This mode decides if KaHPF emulates blocking communication, uses VSCAP, or inserts calls to the shared-memory system functions. (2) The final phase generates the appropriate communication pipelines as explained in Section 2 and inserts them into the program code. The software overhead of these pipelines is very low because no expensive calls to communication libraries exist. The absence of library calls also increases the chance for further optimizations by the Fortran 90 node compiler.

4 Results

This section compares KaHPF compiled programs with the PGI HPF compiler [BMN+97] and the shared-memory system functions. The comparison is done on six application kernels representing six different communication patterns.

The first part of this section presents the benchmarks and their implementation while the second part discusses the results.

4.1 Benchmarks and their Implementations

The parallel algorithms are classified due to their ratio between communication and computation time and their data access patterns. Table 1 gives an overview.

Access Pattern	$T_{\text{Communication}} < T_{\text{Computation}}$	$T_{\text{Communication}} \sim T_{\text{Computation}}$
Reduction	LL3 (64)	LL5 (32)
Indexed Arrays	LL1 (32)	Rotate (32)
Indirectly Indexed Arrays	-	Indirect (32)
2D-grid	Jacobi (64)	-

Table 1: Data access patterns and their representatives in our set of benchmarks.

The columns present the different amount of communication while the rows distinguish communication patterns. Entries show the representatives of the respective class. The number in parentheses show the number of PEs which took part in the benchmark. LL1, LL3, and LL5 are parallelized versions of the corresponding Livermore loops. ROTATE is a cyclic shift of an array. INDIRECT is the example assignment from Section 2, and JACOBI implements the jacobi iteration.

The characteristics of each data access pattern is described below. The reduction of LL3 is a vector product and structured in such a way that each processor performs a local reduction and then collect the results from other processors. Thus, the amount of communication is merely depending on the number of processors and small compared to computation. Reduction in LL5 is not implemented as efficiently. Rather than a local reduction each vector element collects results from $\log n$ neighbors, where n is the size of the vector. Thus, communication grows with the size of the area. LL1 and ROTATE address communication caused by indexing integer arrays. The index offsets are constants. The high communication rate of ROTATE is caused by a cyclic distribution. INDIRECT represents irregular communication, also caused by indexing. The indexes are computed dynamically from other data. The left column is empty because we did not find a suitable application. JACOBI deals with blocked 2D-arrays in which communication is limited to the border of the blocks. There is no benchmark for the right column as the proportion of computation to communication is always high for this kind of communication patterns.

Each kernel is compiled to four different versions.

BLOCK simulates blocking communication on the Cray T3E as described in Section 2.

VSCAP combines 8 E-registers to a vector and operates on them with vector commands that reduce prefetch overhead and increase network bandwidth.

SHMEM uses the shared memory system functions for communication.

PGHPF represents the executables of the PGI HPF compiler. PGI HPF is the commercial HPF compiler on the Cray T3E.

The first three versions are obtained with the KaHPF compiler. The different communication methods are selected on the KaHPF command line. The fourth version is compiled using PGI's HPF compiler using the same HPF source program as for KaHPF. PGI HPF and the Fortran 90 node compiler were tuned with `-O2 -Msmp` and `-O2`, respectively. Time was measured with the real time clock `rtc`. We measured the run-time for different problem sizes while the number of processors was kept constant.

4.2 Measurements

Numbers in Table 2 show maximum relative performance of VSCAP in factors against the remaining three versions. The benchmarks are grouped by their relationship of communication to computation.

Several interesting observation can be made regarding the numbers of Table 2. First, VSCAP is the fastest version in all tested benchmarks. VSCAP outperforms even the shared-memory system functions by at least 20%. Even an improvement of a factor 13.2 can be achieved. The latter factor is reached in the irregular communication benchmark INDIRECT because this kind of communication pattern is not supported by the shared-memory system functions. VSCAP is also at least a factor of 3.9 faster than PGI HPF. The best relative

Benchmark	VSCAP vs. BLOCK	VSCAP vs. SHMEM	VSCAP vs. PGHPF
$T_{\text{Communication}} < T_{\text{Computation}}$			
LL3	2.2	1.5	3.9
LL1	3.0	1.3	4.4
Jacobi	2.7	1.2	11.4
$T_{\text{Communication}} \sim T_{\text{Computation}}$			
LL5	10.8	1.8	9.6
Indirect	9.3	13.2	9.2
Rotate	32.9	2.8	30.4

Table 2: Measurements on the Cray T3E

performance of factor 30.4 in this comparison is reached in our communication benchmark ROTATE.

By comparing the upper half of Table 2 with its lower half it can be noted that the relative performance of VSCAP increases with a higher ratio of communication to computation. This is due to the fact that VSCAP can exploit its latency hiding capabilities in these cases at its best.

5 Conclusions

Previous work showed the good performance of SCAP and VSCAP with simulations and hand coded programs. This paper introduced the HPF prototype compiler of Karlsruhe KaHPF which generates these communication pipelines automatically.

Performance of KaHPF was determined on six benchmarks representing six different communication patterns. The measurements show that data parallel HPF programs compiled by KaHPF are the fastest programs on the Cray T3E. They are even faster than the ones that use shared-memory system functions and they outperform the PGI HPF compiler by a factor of at least 3.9 to 30.4.

VSCAP programs have the lowest possible communication overhead that can not be further decreased by optimizations. The only possible way to achieve further improvements of VSCAP is by data-flow analysis, and by more sophisticated synchronization and locality optimization.

KaHPF generated communication pipelines are therefore a first step towards run-time efficient communication on massively parallel architectures with distributed memory.

References

[BMN+97] Z. Bozkus, L. Meadows, S. Nakamoto, V. Schuster, and M. Young. PGHPF – an optimizing High Performance Fortran compiler for distributed memory machines. *Scientific Programming*, 6(1):29–40, 1997.

[BPA98] Ricardo Bianchini, Raquel Pinto, and Claudio L. Amorium. Data prefetching for software dsms. In *12th International Conference on Supercomputing, Melbourne*, July 13-17, 1998.

[Bra94] Thomas Brandes. Adaptor: A compilation system for data parallel fortran programs. Technical report, German National Center for Computer Science (GMD), St. Augustin, Germany, 1994.

[CB92] Tien-Fu Chen and Jean-Loup Baer. Reducing memory latency via non-blocking and prefetching caches. In *Fifth International Conference on Architectural Support for Programming Languages and Operating Systems*, pages 51–61, Boston, Massachusetts, October 1992. Also available as U. Washington CS TR 92-06-03.

[CKP91] David Callahan, Ken Kennedy, and Allan Porterfield. Software prefetching. In *Fourth International Conference on Architectural Support for Programming Languages and Operating Systems*, pages 40–52, Santa Clara, California, April 1991.

[GGV90] E. Gornish, E. Granston, and A. Veidenbaum. Compiler-directed data prefetching in multiprocessor with memory hierarchies. In *Proceedings 1990 International Conference on Supercomputing*, pages 354–368, 1990.

[KM90] Charles Koelbel and Piyush Mehrotra. Supporting shared data structures on distributed memory architectures. In *Proc. of the 2nd ACM SIGPLAN Symp. on Principles and Practice of Parallel Programming, PPOPP*, pages 177–186, March 1990.

[LCD+97] Honghui Lu, Alan L. Cox, Sandhya Dwarkadas, Ramakrishnan Rajamony, and Willy Zwaenepoel. Compiler and software distributed shared memory support for irregular applications. *ACM SIGPLAN Notices*, 32(7):48–??, July 1997.

[MLG92] Todd C. Mowry, Monica S. Lam, and Anoop Gupta. Design and evaluation of a compiler algorithm for prefetching. In *Fifth International Conference on Architectural Support for Programming Languages and Operating Systems*, pages 62–73, Boston, Massachusetts, October 1992.

[MWT98] Matthias M. Müller, Thomas M. Warschko, and Walter F. Tichy. Prefetching on the Cray-T3E. In *12th International Conference on Supercomputing*, pages 368–375, Melbourne, July 13-17, 1998.

[RL92] Anne Rogers and Kai Li. Software support for speculative loads. In *Fifth International Conference on Architectural Support for Programming Languages and Operating Systems*, pages 38–50, October 1992.

[War97] Thomas M. Warschko. *Effiziente Kommunikation in Parallelrechnerarchitekturen*. PhD thesis, School of Computer Science, University of Karlsruhe, 1997.

A Parallel Object Oriented Framework for Particle Methods*

M. Hipp[1], S. Hüttemann[1], M. Konold[2], M. Klingler[2], P. Leinen[3], M. Ritt[1], W. Rosenstiel[1], H. Ruder[2], R. Speith[2], and H. Yserentant[3]

[1] Wilhem-Schickard-Institut für Informatik, Universität Tübingen
[2] Institut für Astronomie und Astrophysik, Universität Tübingen
[3] Mathematisches Institut, Universität Tübingen

Abstract. A major goal of the Sonderforschungsbereich 382 is the development of different particle methods for solving physical problems. In this paper, we present two different implementations of parallel Smoothed Particle Hydrodynamics codes and a newly developed method called Finite Mass Methode. We follow up with the results of some test simulations which show the pros and cons of the different approaches.

We talk about our experiences with these codes and conclusions for parallelization in general. Our recent work concentrated on object-oriented physical codes and runtime systems. The newly established particle method called Finite Mass Method (FMM), which got developed within the SFB 382 was ported to C++, and parallelized using MPI-1.1. The experiences made with this code led to an object-oriented parallel runtime system, based on message passing and threads. Finally we describe our proposed solutions for higher-level programming abstractions, including design patterns and application-domain specific libraries. Based on this work we currently develop an object-oriented prototype of a QMC simulation.

1 Introduction

There is a need for parallelization in a considerable number of physical research fields. With more computation power, you can compute larger problems (memory-parallelization) or you gain a better resolution for your existing simulations. One major focus in our SFB are particle methods. We assisted several parallelization projects in the last years. One of the first production codes was a parallel implementation of the SPH (Smoothed Particle Hydrodynamics) method, which was done on our own procedural thread-based parallel runtime system (DTS). The system showed its limitations for large-scale distributed memory codes. An implementation of a newer SPH code based on MPI (and SHMEM for the Cray T3E), which achieves very satisfying performance is described in section 2. Included is a comparison of SHMEM and MPI performance on Cray T3E and MPI performance on the SP system.

* This project is funded by the DFG within SFB 382: *Verfahren und Algorithmen zur Simulation physikalischer Prozesse auf Höchstleistungsrechnern* (Methods and algorithms to simulate physical processes on supercomputers).

In the recent time, we noticed the need for a shift towards object-oriented programming methods for reusable parallel algorithms and design patterns in order to cut down development time and to enable the collaborative work of larger group of scientists. For gaining experience we decided to port an existing code of a new approach for a particle-based fluid simulation called Finite Mass Methode (FMM). The object-oriented code was initially parallelized using MPI. The method and some preliminary performance results of the parallel code are presented in section 4.

Based on this work, we propose an object-oriented parallel runtime system and specific algorithm libraries for the parallelization of particle methods in section 5. A new, object-oriented version of the FMM code as well as the distributed SPH code is planned for this year. We are also developing a parallel object-oriented version of a QMC code simulating a neutron star. All three codes are supposed to share a significat amount of library code.

2 SPH on SHMEM and MPI

In [11] we introduced our SPH implementation optimized for Cray SHMEM message passing. We achieved very good speedups on the Cray T3E up to 512 nodes with this code. There was a demand to continue this effort in order to get a library for a wider range of architectures.

From SHMEM to MPI This code seperates the communication part rather strictly from the physical model. In order to compare the implementation for different architectures we ported the communication library of the SHMEM-based code to MPI-1.1. Due to the limitations of the MPI-1.1 standard a few one side communication parts needed a major rewrite but most of the work was simply done by exchanging the SHMEM calls by their equivalent (two-sided) MPI calls. We expect less porting efforts with the availability of a MPI-2.0 implementation on the major platforms. In the exisiting code we wrote wrapper functions around near all communication calls, so we still used only a few direct calls to SHMEM functions and therefor the exchange of the SHMEM calls was quite easy. Not a single line in the main physical application had to be changed.

Due to the dynamic of SPH, the simulation needs a good communication bandwidth and a low latency. The most expensive communication parts are the distribution of particle positions, which is a non parallelized gather/broadcast operation between all nodes. A latency-sensitive part is for example the parallel construction of a particle grid.

We plan to use this application as one of several real-world test application to compare parallel environments. The program has a good mix of different requirements, such as the need for a high communication bandwidth and low latency together with a good floating point performance for some real computations and a considerable integer performance to handle complex data structures. The application should perform well on a good balanced system.

Native SHMEM communication vs. MPI We measured the performance of the MPI code on the Cray T3E in Stuttgart and on the SP system in Karlsruhe. On the SP we used 128 P2SC thin nodes with 120MHz. The tests showed, that the MPI implementation of the Cray T3E is worse compared to the native SHMEM library (see figure 1). Our tests show that CRAY could easily improve the MPI performance by making better wrappers around existing SHMEM calls. For some communication parts, such as gather operations of large arrays, the throughput decreased from about 300MB/s using SHMEM to 120 MB/s using MPI. On the SP we achieved the expected performance of around 50MB/s. The whole code didn't perform this good on the SP system, because the implementation is optimized for Cray T3E and depends heavily on good MPI performance in order to scale beyon 64 Processors.

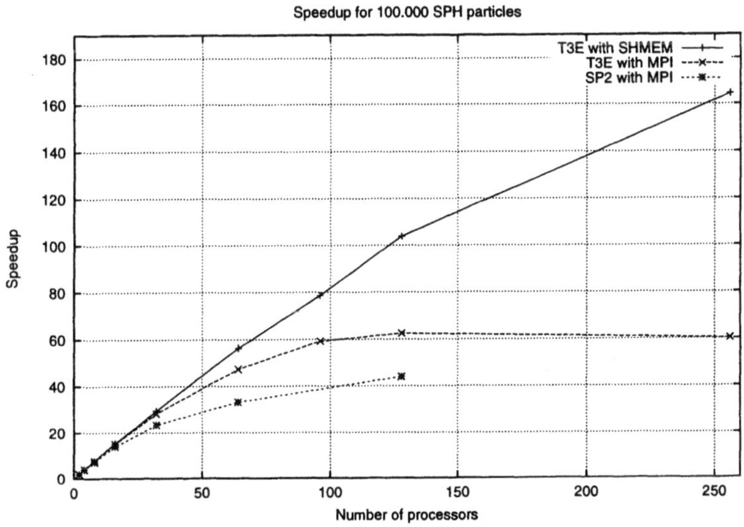

Fig. 1. Speedup of the SPH Simulation on Cray T3E with SHMEM and MPI and on SP with MPI for a mid-size problem with 100000 SPH particles. For larger node numbers the curves are dominated by the non parallelized communication parts such as gather/broadcast operations between all nodes. Please note the effect of the limitations of the MPI implementation on the CRAY T3E beyond 128 nodes.

Results The result of this performance test is, that for existing code it can be useful to exchange performance-critical communication parts in a MPI implementation by SHMEM functions to achieve better performance and better scaling to higher number of nodes. If a high level abstraction of a problem isn't possible you should at least consider to write wrapper functions around your explicit communication calls to simplify the later optimization of the applications hotspots. For new developments it proved to be beneficial to use an abstraction layer, which allows the exchange of low level parts without changing the application to obtain the best performance on a given hardware platform. We chose an object oriented programming model for this abstraction, which is introduced in the next chapter.

3 The need for Object-Oriented techniques

In providing our own runtime system we try to alleviate the error-prone and troublesome process of parallelizing physical simulation code.

Since most existing codes are procedural – written in FORTRAN or C – our first approach for a parallel runtime system was to provide the widely known interface of threads for architectures with distributed memory [2]. The standard model of communication between local threads, data exchange via shared memory, cannot be supported efficiently in a transparent and portable manner, and has been replaced by a distributed shared memory interface [9] for direct thread-thread communication.

From the work of the last years, we found reasons to redesign our simulation environment using object-oriented techniques:

1. using message objects on the lowest level to communicate between concurrent program units on distributed memory computers seems to be most natural and easy to use for our simulation methods and hides the difficult syntax of the communication libraries from the user.
2. the growing complexity of our simulation programs requires structural elements in the programming paradigm not offered by e.g. FORTRAN or C. A special requirement of scientists is ease of understanding and extending of exisiting code due to the high right of fluctuation in the scientific workplace. Also using an object-oriented approach to describe the problem is closer to the physical model used and leads to improvements of the algorithms by making use of more physical properties.
3. to exploit the usual features promised by object-oriented programming (reusability etc.) our project partners programmed in C++; which resulted in code which was reusable and modular within its own scope.

Our goal is to provide a well documented and maintained library of reusable and extensible solutions for astrophysical simulation methods in the near future. This also should give a guideline on how to use object-oriented techniques for our simulation methods (e.g. SPH).

To gain experience, we decided to port a new particle method to C++ and parallelize it using standard MPI.

4 Finite Mass Method

4.1 A new aproach to Lagrangian fluid dynamics

Within the SFB 382 H. Yserentant [13] developed a new approach to solving the continuums mechanics of fluid dynamics which is similiar to finite elements and finite volumes but completely Lagrangean in nature.

In contrast to finite volume and finite element methods this new method is not based on discretization of space but on discretization of mass which is from the point of view of a physicist at least as natural as the former approach.

The demand of astrophysical problems on computational power and high quality algorithms led to the development of this new method for solving fluid dynamics numerically. The method impresses with high accuracy and correctness regarding the conservation of energy, momentum and angular momentum. The later properties and the convergence of the finite mass methode are mathematically prooven which is a major advantage compared to the Smoothed Particle Hydrodynamics method

These mass packets of finite size do have a finite number of interiour degrees of freedom and are under the influence of external and internal forces. The shape and size of the mass packets as well as the translation of the center of mass is controlled by internal and external forces, while obeying the laws of thermodynamics.

It prooved that FMM is highly suited for problems with free boundaries, which are often encountered in astrophysics. The lagrangian approach also avoids the problem of adapting the grid during the simulation.

Arbitrary linear deformations like contraction, expansion, rotation and changes of shape combined with intersection and penetration of the packets leads to a high quality representation of the pysical function with a rather low number of packets. We already obtained very good numerical results with only 20 packets for each dimension for one of our test problems, the viscous gasball.

The higher number of physical values, compared to the Smoothed Particle Hydrodynamics method, stored in each packet and the rather low number of required simulation packets is more easy to scale efficiently on distributed memory machines like the CRAY T3E. The currently available code we used in our numerical experiments shows fourth order convergence and behaves numerically stable while using reasonable amounts of memory.

Originally the very first implementation of FMM was done in PASCAL on a Sun Workstation. This implementation lacked portability, speed and an easy way to do parallization. There is currently no standard way to use MPI safely in a portable manner with PASCAL compilers.

Instead of simply porting the available application to C we decided to use an object oriented approach, which allows the developers of the algorithms to exchanges fundamental parts of the codes without breaking other sections like the physical models. Other major reasons are efficient implementation, which forbids layering and its inheritent copying overhead and ease of parallelization. By putting a large effort in complex classes we could hide very efficent, but hard to understand implementations from the developer of the pysical model while still being able to exchange the underlying algorithms without the need to notify each other. This prooved to be beneficial to heterogenious groups of developers. The object oriented design pattern was helpful in making a cache efficient implementation, which otherwise would have obscurified the code base.

FORTRAN was no option due to missing features in representing the rather complicated data structures employed in the original PASCAL version.

It showed that while the the C++ version employes most complex data structures in order to gain cache efficiency and hiding of the paralization code that the code is optimal for well balanced scalar central processor units like the Alpha 21164 and the Pentium Pro class.

These complex data structures lead to an integer/floating point ratio of nearly one and rather short vector length so that this implementation runs well on the scalar distributed memory machine CRAY T3E in contrast to the NEC SX4 parallel vector processor. Due to the complex data structures it showed that the code runs twice as fast on a single INTEL Pentium II with 450 MHz compared to a single node on the T3E-900 which suffers from branches due to a rather long pipline and missing efficient implementation of branch prediction.

The finite mass method is a purely Lagrangean approach to solve problems in continuum mechanics. It somewhat resembles much more the concepts of finite volume and finite elements methods than pure particle methods like the Boltzmann-like transport equations.

4.2 The particle model of continuum mechanics

Instead of starting from the Euler and Navier Stokes equations the prefered way to introduce FMM is to start from the basic physical principles which finally lead to these well known differential equation. In the trivial case of of an adiabatic, inviscid flow, the equations of motion are straight forwardly derived via the Lagrangian method from the potential and kinetic energies.

The necessary breakage of invariance with respect to time reversal is achieved via frictional forces which tend to zero for the limit of velocity differences of neighbouring packets.

The funktion $\Psi_i(y)$ which describes the mass distribution inside a mass packet i with finite extension in space has to fulfill the following properties.

$$\int \Psi_i(y)dy = m_i, \int \Psi(y)ydy = 0 \qquad (1)$$

One possible choice for the functions Ψ_i as proposed by Yserentant is

$$\Psi_i(y) = \alpha_i \Psi\left(\frac{y}{h_i}\right) \tag{2}$$

with the shape function Ψ.

These properties provide normalization and coincidence of the center of the packet and the center of mass. This Lagrangean approach automatically leads to the conservation of mass.

The fluid is then described as the superposition of the above individual mass packets in space.

It is mathematically proven that constants of motion energy, momentum and angular momentum are conserved within FMM.

We suppose further that with the vector components y_k and y_l of y fullfill the following condition

$$\int \Psi(y) y_k y_l dy = J\delta_{kl}. \tag{3}$$

In our case it prooved to be beneficial to make the function Ψ up from piecewise one dimensional polynomial functions $\tilde{\Psi}(y_k)$. The above constant J is independent on the space dimension and we used J=1/12 for our calculations.

A suitable choice for building up the function $\Psi(y)$ from piecewise polynominal one dimensional functions in our computations was the normalized compact third order B-spline given by

$$\tilde{\Psi}(\xi) = \frac{4}{3}\begin{cases} 2(1+\xi)^3 & , \quad -1 \leq \xi \leq -1/2 \\ 1 - 6\xi^2(1+\xi) & , \quad -1/2 \leq \xi \leq 0 \\ 1 - 6\xi^2(1-\xi) & , \quad 0 \leq \xi \leq 1/2 \\ 2(1-\xi)^3 & , \quad 1/2 \leq \xi \leq 1 \\ 0 & , \quad |\xi| > 1 \end{cases} \tag{4}$$

which then is combined to the tensor product

$$\Psi(y) = \prod_{k=1}^{d} \tilde{\Psi}(y_k). \tag{5}$$

The superposition of the masses of single packets results in the total mass density.

The normalize frictional force for a packet within the fluid is governed by the difference from the average sourounding velocity and leads to dampening of local velocity fluctuations, coupling of particles and production of entropy according to the second law of thermodynamics.

$$F_i^{(r)} = -\frac{1}{2}\int R\Psi_i[\boldsymbol{v}_i - \boldsymbol{v}]dx \qquad (6)$$

The local relaxation time $T = 2/R$ of the system represents for constant R and $\boldsymbol{v} = 0$ the time which is needed to dampen the velocity by a factor of $1/e$. This relaxation time vanishes for smooth flows with the second order of the local packet size but is dominant for shocks.

4.3 Experimental Results

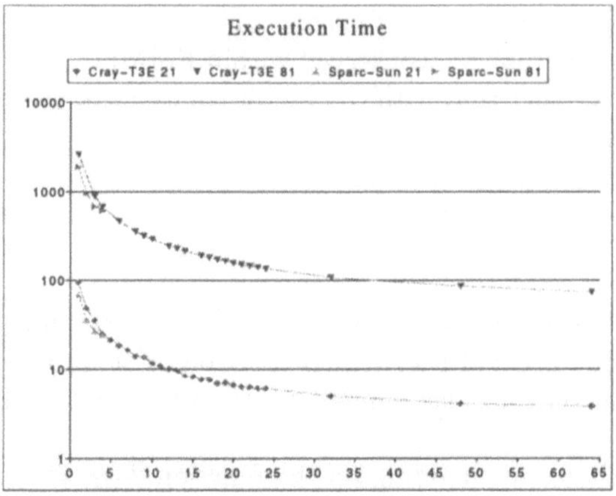

Fig. 2. Execution time over number of processors for a two-dimensional gasball with initial azimutal and radial velocity towards the center.

The current parallel FMM implementation is based on MPI-1.1 and uses a coarse grained parallel tree approach. The parallel tree helps to hide the implementation details of the parallel data structures from the physicist developing new applications and from the numerical mathematicians testing new integration algorithms. It showed to be beneficial that the integration is parallized implicitely via the distributed packets in a parallel tree structure. The users are able to develop new applications and algorithms without beeing concerned about any MPI details.

Figure 2 shows that both the smaller two-dimensional 21×21 packet test problem and the bigger 81×81 test problem. The execution time on the CRAY T3E-900 is compared to a four processor SUN Enterprise Ultra SMP machine running at 300 MHz. It can easily be seen that the SPARC processor family is much better suited to more complex data structures than the ALPHA processor which results in about 40 percent less execution time on the single node 300 MHz SPARC ULTRA compared to the 450 MHz DEC 21164. Some preliminary tests on much cheaper Intel Pentium II/III machines showed the same performance per clock cycle like the SPARC ULTRA. Due to memory bandwidth limitations and unfortunate scheduling algorithms on the SUN ENTERPRISE noticeably performance decradation is already observed for employing only three processor nodes. Due to the excellent and well balanced network interconnect of the CRAY this machine scales very well even for relatively small problems with very high communication overhead. Figure 3 gives an estimate of the possible speedups for the current parallel FMM implementation. It is easily seen that the speedup of the *Sun Enterprise* can not compete with the CRAY T3E. The possible speedup is very much dependent on the problem size. Three dimensional problems are between 20 and 1000 times larger than the current two dimensional test problems so that estimates show that this parallel FMM implementation should scale up to 384 processors on a T3E.

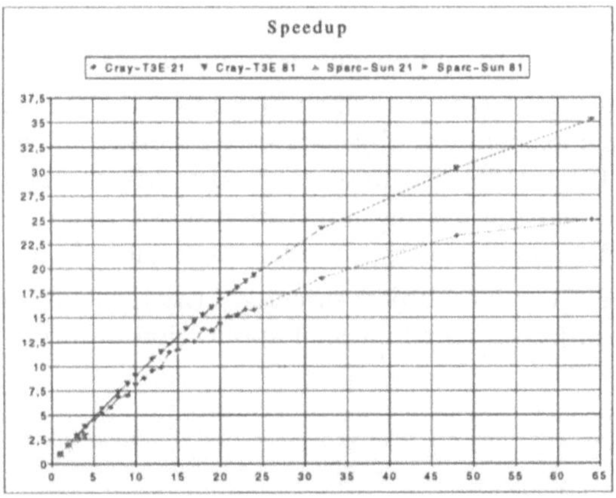

Fig. 3. Speedup over number of processors for a two-dimensional gasball with initial azimutal and radial velocity towards the center

5 An object-oriented parallel runtime system

Based on the experiences with the implementation of FMM, we now introduce our proposal for a object-oriented parallel runtime system. The basic architecture can be seen in figure 4.

On the lowest level of our parallel programming system is an object-oriented message-passing layer with a functionality close to MPI. To keep this layer portable, it is designed to be easily implemented on different low-level communication primitives. There exists a UDP based version for local test runs in a LAN environment and a MPI based version to support nearly every parallel architecture. Due to the disapointing MPI performance on the Cray T3E, we will also implement a Cray SHMEM based version for production runs on this platform. The message passing layer is designed to be thread-safe. A portable object-oriented thread library will be integrated in the near future.

Based on local threads and object-oriented message passing, we propose a second layer with a more abstract interface to hide the message passing and data distribution details from the application programmer. This level will include synchronous and asynchronous remote method invocations as well as distributed shared objects, which are migratable and replicatable.

The second layer will be responsible for load-balancing problems. In our simulation particles move free in space, therefore different particles will interact in each time-step. Access to the particles should be transparent to the application programmer, e.g. a programmer can access a particle as if the particle is in local memory. The access time to the particle of course varies, and will be optimized by the programming system and application library. This layer is scheduled to be implemented by fall this year.

There are numerous reasons, which motivated the redesign of these layers, despite of the existence of object-oriented message passing libraries like MPI++ or MPC++ [12]. The most important are the lack of thread-safe implementations and the missing integration of modern C++ concepts like templates and the support for the standard template library.

Based on the lower-level object-oriented layers, we are developing simulation code and object-oriented parallel application libraries.

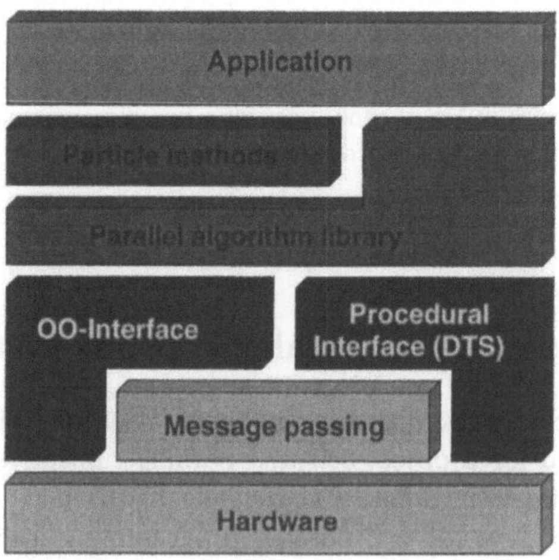

Fig. 4. Layers of the parallel programming system

5.1 The use of Design Patterns

We cannot ignore the demand for programming in C or FORTRAN. To provide just an implementation in C++ will not be accepted by our project partners. We had to find a way to write down our solutions in a "Meta-Language". Using Design Patterns serves this purpose best. We have an easy to understand way to document our solutions, that is not bound to any programming language. Also using UML allows us to use tools to implement the documented Design Patterns in e.g. C++ (almost) automatically.

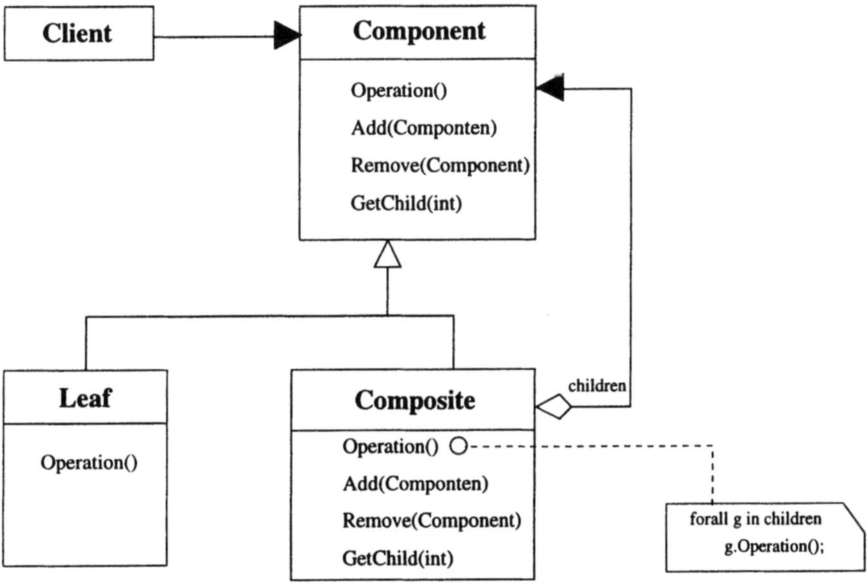

Fig. 5. UML notation of Composite Design Pattern used for Domain-Decomposition

Documenting the Design Patterns in a modern, easy to read way was achieved by using multi-frame HTML documents. Solutions that are easy to read and understand find generally better acceptance, even if there is no direct implementation in the favorite language of the programmer, e.g. FORTRAN. Also documenting the simplicity of our ready to use solutions motivates more physicists to take a look at a new programming paradigm (some even take a second look).

5.2 An object-oriented application library for parallel SPH simulations

As a first step towards an object-oriented SPH program, we used an easy to parallelize Monte Carlo simulation of the pulsar HER-X1. Looking at the problem as a programmer the Monte Carlo simulation and the SPH simulation are similar, because they are both particle simulation methods. We used design patterns to describe the solutions that were implemented in the class library. For algorithmic problems like coordinate transformation or different types of integrators we used a Strategy Design Pattern. The simulation data is created using Factory Design Pattern and the individual sub-domains of our simulation domain are connected with each other by being part of a Composite Design Pattern (see 5).

Prototyping a new simulation using this class library is faster than writing new C or FORTRAN code. Currently we are working on optimizing the performance of our C++-programs. We could get some positive feedback from our physics partners with respect to using these solutions for prototyping.

The special SPH data structures and algorithms are also part of our current work — e.g. nearest neighbor problems.

6 Outlook

Fluid dynamics calculated with modern particle methods remains an important topic not only in the area of astrophysics but increasingly in engineering like combustion engines. A future version of our software will include online visualization on distributed memory parallel machines which requires rather good I/O and interactive use capabilities of the employed hardware.

7 Acknowledgments

We wish to thank our project partners from the SFB 382 for the close and fruitful coorporation. Especially we would like to mention the very helpful staff at the HLRS in Stuttgart which put in a lot of effort in reducing the I/O bottleneck during the recent months.

References

1. On subharmonic solutions of a hamiltonian system. *Comm. Pure Appl. Math.*, 33:609–633, 1980.
2. Tilmann Bubeck. *Distributed Threads System DTS User's Guide.* SFB 382/C6, Universität Tübingen, September 1995.
3. Tilmann Bubeck, Wolfgang Küchlin, and Wolfgang Rosenstiel. Symmetric Distributed Computing with Dynamic Load Balancing and Fault Tolerance. In Boleslaw Szymanski, editor, *Proceedings of the 3rd workshop on Languages, Compilers and Run-Time Systems for Scalable Computers*, Troy, New York, May 1995. Kluwer.

4. F. Clarke and I. Ekeland. olutions périodiques, du période donnée, des équations hamiltoniennes. *Note CRAS Paris*, 287:1013–1015, 1978.
5. F. Clarke and I. Ekeland. Nonlinear oscillations and boundary-value problems for hamiltonian systems. *Arch. Rat. Mech. Anal.*, 78:315–333, 1982.
6. R.A. Gingold and J.J. Monaghan. Smoothed particle hydrodynamics: theory and application to non-spherical stars. *Mon. Not. R. astr. Soc.*, 181:375–389, 1977.
7. Leon B. Lucy. A Numerical Approach to Testing the Fission Hypothesis. *Astron. J*, 82(12):1013–1924, December 1977.
8. R. Michalek and G. Tarantello. Subharmonic solutions with prescribed minimal period for nonautonomous hamiltonian systems. *J. Diff. Eq.*, 72:28–55, 1988.
9. Marcus Ritt. Parallelisierung der berechnung des faltenwurfs von textilien mit partikelsystemen. Master's thesis, Universitt Tbingen, Technische Informatik, 1997.
10. G. Tarantello. Subharmonic solutions for hamiltonian systems via a $\mathbb{Z}_p$ pseudoindex theory. Annali di Matematica Pura, (to appear).
11. T.Bubeck, M.Hipp, S.Hüttemann, S.Kunze, M.Ritt, W.Rosenstiel, H.Ruder, and R.Speith. Parallel SPH on Cray T3E and NEC SX-4 using DTS. In W.Jäger E.Krause, editor, *High Performance Computing in Science and Engineering '98*, pages 396 - 410. Springer, 1999.
12. Gregory V. Wilson and Paul Lu, editors. *Parallel Programming using C++*. The MIT Press, Cambridge, 1996.
13. H. Yserentant. A particle method of compressible fluids. *Numer. Math.*, 76:111–142, 1977.

Parallel solution of Partial Differential Equations with Adaptive Multigrid Methods on Unstructured Grids

Peter Bastian, Klaus Birken, Klaus Johannsen, Stefan Lang, Volker Reichenberger, Christian Wieners, Gabriel Wittum, and Christian Wrobel

Universität Heidelberg, Interdisziplinäres Zentrum für Wissenschaftliches Rechnen
Im Neuenheimer Feld 368, D-69120 Heidelberg, Germany

Abstract. We present new parallel results for the solution of partial differential equations based on the software platform *UG*. State-of-the-art numerical methods have been developed and implemented for the efficient and comfortable solution of real-world problems. *UG* supports distributed unstructured grids, adaptive grid refinement, derefinement/coarsening, robust parallel multigrid methods, various discretizations, dynamic load balancing, mapping and grid partitioning. Here, we give examples for a parallel algebraic multigrid method, for elasto-plastic computations, and for simulations of two-phase flow in porous media.

1 Introduction

The recent decades have seen a substantial improvement in the numerical solution of partial differential equations which is only partly due to the increased computing power. Only by using the combined benefits of both supercomputers and state-of-the-art numerical methods the simulation of real world problems can be achieved. In the field of partial differential equations, four techniques have proven their effectiveness in a wide spectrum of applications: *Unstructured grids*, *Adaptivity*, *Multigrid Methods* and *Parallelism*. Unstructured grids are an essential requisite for the representation of complex geometries and the adaptive refinement of these grids allows for a minimization of the number of grid points to achieve a desired accuracy of the solution. Multigrid methods have proven to be the fastest solvers for the linear systems of equations arising from the discretization of partial differential equations. Finally these methods have to be parallelized for their usage on MIMD machines, the fastest high-performance computers available today.

Despite the wide recognition of their effectiveness, these methods have mostly found their way only into University research codes [2, 4, 10, 16] and some few commercial applications [17]. The reason for the slow acceptance are that robust iterative solvers are difficult to develop and that the combination of the described methods in one software package requires several dozen man-years of work—while the implementation of one component can still be manageable for a single programmer, the strong interconnections of all four produces a considerable increase in complexity.

The software package *UG* was developed with the ultimate goal of providing a toolbox for the development of discretizations, solvers and error estimators based on parallel adaptive multigrid methods on unstructured grids. Constructed around a flexible grid representation data structure which allows the usage of triangular, quadrilateral, tetrahedral, pyramidal, hexahedral and prism elements (in two and three space dimensions, respectively) it supports hierarchical grids, local refinement and coarsening of the grids. The algebraic requirements are satisfied by providing sparse matrix/vector BLAS routines, iterative methods and multigrid methods which can be combined with numerous Krylov subspace methods, nonlinear solvers and time stepping schemes. Degrees of freedom can be associated with nodes, edges, faces and elements of the grid to allow the usage of conforming as well as non-conforming or higher-order finite element methods. Routines for common tasks like the evaluation of shape functions, quadrature formulas, finite volume construction, ... are provided and make it possible for conforming finite element methods and finite volume methods to use adaptive parallel multigrid out of the box—even for systems of partial differential equations.

Applications built with *UG* are controlled by a graphical user interface consisting of a graphic subsystem for visualization and a command line shell with its own simple (yet powerful) scripting language; the object-oriented design of the numerical solution methods allows an interactive assembling of solution methods from basic components. Interfaces for several post-processing tools are part of *UG*, mostly for more sophisticated visualization demands. More information about the *UG* software model can be found in [6].

Much emphasis has been laid on a smooth integration of the parallel capabilities into the project. Based on the programming model DDD (Dynamic Distributed Data) by BIRKEN [8, 9], *UG* has been very successful in achieving this goal and the parallelization has reached a mature state. A portable message passing interface (PPIF) is the basis of the parallelization and has been implemented for PVM, MPI, PARIX, NX and the Cray T3D/T3E. In this stage of development, the usage of parallel computers like the IBM RS/6000 SP is crucial to the development process, because the scalability of the implemented methods has to be proven with real life problems. While relatively small memory-per-processor ratios have often limited the usefulness of otherwise powerful parallel computers, the RS/6000 SP installation in Karlsruhe has been our favorite candidate for several memory-intensive problems. With the rise of inexpensive parallel workstation cluster computers, the methodology of testing and developing parallel programs on comparably small in-house parallel machines/clusters and migrating to powerful high-performance computers only in a late state will certainly grow stronger.

2 Exemplary Applications

In this section we present some new results describing the parallel performance of some exemplary *UG* applications.

2.1 Parallel Algebraic Multigrid FAMG

Besides classical (or geometric) multigrid methods purely algebraic multigrid methods (AMG) are desirable for many reasons; AMG considers only the stiffness matrix but not any geometric information such as elements or coordinates. AMG has the potential to solve complicated problems (e. g. jumping coefficients or highly irregular meshes) better than other methods. Even in a geometric multigrid AMG can be applied namely as the coarse grid solver; geometric multigrid methods refine a given mesh to improve the accuracy of the solution but they still need a quite good solver on the given (coarse) mesh which becomes a hard task for real life problems (cf. the next sections) and can be solved by AMG. A third point of interest is the coupling of multigrid methods with already existing programs. The smallest possible interface—the matrix itself—is already feasible for AMG and the solver has not to interact with the geometry. A major drawback of many existing AMG methods is the missing or even impossible parallelization which enables one first to solve challenging problems.

The starting point for our parallel AMG is the filtering AMG (FAMG) by WAGNER [3, 20]. The main idea is to choose for each vector good parent pairs of vectors for eventual elimination which ensures a certain filtering condition and leads to exactness on a given subspace. The best pairs are selected to eliminate the corresponding vectors; this parent vectors persist on the next coarser grid level and restriction and prolongation matrix's entries with individually calculated values are installed between the vectors and their parents. Recursively applied this process yields a grid hierarchy on which ordinary multigrid solvers can work.

For parallelizing this FAMG one has to do several additional steps. First we must distribute the matrix among the processors. In our applications we use the given domain partition of the load balancing module but this could be done with other (matrix) graph partitioning tools as well. Then a suitable overlap must be generated to perform the calculations necessary for determining the best parent pairs without loss of accuracy and time completely parallel. The most difficult step is the elimination because each eliminated vector influences all its neighbor vectors. Unfortunately the repetition of this process leads to a global dependence between the vectors. But the promising observation is that the order of elimination is not a very crucial one under some constraints. To increase parallel performance that is to employ as many processors at the same time as possible one must break up the sequence of elimination. Our approach is first to eliminate the vectors along the processor interfaces, whereby the processors get separated. This elimination can be accelerated by parallel graph coloring techniques. After that separation each processor can eliminate its inner part of the vectors independently from each other. This results in a good overall parallel performance. Constructing the coarse grid matrix one must ensure the consistency with the *UG* parallel data model. After generating the sequence of coarse grids all available parallel multigrid solvers out of the numerics module can be used.

Now we want to demonstrate some features of the new parallel FAMG. In all examples the multigrid solver uses 1 pre- and 1 postsmoothing Jacobi step. Thus all the efficiency must result from the good coarse grid generation. We examine the Laplace operator on the unit square with a regular tensorproduct mesh and on the shape of Lake St. Wolfgang with an unstructured mesh from a gridgenerator. The structured mesh is distributed on tensorproduct like processor configurations by a special load balancer to conserve the regularity; for this case we compare the results for eliminating the processor interface vectors in the natural order and an optimized order obtained by a parallel graph coloring strategy. The unstructured mesh is distributed by recursive coordinate bisection (RCB).

structured		
unknowns	conv. rate	time [sec]
1089	0.043	5.6
4225	0.044	15.0
16641	0.050	41.7
66049	0.222	98.2
263169	0.252	112.1
1050625	0.254	177.8
4198401	0.255	317.1
16785409	0.237	669.3

unstructured		
unknowns	conv. rate	time [sec]
3617	0.064	7.1
14095	0.095	18.5
55637	0.126	35.3
221065	0.139	91.2
881297	0.158	145.1
3519265	0.180	257.5
14065217	0.212	727.7

Table 1: The convergence rate depends only weakly upon mesh width (structured grid on a 16x8 processor configuration, unstructured grid on 128 processors).

Tab. 1 shows that the convergence rate is rather independent from the meshwidth respectively the number of vectors. This is essential to solve large problems efficiently and strengthens the fact that the order of eliminating the vectors is not very important. But the execution time doesn't scale linearly with the number of unknowns; the responsible effects still must be analyzed.

Next we evaluate the quality of the parallelization. First we consider the so called speedup where the same problem is solved on different numbers of processors; be careful about the meaning of this quantity. Tab. 2 indicates a quite constant convergence rate. But there seems to be a problem with the time of the solving process: it doesn't decrease as expected if the number of processors increases. The bottleneck is the coarsest grid. The coordination between the number of vectors in the coarsest grid and a suitable solver therefore is yet under construction. Presently a direct solver is used and thus its quota increases dramatically as the coarsest grid grows. For many cases

	structured				unstructured	
	without graphcoloring					
PE	conv. rate	time [sec]	conv. rate	time [sec]	conv. rate	time [sec]
4x1					0.291	554.7
8x1	0.129	474.5	0.149	493.9	0.291	374.9
4x2	0.233	374.1	0.212	393.9		
16x1	0.123	369.9	0.147	348.7	0.284	346.2
4x4	0.237	260.5	0.211	222.2		
32x1	0.125	390.4	0.211	307.6	0.238	167.6
8x4	0.230	171.4	0.204	161.3		
64x1	0.180	499.6	0.222	455.9	0.263	149.3
8x8	0.237	133.4	0.230	151.6		
128x1	0.285	1043.0	0.184	551.1	0.158	145.1
16x8	0.254	177.8	0.212	106.5		

Table 2: Speedup on different processor configurations (structured grid with 1 million unknowns, unstructured grid with 880000 unknowns)

the graph coloring keeps the promise to quicken the program. Second we have a look at the scale up in Tab. 3; in practice the scaleup may be more important than the speedup because usually you use a big parallel computer to solve larger problems than prior instead of the same problems quicker. Also here the convergence rate is enjoyably constant and the execution times suffer from the too large baselevel grids.

Fig. 1 shows the given grid (level 0) and two coarsened grid levels for a small example to give you an impression how a grid may look like. Only for visualization purpose the coordinates of the given mesh are used to place the vectors in the picture as small circles; a line between two vectors represents the according matrix entry. The circle color corresponds with the processor number.

	structured		unstructured	
PE	conv. rate	time [sec]	conv. rate	time [sec]
2	0.143	227.1	0.261	175.6
8	0.212	393.9	0.291	374.9
32	0.204	406.6	0.259	483.0
128	0.241	684.0	0.212	727.7

Table 3: Parallel performance for scaled computations (constant load per PE: structured case 131000 unknowns, unstructured case 110000 unknowns).

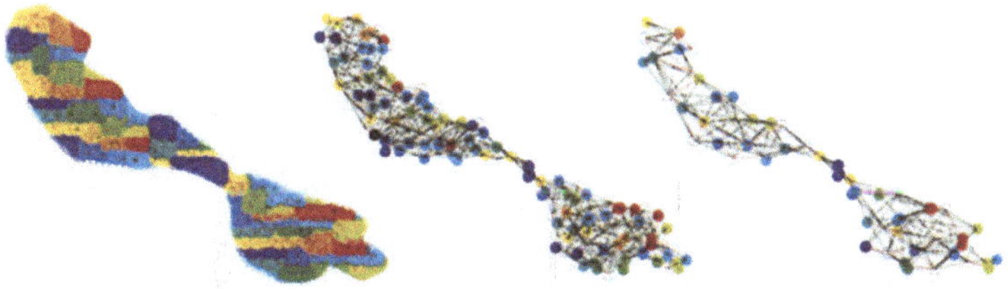

Fig. 1: Sequence of algebraically coarsened grids for a unstructured mesh of Lake St. Wolfgang (64 processors, 3617 unknowns, level 0, 5 and 7).

We have seen very promising features of the new parallel FAMG. The problem with much too large baselevel grids will be tackled soon. And we must explore further model problems (e. g. anisotropy, jumping coefficients) to experience the limitations of this method. For real world problems FAMG must demonstrate its superiority to classical multigrid methods. The advantages of the IBM RS/6000 SP are on the one hand its 2 GB nodes which enable very large serial calculations for several comparisons and on the other hand the quite large memory on the many parallel nodes enables us to do various experiments with not so hard memory limitations.

2.2 Multigrid methods for Prandtl-Reuß-plasticity

The finite element subsystem of *UG* contains a model for elastoplastic computations where the nonlinear problem in every time step is solved by a quasi-Newton method, and in every Newton step the linear system is solved with a multigrid method.

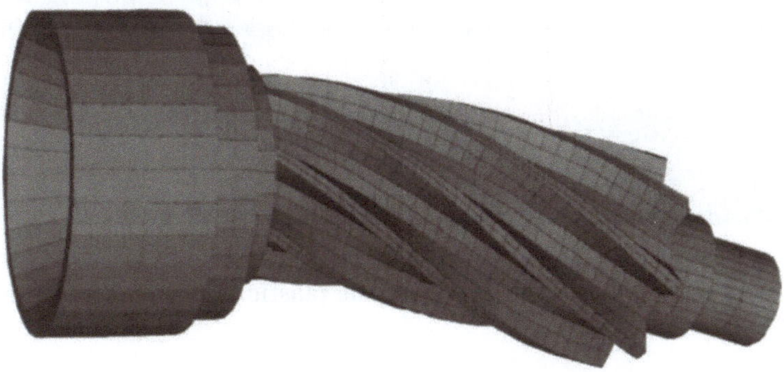

Fig. 2: Deformation of a gear geometry (scaled by a factor of 1000).

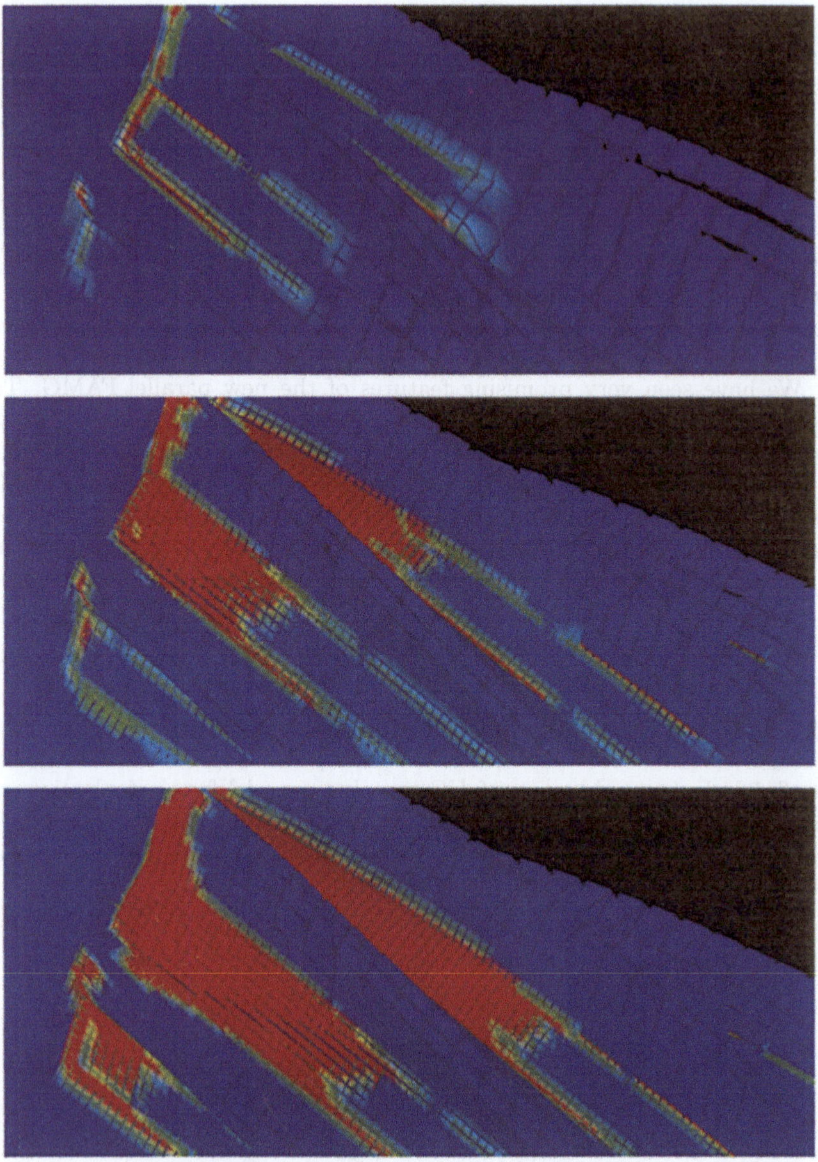

Fig. 3: Local refinement of the gear near the plastic zone (64 processors IBM RS/6000 SP).

We present an example for Prandtl-Reuß-plasticity with nonlinear isotropic hardening. For a detailed description of the model and the material parameters see [22].

For an accuracy of about 10 percent for the problem in Fig. 2, a mesh with uniform refinement requires three refinement levels, i. e. 2000000 unknowns for the displacement vector and 30000000 parameters for the description of the material parameters in this example. It is shown in [22] for a 2D benchmark configuration that nearly the same accuracy can be obtained if only the plastic zone is resolved properly. In particular, the plastification will start near reentrant edges, see Fig. 3.

This example is computed with the static load balancing given in Fig. 4. We give two example for the performance: on an adaptive hierarchy with 1–4 levels, starting with 11456 elements up to 336114 elements after 16 loading steps, the total computation time was ca. 3 hours for 97 Newton steps with 780 linear multigrid cycles; on an adaptive hierarchy with 1–3 levels up to 170675 elements after 21 loading steps, the total computation time was ca. 2 hours for 141 Newton steps with 567 linear multigrid cycles. The next step will be to improve the performance by dynamic load balancing.

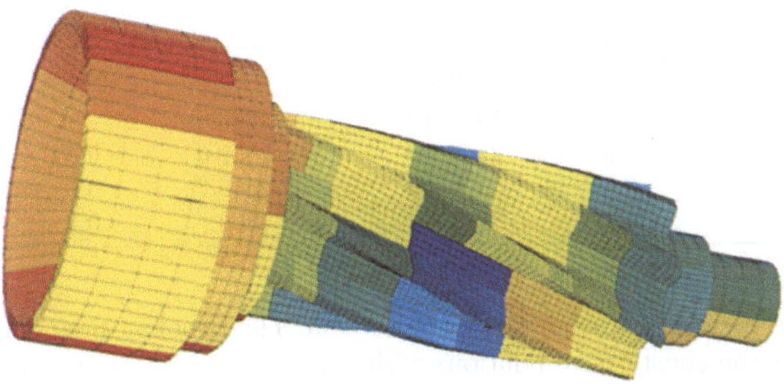

Fig. 4: Load balancing of the gear on 64 processors (IBM RS/6000 SP).

2.3 Two-phase Flow in Porous Media

The flow of two immiscible fluids in a porous medium is described by two coupled non-linear time-dependent partial differential equations; see [1] for an introduction. These equations play an important role in oil reservoir simulation, the development of effective in-situ remediation techniques and the security assessment of underground waste repositories. Due to the hyperbolic/parabolic character of the equations, strong heterogeneities and high non-linearity, they pose a challenging problem for multigrid solution.

A problem class has been developed that solves the two-phase flow equations in a fully implicit/fully-coupled manner using either phase pressure-saturation or a global pressure-saturation formulation [7]. A finite volume

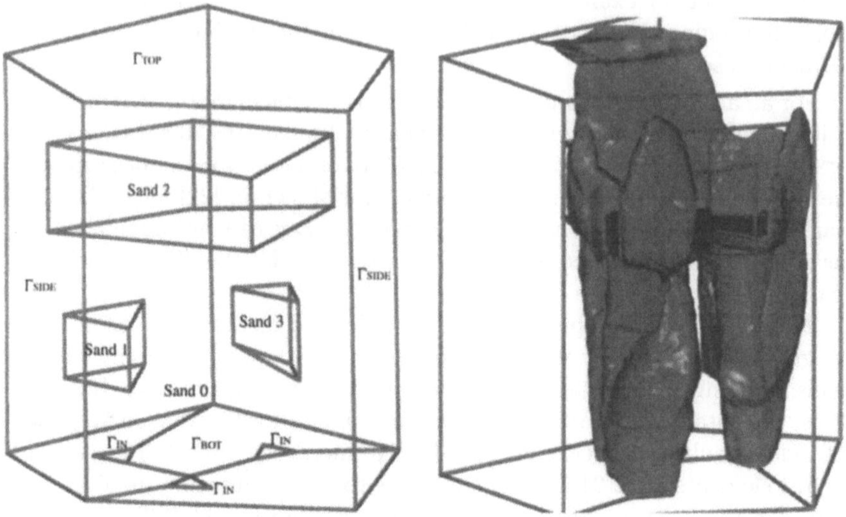

Fig. 5: 3D air sparging in a heterogeneous porous medium. Geometry and 5% saturation isosurface after 640*s*.

	Φ	$k\ [m^2]$	$p_d\ [Pa]$	S_n^*
Sand 0	0.4	$5.04 \cdot 10^{-10}$	1600	–
Sand 1	0.39	$2.05 \cdot 10^{-10}$	1959.6	0.3
Sand 2	0.39	$5.62 \cdot 10^{-10}$	2565.7	0.55
Sand 3	0.41	$8.19 \cdot 10^{-10}$	4800	0.8

Table 4: Parameters for the air sparging problem. The Brooks-Corey function was used for the constitutive relations with $S_{wr} = 0.1$, $S_{nr} = 0$, $\lambda = 2$.

and a control-volume-finite-element discretization with first-order upwinding have been implemented. Entry pressure effects at porous medium discontinuities are handled by incorporating appropriate interface conditions. Both discretizations support all element types in two and three dimensions. Time discretization is fully implicit, resulting in a large set of nonlinear algebraic equations per time step. The nonlinear equations are then solved iteratively by a Newton-Multigrid technique. A line search method is used to achieve global convergence. Several multigrid techniques have been implemented in *UG* to handle coefficient jumps induced by saturation fronts and absolute permeability variations. These jumps are in general *not* aligned with coarse grid lines. In the simulations below, a multigrid method with truncated restriction, point-block ILU smoother and a V-cycle has been used.

The example shown in Figure 5 shows a simulation of the bubbling of air in a three-dimensional heterogeneous medium. The domain has a height of 5

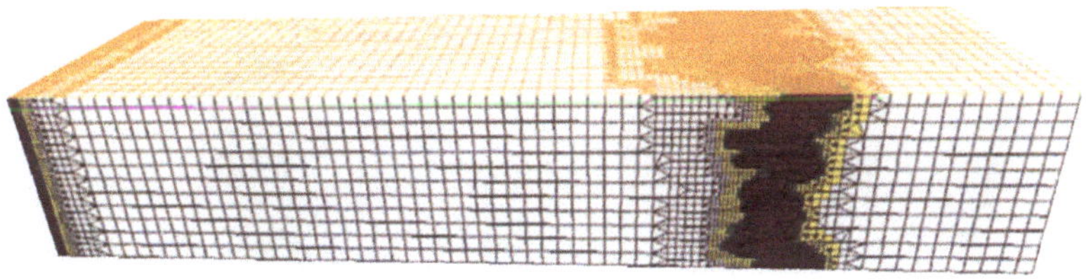

Fig. 6: Immiscible displacement: Adaptively refined multigrid consisting of hexahedra on coarse grid and hexahedra, pyramids and tetrahedra on adaptive refined levels.

Fig. 7: Immiscible displacement: Load balancing of 16 by 16 stripes in flow direction.

meters and an area of 5 by 4 meters. The domain is initially water-saturated and has three lenses with different properties (see Table below). The example was calculated using the phase pressure-saturation formulation with interface conditions and the parameters from table 4.

The coarse grid was created with NETGEN (see [19]), yielding 1492 tetrahedral elements. The finest grid contained 6111232 elements and 1040129 nodes. Figure 5 shows the isosurface of the non-wetting phase saturation $S_n = 0.05$ at time $T = 640[s]$. The visualization was made with the graphics program GRAPE [18].

Figures 6-9 show a recently calculated *UG* application in the area of immiscible displacement. Immiscible displacements are characterized by viscosity and density differences between the fluids and surface tension forces involved [13]. Two phase flow problems in homogenous porous media with phases consisting of immiscible liquids with viscosity differences tend to be unstable at the interface. Small perturbations of the interface, which are on-

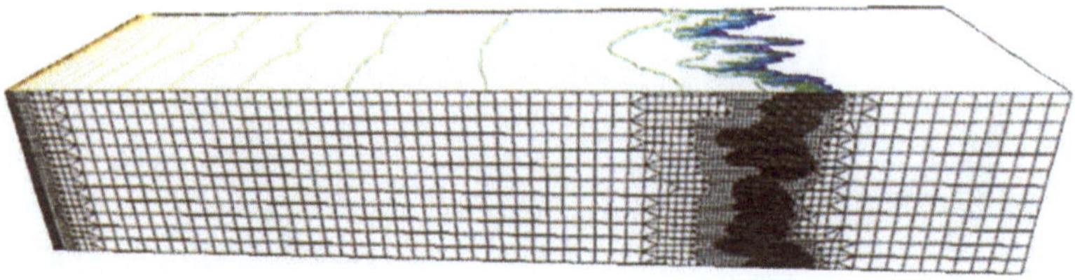

Fig. 8: Immiscible displacement: Horizontal cut through the domain. Isolines of concentration are mapped onto the cutplane.

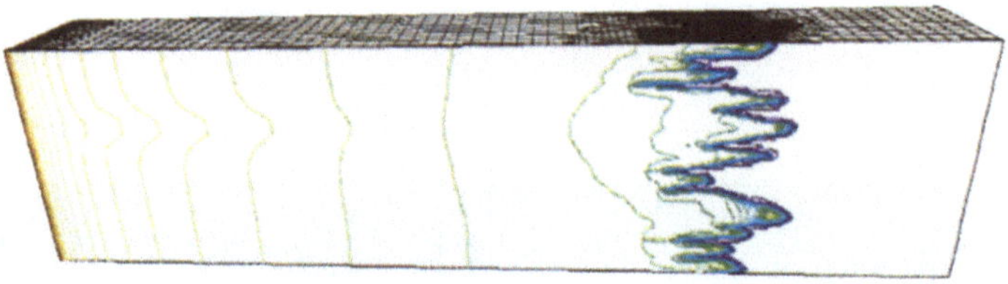

Fig. 9: Immiscible displacement: Vertical cut through the domain. Isolines of concentration are mapped onto the cutplane.

sets of fingers, are subject to microscopic heterogeneity in macroscopically homogenous media. If local velocity is beyond some critical velocity and furthermore surface tension and radius of curvature of the initial perturbation fulfill a specific constraint, then immiscible fingering will develop from perturbations.

The difficulty of simulating such phenomena is a matter of computing power, since the describing equations have to be discretized on a very fine scale to wipe out numerical diffusion leading especially in 3 dimensions to very large equation systems. Moreover since the equations are instationary, many time steps have to be calculated to get immiscible fingering visible.

The problem computed here is a quadratic channel with inflow and outflow boundary conditions, where the wetting phase, water, displaces the nonwetting phase, oil. The combination of several UG inherent features are used to calculate this application: multigrid, adaptivity, parallelism and mixed element types. 209 grid adaptions have been done during the computations of

203 time steps on about 1 million unknowns per timestep. The calculation was performed on 256 T3E processors and needs 12 hours of computing time.

3 Conclusions and Outlook

In this paper, several exemplary applications of the *UG*-package have been shown together with numerical results on the IBM RS/6000 SP and other high-performance computers. The resulting performance shows that it is practical and for many application problems even necessary to combine the computational power and main memory sizes of parallel computers with state-of-the-art numerical techniques for the solution of partial differential equations.

The layered, hierarchical approach of the *UG* software design leads to the following advantages:

- the resulting software is portable on a wide range of platforms;
- for new applications based on *UG* there is only minimal effort for parallelization;
- using the *UG* approach, high-performance computing platforms can be exploited efficiently for a wide spectrum of different application areas.

The software development process for parallel programs based on *UG* supports modularity throughout all components of the software and renders possible the development of parallel applications on a stable basis. Along with this release from the task of debugging low-level parallelization details comes the demand for interactive testing and debugging of real-world application problems, a task that is not well supported on most supercomputer installations. Using a couple of hundred processors in interactive mode is an integral part of parallel program development that can not yet be achieved with workstation clusters and should be considered by Supercomputing Centers.

References

1. K. Aziz and A. Settari. *Petroleum Reservoir Simulation.* Elsevier, 1979.
2. R. Bank. *PLTMG Users Guide Version 7.0*, SIAM, 1994.
3. R. Bank and C. Wagner. *Multilevel ILU decomposition*, Numerische Mathematik, accepted.
4. P. Bastian. *Parallele adaptive Mehrgitterverfahren*, Teubner Skripten zur Numerik, Teubner-Verlag, 1996.
5. P. Bastian, K. Birken, K. Johannsen, S. Lang, N. Neuss, H. Rentz-Reichert, and C. Wieners. UG – a flexible software toolbox for solving partial differential equations. *Computation and Visualization in Science*, (1), 1997.
6. P. Bastian, K. Birken, K. Johannsen, S. Lang, V. Reichenberger, C. Wieners, G. Wittum, and C. Wrobel. *A parallel software-platform for solving problems of partial differential equations using unstructured grids and adaptive multigrid methods*, in W. Jäger and E. Krause, editor: High performance computing in science and engineering, pages 326–339. Springer, 1999.

7. P. Bastian and R. Helmig. *Efficient Fully-Coupled Solution Techniques for Two-Phase Flow in Porous Media.* Advances in Water Resources Research, 1997 (submitted).
8. K. Birken. An efficient programming model for parallel and adaptive CFD-algorithms. In *Proceedings of Parallel CFD Conference 1994*, Kyoto, Japan, 1995. Elsevier Science.
9. K. Birken. Ein Modell zur effizienten Parallelisierung von Algorithmen auf komplexen, dynamischen Datenstrukturen. PhD thesis, Universität Stuttgart, 1998.
10. P. Deuflhard, P. Leinen, and H. Yserentant. *Concepts of an adaptive hierarchical finite element code*, IMPACT of Computing in Science and Engineering, 1 (1989), pp. 3–35.
11. Hendrickson and R. Leland. *The chaco user's guide version 1.0*, Tech. Rep. SAND93-2339, Sandia National Laboratory, October 1993.
12. E. H. Hirschel (ed.). *Flow Simulation with High-Performance Computers II*, Vieweg Verlag, Braunschweig, 1996.
13. B. H. Kueper, E. O. Frind. *An Overview of Immiscible Fingering in Porous Media*, J. of Contaminant Hydrology, 2, 1988, 95-110, Amsterdam.
14. S. Lang. *Lastverteilung für paralle adaptive Mehrgitterberechnungen*, Master's thesis, Universität Erlangen-Nürnberg, IMMD III, 1994.
15. D. J. Mavripilis. *Three-dimensional Multigrid Reynolds-averaged Navier-Stokes solver for unstructured meshes*, AIAA Journal, 33 (1995).
16. L. C. McInnes and B. Smith. *PetSc2.0: A case study of using MPI to develop numerical software libraries*, in Proc. of the MPI Developers Conference, Notre Dame, IN, 1995.
17. M. Raw. *A coupled algebraic multigrid solver for the 3D Navier-Stokes equations*, in Proc. of the 10th GAMM Seminar Kiel, Notes on Numerical Fluid Mechanics, G. W. W. Hackbusch, ed., vol. 49, Vieweg-Verlag, 1995.
18. M. Rumpf, R. Neubauer, M. Ohlberger, R. Schwörer, *Efficient visualization of large-scale data on hierarchical meshes.* Visualization in Scientific Computing '97, Springer, 1997.
19. J. Schöberl, *A rule-based tetrahedral mesh generator.* Computing and Visualization in Science 1, pp. 1–26, 1997.
20. C. Wagner. *Filtering algebraic multigrid*, to appear.
21. C. Wieners. *Robust multigrid methods for nearly incompresible elasticity*, 1998, submitted.
22. C. Wieners. *Multigrid methods for Prandtl-Reuß-plasticity*, Numer. Lin. Alg. Appl., 1999, to appear.
23. G. Wittum. *Multigrid methods for Stokes- and Navier-Stokes equations*, Numer. Math., 54 (1989), pp. 543–563.

Coupling and Parallelization of Grid–based Numerical Simulation Software

P. Adamidis[3], A. Beck[3], U. Becker-Lemgau[1], Y. Ding[3], M. Hackenberg[1], H. Holthoff[2], M. Laux[2], A. Müller[3], M. Münch[3], B. Steckel[1], C. Pospiech[2], and R. Tilch[1]
email: andreas.mueller@informatik.uni-stuttgart.de

[1] GMD SCAI German National Research Center for Information Technology, St. Augustin, Germany
[2] IBM High Performance Support Center, Heidelberg, Germany
[3] IPVR University of Stuttgart, Germany

Abstract. Aims of the GRISSLi project are the parallelization of numerical software for unstructured grids with wide employment of standard tools and the development of the library CI (Coupling Interface) in order to couple different numerical parallel programs based on various kinds of discretization, like structured or unstructured grids. The simulation of steel strip production was chosen as a highly relevant industry application. The chosen simulation uses the block–structured multigrid–solver LiSS and the nonlinear Finite Element package Larstran.

1 Introduction

The production of the bulk material steel strip is traditionally separated by casting and multistage rolling. The rolling process employs a large number of roller pairs and the rolling mill is typically a few hundred meters long. Due to the high costs involved, each simplification of the production process yields tremendous economical savings. The Institute for Metal Forming of the RWTH Aachen maintains one of the few existing pilot systems in cooperation with the Thyssen Krupp Stahl company [1].

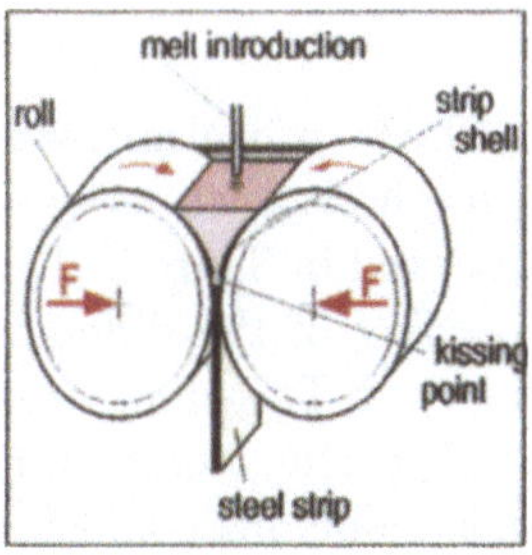

Fig. 1. Principle of twin–roll casting process

Molten steel is introduced between two rotating cooled rolls, which results in an incompressible, turbulent two–phase fluid flow. Due to the intensive heat flux from the fluid steel to the rolls at their surfaces, the steel solidifies in the form of thin shells, which are transported towards the gap between the rolls. Both shells meet at the kissing point and merge into a steel strip. The operational principle is illustrated in figure 1. The fluid flow is simulated by the multigrid–solver LiSS [2] and the metal forming process by the nonlinear Finite Element package Larstran [3].

2 Parallelization of the Finite Element program using domain decomposition methods

Analysis of the computation time spent in different modules of the Finite Element package Larstran shows that more than 90% of the computational time is spent in the assembling and solving modules. According to Amdahl's law, it is clear that the parallelization of the solver of the linear system is not sufficient.

In the GRISSLi project two different strategies to parallelize Finite Element programs are taken into consideration - the method of substructuring and the Schwarz method. The common part of both is the decomposition of the elements into disjunct sets which are later assigned to different processors. In view of the parallelization of the linear solvers, the partitioning of the elements is done topologically. Each partition set is simply connected and the number of common grid points should be as low as possible. For the partitioning of the elements, the software package *metis* is used [4]. A symbolic partitioning with its specific renumbering is depicted in figure 2. The renumbering of the grid points leads to linear systems with arrow shaped matrices which are well suited for parallelization. A linear speed–up for this shape of matrices was achieved despite the overhead for additional matrix operations.

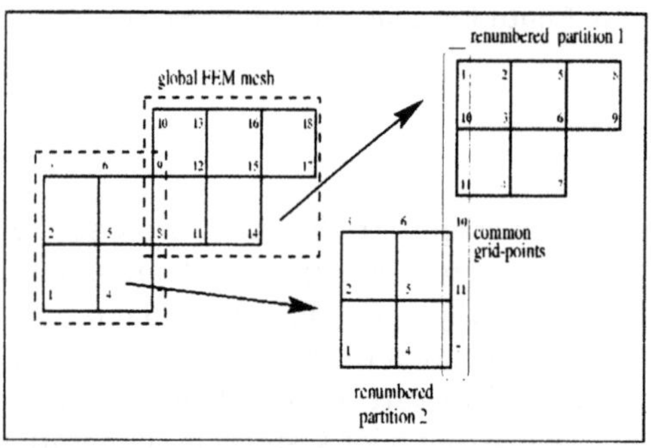

Fig. 2. Disjunct renumbered partitioning

First, the reduced system of the coupled degrees of freedom (DOF) is solved with a Schurcomplement approach. This part of the solution is distributed by message–passing to the processors which have grid–points in common with the respective calculating processor. Afterwards the solution at the inner grid– points can be calculated by a sequential solver in parallel. Results can be found in [5].

The Schwarz methods [6] are iterative methods which utilize an overlapping of the artificially created partitioning using methods such as next neighbor search. The principle of this approach is illustrated in figure 3. The parallelization is attained by creating an external coupling of the different domains at selected grid–points and/or at selected elements. The underlying operations are executed in parallel without communication. Due to the overlapping, there exists for each grid–point laying on a logical boundary another partition in which this grid–point is an inner grid–point.

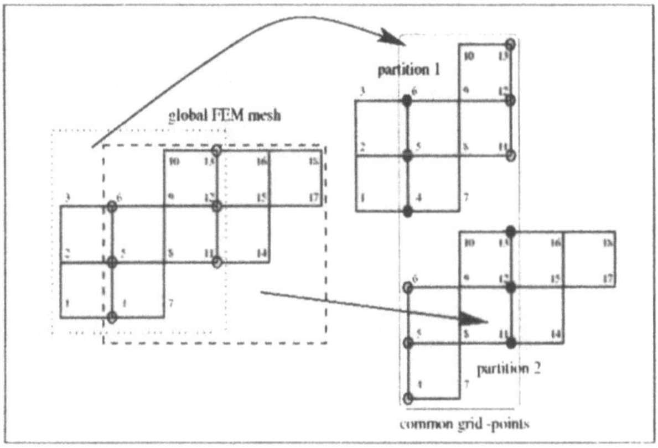

Fig. 3. Overlapping partitioning

After a calculation step, the logical boundary grid–points of each partition are updated with the values of the corresponding inner grid–points of the other partitions. Convergence and efficiency are topics which will be investigated with the queuing implementation.

3 GRISSLi Coupling Interface library (CI)

According to challenges with an increasing complexity of numerical simulation in the recent years, the demand for the integration of different software packages from specialized domains has increased. The CI has been developed to couple different software packages for structured or unstructured grids to unify and simplify the common use of different simulation tools. A new concept to specify the exchange of grids and coupling values, based on grid–based variables, and a communication model have been implemented. Efficient communication is achieved by pre-computing an optimized communication schedule [7].

Each application has its own parallel processes and an instance of the CI is assigned to each of them. The applications start independently on different processors. After reaching a user–defined synchronization point, control is given to the CI, which interpolates values from one grid of an application to a grid of another application and performs the data transfer. Afterwards, the control is given back to the application programs of the coupled calculation (fig. 4).

The CI was successfully engaged in the simulation of steel strip production. The fluid flow is demonstrated at the right hand side of the section between the two rolls in figure 5. The transfered values are interpolated and the metal forming process begins with the initial values shown the left. A more detailed description is given in [8].

4 Acknowledgment

This research was supported by the German Ministry of Education, Science, Research and Technology (BMBF) within the scope of software technology in industry, science and technology (01 IS 512 A-C/GRISSLi). The platform for

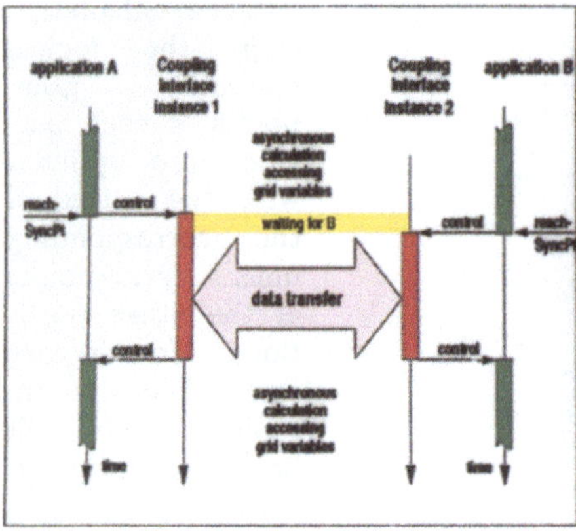

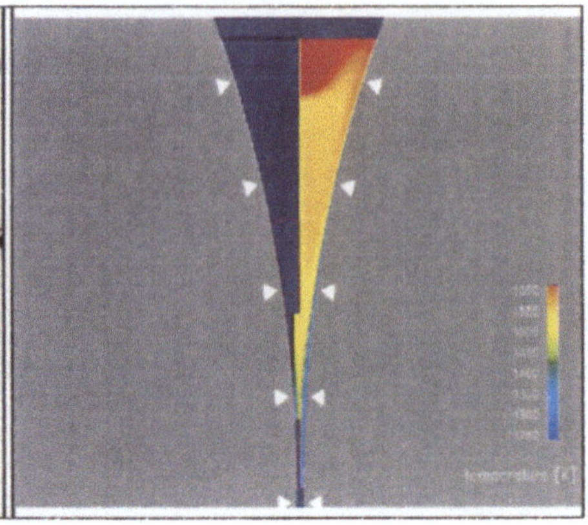

Fig. 4. Coupling of different applications **Fig. 5.** Temperature field of a coupled calculation

implementation was the IBM RS/6000 SP of the Scientific Supercomputing Center (SSC) Karlsruhe, whom we also thank for their kindly support.

References

1. Jürgen Jestrabek. *Stahlbandherstellung nach dem Zweirollenverfahren – Modellierung des Strömungs- und Temperaturfeldes.* PhD thesis, RWTH Aachen, 1995.
2. H. Ritzdorf, A. Schüller, B. Steckel, and K. Stüben. An environment for the parallel multigrid solution of partial differential equations on general 2d domains. *Parallel Comp.*, pages 1559–1570, 1994.
3. Lasso Consulting Engineers. *Larstran User's Manual*, 1993.
4. Vipin Kumar George Karypis. METIS:A Software Package for Partitioning Unstructured Graphs, Partitioning Meshes, and Computing Fill-Reducing Ordering s of Sparse Matrices. Technical report, University of Minnesota,Department of Computer Scie nce / Army HPC Research Center, 1997.
5. A. Müller and P. Adamidis. Parallelization of a metal forming process by substructuring. In *Proceedings of the International Conference on Parallel and Distributed Processing Techniques and Applications (PDPTA'98)*, Las Vegas, Nevada, July 13-16 1998.
6. Barry Smith, Petter Bjørstad, and William Gropp. *Domain decomposition.* Cambridge University Press, 1996.
7. Y. Ding, A. Müller, M. Münch, and A. Reuter. The GRISSLi Coupling Library. In *High Performance Computing on Hewlett-Packard Systems (HiPer'97)*, pages 171–180, 1997.
8. P. Adamidis, A. Beck, U. Becker-Lemgau, Y. Ding, M. Franzke, H. Holthoff M. Laux A. Müller, M. Münch, A. Reuter, B. Steckel, and R. Tilch. Steel strip production - a pilot application for coupled simulation with several calculation systems. In *Metal Forming 98*, Birmingham, September 1–3 1998.

Lecture Notes in Computational Science and Engineering

Vol. 1 D. Funaro, *Spectral Elements for Transport-Dominated Equations.* 1997. X, 211 pp. Softcover. ISBN 3-540-62649-2

Vol. 2 H. P. Langtangen, *Computational Partial Differential Equations.* Numerical Methods and Diffpack Programming. 1999. XXIII, 682 pp. Hardcover. ISBN 3-540-65274-4

Vol. 3 W. Hackbusch, G. Wittum (eds.), *Multigrid Methods V.* Proceedings of the Fifth European Multigrid Conference held in Stuttgart, Germany, October 1-4, 1996. 1998. VIII, 334 pp. Softcover. ISBN 3-540-63133-X

Vol. 4 P. Deuflhard, J. Hermans, B. Leimkuhler, A. E. Mark, S. Reich, R. D. Skeel (eds.), *Computational Molecular Dynamics: Challenges, Methods, Ideas.* Proceedings of the 2nd International Symposium on Algorithms for Macromolecular Modelling, Berlin, May 21-24, 1997. 1998. XI, 489 pp. Softcover. ISBN 3-540-63242-5

Vol. 5 D. Kröner, M. Ohlberger, C. Rohde (eds.), *An Introduction to Recent Developments in Theory and Numerics for Conservation Laws.* Proceedings of the International School on Theory and Numerics for Conservation Laws, Freiburg / Littenweiler, October 20-24, 1997. 1998. VII, 285 pp. Softcover. ISBN 3-540-65081-4

Vol. 6 S. Turek, *Efficient Solvers for Incompressible Flow Problems.* An Algorithmic and Computational Approach. 1999. XVII, 352 pp, with CD-ROM. Hardcover. ISBN 3-540-65433-X

Vol. 7 R. von Schwerin, *Multi Body System SIMulation.* Numerical Methods, Algorithms, and Software. 1999. XX, 338 pp. Softcover. ISBN 3-540-65662-6

Vol. 8 H.-J. Bungartz, F. Durst, C. Zenger (eds.), *High Performance Scientific and Engineering Computing.* Proceedings of the International FORTWIHR Conference on HPSEC, Munich, March 16-18, 1998. 1999. X, 471 pp. Softcover. 3-540-65730-4

Vol. 9 T. J. Barth, H. Deconinck (eds.), *High-Order Methods for Computational Physics.* 1999. VII, 582 pp. Hardcover. 3-540-65893-9

Vol. 10 H. P. Langtangen, A. M. Bruaset, E. Quak (eds.), *Advances in Software Tools for Scientific Computing.* 2000. X, 357 pp. Softcover. 3-540-66557-9

Vol. 11 B. Cockburn, G. Karniadakis, C.-W. Shu (eds.), *Discontinuous Galerkin Methods: Theory, Computation and Applications.* 2000. X, 452 pp. Hardcover. 3-540-66787-3

For further information on these books please have a look at our mathematics catalogue at the following URL: `http://www.springer.de/math/index.html`

MIX
Papier aus verantwortungsvollen Quellen
Paper from responsible sources
FSC® C105338

If you have any concerns about our products,
you can contact us on
ProductSafety@springernature.com

In case Publisher is established outside the EU,
the EU authorized representative is:
Springer Nature Customer Service Center GmbH
Europaplatz 3, 69115 Heidelberg, Germany

Printed by Libri Plureos GmbH
in Hamburg, Germany